ADVANCES IN UNMANNED AERIAL VEHICLES

International Series on

INTELLIGENT SYSTEMS, CONTROL, AND AUTOMATION: SCIENCE AND ENGINEERING

VOLUME 33

Advances in Unmanned Aerial Vehicles

State of the Art and the Road to Autonomy

Edited by
Kimon P. Valavanis

University of South Florida
Tampa, Florida, USA

Springer

A C.I.P. Catalogue record for this book is available from the Library of Congress.

ISBN 978-1-4020-6113-4 (HB)
ISBN 978-1-4020-6114-1 (e-book)

Published by Springer,
P.O. Box 17, 3300 AA Dordrecht, The Netherlands.
www.springer.com

Printed on acid-free paper

...στη μνήμη του πατέρα μου που λάτρευα

...στα δύο 'αγγελάκια' μου,
τη Στελλίτσα και τον Πανούλη

....και
....για την Ντίνα

... to the memory of my father whom I adored

... to my two little "angels"
Stellitsa and Panouli

... and
... for Dina

Table of Contents

Preface

Unmanned Aerial Vehicles (UAVs) have seen unprecedented levels of growth in military and civilian application domains. Fixed-wing aircraft, heavier or lighter than air, rotary-wing (rotorcraft, helicopters), vertical take-off and landing (VTOL) unmanned vehicles are being increasingly used in military and civilian domains for surveillance, reconnaissance, mapping, cartography, border patrol, inspection, homeland security, search and rescue, fire detection, agricultural imaging, traffic monitoring, to name just a few application domains.

When initially introduced during World War I, UAVs were criticized heavily as being unreliable and inaccurate, and only a handful of people recognized at that early stage their potential and (future) impact on changing the battlefield.

To nobody's surprise, about a century later, the total market for UAVs will reach within a few years more than $16 billion, with the US Department of Defense (DOD) being the champion in funding initiatives, research and development, as well as procurement. Europe, as a continent, is a very distant second player, expected to spend about €2 billion in research and development, and procurement.

Having been involved in unmanned systems related research since the very early 1990's, the initial thought of publishing a book on UAVs, subsequently *this* book, dawned on me immediately after Dr. George Vachtsevanos (Professor, Georgia Institute of Technology) and I offered a Tutorial on UAVs during the *11th Mediterranean Conference on Control and Automation*, in June of 2003; the Tutorial was sponsored by the European Aeronautics Defense Systems (EADS) Agency – 3 SIGMA S.A. The response and feedback was so positive, that we decided to offer an expanded Tutorial/Workshop on the subject during the *2004 IEEE International Conference in Robotics and Automation*.

By the end of 2004, that initial and perhaps vague thought had become a very intense idea that 'kept on bothering' me on a daily basis. Contributing factors were the challenges and open questions related to UAV design, control, testing, payloads, sensors, navigation, applications, that demanded 'believable answers', the need for cutting edge technologies, the world wide increasing interest and number of research groups conducting research in this area, the increasing number of submitted and published papers in journals, conferences and technical meetings, the exponentially increasing technical and round table discussions on what UAVs could and should do, what are the bottlenecks to their complete success and acceptance, what is the next step that needs be done, how they can achieve full autonomy, etc.

In short, the momentum and overall attention the UAV field was gaining (in terms of Broad Agency Announcements, funding opportunities, initiatives, development, potential application domains), the major research challenges one had to face and overcome, and the need for some written proof of what may be considered state-of-the-art today, convinced me that it was worth publishing a book on UAVs.

However, my increased load in 2005 slowed me down, and even though I had a plan in my mind, I needed 'a push' to focus on this project. That happened in 2006 in terms of two events: Dr. Paul Y. Oh (Drexel), Dr. Thomas Adams (Boeing Phantom Works) and I, co-organized during the *2006 IEEE International Conference in Robotics and Automation* a Tutorial Workshop on **"UAVs: Payloads and Missions"**, sponsored by Boeing; Dr. George Vachtsevanos and I served as Guest Editors for a Special Issue on UAVs published in September of 2006 at the *IEEE Robotics and Automation Magazine*, Vol. 13, No. 3. All activities in 2006 were also part of coordinated efforts to document research and development in UAVs for the IEEE Robotics and Automation Society *Aerial Robotics and Unmanned Aerial Vehicles* Technical Committee.

In December of 2006, my co-authors and I had completed almost all Chapters of this book. The result of this project is eighteen contributed Chapters from different and/or collaborating groups in the US, Europe, Canada and Australia.

Contributions from US Universities report on research results from representative groups at the Automation and Robotics Research Institute of the University of Texas at Arlington, Brigham Young University, Drexel University, Georgia Institute of Technology, Notre Dame University and University of South Florida.

Contributions from Europe are from research groups in France (CNRS), Greece (Technical University of Crete), Italy (University of Catania, National Institute of Geophysics and Volcano Studies in Palermo), Spain (University of Seville, University Pablo de Olavide), Sweden (Linköping University) and Switzerland (EPFL in Lausanne and ETH in Zurich).

The contribution from Australia is from the Australian National University. The contribution from Canada is from the University of British Columbia.

Even though this is an edited book, I have tried to present it as a unified and complete ensemble as if it were a textbook or a research monograph. For this reason, Chapters have been grouped in five parts according to the subject and topics they present. Summaries are included at the beginning of each Chapter for completeness purposes. The book, as a whole, presents current advances in UAVs and aims at setting the tone of what may come next.

This edited book is suitable for graduate students whose research interests are in the area of unmanned aerial vehicles, for scientists, engineers and practitioners. For better understanding, the interested reader should have knowledge of rigid body kinematics and dynamics, as well as knowledge of advanced graduate level control system theory.

The book may be used as a textbook for a one or two semester graduate level course on UAVs or as a special topics course. The Chapter sequence depends on the potential course emphasis. As such, Chapters 1 and 2 offer a historical perspective on UAVs; Chapters 3 to 6 and Chapter 15 emphasize control; Chapters 7 to 10 discuss aspects of navigation and related techniques; Chapters 11 to 17 emphasize diverse applications, while the road ahead is the topic of the last Chapter, Chapter 18.

I am not sure if my co-authors and I have succeeded in our goal and objectives by publishing this book. I know that our contributions have only 'touched' upon some of the many challenges and open research questions one faces when conducting research in unmanned systems. Perhaps our reported findings reflect personal preferences, research agendas and perspectives we 'think' are important. It is too soon to know and judge. But the unmanned systems field is so widely open, that, regardless, I hope we have made some contribution to its advancement.

Last, but not least, I certainly hope that this project will serve as motivation to our colleagues from academia and industry to publish better, more general or more application specific books, texts or research monographs. As I said, the research area is wide open, the best is yet to come and the sky is the limit.

Kimon P. Valavanis
Tampa, April 2007

Acknowledgements

I wish to acknowledge wholeheartedly several people who have contributed one way or another to publishing this book.

At first, and most important of all, I want to express my profound respect and gratitude to every single author who has participated in completing this book. Without their hard work and contributed Chapters there wouldn't be any book. The credit goes to them for jumping on the train and joining me in this effort. I respect and I value them as colleagues and I hope we will work together again in the very near future.

I owe a lot to George Vachtsevanos who has supported and stood behind me and behind all of my (perhaps crazy) ideas over the years. My long-term collaboration with George in several research areas has taught me how to look at problems not only from the pure theoretical perspective, but also from the application and implementation point of view. My very regular visits to, and interactions with, him, his colleagues and his research group have helped me 'get hooked' in the area of unmanned aerial vehicles and have motivated me to define and develop our research program at USF. It is not an exaggeration if I say that I always run to him every time something new or 'something else' comes to my mind.

I am very thankful to my professionally senior colleagues, Frank Lewis (UT Arlington) and Panos Antsaklis (University of Notre Dame), who have believed in me over the years, encouraged me to complete this project and present the book as I thought it will be best. We have worked together, we have organized conferences together, we have helped establish the *Mediterranean Control Association* and its annual conference, but most important of all, I consider them – on top of pioneering colleagues – as friends. After I moved to USF, Frank and I got closer, and I am very pleased to find out that we have many more things in common than one may think about.

I have had the privilege and honor to interact often with Michael Athans (Professor Emeritus, MIT and now Research Scientist at IST / ISR, Lisbon, Portugal), visit him at his apartment in Clearwater, and talk – among other things - about aerial vehicle control in general and small helicopter control in particular. Michael visited our lab, gave us seminars, lectured our students and offered his help, and opened *my* eyes, and my students' eyes with regards to LQG/LQR based controller designs. Not only is he an excellent 'teacher', but also a great person who has honored me with his friendship.

I am grateful to Dr. Abraham Kandel who played a key role in hiring me at USF in 2003; Abe was Chair of the Computer Science and Engineering Department at that time; he has been and he is one of my very strong supporters, and I am glad we have started working and publishing together. I am also honored that Dr. L. Martin-Vega, Dean of the USF College of Engineering at the time I was hired, went along with Abe's and the Department's recommendation and he agreed to bring me to USF as a tenured faculty. To that extend, Dr. R. Carnahan, Associate Dean for Research who retired in December of 2006, backed-up my ideas to focus on the new research area of unmanned systems. Dr. S. Saigal, Interim Dean, and Dr. R. Kasturi, my Department Chair, have encouraged me with enthusiasm to pursue this research direction and they both went the extra mile to identify and allocate lab space to our group, in essence providing safe heavens for this research program. USF as a University has supported me in my efforts to launch a solid research program and agenda in unmanned systems, and especially in small / miniature unmanned aerial vehicles.

I offer sincere thanks to all of my colleagues within the Computer Science and Engineering Department, and in particular to Miguel Labrador, Wilfrido Moreno and Elias Stefanakos from electrical Engineering, Ali Yalcin from Industrial and Management Systems Engineering, Pei-Sung Lin from the Center for Urban Transportation Research, as well as to Alfredo Weitzenfeld (ITAM, Mexico); we work together and we co-advise graduate students at USF.

My graduate students working at the USF Unmanned Systems Laboratory and the National Institute for Applied Computational Intelligence, N. Aldawoodi, W. Alvis, L. Barnes, C. Castillo, M. Castillo-Effen, C. Dalamagkidis, D. Ernst, R. Garcia, S. Ioannou, M. Kontitsis, S. Murthy, A. Puri, I. Raptis and A. Tsalatsanis - who do everything possible to drive me crazy, and they are being successful at it - have kept me going professionally, they have boosted my energy level and keep me 'young'. Their work speaks by itself and I do thank them from the bottom of my heart.

I also want to thank our Sponsors and Program Managers who have funded our research and they have believed in our capabilities. Without their support, we would not have been able to produce results. At random order, a huge 'thank you' to: Dr. S. Wilkerson and Dr. M-A Fields from the Army Research Lab; Dr. Randy Zachery from the Army Research Office; Dr. G. Toth from the Office of Naval Research; Dr. J. Besser from SPAWAR; Mr. S. Castelin from the Naval Surface Warfare Center in Panama City; Dr. R. Williams from the US SOCOM; Mr. Duane Schultz from Concurrent Technologies Corporation; Dr. J. Michler and Dr. P. Brett from Hillsborough County.

The Publisher, *Springer*, but most importantly Ms. Nathalie Jacobs and her group, have been extremely supportive of this project. Nathalie has gone the extra mile to make it a reality, accommodating us in every single possible way, listening to our concerns. There is not enough I could say about Nathalie's support and how much she wanted this project. She even accepted my recommendation to 'sit on the book' for a couple of weeks before submitting it to Springer because I wanted to look at it once more. With publishers like her, work is fun. I am very pleased with this working relationship and I hope to continue it in the years to come.

All of us who have been part of this book want to thank from the bottom of our hearts Ms. Valerie Mirra not only for formatting the book, but also for proof-reading it, making sure we all follow the same layout and style, looking thoroughly into cosmetic and appearance changes, consistency in equation layout, and also reminding us that we have run out of time. Thanks Valerie.

This project, as a whole, has been supported in part by two Research Grants, ARO W911NF-06-1-0069 and SPAWAR N00039-06-C-0062.

List of Contributing Authors

Wendy Alvis
Unmanned Systems Laboratory
Department of Electrical Engineering
University of South Florida
Tampa, FL 33620, USA
trietley@eng.usf.edu

Panos J. Antsaklis
Department of Electrical Engineering
University of Notre Dame
Notre Dame, IN 46556, USA
antsaklis.1@nd.edu

Gianluca Astuti
Dipartimento di Ingegneria Elettrica Elettronica e dei Sistemi
Università degli Studi di Catania
95125 Catania, Italy
www.robotic.diees.unict.it

D. Blake Barber
MAGICC Laboratory
Brigham Young University
Provo, Utah 84602, USA
d.blake.barber@gmail.com

Randal W. Beard
MAGICC Laboratory
Department of Electrical and Computer Engineering
Brigham Young University
Provo, Utah 84602, USA
beard@ee.byu.edu

Samir Bouabdallah
Autonomous Systems Laboratory
ETH Zurich
8092 Zurich, Switzerland
samir.bouabdallah@ieee.org

Fernando Caballero
Escuela Superior de Ingenieros
Universidad de Sevilla
41092 Sevilla, Spain
caba@cartuja.us.es

Daniele Caltabiano
Dipartimento di Ingegneria Elettrica Elettronica e dei Sistemi
Università degli Studi di Catania
95125 Catania, Italy
www.robotic.diees.unict.it

Carlos Castillo
Unmanned Systems Laboratory
Department of Electrical Engineering
University of South Florida
Tampa, FL 33620, USA
ccastil3@eng.usf.edu

Mauricio Castillo-Effen
Unmanned Systems Laboratory
Department of Electrical Engineering
University of South Florida
Tampa, FL 33620, USA
mcastil7@eng.usf.edu

Andrew Curtis
MAGICC Laboratory
Brigham Young University
Provo, Utah 84602, USA
byudrew@gmail.com

Pritpal Dang
Department of Electrical Engineering
Automation & Robotics Research Institute
University of Texas at Arlington
Fort Worth, TX-76118, USA
pdang@arri.uta.edu

Walter Engel
Mechanical Engineer/Designer
Arvenweg 6a
8840 Einsiedeln, Switzerland
engel.w@bluewin.ch

Lei Fang
Department of Electrical Engineering
University of Notre Dame
Notre Dame, IN 46556, USA
lfang@nd.edu

Joaquín Ferruz
Escuela Superior de Ingenieros
Universidad de Sevilla
41092 Sevilla, Spain
ferruz@cartuja.us.es

Per-Erik Forssen
Laboratory for Computational Intelligence
Department of Computer Science
University of British Columbia
Vancouver, BC V6T 1Z4, Canada
perfo@cs.ubc.ca

Richard D. Garcia
Unmanned Systems Laboratory
Department of Computer Science and Engineering
University of South Florida
Tampa, FL 33620, USA
rdgarcia@cse.usf.edu

Gaetano Giudice
Istituto Nazionale di Geofisica e Vulcanologia
Sezione di Palermo
90146 Palermo, Italy
ggiudice@pa.ingv.it

William E. Green
Drexel Autonomous Systems Laboratory
Department of Mechanical Engineering
Drexel University
Philadelphia, PA 19104, USA
weg22@drexel.edu

Stephen R. Griffiths
Scientific Systems Inc.
500 West Cummings Park, Suite 3000
Woburn, MA 01801, USA
srgriffiths@ssci.com

Tarek Hamel
Laboratoire I3S
UNSA-CNRS
06903 Sophia Antipolis - Cedex
Nice, France
thamel@i3s.unice.fr

Eric N. Johnson
Daniel Guggenheim School of Aerospace Engineering
Georgia Institute of Technology
Atlanta, GA 30332, USA
eric.johnson@ae.gatech.edu

Michael Kontitsis
Unmanned Systems Laboratory
Department of Computer Science and Engineering
University of South Florida
Tampa, FL 33620, USA
mkontits@cse.usf.edu

Frank Lewis
Department of Electrical Engineering
Automation & Robotics Research Institute
University of Texas at Arlington
Fort Worth, TX-76118, USA
lewis@uta.edu

Domenico Longo
Dipartimento di Ingegneria Elettrica Elettronica e dei Sistemi
Università degli Studi di Catania
95125 Catania, Italy
www.robotic.diees.unict.it

Ben T. Ludington
School of Electrical and Computer Engineering
Georgia Institute of Technology
Atlanta, GA 30332, USA
ben.ludington@gatech.edu

Robert Mahony
Department of Engineering
The Australian National University
Canberra, ACT 0200, Australia
Robert.Mahony@anu.edu.au

J. Ramiro Martínez-de-Dios
Escuela Superior de Ingenieros
Universidad de Sevilla
41092 Sevilla, Spain
jdedios@cartuja.us.es

Timothy W. McLain
Department of Mechanical Engineering
Brigham Young University
Provo, Utah 84602, USA
mclain@byu.edu

Donato Melita
Dipartimento di Ingegneria Elettrica Elettronica e dei Sistemi
Università degli Studi di Catania
95125 Catania, Italy
www.robotic.diees.unict.it

Luis Merino
Escuela Politécnica Superior
Universidad Pablo de Olavide
41013 Sevilla, Spain
lmercab@upo.es

Anders Moe
Computer Vision Laboratory
Department of Electrical Engineering
Linköping University
SE-581 83 Linköping, Sweden
moe@isy.liu.se

Wilfrido Alejandro Moreno
Unmanned Systems Laboratory
Department of Electrical Engineering
University of South Florida
Tampa, FL 33620, USA
moreno@eng.usf.edu

Giovanni Muscato
Dipartimento di Ingegneria Elettrica Elettronica e dei Sistemi
Università degli Studi di Catania
95125 Catania, Italy
gmuscato@diees.unict.it

Ioannis K. Nikolos
Department of Production Engineering and Management
Technical University of Crete
Chania 73100, Crete, Greece
jnikolo@dpem.tuc.gr

Klas Nordberg
Computer Vision Laboratory
Department of Electrical Engineering
Linköping University
SE-581 83 Linköping, Sweden
klas@isy.liu.se

André Noth
Autonomous Systems Laboratory
ETH Zurich
8092 Zurich, Switzerland
andre.noth@mavt.ethz.ch

Paul Y. Oh
Drexel Autonomous Systems Laboratory
Department of Mechanical Engineering
Drexel University
Philadelphia, PA 19104, USA
paul@coe.drexel.edu

Aníbal Ollero
Escuela Superior de Ingenieros
Universidad de Sevilla
41092 Sevilla, Spain
aollero@cartuja.us.es

Angelo Orlando
Dipartimento di Ingegneria Elettrica Elettronica e dei Sistemi
Università degli Studi di Catania
95125 Catania, Italy
www.robotic.diees.unict.it

Dan Popa
Department of Electrical Engineering
Automation & Robotics Research Institute
University of Texas at Arlington
Fort Worth, TX-76118, USA
popa@arri.uta.edu

Ioannis A. Raptis
Unmanned Systems Laboratory
Department of Electrical Engineering
University of South Florida
Tampa, FL 33620, USA
iraptis@eng.usf.edu

Jeffrey B. Saunders
MAGICC Laboratory
Brigham Young University
Provo, Utah 84602, USA
saunders.jeff@gmail.com

Roland Siegwart
Autonomous Systems Laboratory
ETH Zurich
8092 Zurich, Switzerland
rsiegwart@ethz.ch

Nikos C. Tsourveloudis
Department of Production Engineering and Management
Technical University of Crete
Chania 73100, Crete, Greece
nikost@dpem.tuc.gr

George J. Vachtsevanos
School of Electrical and Computer Engineering
Georgia Institute of Technology
Atlanta, GA 30332, USA
gjv@ece.gatech.edu

Kimon P. Valavanis
Unmanned Systems Laboratory
Department of Computer Science and Engineering
University of South Florida
Tampa, FL 33620, USA
kvalavan@cse.usf.edu

Johan Wiklund
Computer Vision Laboratory
Department of Electrical Engineering
Linköping University
SE-581 83 Linköping, Sweden
jowi@isy.liu.se

PART I

Background Information

Chapter 1: Introduction[1]

This Chapter justifies the rationale for publishing this edited book. It starts with a non technical, general discussion about unmanned aerial vehicles (UAVs). Then, it presents some fundamental definitions related to UAVs for clarification purposes, and discusses the contents of the book in a very concise way. It paves the way for what is included in subsequent Chapters and how the material, even though it is divided in parts, ties together in a rather unified and smooth way. The goal is to help the potential reader become familiar with the contents of the book and with what to expect reading each Chapter.

1.1 Introduction

UAVs, also called unmanned aircraft systems, have recently reached unprecedented levels of growth in diverse military and civilian application domains.

UAVs were first introduced during World War I (1917), registering the long involvement of the US military with unmanned vehicles [12]. Those early UAVs were very unreliable and inaccurate, and, at that time, their usefulness, their ability to change the battlefield and their overall impact on military applications was not recognized by most military and political leaders.

Only a handful of individuals did envision and predicted their future potential and overall impact on military applications. If it were not for that small group of people who kept alive (over the post World War I years) the concept of an unmanned vehicle pushing for political support and funding, nothing would have been possible today.

Even though UAVs were used in Vietnam, it was only after Operation Desert Storm (1991) and the conflict in the Balkan Peninsula in the early 1990's when interest in UAVs gained momentum. As such, in 1997, the total income of the UAV global market, including the Vertical Take-Off

[1] Written by K. P. Valavanis

Kimon P. Valavanis (ed.), Advances in Unmanned Aerial Vehicles, 3–13.

and Landing (VTOL) segment, reached $2.27 billion dollars [4], a 9.5% increase over 1996. In the middle 1990's the demand for VTOL vehicles was limited, but since then, commercially available products and market share started to increase.

Focusing only on the year 2000, one year before 9/11, Figure 1.1 illustrates the total year funding of the US DOD [3]; as shown in the Figure, 15% of the funding was allocated to VTOL vehicle design.

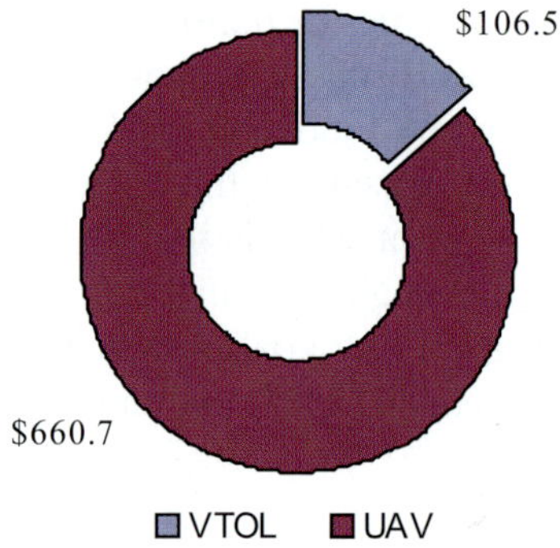

Fig. 1.1. US Government funds ($M) for R&D in UAVs / VTOLs – year 2000.

The critical event that changed completely the perception about UAVs and put them on the everyday life map, on front covers, made them first subject in media coverage and TV documentaries, was the terrorist attack on 9/11. The events on 9/11, coupled with the war in Afghanistan and Operation Iraqi Freedom where UAVs were successfully used in the battlefield and they were deployed successfully for a multitude of missions, resulted in skyrocketed funding and the largest number of production orders [10].

As stated in [10], over the next 8-10 years (until 2015), the UAV market in the US, as a whole, will reach $16 billion, with Europe as a continent playing the role of the second but distant competitor, spending just about €2 billion. US companies hold currently about 63%-64% of the market share, while European companies account for less than 7% [10]. This data is verified in [12] where it is stated that from 1990 to 1999, the US DOD total investment in UAV development, procurement and operations was a bit over $3 billion; but as shown in Table 1.1, the FY03-09 Presidential Budget for related UAV programs reaches $16.2 billion [12]. As a follow up of the data shown in Table 1.1, and just for comparison purposes, Table 1.2 illustrates the revised FY06 President's budget for UAS operations and maintenance [15].

Program	FY03*	FY04*	FY05*	FY06*	FY07*	FY08*	FY09*	Total*
Predator	227.9	358.0	329.4	355.0	241.4	266.5	374.2	2152.4
Pioneer	29.1	36.3	17.7	9.9	10.7	11.2	11.4	126.3
Hunter	33.9	29.2	28.2	28.1	26.0	24.6	25.2	195.2
Global Hawk	510.5	624.2	625.7	688.8	869.6	800.5	730.6	4849.9
Shadow 200	179.0	132.0	124.8	89.0	90.5	93.6	90.2	799.1
ER/MP	0.0	23.1	33.6	71.0	85.7	140.0	168.7	522.1
Fire Scout/VTUAV	38.6	4.0	0.0	0.0	0.0	180.0	270.0	492.6
UCAV (AF & Navy)	171.3	347.8	573.3	530.3	361.7	914.7	1226.1	4125.2
BAMS	0.0	25.1	224.4	187.1	322.1	415.2	440.3	1614.2
GH Maritime Demo	189.4	76.4	57.3	59.2	61.9	56.5	56.9	557.6
UCAR	33.0	49.4	75.9	107.7	84.2	86.2	47.0	483.4
Various Small UAV	51.6	52.4	55.0	55.0	55.0	55.0	55.0	379.0 estimated
Grand Total	**1455.7**	**1761.6**	**2139.1**	**2158.0**	**2199.6**	**3016.9**	**3459.5**	**16190.4**

* All budget figures are given in millions of dollars and roll up RDT&E, procurement, and O&S together.

Table 1.1. Presidential Budget for UAV Programs in $M, FY 04 (Credit: taken from [12], Table 2.4–1).

UAS Program FY06PB ($M)	FY05	FY06	FY07	FY08	FY09	FY10	FY11	Total
Predator (Air Force)	71.9	160.4	175.1	103.0	115.1	116.6	119.2	861.3
Pioneer (USMC)	8.7	10.4	7.7	6.7	3.9	9.5	11.2	58.0
Hunter (Army)	27.9	30.0	30.1	29.8	28.1	9.7	1.1	156.7
Global Hawk (Air Force)	20.0	19.5	68.7	71.3	94.3	108.5	113.5	495.7
Shadow (Army)	29.2	36.2	38.0	34.8	34.0	44.3	45.7	262.3
Fire Scout/VTUA (Navy)	0.0	0.0	0.0	TBD	TBD	TBD	TBD	0.0
BAMS UA (Navy)	0.0	0.0	0.0	0.0	0.0	0.0	31.3	31.3
GH Maritime Demo (Navy)	9.6	18.9	19.1	20.0	20.0	20.0	20.0	127.6
Total	**167.3**	**275.4**	**338.7**	**265.6**	**295.4**	**308.6**	**342.0**	**1992.9**

*Does not include 2005 supplemental request for OIF, OEF, and Operation UNIFIED ASSISTANCE

Table 1.2. FY06 President's Budget for UAS Operations and Maintenance in $M (Credit: taken from [15], Table 2.6–3).

An additional independent study conducted by the Teal Group, a defense and aerospace market analysis firm based in Fairfax, VA [14], claims that UAVs will continue to be the most dynamic growth sector of the world aerospace industry. Their market study that was previewed during the Unmanned Systems North America 2006 Conference, estimates that UAV spending will more than triple over the next decade, totaling close to $55 billion in the next ten years [14]. The same study [14], points out that the US will account for 77% of the worldwide RDT&E spending on UAV technology over the next decade, and about 64% of the procurement. These US expenditures represent higher shares of the aerospace market than for worldwide defense spending in general, with the US accounting for about 67% of total worldwide defense RDT&E spending and 37% of procurement spending, according to forecasts in International Defense Briefing, another Teal Group competitive intelligence service.

Another conclusion that the Teal Group has reached [14] is that a civil UAV market will slowly emerge over the next decade, starting first with government organizations requiring surveillance systems similar to military UAVs such as coast guards, border patrol organizations and similar national security organizations.

A rapidly evolving and dynamic sector of the overall UAV market is the VTOL vehicle segment. America as a continent accounts for 68% of all VTOL vehicles developed worldwide, while Europe and Asia contribute 22% and 10%, respectively, as shown in Figure 1.2 [13]. Moreover, most of VTOL vehicles manufactured in the American continent are contributed by the US. The US alone manufactures 66% of the total number of VTOLs worldwide as shown in Figure 1.3, with most VTOLs being used for military applications.

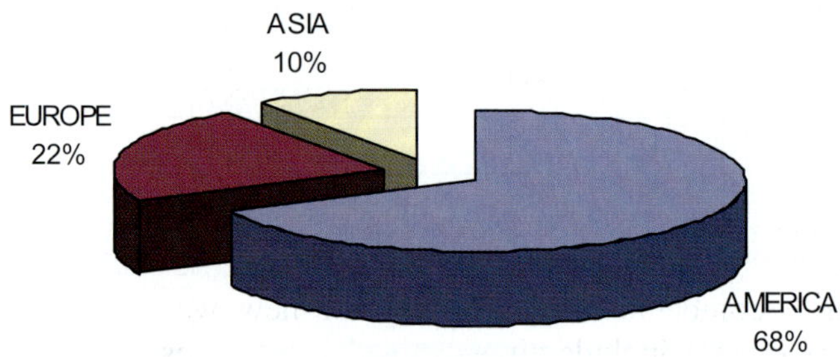

Fig. 1.2. VTOL regional division.

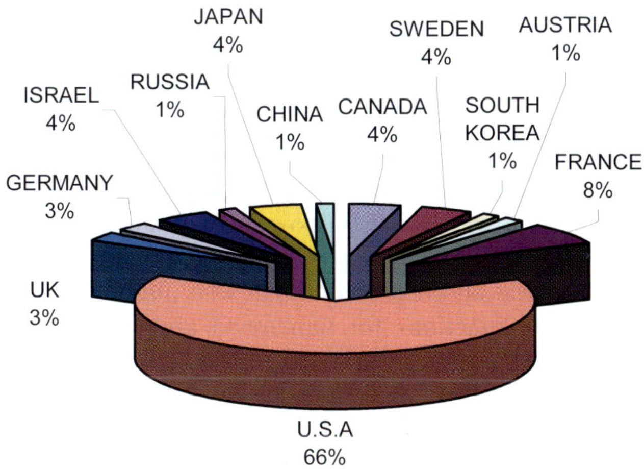

Fig. 1.3. Percentages of VTOL models produced over the world.

It is essential to state that unmanned airplanes are basically used for military applications; however VTOL applications extend to the non-military domains as well. VTOL military applications include surveillance and reconnaissance, combat uses and testing for new weapon systems. Non-military applications include pipelines and power lines inspection and surveillance, border patrol, rescue missions, region surveillance, oil and natural gas search, fire prevention, topography and natural disasters, as well as agricultural applications (mostly in Japan).

As the field matures, the tendency shifts to smaller, more flexible and versatile UAVs. From that perspective, and regardless of application and type of UAV under consideration, the US Army states that "the role of small unmanned aerial vehicles as a critical component for providing unprecedented situational awareness, is rapidly increasing" [11].

Certainly, this brief introduction supports the claim that the future of UAVs is bright and that this area will continue to grow. Therefore, it is important to concentrate in thrust areas related to the current state of the art in research and development, register application domains but also discuss challenges and limitations that need be overcome to improve functionality and utilization of unmanned aerial systems.

Before any further discussion, it is necessary to provide clarifications related to the UAV terminology.

1.2 Clarifications and Related Definitions

In general, an aircraft is any flying vehicle/machine in all possible configurations: fixed-wing, rotary-wing or rotorcraft, helicopters, VTOL vehicles, or short take-off and landing (STOL).

As stated in [8] [9], an aircraft may be either heavier or lighter than air, with balloons and airships belonging to the latter category. Moreover, the term unmanned aerial vehicle (also known as a drone) refers to a pilotless aircraft, a flying machine without an on-board human pilot. As such, 'unmanned' refers to total absence of a human who directs and actively pilots the aircraft. Control functions for unmanned aircraft may be either onboard or off-board (remote control).

A fixed-wing UAV refers to an unmanned airplane that requires a runway to take-off and land, or catapult launching.

A helicopter refers to an aircraft that takes off and lands vertically; it is also known as a rotary aircraft with the ability to hover, to fly in very low altitudes, to rotate in the air and move backwards and sideways. It is capable of performing non-aggressive or aggressive flights.

A helicopter may have different configurations, with a main and a tail rotor (most common), with only a main rotor, with tandem configuration, with coaxial but opposite rotary rotors, as well as with one, two or four rotors.

1.3 Objectives and Outline of the Book

The main objective of the book is to register current research and development in small / miniature unmanned aerial vehicles, fixed- or rotary-wing ones discussing integrated prototypes developed within research laboratories. It aims at describing advances in UAVs, highlighting challenges that need be overcome when dealing with such flying machines, as well as demonstrating their wide applicability to diverse application domains.

Even though this is not a comprehensive edited Volume of contributed Chapters (since it does not include research results from every group working in this area), it does offer a wide perspective of important problems and research questions that need be addressed and solved.

The book is unique in at least one aspect: even though it consists of contributed Chapters from different individuals and groups, material is presented in a rather unified way, classified per topic discussed, assuring continuity in reading.

The book is divided in five Parts:

- Part I consists of Chapters 1 and 2. Both Chapters are introductory motivating and guiding the reader gradually in to the field of UAVs. A historical overview of the evolution of such vehicles, starting from Ancient Greece to the most recent models shows that the idea of a flying machine is a very old one, and provides proof of the tremendous progress in the field.

- Part II focuses on modeling and control fundamentals of small fixed-wing airplanes and small rotorcraft. It includes four Chapters:

    - Chapter 3 provides fundamental background information related to the derivation of the basic equations of motion of a traditional airplane. It explains how the airplane's position and orientation are determined with respect to an Earth-fixed inertia reference frame, derives the aerodynamic forces that act on the airplane, defines the corresponding control angles, and concludes with derivation of the open-loop dynamics. This Chapter is the basic one a designer or control engineer needs to understand before proceeding in controller design, testing and implementation.
    - Chapter 4 focuses on low-level controller design of miniature helicopters for autonomous flights. After summarizing major contributions to small helicopter control, the Chapter describes a general model suitable for small / miniature helicopter non-aggressive flights and compares three different controllers, a PID, a Linear Quadratic Regulator (LQR) and an $\boldsymbol{H}_{\infty}$ controller in terms of their practical implementation to achieve autonomous, self-governing flights.
    - Chapter 5 presents a tutorial–like approach to studying, designing, implementing and testing controllers for small unmanned helicopters performing autonomous non-aggressive flights, putting emphasis on hovering and cruising. It describes simplified, decentralized single input single output, PID and PID-like fuzzy logic controller designs with optimized gains, and a complete design of a multiple inputs multiple outputs linear quadratic regulator (LQR) controller. The presented approach is general enough to be applicable to a wide range of small unmanned helicopters. Chapters four and five are complementary and 'loosely coupled'. Taken together, they offer a comprehensive perspective to small helicopter controller design.

- ➢ Chapter 6 takes advantage of progress in low-power processors and miniature sensors to design and control a miniature quadrotor. This is a rather difficult problem in the field of miniature flying robots (MFR) that are used in search and rescue missions, after earthquakes, explosions, collapsed buildings, etc, since such a MFR should fit through small openings, maneuver around pillars and destructed wall structures.

- Part III is devoted to autonomous navigation, discussing approaches that contribute to improving UAV autonomicity, a key requirement dictated by the US DOD [12] [15]. This part is composed of four Chapters:

  - ➢ Chapter 7 concentrates in micro air vehicle (MAV) obstacle and terrain avoidance building on the notion of utilizing useful but imperfect map information to plan nominal paths through city or mountain terrain. The focal point is that MAVs utilize sensory information to detect and avoid obstacles unknown to the path planner (due to maps being outdated, inaccurate, etc.).
  - ➢ Chapter 8 focuses on UAV vision-based navigation and target tracking, demonstrating that the addition of a camera to a UAV allows the vehicle to perform a variety of tasks autonomously. This Chapter presents vision systems developed and tested at the Georgia Institute of Technology using the GTMax unmanned research helicopter. On top of discussing the vision-based navigation system, the Chapter includes demonstrations of an automated search routine for stationary ground targets, as well as a ground target tracking architecture for mobile targets.
  - ➢ Chapter 9 describes how vision-based techniques for single UAV localization may be extended to deal with the problem of multi-UAV relative position estimation. The approach is built on the assumption that if different UAVs identify using their cameras common objects in the scene, then, the relative pose displacement between the UAVs can be computed from these correspondences.
  - ➢ Chapter 10 derives and tests an evolutionary algorithm based path planner for cooperating UAVs. The scenario under consideration assumes that several UAVs are launched from the same or different but known initial locations. Then, the main goal is to produce 3-D trajectories that ensure a collision free operation with respect to mission constraints. The path planner produces curved routes that are represented by 3-D B-Spline curves. An off-line and an on-line path planner are derived. Both off-line

and on-line path planning problems are formulated as optimization problems, with a differential evolution algorithm serving as the optimizer.

- Part IV refers to diverse applications using UAVs; it includes seven Chapters:

  - Chapter 11 talks about robust non-linear filters for attitude estimation of micro UAVs. It proposes a suite of non-linear attitude observers that fuse angular velocity and orientation measurements in an analogous manner to that of a complementary filter for a linear system. By exploiting the natural geometry of the group of rotations an attitude observer is derived that: requires only accelerometer and gyro outputs; it is suitable for implementation on embedded hardware, and, provides robust attitude estimates as well as estimating the gyro biases on-line.
  - Chapter 12 refers to autonomous solar UAV for sustainable flights. A methodology is presented that is suitable for the global design of a solar powered airplane intended to achieve continuous flight on Earth.
  - Chapter 13 illustrates how integrating optic flow sensing for lateral collision avoidance with a novel MAV platform results in a vehicle that is well suited for flight in near-Earth environments. A novelty is a fixed-wing MAV with hovering capabilities.
  - Chapter 14 is on the topic of dynamic localization of air-ground wireless sensor networks. It presents a method for relative and absolute localization based on potential fields. The relative localization algorithm assumes that distance measurements between sensor nodes are available. For absolute localization, it is assumed that some nodes have GPS absolute position information.
  - Chapter 15 focuses on the problem of decentralized formation tracking of multi-vehicle systems with consensus-based controllers. The problem is stated as multiple vehicles are required to follow spatial trajectories while keeping a desired inter-vehicle formation pattern in time. The Chapter considers vehicles with nonlinear dynamics to follow very general trajectories that can be generated by some reference vehicles. The key idea is to combine consensus-based controllers with the cascaded approach to tracking control, resulting in a group of linearly coupled dynamical systems. The method is general and may be used for both unmanned ground and unmanned aerial vehicles.

- ➢ Chapter 16 describes a complete system including hardware in the loop tuning for a volcanic gas sampling UAV developed at the University of Catania, Italy.
- ➢ Chapter 17 presents two detailed designs on on-board processing systems for small / miniature helicopters with very strict payload limitations. Designs are general and generic enough that may be used across aerial and ground platforms.

- Part V concludes the book. It includes only one Chapter:

  - ➢ Chapter 18 summarizes the book, gives the road map to future developments and designs, talks about the road to complete autonomy and highlights what may be next.

The contributed Chapters reflect mostly current research findings, with the background information needed for completeness purposes. References are included at the end of each Chapter for additional information.

References

1. Van Blyenburgh P., "*UAVs: an Overview*", Air & Space Europe, Vol. 1, No 5/6, pp. 43-47, 1999.
2. *Unmanned Vehicles Handbook 2002*, The Shepard Press, 2002.
3. *Unmanned Aerial Vehicles Roadmap*, Office of the Secretary of Defense, April 2001.
4. *World Markets for Military, Civil and Commercial UAVs: Reconnaissance UAVs and Aerial Targets*, Frost & Sullivan, 1998.
5. UAVForum Internet page, "*Vehicle Overview*", www.uavforum.com.
6. Castillo P., Lozano R., Dzul A. E., *Modeling and Control of Mini-Flying Machines*, Springer 2005.
7. Mettler B., *Identification Modeling and Characteristics of Miniature Rotorcraft,* Kluwer Academic Publishers 2003.
8. Wikipedia, The free encyclopedia, http://en.wikipedia.org/wiki/Main_Page.
9. National Air and Space Museum, *Centennial of Flight*, July 2004. Available at: http://www.centennialofflight.gov/index.htm.
10. Dickerson L., "*UAVs on the Rise*", Aviation Week & Space Technology, Aerospace Source Book 2007, Vol. 166, No. 3, January 15 2007.
11. Lyon D. H., "*A Military Perspective on Small Unmanned Aerial Vehicles*", IEEE Instrumentation & Measurement Magazine, pp: 27-31, September 2004.
12. *OSD UAV Roadmap 2002-2027*, Office of the Secretary of Defense (Acquisition, Technology, & Logistics), Air Warfare, December 2002.

13. Spanoudakis P., Doitsidis L., Tsourveloudis N. C., Valavanis K. P., "*Vertical Take-Off and Landing Vehicle Market Overview*", Unmanned Systems , Vol. 21, No. 5, pp: 14-18, September/October 2003.
14. http://www.roboticstrends.com/displayarticle880.html, September' 06, Robotics Trends.
15. *Unmanned Aircraft Systems Roadmap 2005-2030*, Office of the Secretary of Defense, August 2005.

Chapter 2: A Historical Perspective on Unmanned Aerial Vehicles[1]

This 'pictorial' Chapter presents a historical perspective on unmanned aerial vehicles (UAVs) starting from Ancient Greece to the beginning of the 21st Century. The UAV history, from a very early dream to today's reality is illustrated through a series of figures with detailed legends that are arranged mostly chronologically; they reveal the unmanned vehicle evolution and designs over a period of almost 2,500 years. The Chapter, even though it is non-technical, offers an accurate glimpse of history and helps the reader understand the tremendous level of growth in the unmanned systems area. Almost all figures have been taken from archives and web sites available on-line. The list is by no means complete, but it is very informative. The Chapter layout and contents are similar to Chapter 1 of reference [10].

2.1 UAVs: A Journey through History

In modern times, UAVs appeared during the World War I (1917). However, the idea for a 'flying machine' originated and it was first conceived about 2,500 ago!

2.1.1 Early Designs

It has been documented that the first major breakthrough contribution to autonomous mechanisms occurred during the era of Pythagoras, who was Thales' student for a few years, and the Pythagorean Mathematicians. The first breakthrough on autonomous mechanisms is attributed to Archytas from the city of Tarantas in South Italy, known as Archytas the Tarantine, also referred to as Leonardo Da Vinci of the Ancient World. Archytas was not only the inventor of the number 'one', 'the father of 1' in number theory, but he was also the first engineer. By applying a series of geometric

[1] Written by K. P. Valavanis, M. Kontitsis

Kimon P. Valavanis (ed.), Advances in Unmanned Aerial Vehicles, 15–46.

notions and observations to the study of structures, links and joints, he created Mechanics (in Greek *Μηχανική*). He was drawing mechanisms and he was building them. As such, in 425 B.C. he created the first UAV of all times by building a mechanical bird, a pigeon (in Greek *περιστέρι*) that could fly by moving its wings getting energy from a mechanism in its stomach, see Figure 2.1. It is alleged that it flew about 200 meters before falling to the ground, once all energy was used. The pigeon could not fly again [9], unless the mechanism was reset. As reported by Latin author Aulus Gellius, it is believed to be the first artificial, self propelled flying machine - allegedly propelled by streams of water and vapor.

Fig. 2.1. An artist's depiction of the flying pigeon, the first documented UAV in history. It is reported that it flew for about 200 meters.

During the same era of the Pythagorean Mathematicians, at another part of the Ancient World, in China, at about 400 B.C., the Chinese were the first to document the idea of a vertical flight aircraft. The earliest version of the Chinese top consisted of feathers at the end of a stick. The stick was spun between the hands to generate enough lift before released into free flight.

More than seventeen centuries later, the initial idea attributed to Archytas surfaced again: a similar 'flying bird', credited to some unknown engineer of the Renaissance was documented, see Figure 2.2. It is not known whether this new design was based on Archytas' idea; however, the concept was very similar.

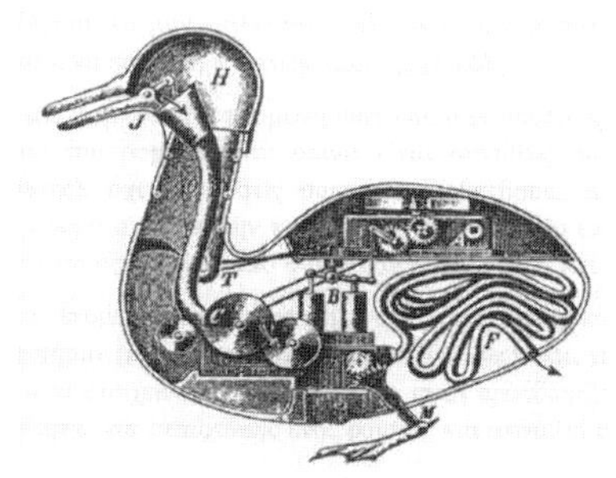

Fig. 2.2. A similar 'flying bird' with a mechanism in its stomach, attributed to an engineer during the Renaissance.

Leonardo Da Vinci, in 1483, designed an aircraft capable of hovering, called aerial screw or air gyroscope, shown in Figure 2.3. It had a 5 meter diameter and the idea was to make the shaft turn and if enough force were applied, the machine could spun and fly. This machine is considered by some experts as the ancestor of today's helicopter [1] [2].

Fig. 2.3. Leonardo Da Vinci's air screw (Credit, Hiller Aviation Museum [2]).

Further, Da Vinci devised a mechanical bird in 1508 that could flap its wings by means of a double crank mechanism as it descended along a cable [13].

Two additional designs based on the initial Chinese top idea were documented in 1754 and 1783, respectively. The first is credited to Mikhail Lomonosov who designed a coaxial rotor powered by a wound-up spring device. The second is credited to Launoy and Bienvenue whose model consisted of a counter rotating set of turkey feathers [1] [2].

Figure 2.4 illustrates George Cayley's aerial carriage that was designed in 1843; it is a converti-plane capable of hovering, which remained an idea due to the fact that the only available power plants at that time were steam engines that could not be used for powered flight [1] [2].

Fig. 2.4. Aerial carriage (Credit, Hiller Aviation Museum [2]).

A vertical flight machine was also designed in the 1840's by Horatio Phillips. A miniature boiler was used to generate steam that was ejected out of blade tips [1]. However, it was Ponton d' Amecourt in the 1860's who flew small helicopter models powered by steam [1] [2], shown in Figure 2.5. It was at that time the term '*helicopteres*' was first coined, based on the complex Greek word '*ελικόπτερο*' that is composed of two words, '*έλικας*' referring to something that spins (spiral) and '*πτερόν*' that means feather (like a bird feather) or wing (like an airplane wing).

Additional helicopter models were introduced between 1860 and 1907. The one standing out was introduced by Thomas Alva Edison who in the 1880's experimented with different rotor configurations, eventually using an electric motor for power [1] [2]. Through his experiments it was revealed that for best hovering abilities, a large diameter rotor was needed with low blade area. In 1907, Paul Cornu developed a two-rotor vertically flying machine, see Figure 2.6, that presumably carried the first human off the ground for the first time. Rotors rotated in opposite directions, the machine flew for about 20 seconds and was merely lifted off the ground.

Fig. 2.5. Ponton d'Amecourt's helicopters (Credit, Hiller Aviation Museum [2]).

Fig. 2.6. Paul Cornu's helicopter (Credit, Hiller Aviation Museum [2]).

The major breakthrough of modern times in helicopter history was the Igor Ivanovitch Sikorsky helicopter, even though his first prototype built by 1909, a non-piloted coaxial helicopter, never flew because of vibration problems and lack of a powerful engine. Russia's contribution came in 1912; Boris Yur'ev's design included a main rotor and a tail rotor (used for the first time), see Figure 2.7, while he was the first to propose cyclic pitch for rotor control.

Fig. 2.7. Boris Yur'ev's aircraft (Credit [1]).

2.1.2. Post World War I Designs

UAVs entered the military applications arena during the First World War. Figures 2.8 to 2.11 depict post war major efforts to design and test manned flying machines with different levels of success.

Fig. 2.8. Stephan Petroczy and Theodore von Karman's helicopter consisting of two superimposed lifting propellers (Credit, Hiller Aviation Museum [2]).

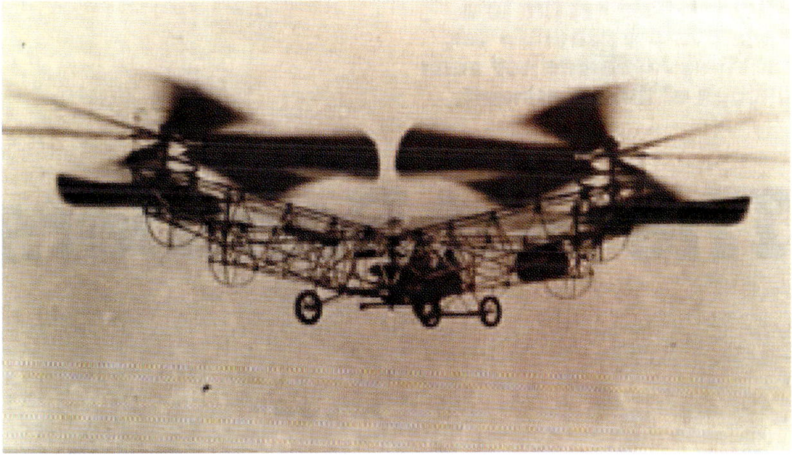

Fig. 2.9. Bothezat's helicopter with four six-bladed rotors (1922). Design was sponsored by the US Army (Credit, National Museum of the United States Air Force [3]).

Fig. 2.10. Cierva's autogyro (1923); it is a hybrid aircraft with fixed-wings and tail but also with a rotor on a vertical shaft above the fuselage (Credit, Hiller Aviation Museum [2]).

Fig. 2.11. The Fa-61 helicopter (1936). It is a side-by-side two rotor machine. It is the first helicopter that demonstrated fully controlled flight and successful autorotations (Credit [1]).

However, as previously mentioned, the field of rotary-wing aviation owes its success almost entirely to Sikorsky, who built in 1939 the classical modern helicopter shown in Figure 2.12. Advances continued with Sikorsky dominating the market, and building during the 1950's the first commercial transport helicopter, the S-55 Chickasaw (H-19).

Fig. 2.12. Sikorsky's modern helicopter, Sikorsky Aircraft Corporation [4].

Of course, in parallel with building vertically flying machines and helicopters, fixed wing aircraft started to evolve over the last one hundred plus years, with the first flight demonstrated by the Wright brothers in 1903.

Focusing on unmanned fixed-wing aircraft, major breakthroughs happened over the past thirty years; therefore, a new Section is devoted to summarize progress. Included in new era designs are also modern rotorcraft configurations.

2.1.3 The New Era in UAVs

This Section presents modern UAV designs and prototypes, what may be termed as 'new UAV configurations'. It includes giant scale, large, small, miniature and mini scale UAVs (compared to their manned counterparts), as well as airship models. Figures have been retrieved mostly from references [5], [11] and [12].

Some of the most well known models are the ones being used by the military around the world. Their advantages over manned aircraft in terms of endurance (due to pilot fatigue) and expendability have been demonstrated and proven in recent conflicts. UAV types in service are shown in Figures 2.13 to 2.20

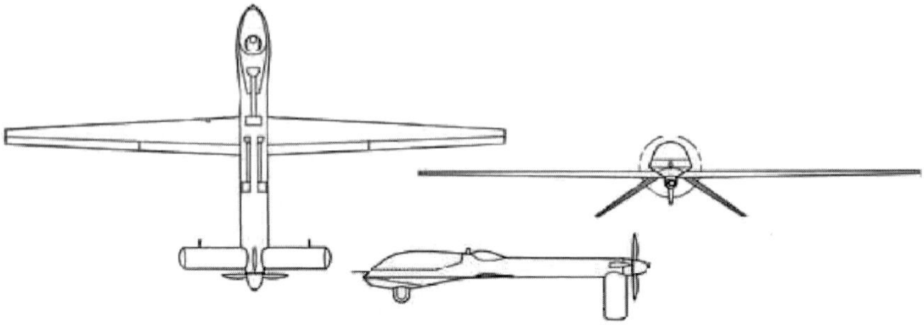

Fig. 2.13. The MQ-1 Predator built by General Atomics Aeronautical Systems Inc. [12].

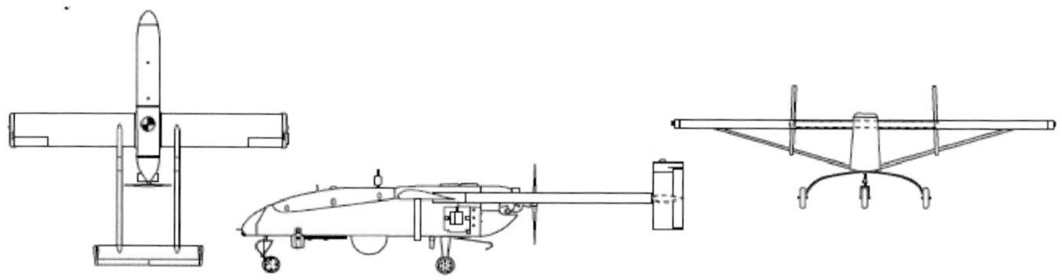

Fig. 2.14 The RQ-2B Pioneer designed by Pioneer UAV Inc. and operated by the US Marine Corps [12].

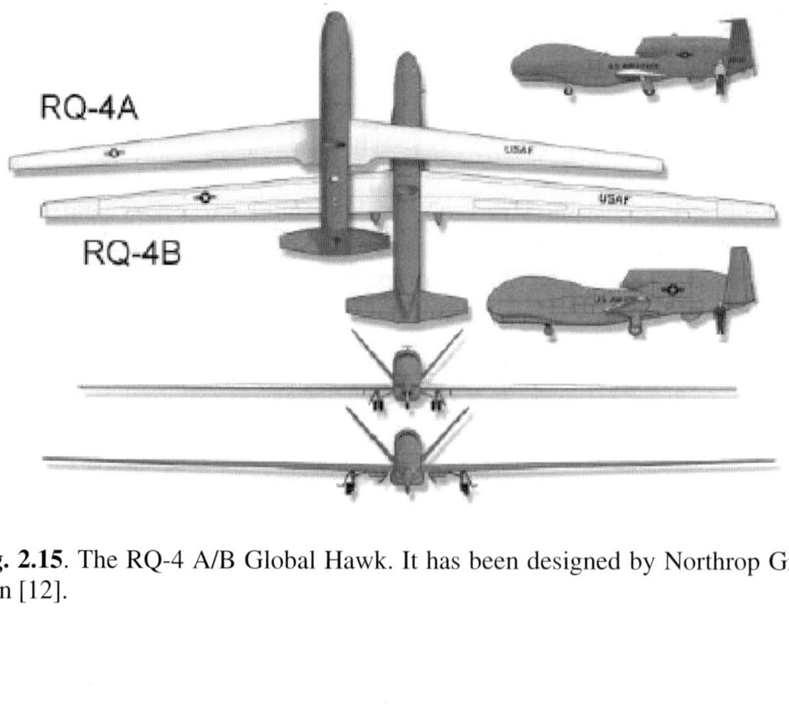

Fig. 2.15. The RQ-4 A/B Global Hawk. It has been designed by Northrop Grumman [12].

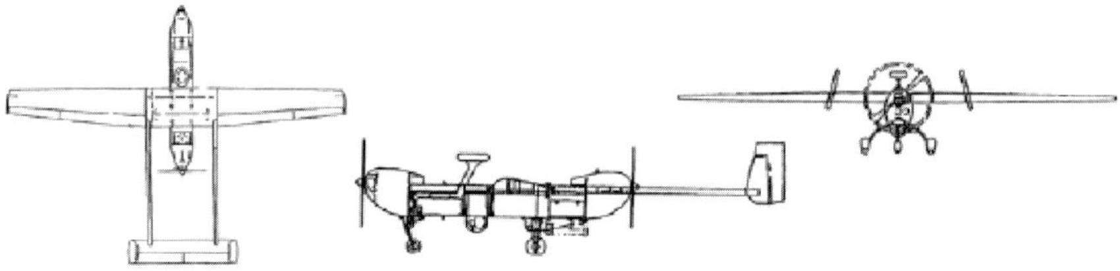

Fig. 2.16. The UAV designated as RQ-5A / MQ-5B Hunter is in use by the US Army. It is capable of delivering munitions. It has been designed by Northrop Grumman [12].

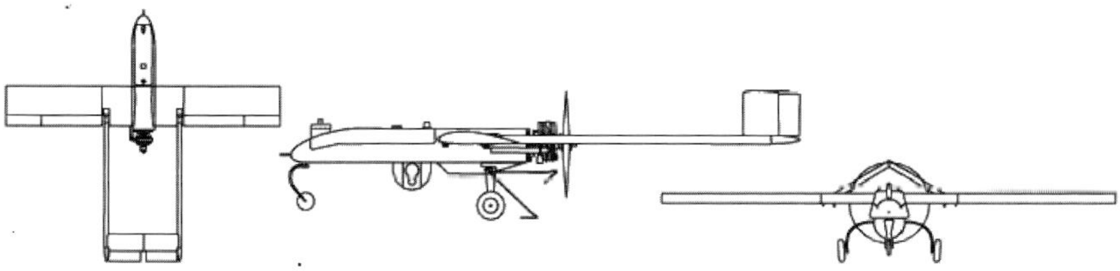

Fig. 2.17. The RQ-7A/B Shadow 200 manufactured by AAI. It is used mainly for reconnaissance [12].

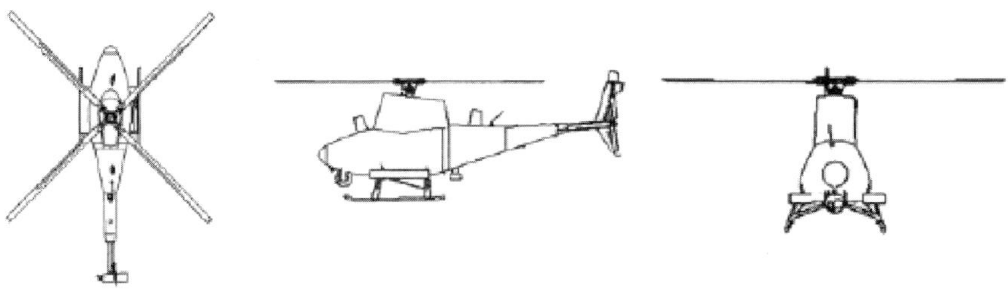

Fig. 2.18. The RQ-8A/B FireScout. It is designed by Northrop Grumman and it has demonstrated autonomous flight capabilities [12].

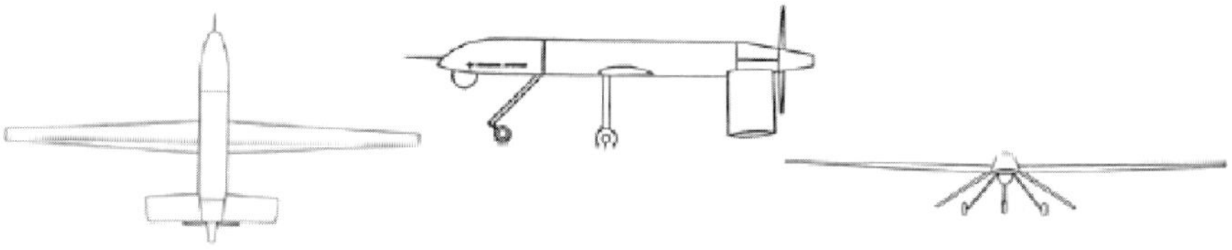

Fig. 2.19. The I-Gnat-ER manufactured by General Atomics Aeronautical Systems Inc. It was used during the G-8 Heads of State Meeting in Alberta Canada to augment security measures.

Fig. 2.20. The X-45 UCAV aircraft built by Boeing Corp, Technology demonstrator for strike missions.

The next set of Figures, up to and including Figure 2.43, presents designs that attempt to explore new and somewhat unconventional configurations. Representative UAVs of this group are the Seagull- Elbit UAV that consists of a single wing carrying the fuselage over it, powered by a rear mounted propeller. Similarly, the Dragoneye by AeroViroment has no tail wings but maintains its tail rudder. The Mikado Aircraft-EMT follows the flying wing configuration with a single tail rudder. Duct-shaped rotorcrafts like Golden Eye, iSTAR, Kestrel are also shown. The Sikorski Cypher II is a duct-shaped rotorcraft that also has fixed wings. The X-50 experimental UAV explores the canard rotating wing configuration. UAVs that belong to this category are shown next.

Fig. 2.21. The Seagull built by Elbit Systems, Israel (Credit, Defense Update [5]).

Fig. 2.22. The Dragoneye built by AeroViroment, Inc. USA (Credit, Defense Update [5]).

Fig. 2.23. The Skylite built by RAFAEL, Israel (Credit, Defense Update [5]).

Fig. 2.24. The Skylark built by Elbit Systems, Israel (Credit, Defense Update [5]).

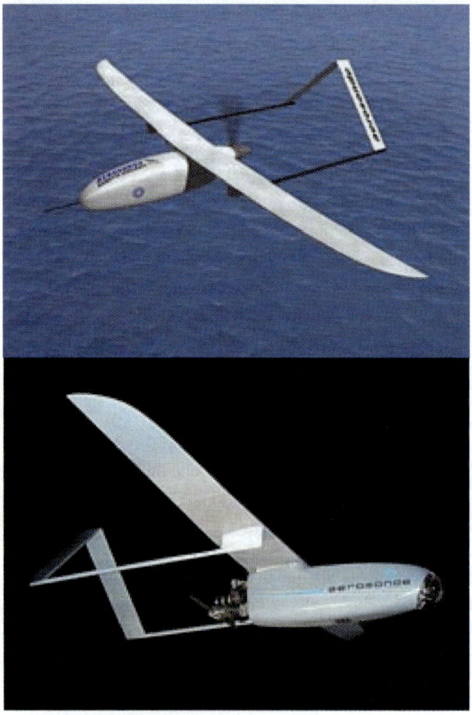

Fig. 2.25. The Aerosonde aircraft built by Aerosonde Robotic Aircraft; Designed for surveillance missions (Credit [6]).

Fig. 2.26. The Mikado Aircraft, EMT, Germany (Credit, Defense Update [5]).

Fig. 2.27. Sikorsky Cypher II, Sikorsky Aircraft Corp [4].

Fig. 2.28. The Golden Eye 100 built by Aurora Flight Systems Corp [12].

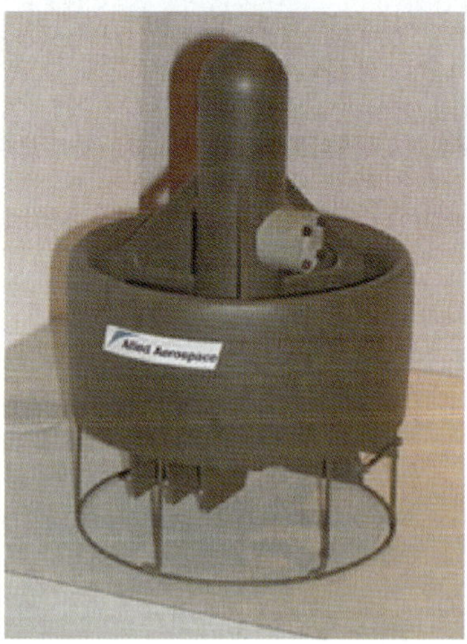

Fig. 2.29. The iSTAR MAV aircraft built by Allied Aerospace (Credit, Defense Update [5]).

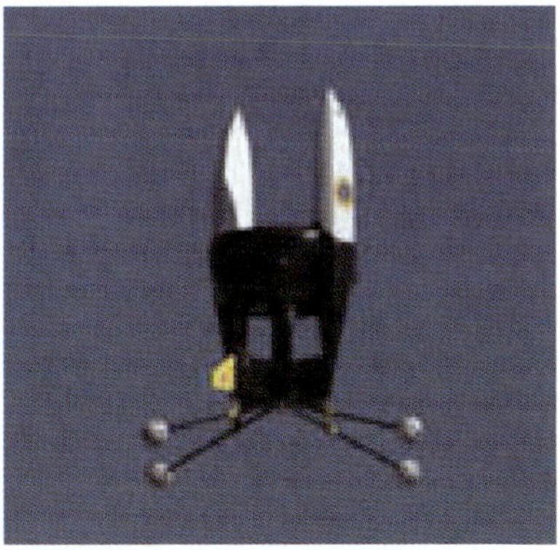

Fig. 2.30. The Kestrel Aircraft built by Honeywell [10].

Fig. 2.31. The X-50 aircraft built by Boeing Corp. It is a technology demonstrator for the Canard Rotor Wing (CRW) configuration [12].

Fig. 2.32. The Guardian CL-327 aircraft built by Bombardier Services Corp [10].

Fig. 2.33. T-Wing aircraft, University of Sydney Australia [7].

Fig. 2.34. Four Rotor configuration. It is designed by Draganfly Innovations Inc. [8].

Fig. 2.35. The A-160 Hummingbird built by Boeing/Frontier. It is a demonstrator for improvements in range endurance and controllability [12].

Fig. 2.36. The Cormorant built by Lockheed-Martin. An immersible vehicle demonstrating launch, recovery and re-launch from a submerged submarine or surface ship [12].

Fig. 2.37. The DP-5X by Dragonfly Pictures. It is designed to serve as a tactical Reconnaissance, Surveillance, and Target Acquisition (RSTA) and Communication Relay platform [12].

Fig. 2.38. The Long Gun by Titan Corporation. It is designed as a reusable, low cost alternative to cruise missiles [12].

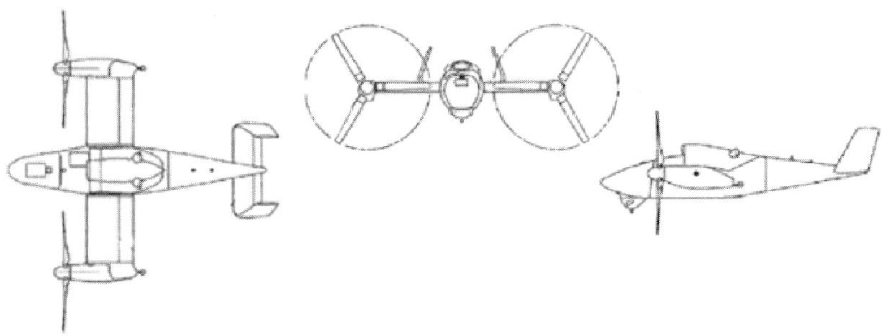

Fig. 2.39. The Eagle Eye by Bell Textron. The tilt-rotor configuration is to be evaluated in 2007 [12].

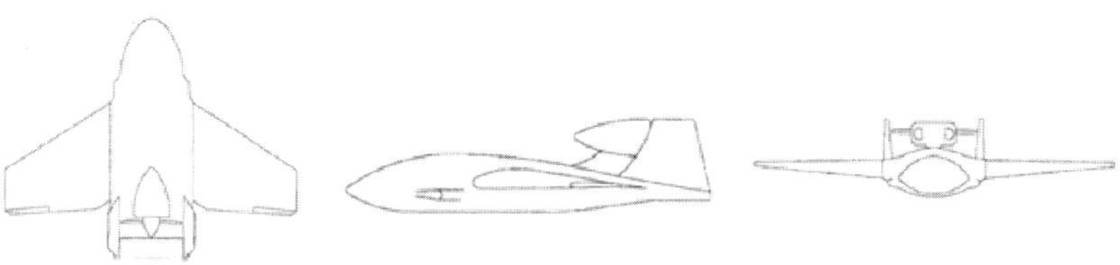

Fig. 2.40. The Neptune built by DRS Unmanned Technologies. Surveillance vehicle designed for sea-launch and recovery from small vessels [12].

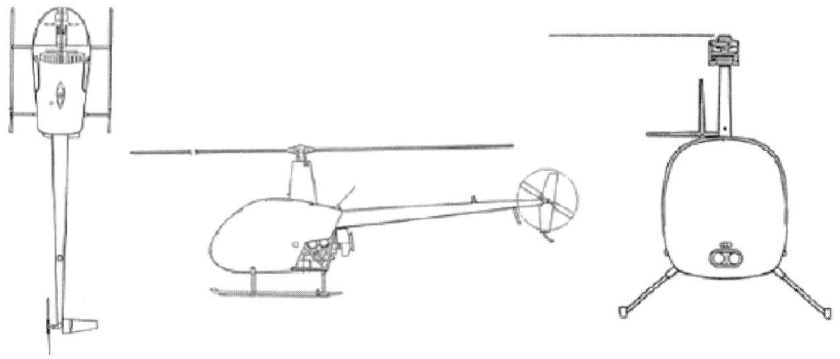

Fig. 2.41. The Maverick built by Boeing/Frontier/Robinson utilized as a testbed for development of control logic [12].

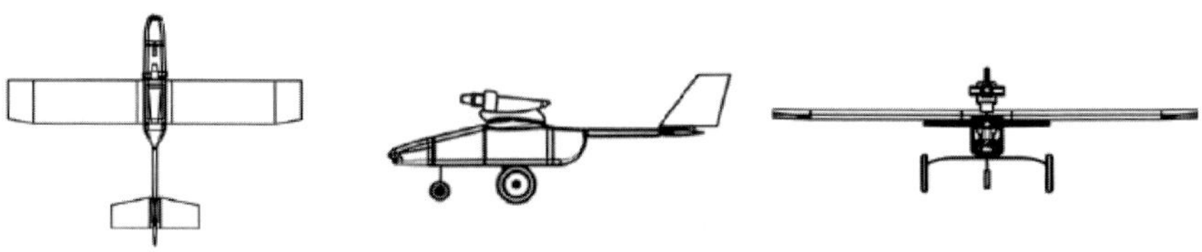

Fig. 2.42. The XPV-1 built by BAI Aerosystems. It is developed for force protection and ground sensor dispersion missions [12].

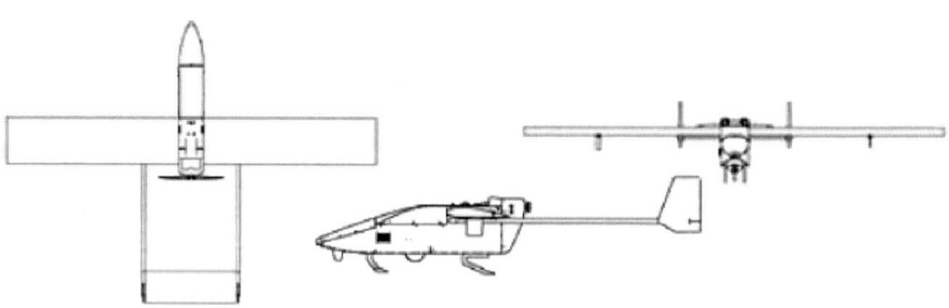

Fig. 2.43. The XPV-2 Mako built by NAVMAR Applied Sciences Corporation/BAI Aerosystems. It is designed as a low cost multi-role UAV [12].

The mainstream of UAV applications has been surveillance, monitoring, and even delivering munitions is some cases. The following vehicles differ both in terms of their configuration and their mission. These parafoils have been used to deliver cargo to otherwise inaccessible areas or propaganda leaflets to enemy troops. Figures 2.44 and 2.45 depict two such vehicles.

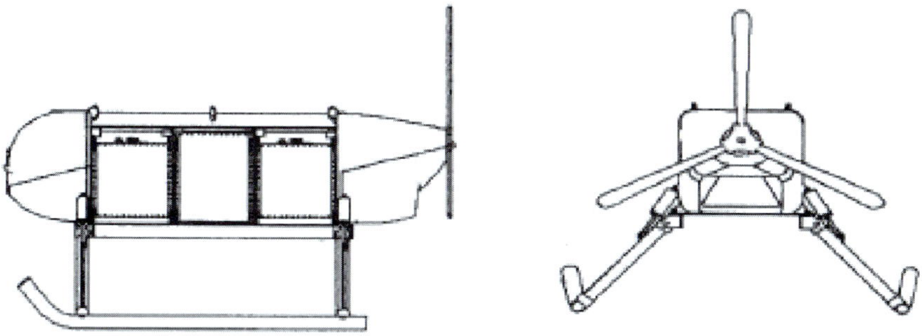

Fig. 2.44. The CQ-10 SnowGoose built by MMIST Inc. It is designed as a powered, GPS guided parafoil for delivery of propaganda leaflets [12].

Fig. 2.45. The Onyx Autonomously Guided Parafoil System by Atair Aerospace Inc. It is designed to deliver cargo for ground and special operation forces [12].

The UAVs depicted below are noteworthy because of their small size. They are versatile, portable, and easy to maintain; they can be employed for the same applications as larger UAVs on a smaller scale and at a lower cost. Representatives are shown in Figures 2.46 to 2.55.

Fig. 2.46. Force Protection Aerial Surveillance System (FPASS) developed by the Air Force Electronics Systems Center to enhance the security of its bases [12].

Fig. 2.47. The FQM-151 Pointer by AeroVironment has been used to test several miniaturized sensors [12].

Fig. 2.48. The Raven by AeroVironment. This UAV is light enough to be hand-launched by soldiers [12].

Fig. 2.49. The BUSTER built by the U.S. Army Night Vision Laboratories. It is being utilized as a testbed for various sensors [12].

Fig. 2.50. A picture of the Silver Fox. It is being developed by the Office of Naval Research for ship security and harbor patrol [12].

Fig. 2.51. The Scan Eagle provides force protection for elements of the Marine Corps [12].

Fig. 2.52. The Battlefield Air Targeting Camera Micro Air Vehicle (BATCAM) is designed as an autonomous, covert, reconnaissance tool [12].

Fig. 2.53. Micro Aerial Vehicle (MAV) built by Honeywell [12].

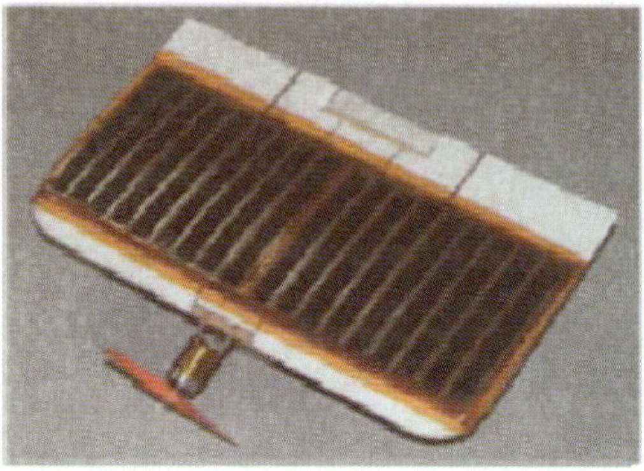

Fig. 2.54. The Hornet built by AeroVironment uses fuel cells for power [12].

Fig. 2.55. The Wasp built by AeroViroment is shown with a pencil for scale [12].

A distinct category of unmanned systems is the airships. Their main advantage over fixed wing or rotary configurations is their unparallel endurance. Many of the models can stay aloft for days or even months. Applications include surveillance, monitoring and communications relay.

Fig. 2.56. Advanced Airship Flying Laboratory developed by the American Blimp Corporation as a testbed for improving airship systems technologies, sensors, communications etc [12].

Fig. 2.57. Tethered Aerostat Radar System (TARS) by ILC Dover is being used as a surveillance platform [12].

Fig. 2.58. Joint Land Attack Elevated Netted Sensor (JLENS) by Raytheon/TCOM capable of providing over-the-horizon surveillance [12].

Fig. 2.59. Rapidly Elevated Aerostat Platform (REAP) by Lockheed Martin/ ISL-Bosch Aerospace [12].

Fig. 2.60. High Altitude Airship (HAA) developed by Lockheed Martin. It is a solar powered, untethered, long endurance, high altitude demonstrator [12].

Fig. 2.61. Marine Airborne Re-Transmission System (MARTS) by SAIC/ TCOM LP, provides over-the-horizon communications relay [12].

All figures are very representative of the current state-of-the-art in UAV models, designs, and applications. Most of such UAVs have been used for military missions. Civilian applications gain momentum, but the consensus is that much more cost effective UAVs need be utilized.

References

1. Helicopter History Site, History of Helicopters, June 2004; Available at http://www.hiller.org.
2. Hiller Aviation Museum; Available at http://www.hiller.org/.
3. National Museum of the United States Air Force; Available at http://www.wpafb.af.mil/museum/.
4. Sikorsky Aircraft Corporation; Available at http://www.sikorsky.com/sac/Home/0,9746,CLI1_DIV69_ETI541,00.html.
5. Defense Update. International Online Defense Magazine; Available at http://www.defense-update.com/.

6. Aerosonde Robotic Aircraft, March 2007; Available at http://www.Areosonde.com/index.php.
7. Stone H., *Configuration design of a canard configured tail-sitter unmanned vehicle using multidisciplinary optimization*, PhD Thesis, University of Sydney, Sydney, Australia, 1999.
8. Dragonfly Innovations, March 2007; Available at http://www.rctoys.com/.
9. Guedj D., *Le Theoreme du Perroquet*, Editions du Seuil, 1998.
10. Castillo P., Lozano R., Dzul A. E., *Modeling and Control of Mini-Flying Machines*, Springer, 2005.
11. *OSD UAV Roadmap 2002-2027*, Office of the Secretary of Defense (Acquisition, Technology, & Logistics), Air Warfare, December 2002.
12. *Unmanned Aircraft Systems Roadmap 2005-2030*, Office of the Secretary of Defense, August 2005.
13. Rosheim, M. E., *Leonardo's Lost Robots*, Springer 2006.

PART II

Modeling and Control Fundamentals

Chapter 3: Airplane Basic Equations of Motion and Open-Loop Dynamics[1]

The goal of this Chapter is to present fundamental background information related to the derivation of the basic equations of motion of a traditional airplane, explain how the airplane's position and orientation are determined with respect to a reference frame (Earth-fixed inertia reference frame), derive the aerodynamic forces that act on the airplane, define the corresponding control angles, and conclude with derivation of the open-loop dynamics. The material included in this Chapter is a very concise version of what may be found in any related textbook, and follows the same notation and derivation approach described in the references.

3.1 Introduction

The overall objective of this Chapter is to discuss the fundamental behavior of a traditional airplane in flight. It describes the kinematics properties and basic equations of motion of a generic airplane, where the term generic is used to emphasize that the airplane's structural components and flight control systems may be found in every 'traditional' airplane design.

Equations of motion are derived by implementing Newton's second law that deals with vector summations of all forces and moments as applied to the airplane relative to an inertial reference frame. However, for practical reasons, analysis may be significantly simplified if motion is described relative to a body-fixed reference frame attached to the airplane. When this is the case, the equations of motion are derived relative to this non-inertial frame. Further, Euler angles are used to define the airplane orientation relative to a general Earth-fixed inertial frame.

---

[1] Written by I. A. Raptis, K. P. Valavanis. This work has been supported partially by two Research Grants, ARO W911NF-06-1-0069 and SPAWAR N00039-06-C-0062.

Kimon P. Valavanis (ed.), Advances in Unmanned Aerial Vehicles, 49–72.

The equations of motion are nonlinear. Aerodynamic forces and moments are also nonlinear functions of motion characteristics and airplane controls. Linearization of the nonlinear equations is based on considering a specific configuration of airplane non-accelerating motion that is subject to small perturbations of linear and angular velocities from the reference non-accelerating steady flight. Under such constraints, the resulting perturbed aerodynamic forces and moments may be considered as linear functions of the perturbed linear and angular velocities, the airplane control angles, and their associated derivatives. This is a common practical approximation of real flight behavior, despite the fact that it is not based on a rigorous mathematical background.

This linearization results in obtaining a set of linear differential equations (for the perturbed model). Using Laplace transform one may obtain a set of algebraic equations for controller design purposes. This controller may be used for disturbance rejection. Subsequently, closed-loop controllers may be designed that meet set performance criteria and stability of flight. However, this is beyond the scope of this Chapter.

3.2 Equations of Motion

The equations of motion include derivation of the respective equations with respect to the body-fixed reference frame that is attached to the airplane, as well as position and orientation of the airplane relative to an Earth-fixed inertial frame.

The first step towards dynamic modeling of an airplane is to consider it as a rigid body with six degrees of freedom (DOF), followed by application of Newton's laws to the rigid body (airplane). As previously mentioned, an Earth-fixed inertial frame makes analysis impractical since moments and products of inertia vary with time. This is not the case when a body-fixed reference frame is considered, where moments and products of inertia are constant.

Figure 3.1 depicts the body-fixed reference frame (moving frame) that is attached to the airplane. The center C of the body-fixed reference frame C_{xyz} coincides with the center of gravity (CG) of the airplane. The C_{xz} plane coincides with the plane of symmetry of the airplane with the C_x and C_z axes pointing forward and downward, respectively. The C_y axis is perpendicular to the plane of symmetry in the direction of the right wing. The C_{xyz} body-fixed reference frame is a right-handed Cartesian coordinate system.

The linear velocity components of the CG along the C_x, C_y and C_z axes are defined as *U*, *V* and *W*, respectively. The angular velocity components about the axes of the body-fixed reference frame are defined as *P*, *Q* and *R*, respectively.

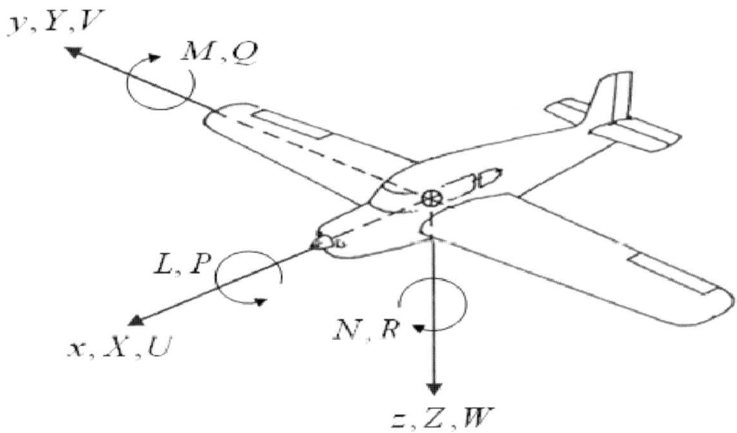

Fig. 3.1. Body-fixed coordinate system.

However, it is important to clarify that the linear and angular velocity vectors of the CG of the airplane are vectors relative to the Earth-fixed inertial frame, that is, vectors viewed by a stationary observer in the Earth-fixed inertial frame. The values of *U*, *V* and *W* are the instantaneous components of that vector relative to the body-fixed reference frame. The same holds for the angular velocities as well.

External aerodynamic forces components along the axes are denoted by *X*, *Y* and *Z*. The components about the axes of the external aerodynamic moments are denoted by *L*, *M* and *N* as shown in Figure 3.1.

Positive direction of the angular velocity components and of the moment components refers to the clockwise direction about the respective axis.

Basic concepts of kinematics analysis for rotating frames are used to derive the equations of motion. A more detailed presentation may be found in [8]. The first step is to define an Earth-fixed reference frame. It is a right handed Cartesian system denoted by $O_{x'y'z'}$. The underlying assumption is that the Earth is fixed in space, so $O_{x'y'z'}$ is an inertia frame.

As illustrated in Figure 3.2, $\vec{R}_0$ is the position vector of the origin *C* relative to the Earth-fixed reference frame. The set of the unit vectors for the body-fixed reference frame is denoted by $\{\hat{I}, \hat{J}, \hat{K}\}$. Point P is the position in space of a mass element dm of the airplane. Point P is rigidly attached to the body-fixed reference frame. The position vector of point P relative to the body-fixed reference frame is denoted by $\vec{r}$. If the coordinates of P relative to the body-fixed reference frame are (x, y, z) then:

$$\vec{r} = x\hat{I} + y\hat{J} + z\hat{K} \tag{3.1}$$

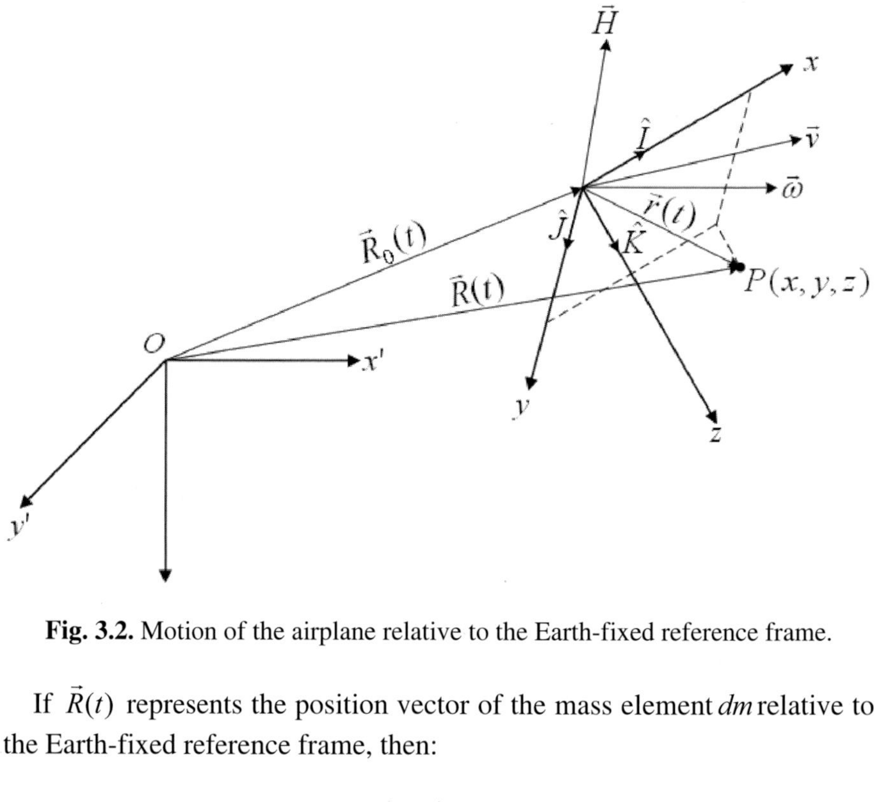

Fig. 3.2. Motion of the airplane relative to the Earth-fixed reference frame.

If $\vec{R}(t)$ represents the position vector of the mass element dm relative to the Earth-fixed reference frame, then:

$$\vec{R} = \vec{R}_0 + \vec{r} \tag{3.2}$$

The velocity of the mass element at point P relative to the Earth-fixed reference system is given by:

$$\vec{v}_P = \left.\frac{d\vec{R}}{dt}\right|_E = \left.\frac{d\vec{R}_0}{dt}\right|_E + \left.\frac{d\vec{r}}{dt}\right|_E \tag{3.3}$$

where $\left.\frac{d(\circ)}{dt}\right|_E$ denotes the time derivative of a vector in space relative to the Earth-fixed reference frame, as viewed by an observer in the Earth-fixed reference frame. The derivative of the position vector $\vec{R}_0$, relative to the Earth-fixed reference frame equals the velocity of the CG. The linear velocity of the airplane's CG is measured with respect to the Earth-fixed frame. Since the components of the linear velocity along the axes of the body-fixed reference frame are U, V and W, it follows that:

$$\vec{v} = \left.\frac{d\vec{R}_0}{dt}\right|_E = U\hat{I} + V\hat{J} + W\hat{K} \tag{3.4}$$

and $\hat{v}$ denotes the instantaneous velocity of the CG of the airplane relative to the Earth-fixed reference frame. The vector $\vec{r}$ is a position vector of the rotating body-fixed reference frame. According to [8], the time derivative of $\vec{r}$ with respect to the Earth-fixed reference frame is:

$$\left.\frac{d\vec{r}}{dt}\right|_E = \left.\frac{d\vec{r}}{dt}\right|_B + \vec{\omega}\times\vec{r} \tag{3.5}$$

where $\vec{\omega} = P\hat{I} + Q\hat{J} + R\hat{K}$ denotes the angular velocity of the body-fixed frame with respect to the Earth-fixed reference frame. The operator $(\times)$ is the vector cross product. The term $\left.\frac{d\vec{r}}{dt}\right|_B$ denotes the time derivative of the position vector $\vec{r}(t)$ with respect to the body-fixed reference frame. In general, $\left.\frac{d(\circ)}{dt}\right|_B$ denotes the derivative of a vector from the viewpoint of an observer in the body-fixed reference frame. Since point P is rigidly attached to the body-fixed reference frame, it follows that $\left.\frac{d\vec{r}}{dt}\right|_B = \vec{0}$. Hence, the velocity of the airplane's arbitrary element mass placed at the point P is given by:

$$\vec{v}_P = \left.\frac{d\vec{R}_O(t)}{dt}\right|_E + \vec{\omega}\times\vec{r}(t) \tag{3.6}$$

If u_{Px}, u_{Py} and u_{Pz} are the velocity components of the element mass dm along the axes of the body-fixed frame, then by equating both sides of (3.6) one obtains:

$$\begin{aligned} u_{Px} &= U + Qz - Ry \\ u_{Py} &= V + Rx - Pz \\ u_{Pz} &= W + Py - Qx \end{aligned} \tag{3.7}$$

The acceleration vector $\vec{a}$ of the airplane's CG is:

$$\vec{a} = \left.\frac{d\vec{v}(t)}{dt}\right|_E \tag{3.8}$$

Since $\vec{v}(t)$ is expressed in terms of the body-fixed frame unit vectors, and the body-fixed frame is rotating, following analysis presented in [3] and [8], the acceleration vector of the CG is given by the following equation:

$$\vec{a} = \left.\frac{d\vec{v}(t)}{dt}\right|_E = \left.\frac{d\vec{v}(t)}{dt}\right|_B + \omega \times \vec{v}(t) \tag{3.9}$$

But $\vec{v} = U\hat{I} + V\hat{J} + W\hat{K}$, therefore, $\left.\frac{d\vec{v}}{dt}\right|_B = \dot{U}\hat{I} + \dot{V}\hat{J} + \dot{W}\hat{K}$ since $\left.\frac{d\vec{v}}{dt}\right|_B$ is the time derivative of the velocity with respect to the body-fixed frame. It is clarified that the vector $\vec{a}$ is the instantaneous acceleration of the airplane's CG with respect to the Earth-fixed inertia frame. If a_x, a_y and a_z denote the instantaneous components of the vector $\vec{a}$ along the axis of the body-fixed reference frame, then from (3.9) the following algebraic equations are derived:

$$\begin{aligned} a_x &= \dot{U} - RV + QW \\ a_y &= \dot{V} - PW + RU \\ a_z &= \dot{W} - QU + PV \end{aligned} \tag{3.10}$$

If the vectors of all forces acting on the airplane are expressed in terms of their components $\sum X, \sum Y$ and $\sum Z$ along the respective axes of the body-fixed reference frame, then:

$$\begin{aligned} \sum X &= m(\dot{U} - RV + QW) \\ \sum Y &= m(\dot{V} - PW + RU) \\ \sum Z &= m(\dot{W} - QU + PV) \end{aligned} \tag{3.11}$$

To conclude derivation of the equations of motion, Newton's second law is applied to all moments that act on the CG. Let $\vec{H} = h_x\hat{I} + h_y\hat{J} + h_z\hat{K}$ be the vector of the airplane's angular momentum expressed in the body-fixed frame unit vectors. From [8], the angular momentum components of the body-fixed reference frame are expressed as a function of moments of inertia and products of inertia as:

$$\begin{aligned} h_X &= I_{XX}P - I_{XY}Q - I_{XZ}R \\ h_Y &= -I_{YX}P + I_{YY}Q - I_{YZ}R \\ h_Z &= -I_{ZX}P - I_{ZY}Q + I_{ZZ}R \end{aligned} \tag{3.12}$$

where $I_{XX} = \sum dm(y^2 + z^2)$, $I_{YY} = \sum dm(x^2 + z^2)$, $I_{ZZ} = \sum dm(x^2 + y^2)$ and the products of inertia are $I_{XY} = \sum dmxy = I_{YX}$, $I_{XZ} = \sum dmxz = I_{ZX}$, $I_{YZ} = \sum dmyz = I_{ZY}$.

The above sums apply to all elementary masses of the airplane, and x, y and z are the distances of each elementary mass from the origin (the CG). Moreover, since C_{xz} is a plane of symmetry for the airplane, it follows that $I_{XY} = I_{YX} = 0$ and $I_{YZ} = I_{ZY} = 0$.

The external moments equal the time rate of change of the angular momentum with respect to the Earth-fixed reference frame. Since the angular momentum is described by the unit vectors of the body-fixed frame, the following is true:

$$\left.\frac{d\vec{H}}{dt}\right|_E = \left.\frac{d\vec{H}}{dt}\right|_B + \vec{\omega} \times \vec{H} \tag{3.13}$$

The term $\left.\frac{d\vec{H}}{dt}\right|_E$ is the time rate of change of the angular momentum with respect to the Earth-fixed reference frame. Regarding $\left.\frac{d\vec{H}}{dt}\right|_B$, the time derivative of the angular momentum with respect to the body-fixed reference frame is derived as:

$$\begin{aligned}\dot{h}_X &= I_{XX}\dot{P} - I_{XZ}\dot{R} \\ \dot{h}_Y &= I_{YY}\dot{Q} \\ \dot{h}_Z &= -I_{ZX}\dot{P} + I_{ZZ}\dot{R}\end{aligned} \tag{3.14}$$

Let $\sum L, \sum M$ and $\sum N$ denote the moments of all forces about the axes of the body-fixed reference frame. Then:

$$\begin{aligned}\sum L &= \frac{dH_X}{dt} = I_{XX}\dot{P} + QR(I_{ZZ} - I_{YY}) - I_{XZ}(\dot{R} + PQ) \\ \sum M &= \frac{dH_Y}{dt} = I_{YY}\dot{Q} + PR(I_{XX} - I_{ZZ}) + I_{XZ}(P^2 - Q^2) \\ \sum N &= \frac{dH_Z}{dt} = I_{ZZ}\dot{R} + PQ(I_{YY} - I_{XX}) + I_{XZ}(QR - \dot{P})\end{aligned} \tag{3.15}$$

Therefore, the final form of the equations of motion with respect to the Earth-fixed frame but expressed in the body-fixed frame unit vectors is given by (3.11) for the forces and (3.15) for the moments.

3.3 Position and Orientation of the Airplane

The main disadvantage of using a body-fixed reference frame C_{xyz} attached to the airplane relates to the inability to express the airplane's position and orientation with respect to this body-fixed frame. Position and orientation of rigid bodies is defined with respect to fixed, inertial reference frames. Therefore, the airplane position and orientation equations should and will be derived relative to a generic Earth-fixed inertial reference frame. Derivation follows [4] but with additional details for clarification purposes.

A right-handed Cartesian system $O_{x'y'z'}$ is first defined as the Earth-fixed reference frame. Airplane directions at specific time instances are described by the orientation of body-fixed frames relative to the Earth-fixed reference frame. The origin of those frames is the CG of the airplane. At time instant $t = 0$ the CG of the airplane coincides with the origin of the frame $O_{x'y'z'}$.The initial position of the airplane is described by the frame $C_{x_1y_1z_1}$ that is aligned with $O_{x'y'z'}$. The final orientation of the airplane at time t is described by the body-fixed frame C_{xyz}. Figure 3.3 shows the schematics of deriving the orientation of the airplane.

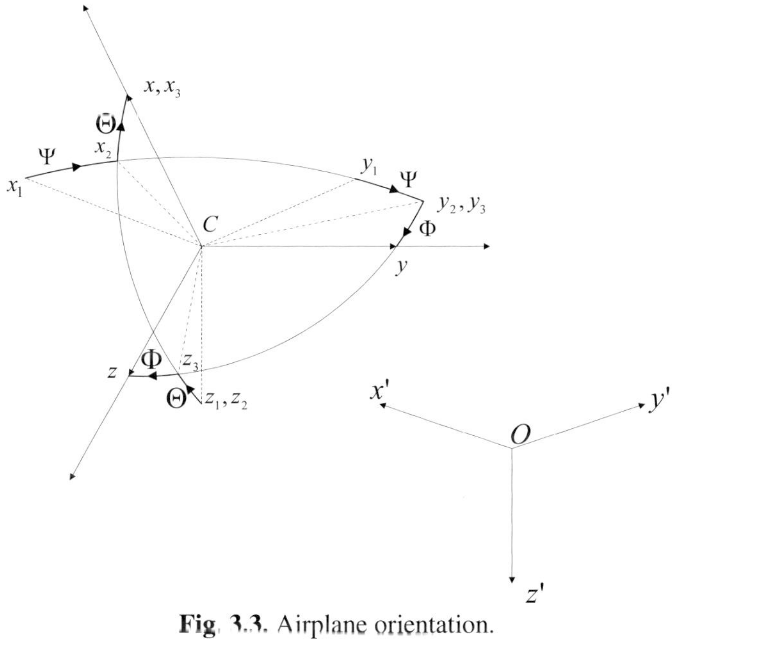

Fig. 3.3. Airplane orientation.

The airplane orientation at any time instant may be obtained by performing three consecutive rotations relative to the Earth-fixed frame; rotations are performed at a specific order, they cannot be considered as vectors and they are not commutative [8]. Therefore, the rotation order is important for consistency, as follows:

- A counterclockwise rotation of an angle Ψ about axis C_{z_1}. This rotation moves the airplane to the position described by $C'_{x_2y_2z_2}$, bringing C_{x_2} parallel to the plane C_{xz_2}.

- A counter clockwise rotation of an angle Θ about axis C_{y_2} .This rotation moves the airplane to the position described by $C_{x_3y_3z_3}$, aligning C_{x_3} with the C_x axis.
- A counterclockwise rotation of an angle Φ about axis C_{x_3} bringing the axes to their final direction C_{xyz} .

3.3.1 Airplane Position

The position of the airplane may be calculated by integrating velocity components at any time instant. Let $\hat{I},\hat{J},\hat{K}$ denote unitary vectors of C_{xyz}, $\hat{I}',\hat{J}',\hat{K}'$ unitary vectors of $O_{x'y'z'}$, and $\hat{I}_i,\hat{J}_i,\hat{K}_i$ unitary vectors of frames $Cx_iy_iz_i$, where $i=1,2,3$ .The linear velocity of the airplane relative to the C_{xyz} frame and relative to the Earth-fixed frame is, respectively:

$$\hat{v}=U\hat{I}+V\hat{J}+W\hat{K} \tag{3.16a}$$

$$\hat{v}=\frac{dx'}{dt}\hat{I}'+\frac{dy'}{dt}\hat{J}'+\frac{dz'}{dt}\hat{K}' \tag{3.16b}$$

The unit vectors of the body-fixed reference frame C_{xyz} are written relative to the frame $C_{x_3y_3z_3}$ as:

$$\begin{aligned}\hat{I}&=\hat{I}_3\\ \hat{J}&=\cos\Phi\hat{J}_3+\sin\Phi\hat{K}_3\\ \hat{K}&=-\sin\Phi\hat{J}_3+\cos\Phi\hat{K}_3\end{aligned} \tag{3.17}$$

The unit vectors of the frame $C_{x_3y_3z_3}$ are expressed relative to the frame $C_{x_2y_2z_2}$ as:

$$\begin{aligned}\hat{I}_3&=\cos\Theta\hat{I}_3-\sin\Theta\hat{K}_2\\ \hat{J}_3&=\hat{J}_2\\ \hat{K}_3&=\sin\Theta\hat{I}_2+\cos\Theta\hat{K}_2\end{aligned} \tag{3.18}$$

Finally, the unit vectors of the frame $C_{x_2y_2z_2}$ relative to $C_{x_1y_1z_1}$ are expressed as:

$$\begin{aligned}\hat{I}_2 &= \cos\Psi\hat{I}_1 + \sin\Psi\hat{J}_1 \\ \hat{J}_2 &= -\sin\Psi\hat{I}_1 + \cos\Psi\hat{J}_1 \\ \hat{K}_2 &= \hat{K}_1\end{aligned} \tag{3.19}$$

Substituting (3.17) to (3.19) to (3.16a) and equating with (3.16b) the following algebraic equations are obtained for the velocity components of the airplane relative to the Earth-fixed frame:

$$\begin{aligned}\frac{dx'}{dt} &= U\cos\Theta\cos\Psi + V(\sin\Phi\sin\Theta\cos\Psi - \cos\Phi\sin\Psi) + W(\cos\Phi\sin\Theta\cos\Psi + \sin\Phi\sin\Psi) \\ \frac{dy'}{dt} &= U\cos\Theta\sin\Psi + v(\sin\Phi\sin\Theta\sin\Psi + \cos\Phi\cos\Psi) + W(\cos\Phi\sin\Theta\sin\Psi - \sin\Phi\cos\Psi) \\ \frac{dz'}{dt} &= -U\sin\Theta + V\sin\Phi\cos\Theta + W\cos\Phi\cos\Theta\end{aligned} \tag{3.20}$$

Therefore, the position coordinates of the CG of the airplane are obtained by integrating (3.20). Due to integration complexity, (3.20) is usually linearized for simplification purposes.

3.3.2 Airplane Orientation

The orientation of the airplane is expressed with respect to the angular velocity components (P, Q, R) of the body-fixed reference frame.

Consider that during an infinitesimal time interval dt the airplane is subjected to three infinitesimal rotations $d\Psi, d\Theta$ and $d\Phi$ resulting in a position defined by angles $\Psi + d\Psi$, $\Theta + d\Theta$ and $\Phi + d\Phi$. Although finite rotations cannot be treated as vectors, infinitesimal rotations may be treated as such, thus, according to [4] the vector that represents the above rotation is:

$$\hat{n} = d\Phi\hat{I} + d\Theta.\hat{J}_3 + d\Psi\hat{K}_2 \tag{3.21}$$

The angular velocity is:

$$\vec{\omega} = \frac{d\vec{n}}{dt} = \dot{\Phi}\hat{I} + \dot{\Theta}.\hat{J}_3 + \dot{\Psi}\hat{K}_2 \tag{3.22}$$

The unit vectors of the frame $C_{x_3y_3z_3}$ are expressed with respect to the unit vectors of the reference frame C_{xyz} as:

$$\begin{aligned} \hat{I}_3 &= \hat{I} \\ \hat{J}_3 &= \cos\Phi\hat{J} - \sin\Phi\hat{K} \\ \hat{K}_3 &= \sin\Phi\hat{J} + \cos\Phi\hat{K} \end{aligned} \tag{3.23}$$

The unit vectors of the frame $C_{x_2y_2z_2}$ are expressed with respect to the unit vectors of the reference frame $C_{x_3y_3z_3}$ as:

$$\begin{aligned} \hat{I}_2 &= \cos\Theta\hat{I}_3 + \sin\Theta\hat{K}_3 \\ \hat{J}_3 &= \hat{J}_2 \\ \hat{K}_2 &= -\sin\Theta\hat{I}_3 + \cos\Theta\hat{K}_3 \end{aligned} \tag{3.24}$$

After substituting equations (3.23) and (3.24) to (3.22), equating with $\vec{\omega} = P\hat{I} + Q\hat{J} + R\hat{K}$ and then solving for $\dot{\Psi}, \dot{\Theta}, \dot{\Phi}$, one obtains the differential equations that relate the orientation angles of the airplane relative to the Earth-fixed frame with the components of the angular velocity expressed in the body-fixed frame unit vectors:

$$\begin{aligned} \dot{\Psi} &= Q\sin\Phi\sec\Theta + R\cos\Phi\sec\Theta \\ \dot{\Theta} &= Q\cos\Phi - R\sin\Phi \\ \dot{\Phi} &= P + Q\sin\Phi\tan\Theta + R\cos\Phi\tan\Theta \end{aligned} \tag{3.25}$$

Integration leads to the final orientation equations:

$$\begin{aligned} \Psi &= \int_0^t (Q\sin\Phi\sec\Theta + R\cos\Phi\sec\Theta)ds \\ \Theta &= \int_0^t (Q\cos\Phi - R\sin\Phi)ds \\ \Phi &= \int_0^t (P + Q\sin\Phi\tan\Theta + R\cos\Phi\tan\Theta)ds \end{aligned} \tag{3.26}$$

3.4 Aerodynamic Forces Acting on the Airplane

An important and very common term in flight terminology is *trimmed flight* that refers to the airplane being at equilibrium or in a non-accelerating motion. An airplane is at trim when the vector sum of all aerodynamic and gravitational forces acting on the CG equals zero. At trim, the sum of all aerodynamic moments acting on the CG is zero, too. Subsequent analysis follows the method found in [5].

Figure 3.4 depicts an airplane in longitudinal trim flight. V is the total velocity of the airplane CG that lies in its plane of symmetry. The angle between the body-fixed frame axis that points forward and the Earth-fixed frame plane $O'_{x'y'}$ is defined as Θ. In general, aerodynamic forces are not applied to the CG of the airplane [5]. However, since the airplane is performing a trimmed flight, all acting forces are in equilibrium. Since weight is acting on the CG, it may be considered, for that specific case, that aero dynamic forces act on the CG as well.

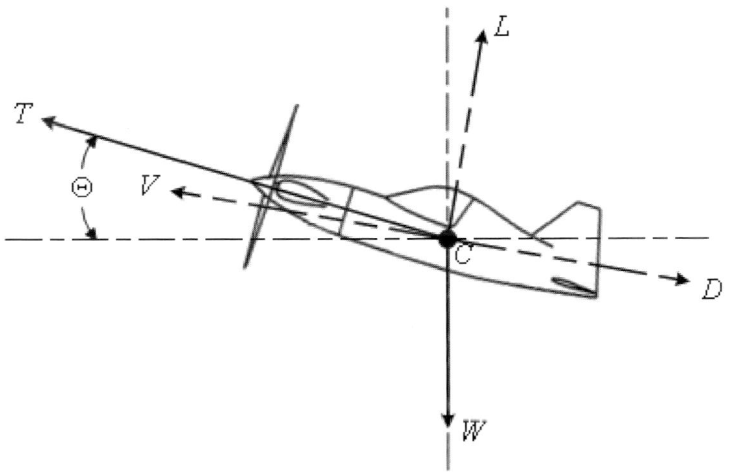

Fig. 3.4. Acting forces on the airplane during a trimmed longitudinal flight.

The first aerodynamic force is the thrust T acting on the body-fixed frame axis that points to forward direction. Thrust is produced by a propeller, a turbojet, or a rocket engine.

The second force is the lift L, normal to the velocity vector. The main component of the lift is the vertical force acting on the wings, facing upwards. The actual lift is the sum of all lifts produced in the wings, tail, even in the propellers.

The third force is the drug D, collinear with the velocity vector of the airplane. Its direction opposes the motion of the airplane. Drug is mainly composed of the parasite and the induced drug. The induced drug is a side effect of the lift generation. Parasite drug is created by airplane surfaces exposed to air.

The fourth force is the airplane weight W pointing downwards.

3.5 Angle of Attack and Sideslip Angle

The *angle of attack* and the *sideslip angle* shown in Figure 3.5 affect the aerodynamic forces of the airplane. Both angles are useful for describing the airplane dynamics and they are characteristics of its dynamic behavior.

The accurate definition of the angle of attack is given in [3] as the angle between the velocity vector of the relative wind and the wing cord. However, since this Chapter does not focus on specifics and details of aerodynamic forces, for simplicity purposes, the angle of attach will be defined in terms of the linear velocity components of the body-fixed reference frame as the angle between the velocity components of the C_x and the C_z axes:

$$\alpha = \tan^{-1}\frac{W}{U} \tag{3.27}$$

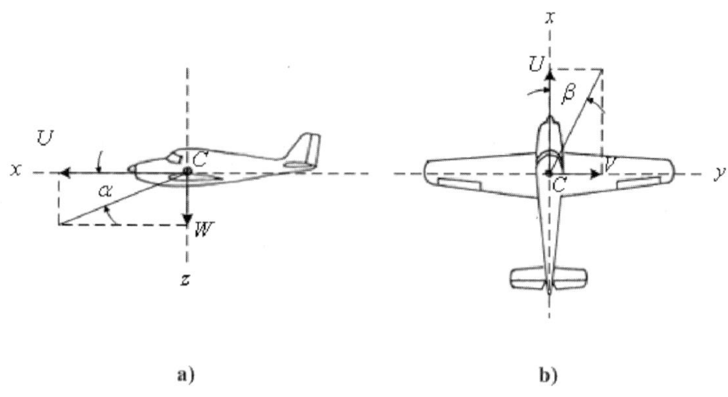

Fig. 3.5. a) Angle of attack; b) Sideslip angle.

The sideslip angle is defined for the lateral motion as:

$$\beta = \tan^{-1}\frac{V}{U} \tag{3.28}$$

The above definition is simplified; the strict definition may be found in [3] and [6]. The sideslip angle is not the same as the yaw angle. The definition of this angle is different depending on the type of motion the airplane performs. The first definition corresponds to longitudinal steady motion that experiences a small disturbance and moves the airplane about the C_z axis. The second definition is a result of asymmetric flight.

Due to the influence of the angle of attack on the lift force, it is an important parameter to control. On the contrary, the sideslip angle is considered a negative effect for airplane stability.

3.6 Weight Components in the Body-Fixed Frame

Weight along the body-fixed reference frame is calculated by the following equations:

$$\vec{W} = mg\hat{K}_1 = -mg\sin\Theta\hat{I} + mg\cos\Theta\sin\Phi\hat{J} + mg\cos\Theta\cos\Phi\hat{K}$$

$$\begin{aligned} X_g &= -mg\sin\Theta \\ Y_g &= mg\cos\Theta\sin\Phi \\ Z_g &= mg\cos\Theta\cos\Phi \end{aligned} \tag{3.29}$$

3.7 Airplane Control Angles

Airplane behavior is controlled by manipulating the angles of several movable parts on its surface. The airplane structural components associated with control actions are depicted in Figure 3.6. Control actions manipulate mainly the three moments about the CG; a secondary action of this type of control changes the external aerodynamic forces.

The horizontal tail is composed of: the forward part, which is fixed, called the *horizontal stabilizer*, and the rare part of the tail wing, which is a movable flap, called the *elevator*. The elevator angle, denoted as δ_e, affects longitudinal motion. Changes of elevator angle affect the equilibrium lift force and, more importantly, the pitching moment about the CG. When the elevator is deflected downwards, the elevator angle has a positive sign; when the deflection is upwards, the elevator angle has a minus sign.

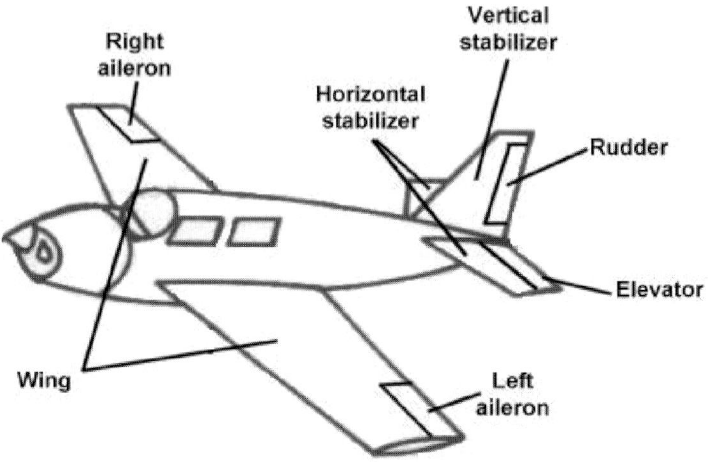

Fig. 3.6. Airplane configuration associated with control actions.

The movable aft portion of the vertical stabilizer is called the *rudder*. By controlling the rudder angle, denoted by δ_r, one manipulates the yawing moment about the C_z moving axis of the body-fixed frame. The effect of the rudder is useful for lateral directional motion and maneuvering. The rudder angle has a positive sign when the rudder is deflected at the left side of the vertical stabilizer's symmetry axis, and a negative sign when it is deflected on the right side.

The *ailerons* are movable parts hinged in the rear side of the wing. By controlling the ailerons angle one can change the airplane roll. The angles of the two ailerons have opposite directions in the two wings and they are not necessarily the same. The total aileron angle is considered to be the sum of the two, left and right aileron angle. The aileron is denoted as δ_a and $\delta_a = \delta_L + \delta_R$ where δ_R and δ_L are the absolute values of the left and right aileron angle, respectively. The aileron angle is considered positive when the right aileron is deflected downwards.

3.8 Open-Loop Dynamics

Equations describing the airplane motion are nonlinear differential equations. Linearizing these equations, under specific assumptions, is a common practice that simplifies greatly calculations and at the same time provides an accurate description of the actual behavior of the airplane. Derivations follow work described in [4] and [5].

Model linearization is based on small disturbance theory. According to the theory, analysis is done under small perturbations of motion characteristics (related to forces, momentums, velocities, angular velocities, etc.) from a steady non-accelerating reference flight. The rationale behind this approach is the fact that external aerodynamic forces and moments acting on the CG depend mainly on control angles and flight characteristics such as linear and angular velocities, angle of attack and sideslip angle. When this is the case, the perturbed aerodynamic forces and moments may be considered as linear functions of the disturbances [4].

Notation wise, parameter values with a zero subscript refer to the trimmed value of the variable, while lower case letters denote the perturbed value of that variable. The airplane is assumed to perform a reference trimmed flight when the disturbances occur. For simplicity purposes it is assumed that the body-fixed frame axes are identical to the airplane *stability axes.* When the airplane performs a longitudinal motion and the C_x axis is collinear and in the same direction with the total velocity of the airplane, with no sideslips, then those axes are termed as stability axes. It is obvious that for the stability axes the trim values of V and W are zero. Moreover, the trim values of the angular velocities and the moments about the CG will be zero as well. Summarizing the above leads to:

$$v_0 = w_0 = p_0 = q_0 = r_0 = \phi_0 = L_0 = M_0 = N_0 = 0$$

The motion of the airplane before the effect of the disturbances can be seen in Figure 3.7. The perturbation quantities and their derivatives will have very small values; therefore, their products are negligible. Without loss of generality, it is assumed that the trigonometric quantities of the perturbed variables, for example θ, will be $\cos\theta = 1$ and $\sin\theta = \theta$. Therefore:

$$\sin(\theta_0 + \theta) = \sin\theta_0 \cos\theta + \cos\theta_0 \sin\theta = \sin\theta_0 + \theta\cos\theta_0$$
$$\cos(\theta_0 + \theta) = \cos\theta_0 \cos\theta - \sin\theta_0 \sin\theta = \cos\theta_0 - \theta\sin\vartheta_0$$

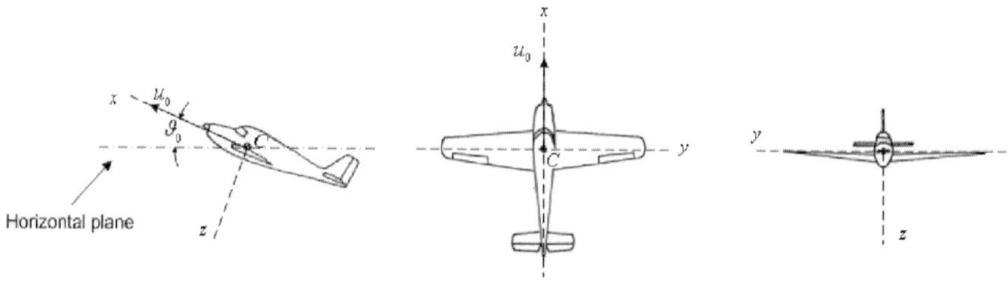

Fig. 3.7. Airplane motion before the effect of the disturbances.

Based on the above assumptions, substitutions into (3.11) and (3.15) results in the following perturbed equations:

$$\begin{aligned} &X_0 + \Delta X - mg(\sin\theta_0 + \theta\cos\theta_0) = m\dot{u} \\ &Y_0 + \Delta Y + mg\phi\cos\theta_0 = m(\dot{v} + u_0 r) \\ &Z_0 + \Delta Z + mg(\cos\theta_0 - \theta\sin\theta_0) = m(\dot{w} - u_0 q) \end{aligned} \tag{3.30}$$

$$\begin{aligned} &L_0 + \Delta L = Ixx\dot{p} - Ixz\dot{r} \\ &M_0 + \Delta M = I_{YY}\dot{q} \\ &N_0 + \Delta N = I_{ZZ}\dot{r} - I_{XZ}\dot{p} \end{aligned} \tag{3.31}$$

$$\begin{aligned} &\dot{\vartheta} = q \\ &\dot{\phi} = p + r\tan\theta_0 \\ &\dot{\psi} = r\sec\theta_0 \\ &p = \dot{\phi} - \dot{\psi}\sin\theta_0 \end{aligned} \tag{3.32}$$

$$\begin{aligned} &\frac{dx'}{dt} = (u_0 + u)\cos\theta_0 - u_0\theta\sin\theta_0 + w\sin\theta_0 \\ &\frac{dy'}{dt} = u_0\psi\cos\theta_0 + r \\ &\frac{dz'}{dt} = -(u_0 + u)\sin\theta_0 - u_0\theta\cos\theta_0 + w\cos\theta_0 \end{aligned} \tag{3.33}$$

In the above equations $\Delta X, \Delta Y, \Delta Z$ denote the perturbed values of the external aerodynamic forces and $\Delta L, \Delta M, \Delta N$ denote the perturbed values of the moments about the CG. When the airplane is at trim, the vector sum of all forces applied to the airplane is zero. Moreover the momentums about CG will be zero as well. Hence at trim:

$$\begin{aligned} &X_0 - mg\sin\theta_0 = 0 \\ &Y_0 = 0 \\ &Z_0 + mg\cos\theta_0 = 0 \\ &L_0 = M_0 = N_0 \end{aligned} \tag{3.34}$$

Substituting (3.34) in the perturbed model, the following two sets of equations are obtained:

$$\begin{aligned} &\Delta X - mg\theta\cos\theta_0 = m\dot{u} \\ &\Delta Z - mg\theta\sin\theta_0 = m(\dot{w} - u_0 q) \quad \text{(a)} \\ &\Delta M = I_{YY}\dot{q} \\ \\ &\Delta Y + mg\phi\cos\theta_0 = m(\dot{v} + u_0 r) \\ &\Delta L = Ixx\dot{p} - Ixz\dot{r} \qquad\qquad \text{(b)} \\ &\Delta N = I_{ZZ}\dot{r} - I_{XZ}\dot{p} \end{aligned} \tag{3.35}$$

The first set, (3.35a) includes terms and moments which lie in the C_{xy} plane and they describe the longitudinal motion of the airplane. The second set (3.35b) includes terms which tend to move the plane of symmetry itself. Those equations are the equations of lateral-directional motion, and they may be solved independently [5].

3.8.1 Stability Derivatives

Definition of the stability derivatives is based on [4] and [5]. Analysis of the perturbed aerodynamic forces and moments follows the assumption that the latter are functions of the angle of attack, the sideslip angle, the linear perturbed and the angular perturbed velocities, the control angles, and their respective derivatives. An additional assumption is that the disturbed forces and moments can be described in a linear manner. An example of the aerodynamic force in the C_x axis is:

$$\Delta X = (\frac{\partial X}{\partial u})_0 u + (\frac{\partial X}{\partial v})_0 v + (\frac{\partial X}{\partial \alpha})_0 \alpha + (\frac{\partial X}{\partial \dot{\alpha}})_0 \dot{\alpha} + (\frac{\partial X}{\partial p})_0 p + (\frac{\partial X}{\partial q})_0 q + (\frac{\partial X}{\partial r})_0 r +$$
$$(\frac{\partial X}{\partial \delta})_0 \delta + (\frac{\partial X}{\partial \phi})_0 \phi + (\frac{\partial \vartheta}{\partial \vartheta})_0 \vartheta + (\frac{\partial X}{\partial \psi})_0 \psi$$

where $(\frac{\partial X}{\partial u})_0$ denotes the partial derivative of the external aerodynamic force in the C_x axis, with respect to u and calculated under the trim flight condition. Stability derivatives are obtained by dividing the above partial derivatives by the airplane mass when the partial derivative corresponds to an external force. When the partial derivative corresponds to a moment, then the stability derivative is obtained by dividing by the appropriate mass inertia. An example is:

$$X_u = \frac{1}{m}(\frac{\partial X}{\partial u})_0$$
$$M_\alpha = \frac{1}{I_{YY}}(\frac{\partial M}{\partial \alpha})_0$$

The calculation of the stability derivatives is beyond the scope of this Chapter; however, details may be found in [2], [4] and [5], [6], [7].

In general not all stability derivatives are necessary for linearization of the forces or moments. The significance of each stability derivative in approximating forces and moments is summarized in Table 3.1, duplicated from [5]. The (+) entry indicates that the value of the specific stability derivative can be significant for calculation of force or moment. A (-) entry indicates that the value of the corresponding stability derivative is zero or insignificant for calculation of force or moment.

Independent variable	Dependent variable					
	X	Y	Z	L	M	N
u	-	-	+	-	-	-
v	-	+	-	+	-	+
α	+	-	+	-	+	-
$\dot{\alpha}$	-	-	+	-	+	-
p	-	-	-	+	-	+
q	-	-	+	-	+	-
r	-	+	-	+	-	+
δ_e	-	-	+	-	+	-
δ_r	-	+	-	+	-	+
δ_a	-		-	+	-	-
ϕ	-	+	-	-	-	
θ	+	-	-	-	-	-
ψ	-	-	-	-	-	-

Table 3.1. Stability derivatives significance (duplicated from [5]).

3.8.2 Longitudinal Motion

Due to perturbation in the velocity of the C_z axis a small angle of attack is generated, which may be approximated by:

$$\alpha = \frac{w}{u_0 + u} \approx \frac{w}{u_0} \tag{3.36}$$

since $u << u_0$. Under this assumption, and observing Table 3.1, the equations describing the longitudinal motion are written as:

$$\begin{aligned}
&X_u u + X_\alpha \alpha - g\theta\cos\theta_0 = \dot{u} \\
&Z_u u + Z_\alpha \alpha + Z_{\dot{\alpha}}\dot{\alpha} + Z_q \theta - g\theta\sin\theta_0 + Z_{\delta_e}\delta_e = u_0(\alpha - \dot{\theta}) \\
&M_\alpha \alpha + M_{\dot{\alpha}}\dot{\alpha} + M_q \dot{\theta} + M_{\delta_e}\delta_e = \ddot{\vartheta}
\end{aligned} \tag{3.37}$$

The value of the angular velocity q has been replaced by $\dot{\vartheta}$. The Laplace transform of the above equations is:

$$\begin{bmatrix} g\cos\theta_0 & s - X_u & -X_a \\ -s(Z_q + u_0) + g\sin\theta_0 & -Z_u & s(u_0 - Z_{\dot{\alpha}}) - Z_\alpha \\ s^2 - sM_q & 0 & -(sM_{\dot{\alpha}} + M_\alpha) \end{bmatrix} \begin{bmatrix} \theta(s) \\ u(s) \\ \alpha(s) \end{bmatrix} = \begin{bmatrix} 0 \\ Z_{\delta_e} \\ M_{\delta_e} \end{bmatrix} \delta_e(s) \tag{3.38}$$

3.8.3 Lateral-Directional Motion

Due to perturbation in the velocity of the C_y axis, a small sideslip angle is generated, which may be approximated by:

$$\beta = \frac{v}{u_0 + u} \approx \frac{v}{u_0} \tag{3.39}$$

since $u << u_O$. Under this assumption, and from Table 3.1, the equations describing the longitudinal motion are written as:

$$\begin{aligned} &Y_\beta\beta + Y_r r + Y_p p + Y_{\delta_r}\delta_r + g\phi\cos\theta_0 = u_0(\dot{\beta} + r) \\ &L_\beta\beta + L_r r + L_p p + L_{\delta_r}\delta_r + L_{\delta_a}\delta_a = \dot{p} - \frac{Ixz}{I_{XX}}\dot{r} \\ &N_\beta\beta + N_r r + N_p p + N_{\delta_r}\delta_r = \dot{r} - \frac{I_{XZ}}{I_{ZZ}}\dot{p} \end{aligned} \tag{3.40}$$

by substituting $r = \dot{\psi}\cos\theta_0$ and $p = \dot{\phi} - \dot{\psi}\sin\theta_0$ (3.40) takes the form:

$$\begin{aligned} &Y_\beta\beta + \left(Y_r\cos\theta_0 - Y_p\sin\theta_0 - u_0\cos\theta_0\right)\dot{\psi} + Y_p\dot{\phi} + Y_{\delta_r}\delta_r + g\phi\cos\theta_0 = u_0\dot{\beta} \\ &L_\beta\beta + \left(L_r\cos\theta_0 - L_p\sin\theta_0\right)\dot{\psi} + L_p\dot{\phi} + L_{\delta_r}\delta_r + L_{\delta_a}\delta_\alpha = \ddot{\phi} - \left(\sin\theta_0 + \frac{Ixz}{I_{XX}}\cos\theta_0\right)\ddot{\psi} \\ &N_\beta\beta + \left(N_r\cos\theta_0 - N_p\sin\theta_0\right)\dot{\psi} + N_p\dot{\phi} + N_{\delta_r}\delta_r = \left(\cos\theta_0 + \frac{I_{XZ}}{I_{ZZ}}\sin\theta_0\right)\ddot{\psi} - \frac{I_{XZ}}{I_{ZZ}}\ddot{\phi} \end{aligned} \tag{3.41}$$

The Laplace transform of the above equations will be:

$$\begin{bmatrix} su_0 - Y_\beta & \left(u_0\cos\theta_0 + Y_p\sin\theta_0 - Y_r\cos\theta_0\right)s & -(Y_p s + g\cos\theta_0) \\ -L_\beta & -\left\{\left(\sin\theta_0 + \frac{I_{XZ}}{I_{XX}}\cos\theta_0\right)s^2 + \left(L_r\cos\theta_0 - L_p\sin\theta_0\right)s\right\} & s^2 - L_p s \\ -N_\beta & \left(\cos\theta_0 + \frac{I_{XZ}}{I_{ZZ}}\sin\theta_0\right)s^2 + (N_p\sin\theta_0 - N_r\cos\theta_0)s & -\frac{I_{XZ}}{I_{ZZ}}s^2 - N_p s \end{bmatrix} \begin{bmatrix} \beta(s) \\ \psi(s) \\ \phi(s) \end{bmatrix} =$$

$$\begin{bmatrix} Y_{\delta_r} \\ L_{\delta_r} \\ N_{\delta_r} \end{bmatrix} \delta_r(s) + \begin{bmatrix} 0 \\ L_{\delta_a} \\ 0 \end{bmatrix} \delta_a(s) \tag{3.42}$$

Equations (3.38) and (3.42) are the two sets of equations that are used for controller design.

3.9 Conclusions

This Chapter presented background information related to how an airplane behaves in flight. Control of the airplane is accomplished by manipulating the angles of the airplane moving surfaces in the rare of the wings, the horizontal and vertical stabilizers.

The first challenge was to express equations with respect to the body-fixed reference frame attached to the airplane. By doing so, the moments and products of inertia have constant values over time. The difficulty in this derivation is that Newton's second law applies to inertial frames. Therefore, the idea is to express appropriately the inertia frame instantaneous linear and angular velocities and their derivatives relative to the body-fixed reference frame; then, Newton's second law is applied.

The derivation of the equations of motion relative to the body-fixed reference frame is not sufficient to describe the position and orientation of the airplane in space. Therefore, still using an Earth-fixed reference frame, the orientation of the airplane is given in terms of three Euler angles. More specifically, the airplane's orientation at any instant is determined by three sequential finite rotations, starting with a frame initially aligned with the Earth-fixed frame. The final position and orientation equations relative to the Earth-fixed frame are nonlinear and coupled. Since integration of those equations is highly complicated, numerical integration may be followed, or analysis should be restricted to special motion case studies.

The next step was the analysis of the aerodynamic forces and moments acting on the airplane under the assumption of a trimmed longitudinal flight subjected to small perturbations. Perturbations are small changes in orientation, linear and angular velocities and their derivatives. The perturbed aerodynamic forces may be expressed as linear combinations of the perturbed position, linear and angular velocities their derivatives and control angles. The resulting equations are linearized around conditions of trimmed longitudinal flight. Apart from the linearity advantage, it appears that the equations of motion may be split in two uncoupled subsets. The first set describes the longitudinal motion while the second set describes the lateral-directional motion. The two sets may be treated separately.

Since those equations are linear differential equations, Laplace transform is used to obtain the equations for designing closed-loop controllers.

References

1. Anderson J. D., *Introduction to Flight*, McGraw-Hill, 1989.
2. Babister A. W., *Aircraft Dynamic Stability and Response*, Pergamon Press, 1980.
3. Blakelock J. H., *Automatic Control of Aircraft and Missiles*, John Wiley & Sons, 1991.
4. Etkin B., *Dynamics of Flight: Stability and Control*, John Wiley & Sons, 1959.
5. McCormick B. W., *Aerodynamics Aeronautics and Flight Mechanics*, John Willey & Sons, 1995.
6. Perkins C. D., Hage R. E., *Airplane Performance Stability and Control*, John Wiley & Sons, 1965.
7. Seckel E., *Stability and Control of Airplanes and Helicopters*, Academic Press, 1964.
8. Williams J. H., *Fundamentals of Applied Dynamics*, John Wiley & Sons, 1996.

Chapter 4: Control Fundamentals of Small / Miniature Helicopters - A Survey[1]

Miniature helicopters are increasingly used in military and civilian applications, mainly due to their ability to hover, fly in very low altitudes and within confined spaces. However, due to model nonlinearities and inherent instabilities, low-level controller design for autonomous flights is a challenge. This Chapter presents an overview of major accomplishments in the area of unmanned helicopter control by several research groups, and focuses on techniques used for low-level control. It then describes a general model suitable for small or miniature helicopter non-aggressive flights and compares three different controllers, a PID, a Linear Quadratic Regulator (LQR) and an H_∞ controller in terms of their practical implementation to achieve autonomous, self-governing flights.

4.1 Introduction

The specific type of unmanned helicopter that is considered in this Chapter is the 'miniature version' of a conventional helicopter with a main rotor and a tail rotor as shown in Figure 4.1. This configuration offers many advantages, but at the same time, it imposes challenges on the design of low-level controllers.

Self-governing flights require generation of low-level signals sent to actuators as well as decision making related to guidance, navigation, path planning, motion planning, mission planning, communications, fault tolerance and emergency landing. When focusing only on sensors and actuators, a thorough understanding of helicopter control issues is needed before designing a controller.

Most control design techniques require at least availability of a system model describing plant dynamics, with higher levels of model details and

[1] Written by M. Castillo-Effen, C. Castillo, W. Moreno, K. P. Valavanis. This work has been supported partially by two Research Grants, ARO W911NF-06-1-0069 and SPAWAR N00039-06-C-0062.

Kimon P. Valavanis (ed.), Advances in Unmanned Aerial Vehicles, 73–118.

complexity implying a better approximation. Highly complex models are not practical for most advanced control design methods. For example, controllers designed assuming linear time-invariant models are usually of high order, giving rise to issues related to on-board processing requirements. Model reduction techniques are used to overcome such problems [42] [13]. Concentrating in non-aggressive flights, it is sufficient to evaluate controller design under forward flight and hovering [45].

Fig. 4.1. Small helicopter with a custom-made controller and pan-tilt camera.

Helicopter models are derived from first principles where most parameters have a specific interpretation, or using a 'black box' model where parameters do not have a physical meaning, or using 'hybrid models' that combine the previous two. Regardless, experimental data are required to tune model parameters via system identification techniques.

Models derived from the system's physics, like momentum and blade element theory, provide detailed information related to the physics and plant behavior [44] [45]; when combined, they offer a good balance between complexity and accuracy. The momentum method helps assessing thrust, forces and moments required to keep a helicopter airborne. The blade element method helps computing moments and forces by observing the rotor blades from the aerodynamic perspective. However, derivation of simple and accurate models for model-based control design requires careful judgment.

There is considerable published research in modeling and identification of miniature rotorcraft. In [36], a standard rotorcraft system identification tool called ***C****omprehensive* ***I****dentification from* ***F****requency* ***R****esponses* (CIFER) [61] was applied to the Yamaha R-50 and X-Cell 60 miniature helicopters. Scaling rules applied for predicting flying qualities of model helicopters based on known parameters of full-size helicopters were initially studied in [35] and extended in [27] by integrating under the same framework first principles and system identification techniques to a Yamaha R-50 helicopter. This approach was named ***Mo****deling for Flight* ***S****imulation and* ***C****ontrol* ***A****nalysis* (MOSCA). An accurate model of the smaller X-Cell helicopter with significantly different dynamics covering a broad spectrum of its flight envelope was proposed in [14].

The most basic modeling technique requires derivation of equations of motion of the fuselage, assuming that it is a rigid body, following the simple Newton-Euler equations [24] [36] using the concise wrench notation and quaternion algebra [19], or the energy oriented approaches such as the Lagrange formulation [4].

The next Section summarizes efforts in control of miniature rotorcraft, followed by introduction of concepts and specifications before actual controller design to achieve autonomous stable flight. This is followed by modeling fundamentals and an overview of control design techniques suitable for small rotorcraft.

4.2 Contributors to Miniature Helicopter Control

A sample list of major accomplishments on small / miniature helicopter control and research groups that have had an impact on autonomous flight are presented below. The list is neither comprehensive nor the most recent one as the number of groups worldwide conducting research in this area keeps on increasing.

The IARC/AUVSI Competition

The **I**nternational **A**erial **R**obotics **C**ompetition (IARC) has been organized by the **A**ssociation for **U**nmanned **V**ehicle **S**ystems **I**nternational (AUVSI) since 1991. Four missions have been proposed with increasing level of difficulty and realism, being autonomy the main mission tasks requirement.

In its current version, the fourth mission, introduced in 2001, involves several behaviors that must be completed in four qualification levels: i) *L1:* Autonomous flight over a distance of 3 km, waypoint following, hovering over the final waypoint; ii) *L2:* Identifying a target building and open portals autonomously; iii) *L3:* Entering the target structure and relaying reconnaissance information from within; iv) *L4:* While previous levels can be completed in stages, level 4 requires execution of the three previous level behaviors in less than 15 minutes.

No team has completed successfully the fourth mission. According to [18], the mission is 'beyond the capability of any system in existence today, including those of the super-power's military machines'.

The Sugeno Laboratory

Sugeno's research towards achieving autonomous flight control of miniature helicopters dates back to 1988 [41]. Most difficulties associated with autonomous flight were identified and solved to some extent by designing, testing and implementing Fuzzy Logic controllers.

The Carnegie Mellon University Robotics Institute

The CMU Robotics Institute group has reported results on autonomous flight control of helicopters since the early 1990's [3]. The CMU group was the first one that implemented vision-based techniques for navigation, in addition to other traditional sensors commonly used for the same purpose. Besides major accomplishments found in [36] and references therein, recent research in robust control for full-envelope flight control is reported in [26].

Draper Laboratories / MIT

The UAV designed and built at Draper Laboratories, called the ***D**raper **S**mall **A**utonomous **A**erial **V**ehicle* (DSAAV), was the only entry in the 1996 AUVSI competition that achieved fully autonomous flight [20].

In 2000, a new approach to UAV aggressive maneuvering using hybrid control techniques was introduced [32]. The main idea was based on incorporating a 'maneuver automaton' that selected optimally different control laws according to the motion primitive that needed be executed. This maneuver automaton concept was developed further and tested in simulation within the framework of the **S**oftware **E**nabled **C**ontrol (SEC) program [46]. An original avionics system and its application to a miniature acrobatic helicopter are also presented in [15] and [58].

The BEAR Project at UC Berkeley

The *aerobot* research team at UC Berkeley has consistently contributed to the field of VTOL type UAVs since 1996 [54]. Recent research deals with control of multiple UAVs [53] and the incorporation of obstacle avoidance strategies for navigation in urban environments [51].

The AVATAR Project at USC

Research at the University of Southern California started in 1991 with the first version of an **A**utonomous **F**lying **V**ehicle (AFV), winning the IARC competition in 1994 with the first generation of ***A****utonomous* ***V****ehicle* ***A****erial* ***T****racking* ***a****nd* ***R****etrieval* (AVATAR) helicopters [39]. The AVATAR software and control architecture is further explained in [48] along with other research efforts in autonomous landing and vision-based state estimation. The AVATAR main feature is its hierarchical behavior-based control architecture with all behaviors acting in parallel at different levels. An autonomous landing approach on a moving target and visual servoing in urban areas are the topics discussed in [34] and [50]. Behavior-based architectures for helicopter control have also been reported in [39] [11].

The SEC Program

The Software Enabled Control (SEC) program started in late fiscal year 1999 under Defense Advanced Research Projects Agency (DARPA) funding and sponsorship. Its goal, among other issues, is the search for solutions that would lead to greater levels of autonomy in man-made systems. Realization of complex controls for such systems involves major computational complexity concerns and requires computationally efficient techniques that can be implemented in real-time. Thus, computing plays a prominent role when dealing with such man-made systems and complex controls.

The main focus of the SEC program was to advance control technologies that improve UAV performance, reliability and autonomy. One of the main results was derivation and implementation of an ***O****pen* ***C****ontrol* ***P****lat form* (OCP), which enables development and deployment of control functions in terms of objects. In this way, 'object-oriented' control components are distributed across embedded platforms and enable coordination and cooperation among UAVs [46].

Further, a component based design environment called *Ptolemy* was developed, integrated with OCP, to allow for model based control design of heterogeneous systems taking into account the hybrid nature of most technical systems, as well as different models of computation.

Major contributions of the SEC program in the field of low-level VTOL vehicle control are in deriving several Model Predictive Control

(MPC) strategies and the so called *mode transition controller* that blends different linear controllers according to the corresponding appropriate flight mode. To this date, the SEC program has been the most comprehensive effort involving major companies and Universities across the US.

The Georgia Institute of Technology Program

The Georgia Institute of Technology research group is arguably the one group that has contributed the most to the UAV field, not only winning most IARC competitions, but also playing the crucial role in developing and implementing under the DARPA SEC program. Among the main contributions to the unmanned helicopter community the following stand out: i) A prototype implementation of OCP in the form of a fully rigged autonomous helicopter incorporating a fault detection and identification module to compensate for collective actuator failures; ii) A control design methodology for accommodating different flight modes and limit avoidance through mode transition controllers; iii) A helicopter testbed called GTMax, which includes a simulation model of a Yamaha R-Max helicopter with sensors, a ground control station and all baseline on-board routines. The first two components run on Windows platforms and the on-board routines run under QNX.

4.3 Challenges and Specifications

Before any control design takes place, it is essential to define a set of specifications that relate to the mission type and tasks an unmanned helicopter is required to fulfill, which in turn define maneuvers, agility, tracking accuracy and optimality criteria that need be satisfied. Design constraints may depend on physical limitations imposed by the flight mechanics and instrumentation. Availability of information regarding types of disturbances the helicopter will be exposed to, model accuracy and parameter uncertainty are also crucial in defining robustness measures. A limiting factor is the on-board computing capability that will determine the allowable level of software complexity of the control system to be finally implemented.

In manned aerial vehicles, control systems perform different levels of augmentation, while the pilot assesses the ease and precision with which tasks are performed; these are the so called 'handling qualities' or 'flying qualities' [44]. In unmanned rotorcraft, handling qualities involve artifacts that need be considered as these artifacts will operate at different levels of autonomy [17]. A valid set of goals and identification of critical missions and payloads has been defined in [64].

Regardless, the complete design cycle of control systems for autonomous unmanned helicopters is subject to severe challenges [57] with the most important listed below:

- Understanding helicopter traits and flight dynamics is a difficult task that requires, almost unavoidably, the use of flight simulators and learning how to pilot small model helicopters in order to acquire an understanding of the process.
- Most system identification techniques cannot be applied directly; hence a well trained pilot is needed to excite the most dominant rotorcraft modes. That is, perform maneuvers that are almost acrobatic while preventing the rotorcraft from crashing
- Data collection for system identification procedures is not a straight forward task, since experiments are limited by fuel autonomy constraints.
- Due to inherent instabilities, it is practically impossible to perform open-loop tests, the exception being hinged test-stands to perform preliminary tests.
- The configuration of the rotors forces a helicopter to act as a gyroscope creating high degrees of coupling between dynamic modes. Modes cannot be easily isolated for analysis usually performed on fixed-wing aircraft, where heave, longitudinal and lateral dynamics have very distinct features.
- Helicopters can perform several types of flights like hovering, flying forward, flying sideward, hovering turn, etc. Flight dynamics are different in each flight mode, so it is very unlikely that a single controller will handle all flight modes.
- Rotors are subject to highly nonlinear quasi-chaotic dynamics due to their turbulent interaction with air and wind gusts. Blades are flexible causing the system to have non-negligible high frequency dynamic components. This is a significant hurdle for deriving a high fidelity lumped model appropriate for control design. Models for high-fidelity simulation are easier to realize through application of computational fluid dynamics (CFD) techniques, albeit not applicable to control design methods.
- Vibration and noise affect main avionic instruments like inertial measurement units, compasses, GPS, etc. This problem may be alleviated, but not completely solved by finding appropriate locations for different instruments, shielding, and applying mechanical damping measures like shock absorbing joints. More recent helicopter designs based on

using electric motors offer improved vibration behavior. However, they usually offer less flight endurance than their combustion engine based counterparts.

- Payload limitations may be the main obstacles in miniature helicopters preventing the use of higher quality instrumentation, vision systems or computing platforms. This factor is tightly coupled to fuel autonomy limitations. The more the payload, the less time the helicopter is able to maintain autonomous flight.

4.4 Modeling and Simulation

This Section begins with a brief introduction to how a helicopter functions, followed by derivation of a model that will be used for simulation and controller design.

4.4.1 Basic Helicopter Function

A helicopter may be studied as a vehicle with six Degrees of Freedom (DOF) that is subject to non-holonomic constraints. It changes its position along three coordinate axes, longitudinal *X*, lateral *Y* and vertical *Z*; its attitude is described by three angles, pitch (θ), roll (ϕ), and yaw (ψ) as shown in Figure 4.2. The configuration space is $SE(3) = R^3 \times SO(3)$. For motion planning purposes, the helicopter configuration space may be reduced to $R^3 \times S^1$ that corresponds to changing the position and heading, since the pitch and roll cannot be chosen arbitrarily.

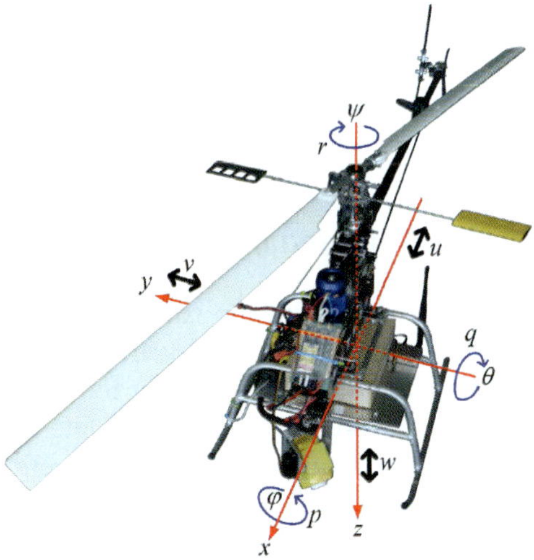

Fig. 4.2. Helicopter and its six degrees of freedom.

Components that produce moments and forces are the helicopter main rotor, the tail rotor, the vertical fin and the horizontal stabilizer. The main rotor blades can change their pitch simultaneously creating different levels of lift. Under manual control mode the pilot may command the pitch of the blades by the *collective* control (u_{col}), used to control heave motion. The main rotor is affected by the *cyclic* control that varies the pitch of the blades cyclically creating different horizontal propulsive forces at different angles. This in turn causes the helicopter to move in the lateral and longitudinal directions. For this reason, cyclic control is usually decomposed into lateral and longitudinal components, u_{lat} and u_{lon}, respectively. Given the rate gyro stabilization mechanism, the pitch of the rear rotor is used to counteract the moment created by the main rotor that can make the helicopter spin. However, the pitch of the rear rotor can also be commanded around the stabilization point allowing the helicopter to change its heading. In real size helicopters this control is accessible through pedals, similar to the rudder in fixed-wing aircraft, thus the name *pedal* control, u_{ped}. The speed of the rotors is usually kept constant through a governor stabilization mechanism within the throttle. When not constant, it creates variable amounts of vertical thrust with a similar effect as the collective control within certain limits.

A component that helps stabilizing longitudinal and lateral dynamics improving ability to fly a model helicopter is the *flybar*. The Bell-Hiller stabilizer bar provides mechanical rate feedback (damping), but it also increases the dynamic complexity of the main rotor and adds aerodynamic drag. It has been shown that this may be safely eliminated improving closed-loop performance under computer control as well as energy savings, which results in an increase of payload capacity or flight endurance as shown in [22].

4.4.2 Modeling, System Dynamics, System Identification

Considering only the fuselage and the rotors as sources of moments and forces it is possible to derive low-fidelity models used for basic assessment of helicopter handling qualities, but not for simulation or control design. Higher fidelity models require additional state variables that consider model dynamics in higher frequency regions, as well as dynamic features of the rotor such as flapping and coupling between body and main rotor. In most cases, actuators add up additional lags that need be taken into account as first or second order elements.

A key component of a helicopter model is transformation between reference frames. Rigid body equations are derived with respect to the body frame of reference that is fixed on the helicopter. However, to simulate the helicopter in the inertial reference frame, a transformation is needed. If the helicopter attitude is parameterized in terms of Euler angles, the resulting transformations may be given as rotation matrices that are functions of roll, pitch and yaw. Using s , c and t as abbreviations for *sin(.)*, *cos(.)* and *tan(.)*, the respective transformations with variables in the inertial frame on the left hand side and variables in the body frame on the right hand side are:

$$\begin{bmatrix} v_x \\ v_y \\ v_z \end{bmatrix} = \begin{bmatrix} c\theta c\psi & s\phi s\theta c\psi - c\phi s\psi & c\phi s\theta c\psi + s\phi s\psi \\ c\theta s\psi & s\phi s\theta s\psi - c\phi c\psi & c\phi s\theta s\psi - s\phi c\psi \\ -s\theta & s\phi c\theta & c\phi c\theta \end{bmatrix} \begin{bmatrix} u \\ v \\ w \end{bmatrix} \quad (4.1)$$

$$\begin{bmatrix} \dot{\phi} \\ \dot{\theta} \\ \dot{\psi} \end{bmatrix} = \begin{bmatrix} 1 & \sin(\phi)\tan(\theta) & \cos(\phi)\tan(\theta) \\ 0 & \cos(\phi) & -\sin(\theta) \\ 0 & \sin(\phi)\sec(\theta) & \cos(\phi)\sec(\theta) \end{bmatrix} \begin{bmatrix} p \\ q \\ r \end{bmatrix} \quad (4.2)$$

For system identification and control design purposes, the helicopter nonlinear model may be linearized for specific flight regimes [27] [36] [52]. Parameters are extracted such that the model may be written in the state-space form:

$$\dot{\mathbf{x}} = \mathbf{A}\mathbf{x} + \mathbf{B}\mathbf{u} \tag{4.3}$$

with the state vector $\mathbf{x}$ and the control $\mathbf{u}$ given by:

$$\mathbf{x} = \begin{bmatrix} u & v & p & q & \phi & \theta & a & b & w & r & r_{fb} \end{bmatrix}^T \tag{4.4}$$

$$\mathbf{u} = \begin{bmatrix} \delta_{lat} & \delta_{lon} & \delta_{ped} & \delta_{col} \end{bmatrix} \tag{4.5}$$

The state vector includes 11 linear velocity variables[2]: linear velocities in x, y, and z directions, u, v and w, respectively; attitude variables: roll ϕ, and pitch θ; angular rates for roll, pitch and yaw: p, q, and r, respectively; rotor longitudinal and lateral flapping: a and b, respectively; yaw rate gyro feedback r_{fb}.

The control vector is composed from deviations from trim conditions of the four control variables cyclic ($\delta_{lat}, \delta_{lon}$), collective ($\delta_{col}$), pedal ($\delta_{ped}$).

The specific state matrices ($\mathbf{A}, \mathbf{B}$) for different flying conditions and different types of helicopters, determined through system identification procedures performed before the control system design, have been obtained and tabulated in [52] and [36]. In this Chapter, the parameters obtained for the linearized model of a Yamaha R-50 in hovering flight [36] will be used for assessing application of commonly used techniques for small helicopter control analysis and design. Since parameters are associated with linear models, it is expected that they change according to the current flight mode. The specific parameter values used in [36] are also shown below:

[2] A more basic version of the helicopter model may include only eight state variables, neglecting the last three flapping variables and the yaw rate gyro feedback [51].

$$\mathbf{A} = \begin{bmatrix} -0.0505 & 0 & 0 & 0 & 0 & -32.2 & -32.2 & 0 & 0 & 0 & 0 & 0 & 0 \\ 0 & -0.154 & 0 & 0 & 32.2 & 0 & 0 & 32.2 & 0 & 0 & 0 & 0 & 0 \\ -0.144 & 0.143 & 0 & 0 & 0 & 0 & 0 & 166 & 0 & 0 & 0 & 0 & 0 \\ -0.0561 & -0.0585 & 0 & 0 & 0 & 0 & 82.6 & 0 & 0 & 0 & 0 & 0 & 0 \\ 0 & 0 & 1 & 0 & 0 & 0 & 0 & 0 & 0 & 0 & 0 & 0 & 0 \\ 0 & 0 & 0 & 1 & 0 & 0 & 0 & 0 & 0 & 0 & 0 & 0 & 0 \\ 0 & 0 & 0 & -1 & 0 & 0 & -21.74 & -4.109 & 0 & 0 & 0 & 14 & 0 \\ 0 & 0 & -1 & 0 & 0 & 0 & 8 & -21.74 & 0 & 0 & 0 & 0 & 15.43 \\ 0 & 0 & 0 & 0 & 0 & 0 & -9.75 & -131 & -0.614 & 0.93 & 0 & 0 & 0 \\ 0 & 0.0301 & -3.53 & 0 & 0 & 0 & 0 & 0 & 0.0857 & -4.13 & -33.1 & 0 & 0 \\ 0 & 0 & 0 & 0 & 0 & 0 & 0 & 0 & 0 & 2.16 & -8.26 & 0 & 0 \\ 0 & 0 & 0 & -1 & 0 & 0 & 0 & 0 & 0 & 0 & 0 & -2.924 & 0 \\ 0 & 0 & -1 & 0 & 0 & 0 & 0 & 0 & 0 & 0 & 0 & 0 & -2.924 \end{bmatrix}$$

$$\mathbf{B} = \begin{bmatrix} 0 & 0 & 0 & 0 \\ 0 & 0 & 0 & 0 \\ 0 & 0 & 0 & 0 \\ 0 & 0 & 0 & 0 \\ 0 & 0 & 0 & 0 \\ 0 & 0 & 0 & 0 \\ 0.6804 & -2.174 & 0 & 0 \\ 3.043 & 0.3 & 0 & 0 \\ 0 & 0 & 0 & -45.8 \\ 0 & 0 & 33.1 & -3.33 \\ 0 & 0 & 0 & 0 \\ 0 & -0.7573 & 0 & 0 \\ 0.7982 & 0 & 0 & 0 \end{bmatrix}$$

Given the availability of experimental input - output data and a model as the one in (4.3), there are readily available numerical implementations of Prediction Error Identification Methods (PEM) that allow for optimally determining unknown parameters [29] [30]. Following approaches that resemble Maximum Likelihood Estimation (MLE), PEM methods find estimated parameters by seeking minimization of a function which is a norm of the prediction error sequence, according to:

$$\hat{\theta}_N = \underset{\theta \in D_M}{\operatorname{argmin}} \frac{1}{N} \sum_{t=1}^{N} \ell(\varepsilon_F(t,\theta)) \tag{4.6}$$

The prediction error sequence is a function of the data, the parameters of the assumed model and a stable linear filter $L(q)$:

$$\varepsilon_F(t,\theta) = L(q)\left[y(t) - \hat{y}(t \mid \theta_*)\right] \tag{4.7}$$

Knowing the parameters for a specific flight regime, the main modes that define the key dynamics of the helicopter may be obtained via eigenvalues analysis and exciting the system with the corresponding eigenvectors. Given (4.3), there exist seven basic modes under hovering shown in Table 4.1.

Mode	Natural Frequency (rad/sec)	Damping	Time Constant (seconds)
1	0.32	-0.96	3.27
2	0.41	0.98	2.50
3	0.61	1.00	1.64
4	8.36	0.20	0.59
5	11.86	0.22	0.38
6	10.28	0.60	0.16
7	20.86	0.97	0.05

Table 4.1. Modes in hovering flight.

Mode 1 is an unstable slow phugoid mode that affects mainly the lateral and longitudinal linear velocities. Mode 2 is a barely under damped stable phugoid mode involving the horizontal and vertical linear velocities. Mode 3 is a damped mode affecting the heave velocity. Mode 4 is an under damped mode involving roll, pitch and yaw angular rates, as well as the heave velocity. This mode shows clearly the gyroscopic coupling between rotational speeds. It illustrates the fact that attitude dynamics is faster than translational dynamics. Mode 5 is an under damped mode affecting the pitch angular rate and the heave velocity. Mode 6 is a relatively fast under damped mode showing coupling between yaw and heave velocity. Mode 7 is a fast critically damped mode that shows strong coupling between fast changes in pitch and heave velocity.

This modal analysis illustrates the degree of coupling among different state variables at specific frequencies and helps obtaining a better understanding of plant dynamics. It may be used to verify qualitatively correctness of system identification procedures.

Figure 4.3 shows the plot of singular values against frequencies for the used linearized model and obtained modes. The plot provides a graphical tool for obtaining a measure of the magnitude of the gain of a multivariable system.

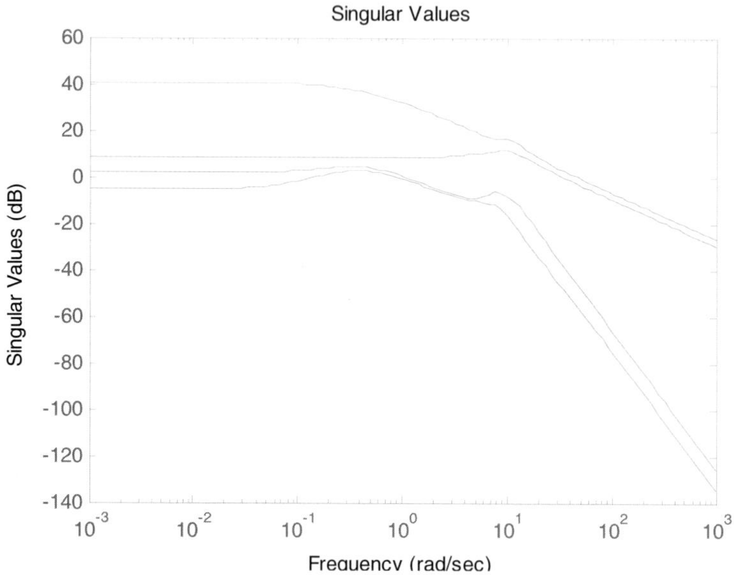

Fig. 4.3. Singular values of Yamaha R-50 in hovering mode.

The largest singular value starts declining at a corner frequency of 0.4rad/sec at -20dB/Dec. The smallest singular value decays rapidly after 10rad/sec at -60dB/Dec. A resonant peak appears to take place around 10rad/sec. The frequency response shows that in general, most relevant features are found between 0.1rad/sec and 100rad/sec.

One attribute of helicopter dynamics with important implications for control design is the non-minimum phase characteristic. This means that when linearized, the model has transmission zeros in the right half plane. This fact has been proven in [25], used in [54], and confirmed in [12] and [37]. Non-minimum phase systems present and additional challenge for application of certain control methods such as in Linear Quadratic Control / Loop Transfer Recovery (LQR/LTR) or for dynamic inversion procedures. It has been shown that under certain assumptions, the system may be analyzed as being minimum-phase as highlighted in [54] by using the term 'slightly non-minimum phase'.

4.5 Overview of Control Techniques

Controller design for small unmanned helicopters requires addressing and overcoming at least the following characteristics that are specific to these flying machines:

- High degree of coupling among different state vector different variables.
- Nonlinear behavior that makes the linear model valid only for certain regions of the flight envelope.
- Difficulties associated with obtaining accurate models due to issues related to application of standard system identification procedures and methodologies, resulting in models with high parameter uncertainty.
- Open-loop instability.
- Dynamics spanning a wide range of frequencies.
- Very fast dynamics, especially in the case of very small model helicopters.
- Diverse sources of noise and disturbances. The helicopter is subject to effects of rotor wake, wind gusts, and vibrations. To reduce payload the avionics system incorporates lower grade instruments and sensors that increase measurement noise. Light structures imply flexibility, a major source of unmodeled dynamics.

To overcome such difficulties, modern ***Computer-Aided Control System Design*** (CACSD) technologies offer tools to design controllers that satisfy conditions imposed by different applications. Basic approaches and commonly used techniques for design of helicopter control systems are elaborated next, putting emphasis on simplified techniques with practical relevance.

4.5.1 Low-Level Control Architecture

There exist low-level control structures that may serve as a foundation for new approaches to designing small helicopter controllers, or for application of already well-known methodologies.

Cascaded Control

Based on prior experience in piloted aeronautical vehicles [1], the cascaded control approach [21] [47] considers that the helicopter control system is decomposed into two loops where reference signals for the inner-loop are produced by controllers in the outer-loop as shown in Figure 4.4.

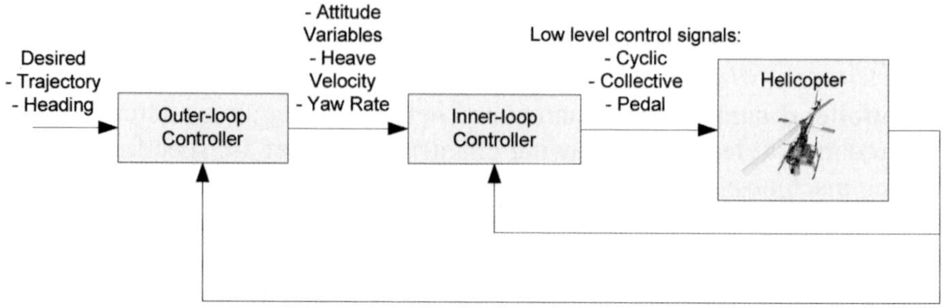

Fig. 4.4. Inner- and outer- loop cascaded control.

The inner-loop stabilizes the helicopter and decouples state variables controlling attitude. Typically, design of the inner-loop aims at achieving high bandwidth and robust stability. The outer-loop is used for guidance, generating velocity commands for the inner-loop. It operates at a slower time scale than the inner-loop.

Further, one of the main tasks of the inner-loop is to provide adequate decoupling such that outer-loop controllers may control each variable independently. Hence, a coupling metric is used to evaluate the degree of success with which the inner-loop controllers achieves decoupling.

Viewed from the behavior-based robotics perspective, the inner-loop may be considered as a set of reactive behaviors that act at the lowest level of a 'subsumption architecture' [7], while the outer-loop includes behaviors with higher levels of competence.

Sensor Fusion in Miniature Helicopters

Sensor fusion is essential for estimating non-measurable state variables and filtering, and also for performing system identification procedures. The main task of the data fusion module is to yield optimal estimates for key state variables using sensor measurements. State estimation techniques are also needed since almost all most small helicopters are equipped with low cost light-weight sensors prone to noisy measurements at different data rates and they are subject to considerable vibration and turbulent gusts. Reference [49] offers an excellent comparative overview of avionic systems and sensor fusion algorithms of some of the early autonomous helicopter projects. However, small helicopters are equipped at a minimum with the following sensors:

- An on-board Inertial Measurement Unit (IMU) with three angular rate gyros, three orthogonal accelerometers and three orthogonal magnetometers to measure the Earth's magnetic field, measuring dynamic and static attitude of the rotorcraft. The challenge is in estimating and eliminating bias drift in IMU measurements. This may be accomplished by including biases as states, and by applying a period of pre-flight warm up. Attitude measurements may be obtained at relatively short sampling intervals, 10ms or 100Hz. When the IMU does not include magnetometers, they need be added externally.

- A Global Positioning System (GPS) receiver that supplies position estimates. GPS receivers have several sources of errors such as atmospheric and multi path effects. Four satellites need be in the Line of Sight (LOS) of the receiver – 'GPS lock'. GPS are relatively slow measurement devices and samples can generally obtained at 1Hz, with temporary dropouts when GPS lock is lost. A similar technology called differential GPS may provide very accurate position estimates (2 cm) with the main disadvantage being the need of a base station with a precisely known location and higher cost of the receiver.
- A barometric altimeter that is used to provide altitude information. This measurement may be considered complementary to GPS readings. Measurements are relatively slow (20Hz) and noisy. However, sensor fusion filters can successfully eliminate it. For lower altitudes, in particular for enabling autonomous take-off and landing maneuvers, ultrasound sensors and radar may also provide accurate altitude readings.

The Kalman filter (KF) calculates optimal estimates provided that two main conditions are met: the system is linear and noise is Gaussian [8] [56] [65]. Since helicopter motion equations are evidently nonlinear and sensor noise is not necessarily Gaussian, the extended Kalman filter (EKF) relaxes the linearity condition by applying linearization of the model, implemented in two steps of prediction and update. The EKF computational complexity is not greater than the KF; therefore, it is best to follow the EKF recursive estimation equations.

To derive the EKF equations, it is necessary to know the magnitude and model sensor noise and system noise associated with the helicopter model. The helicopter model may not incorporate the full state vector of (4.4) but only the state variables of the kinematics model. However, the state vector may be augmented with the gyro biases to account for the required assumption of Gaussian noise. The EKF state vector $\mathbf{x}$ usually includes the inertial position x, y, z, the body frame velocities: u, v, w, the attitude quaternion: e_0, e_1, e_2, e_3, and the gyro biases: p_{bias}, q_{bias} and r_{bias}. The system equation based on kinematics has the form:

$$\dot{\mathbf{x}} = f(\mathbf{x}, \mathbf{a}, \omega) + vN(0, Q) \tag{4.8}$$

In (4.8), **a** and ω are measured accelerations and angular rates; its discrete-time version yields the state estimate $\hat{\mathbf{x}}$ according to the kinematics model, while the system matrix $\mathbf{F}$ is obtained (before obtaining the Kalman gain) calculating the Jacobian of the function $f(\cdot)$:

$$\mathbf{F} = \left. \frac{\partial f}{\partial \mathbf{x}} \right|_{\mathbf{x}=\hat{\mathbf{x}}} \tag{4.9}$$

Additional sensing information from cameras can be added to the standard instrumentation list shown above. Successful incorporation of vision to the instrumentation system of autonomous helicopters has been documented in [9] and [34].

4.5.2 Control Techniques Applied to Helicopter Flight Regimes

Linear controllers allow for determining exact measures of closed-loop performance, stability and robustness. Nonetheless, these techniques rely on the strong assumption of the plant being linear, which holds true only for a region of the state-space around an operating point. When a nonlinear system traverses different regions of the state space, several linear models may be used to approximate the overall nonlinear dynamics. Thus, several controllers need be designed and 'activated' according to the linear model that approximates best local dynamics. Since the controllers do not necessarily have different structures, only parameters may need to change giving rise to a gain-scheduling scheme. If controllers with different structures are needed, blending techniques may be necessary to achieve smooth transitions between them, as illustrated in Figure 4.5.

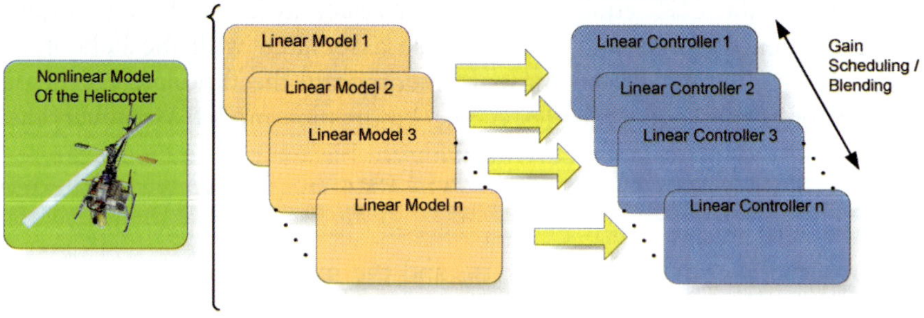

Fig. 4.5. Divide and conquer approach - linear control.

Coupling Analysis

The helicopter as a whole may be considered as a square system with four inputs and outputs as shown in Figure 4.6, with the best input / output pairing shown in Table 4.2.

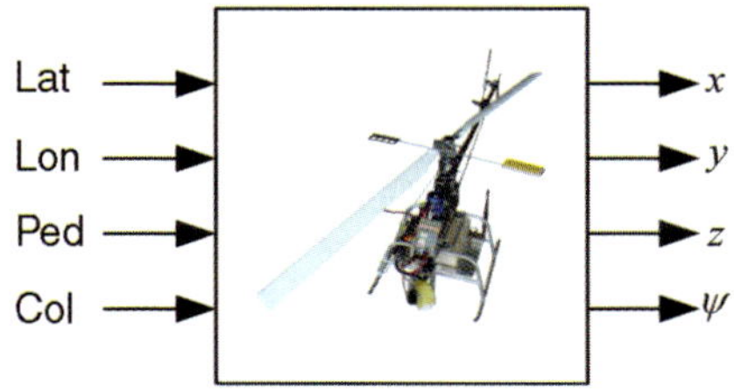

Fig. 4.6. Helicopter viewed as a square system.

Input	Pairing
Lateral Cyclic (u_{lat}), Longitudinal Cyclic (u_{lon})	Position in Horizontal Plane (x, y)
Collective (u_{col})	Altitude (z)
Pedal (u_{ped})	Yaw (ψ)

Table 4.2. Input/output pairing.

However, following decomposition in inner- and outer- loop, internal variables to be paired with inputs are not as unequivocal. The most commonly used input / output pairs for the inner-loop controller are shown in Table 4.3, with each pair often referred to as 'channel', sometimes treated in isolation for tuning purposes.

Input	Pairing 1	Pairing 2
Lateral Cyclic (u_{lat})	Roll (ϕ)	Lateral speed (v)
Longitudinal Cyclic (u_{lon})	Pitch (θ)	Longitudinal speed (u)
Pedal (u_{ped})	Yaw rate (r)	Yaw rate (r)
Collective (u_{col})	Heave velocity (w)	Heave velocity (w)

Table 4.3. Most commonly used input/output pairings.

Two main metrics are considered to quantify the degree of interaction among channels justifying a particular pairing: the relative gain array (RGA), or gain array number, RGAN, and diagonal dominance. The $(i,j)^{\text{th}}$ element of the RGA $\Lambda(\omega)$ [16] [57] is calculated for all relevant frequencies as:

$$\lambda_{ij} = [G(\omega)]_{ij}\left[G^{-1}(\omega)\right]_{ji} \tag{4.10}$$

The RGAN summarizes results obtained from the RGA into one number according to:

$$\text{RGA number} \overset{\Delta}{=} \left\|\Lambda(G) - I\right\|_{\text{sum}} \tag{4.11}$$

Here $\left\|A\right\|_{\text{sum}} = \sum_{i,j} |a_{ij}|$. It penalizes the off-diagonal elements; therefore smaller RGAN implies better decoupling.

For a square system with n inputs and outputs, diagonal dominance is also used to verify decoupling, its elements being computed as a function of frequency according to:

$$\Phi_j(j\omega) = \frac{\sum_{i=1, i\neq j}^{n} |z_{ij}(j\omega)|}{|z_{jj}(j\omega)|} \tag{4.12}$$

Values that are smaller than 1 indicate a level of diagonal dominance that allows for decoupling. Values greater than 1 indicate high degree of coupling.

Considering the linear model of the Yamaha R-50 in hovering flight, Figures 4.7 and 4.8 depict the RGAN and diagonal dominance for the two pairings of Table 4.3.

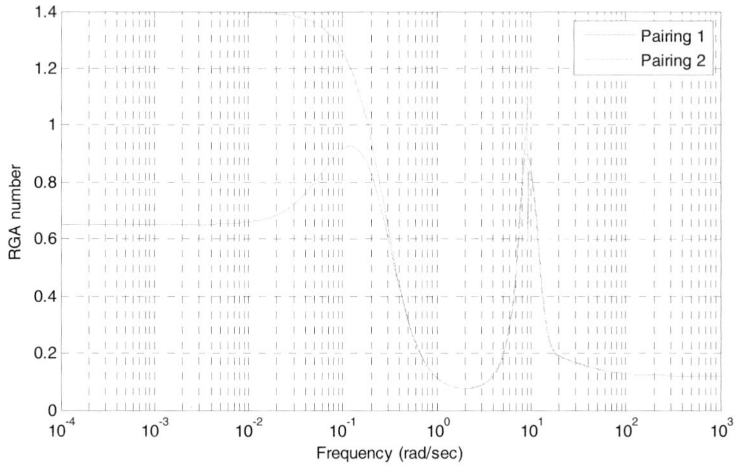

Fig. 4.7. RGAN; Yamaha R-50 model, hover condition.

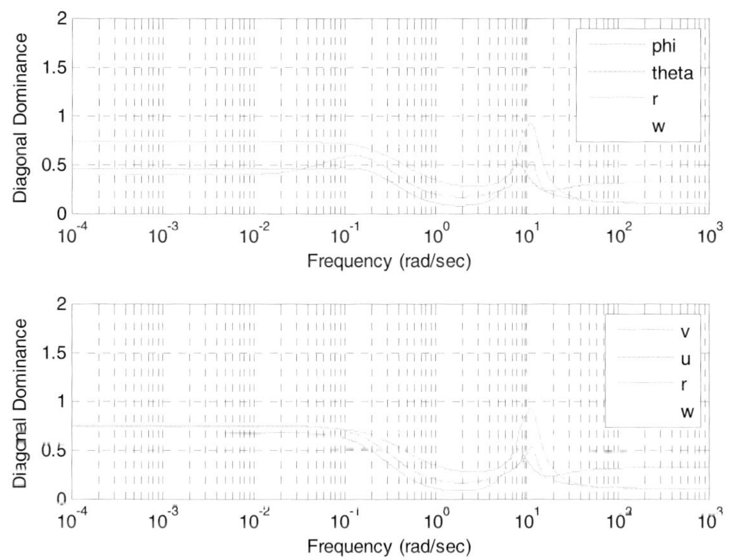

Fig. 4.8. Diagonal dominance; Yamaha R 50 model, hover condition

In general, it may be concluded that the helicopter behaves strongly as a set of two Two-Input / Two-Output (TITO) system given the coupling between the lateral and longitudinal channels, as well as the coupling between the pedal and collective channels. The latter two may be also approximated by two Single-Input / Single-Output, SISO, channels. The output affected by most of the inputs is the heave velocity. The pedal and collective inputs do not have much influence on the upper TITO system containing the longitudinal and lateral cyclic. From Figures 4.7 and 4.8 it may be concluded that the first pairing represents a slightly better choice since values are smaller.

Robustness Analysis

Robustness analysis helps quantify control system stability and performance under all possible sources of uncertainty and disturbances. UAV deployment does involve factors contributing to unexpected behavior. Therefore, plant modeling but also the uncertainty resulting from the mismatch between a nominal model and the real plant need be considered. Since a 'perfect' physical plant model is unachievable, the helicopter dynamic behavior is described by a set of possible linear time-invariant (LTI) models Π [57]. During flight any perturbed model $G'(s) \in \Pi$ may be valid as in Figure 4.5 where each linear model may be seen as a model incorporating uncertainty. A robust controller must guarantee stability and predefined performance.

Considering SISO systems, phase and gain margins are usually a good measure of 'slack' before reaching instability. Robustness analysis in MIMO systems requires more elaborate development as gain and phase margins of the transfer function matrix elements do not provide sufficient information. The generalized configuration depicted in Figure 4.9 will be studied.

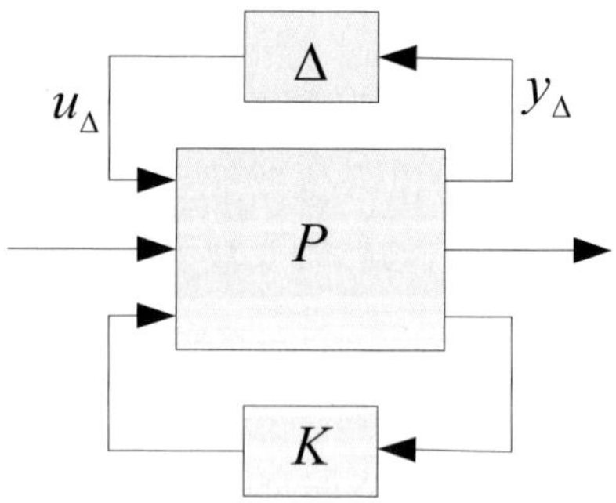

Fig. 4.9. Generalized control configuration.

Blocks P, K and Δ represent the plant, the controller, and source of uncertainties. w models all exogenous inputs to the system. z represents all variables that need be controlled, including performance variables with small desired value. u represents controller inputs and v all sensor measurements. The transfer function from inputs u_Δ, w to output z has to be small in all relevant frequencies and directions. The system of Figure 4.9 is described by [57]:

$$\begin{bmatrix} z \\ v \end{bmatrix} = P(s)\begin{bmatrix} w \\ u \end{bmatrix} = \begin{bmatrix} P_{11}(s) & P_{12}(s) \\ P_{21}(s) & P_{22}(s) \end{bmatrix}\begin{bmatrix} w \\ u \end{bmatrix} \tag{4.13}$$

with
$$z = [P_{11} + P_{12}K(I - P_{22}K)^{-1}P_{21}]w = F_l(P,K)w \tag{4.14}$$

The block diagram of Figure 4.10 shows a one degree-of-freedom controller that will be considered to connect robustness with the traditional frequency shaping followed in SISO system design and understand the importance of *sensitivity functions*. Design constraints and closed-loop requirements will be met by 'playing' with the singular values of the open-loop system.

Observing Figure 4.10, relevant multivariable transfer functions may be defined [66] [5] [57]: the loop transfer matrix $L = GK$, the sensitivity function $S = (I + L)^{-1}$ and the complementary sensitivity function $T = L(I + L)^{-1}$ such that $S + T = I$.

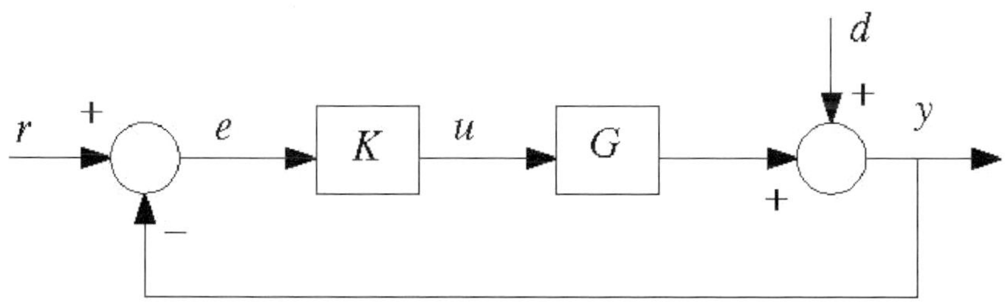

Fig. 4.10. One degree-of-freedom control configuration.

Therefore, the closed-loop response, the control error and control effort may be obtained, respectively, in terms of L, S and T, as follows:

$$y = Tr + Sd \tag{4.15}$$
$$e = S(d - r) \tag{4.16}$$
$$u = KS(r - d) \tag{4.17}$$

The greatest and smallest singular value of the particular transfer function at a particular frequency is utilized to quantify their magnitude. For example, the condition for desentization of y with respect to disturbances in a specific frequency band (where the attenuation factor $W_1 \geq 1$ varies with frequency) can be given by:

$$\bar{\sigma}(S(j\omega)) \leq W_1^{-1}(j\omega) \tag{4.18}$$

For reference tracking, the condition $\bar{\sigma}(T(j\omega)) \approx \underline{\sigma}(T(j\omega)) \approx 1$ should be fulfilled. The control effort can also be reduced by keeping $\bar{\sigma}(KS(j\omega))$ bounded. The same condition helps achieving robust stability when plant uncertainty is modeled as an additive perturbation, that is, $G_p = G + \Delta$. When plant uncertainty is modeled as a multiplicative output perturbation $G_p = (I + \Delta)G$, then $\bar{\sigma}(T(j\omega))$ needs be kept small:

$$\bar{\sigma}(T(j\omega)) \leq W_2^{-1}(j\omega) \tag{4.19}$$

It is easily verified that for low /high frequencies, respectively:

$$\frac{1}{\bar{\sigma}(S(j\omega))} \approx \underline{\sigma}(L(j\omega)) \tag{4.20}$$

$$\bar{\sigma}(T(j\omega)) \approx \bar{\sigma}(L(j\omega)) \tag{4.21}$$

Loop shaping constraints are shown in Figure 4.11.

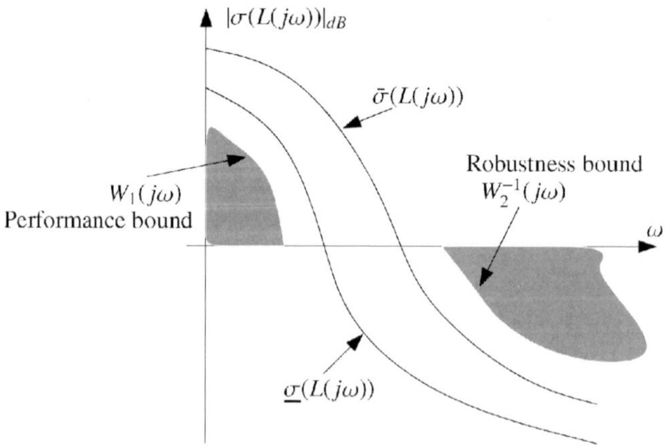

Fig. 4.11. Loop shaping constraints.

As previously mentioned, the Yamaha R-50 parameters obtained via system identification procedures have tolerances, and the parameters presented in Section 4.4.2 represent only dynamics of the nominal model. Further, the state variables of (4.3) are handled in the body-fixed frame, but the actual rotorcraft needs be controlled in the earth-fixed frame. The transformation (4.1) between the two reference frames introduces a nonlinearity that may be accounted for as 'uncertainty factor'. Figure 4.12 shows a similar frequency response as Figure 4.3 with variations that result from uncertainties; this singular value plot includes uncertainties due to variations of pitch and roll of $\pm 20^\circ$.

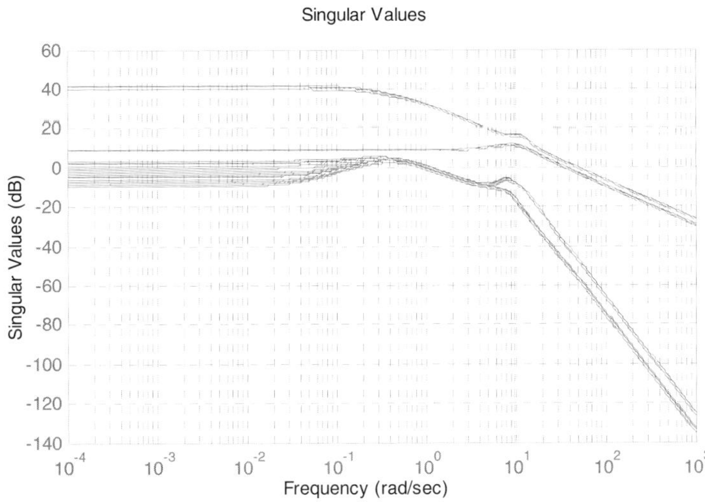

Fig. 4.12 Singular values of Yamaha R-50 considering uncertainties.

PID Controllers

Most initial attempts to achieve autonomous helicopter flight have been based on PID controller design [2] [23] [36]. Most commercial autopilots such as Micropilot's MP2128-Heli [38] or Rotomotion's AFCSV2.5 UAV Helicopter Controller [62] are based on PID controllers.

PID controller parameters may be easily adjusted allowing for on-line tuning when the model is not known. On the other hand when a sort of accurate model is given, classical design or optimization-based tuning strategies may be attempted [10]. The main disadvantage of SISO PID controllers relates to overlooking coupling among controlled variables, limiting bandwidth and agility that could be achieved. However, since standard

PID controllers are of the SISO type, it is assumed implicitly that controlled variables are not strongly coupled. Figure 4.13 shows the helicopter decentralized inner-loop controller structure based on PID controllers.

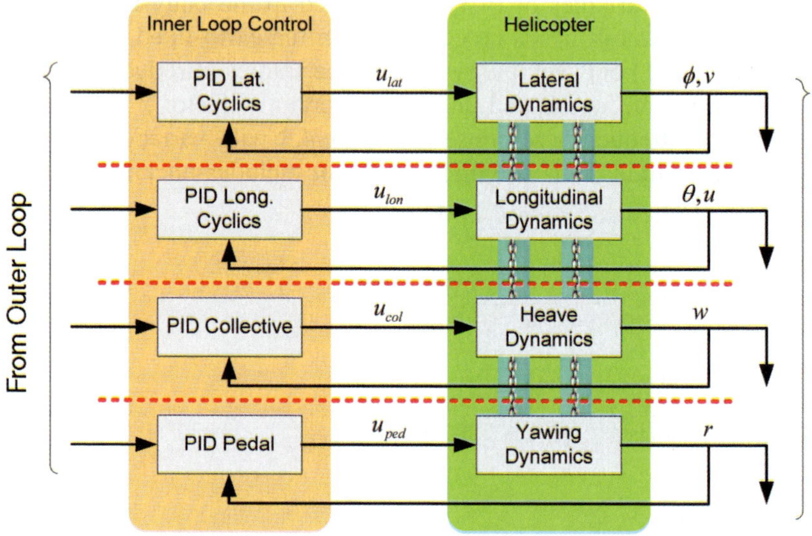

Fig. 4.13. Inner-loop PID controllers.

For the Yamaha R-50 PID controllers are assigned to each channel of the plant. The best flexibility is obtained when using two-degree-of-freedom PID controllers as shown in Figure 4.14 with an anti-windup term at the output preventing saturation.

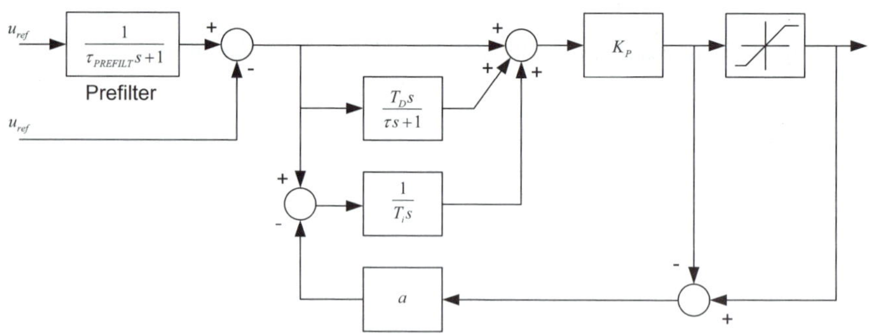

Fig. 4.14. Structure of a 2-DOF PID controller.

The lateral and longitudinal channels controlled by the PID compensators cannot be tuned individually since the diagonal terms of the matrix transfer function present unstable non-minimum phase dynamics. However, iterative tuning of the controllers, starting from rough hand-tuning, gives very good results as validated through experiments conducted with the controlled plant. PID controllers have been parameterized using the following standard form given by [16]:

$$C_{\text{PID}}(s) = K_p\left(1+\frac{1}{T_i s}+\frac{T_d s}{\tau s+1}\right) \tag{4.22}$$

Parameters for all four Yamaha R-50 PID controllers are listed in Table 4.4.

Channel	K_p	T_i	T_d	τ
Lateral	1.63	0.48	0.096	0.02
Longitudinal	2.17	0.333	0.058	0.017
Pedal	3.74	0.374	0	0
Collective	9.83	0.983	0.120	0.05

Table 4.4. PID parameters for inner-loop controllers.

Figure 4.15 illustrates the behavior of the PID controllers with respect to decoupling using diagonal dominance. The values for the open-loop plant are also included for comparison purposes.

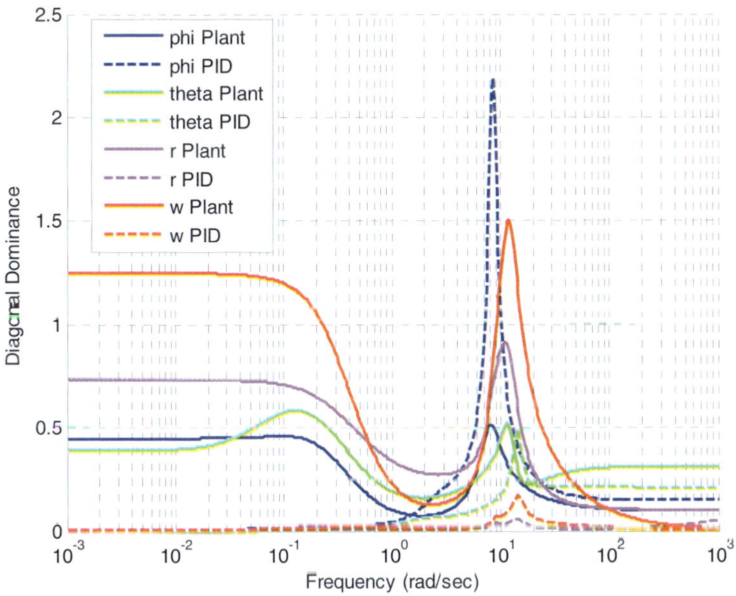

Fig. 4.15. Inner-loop decoupling effect using PID controllers.

The singular values of the open-loop transfer function matrix are shown in Figure 4.16. Singular values of the plant alone are also included to observe the effect of the control on the loop shape.

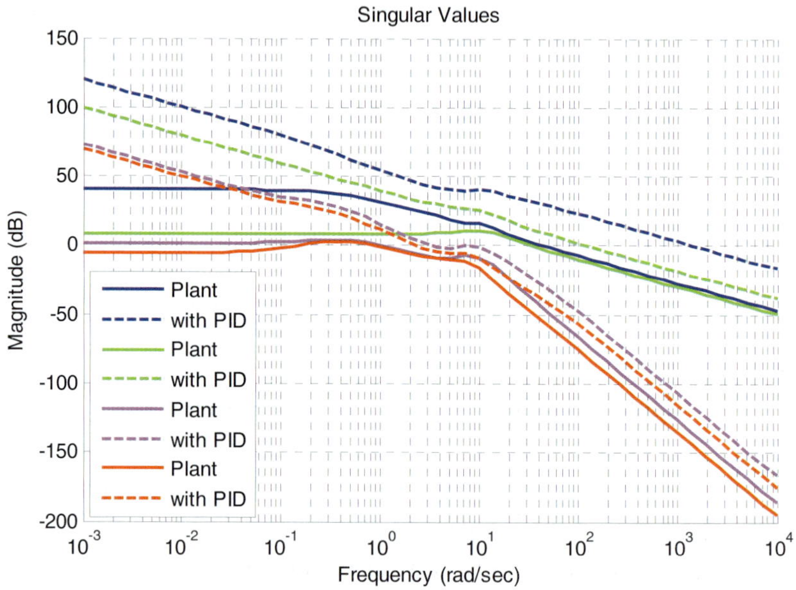

Fig. 4.16. Singular values of the PID control loop.

The singular values show that the integral action provides high gain at lower frequencies, guaranteeing good decoupling and low steady state errors. High gain at lower frequencies contributes to achieving a high level of agility. High gains at high frequencies indicate potential problems with respect to noise sensitivity, especially sensor noise, and high gains at higher frequencies points to potential robustness problems.

Linear Quadratic Control Techniques

When the plant model is considered as a full MIMO system, the first systematic technique to control the helicopter is the linear quadratic regulator (LQR) design methodology [13] [55] [58]. LQR control has been applied successfully to aeronautical control problems [60] mainly due to guaranteed robust asymptotic stability of the closed-loop. However, stability measures are only valid under the assumption of full state noise-free feedback and perfect knowledge of the model. In reality, the helicopter's nonlinearity makes the model valid within a region around a specific operating point used for design. Furthermore, not all state variables can be measured, therefore, applying a filter-observer as previously discussed may be required. Even in this case, robustness properties do not hold anymore, unless observer dynamics are much faster than rotorcraft dynamics.

In any case, noise presence in measured signals requires studying the problem from the stochastic systems perspective. Under this perspective, the linear quadratic Gaussian (LQG) control methodology (design) requires obtaining the linear quadratic controller and determining an optimal observer. Implementation, design procedures and experimental results of LQG control applied to miniature helicopters have been documented in [6], [40] and [64]. Designs may be obtained using numerical routines included in standard CACSD packages [31].

In detail, LQR is a full-state feedback control where x is the state vector and u is the control effort designed to minimize the quadratic performance measure described by:

$$V = \int_0^T (x'Qx + u'Ru)dt \tag{4.23}$$

The weighting matrices Q and R are design parameters varied to obtain a specified closed-loop dynamic behavior, pole placement, penalizing control accuracy and control effort, respectively. Since (4.23) has Q and R in two bilinear forms, they need be positive semi-definite and positive definite, respectively, $Q = Q^T \geq 0, R = R^T > 0$. This minimization problem is translated to the solution of an algebraic Riccati equation given by:

$$A^T X + XA - XBR^{-1}B^T X + Q = 0 \tag{4.24}$$

with a positive semi-definite solution X and control law:

$$u = -K_r x, \; K_r = R^{-1}B^T X \tag{4.25}$$

Design of an optimal observer under the assumption of Gaussian noise leads also to an Algebraic Riccati equation whose solution yields a Kalman filter. The combination of optimal estimator and linear quadratic regulator gives the LQG controller. The two components plus integral action for tracking reference signals are shown in Figure 4.17 applicable to helicopter control.

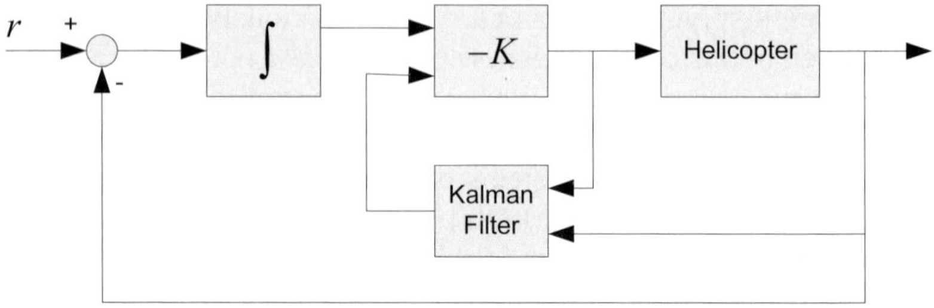

Fig. 4.17. LQG controller as combination of a KF and state feedback [57].

In order to guarantee certain robustness properties, particularly considering the frequency behavior of sensitivity functions over specific frequency ranges, frequency dependent Q and R weight matrices need be considered in (4.23). To achieve this, it is possible to augment the open-loop system with fictitious states and apply the standard LQG design procedure with the help of computerized tools. This method results in gains that are not static but have the form of dynamic compensators. This extension of the LQG methodology is also known as Wiener-Hopf design or H_2 control. In the case of the H_2 controller, the energy of the impulse response of the fictitious plant G is minimized:

$$\|G\|_2 = \sqrt{\frac{1}{2\pi}\int_{-\infty}^{\infty} \mathrm{tr}\left[G(j\omega)G^*(j\omega)\right]d\omega} \tag{4.26}$$

Although H_2 optimal control design is easily accomplished using CACSD tools, considering that the plant is open-loop unstable in some of its flight regimes, one should take into account (in order to avoid unexpected behavior) measures such as the integrated log of sensitivity magnitude [59]:

$$\int_0^{\infty} \ln|\det S(j\omega)|\, d\omega = \pi \sum_{i=1}^{N_p} \mathrm{Re}(p_i) \tag{4.27}$$

Here, $S(j\omega)$ represents the frequency dependent sensitivity function considered in its MIMO form and the summation in the right term is carried over the poles (controller and plant) that lie on the right half plane (RHP).

For the Yamaha R-50, an output feedback LQG controller with integral action has been designed assuming measurement availability of the inner-loop controlled variables. The plant has been augmented with free integrators for high open-loop gain at lower frequencies, achieving good tracking accuracy and mainly good decoupling. An LQR regulator has been designed assuming knowledge of the complete state vector with $Q = 0_{17\times17}$ and $R = \begin{bmatrix}0.5 & 0.02 & 0.025 & 10\end{bmatrix} \times I_{13\times13}$. The Q matrix has four more terms since it includes the states associated with the augmented integrators.

Design of the KF involves selection of weights for process and measurement noise, W and V, respectively, both set to identity matrices for simplicity purposes. The states of the integrators do not need be estimated, hence $W = I_{13\times13}$ and $V = I_{4\times4}$.

The resulting design can be seen as a 2-DOF controller shown in Figure 4.18. The integrators have been isolated from the controller to enable an anti-windup scheme by using limited outputs.

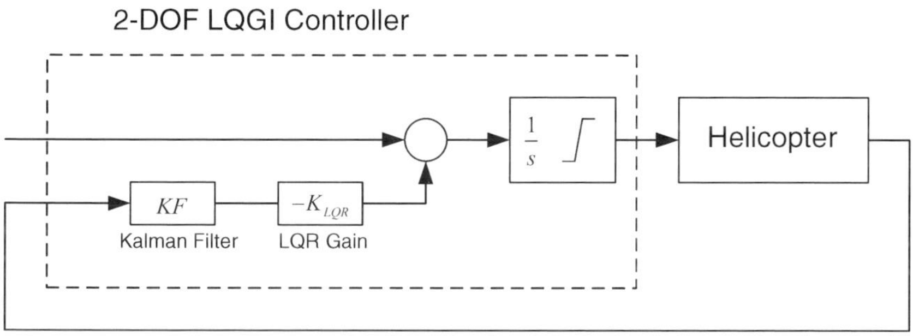

Fig. 4.18. LQG with integrator.

Figure 4.19 presents the resulting diagonal dominance values along with those of the plant.

Singular values of the open-loop system are shown in Figure 4.20 and compared against the plant. In general, the loop presents lower gains at lower frequencies in comparison to the PID design. Much lower gains at higher frequencies show a very good behavior concerning sensor noise. Low noise sensitivity is achieved at the expense of slightly lower bandwidth, thus agility is not as good as in the case of the PID design

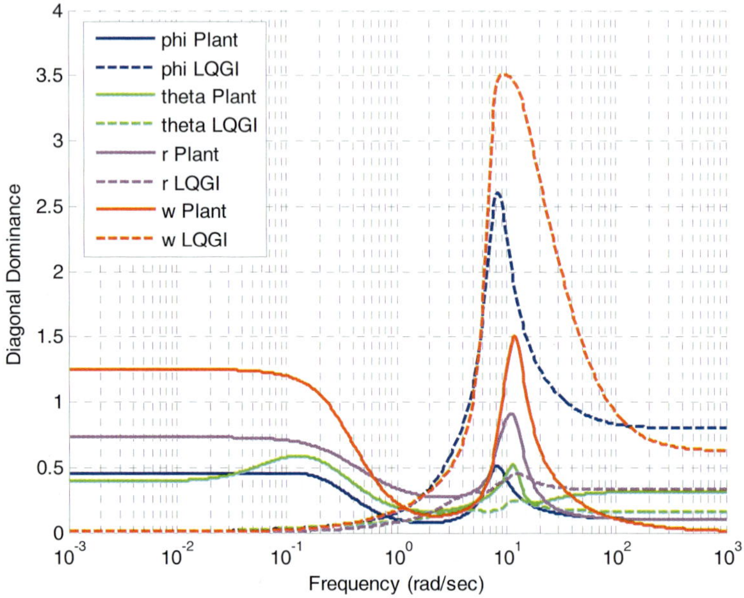

Fig. 4.19. Inner-loop decoupling using LQG control with integration.

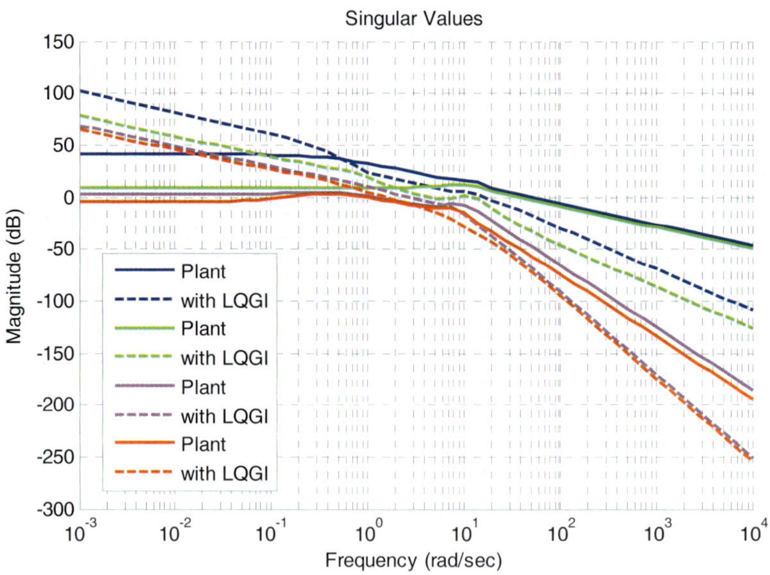

Fig. 4.20. Singular values of the LQGI control loop.

Robust control design

Robust control represents the most widely used and proven design technique for small-scale helicopters allowing for aggressive maneuvers exploiting their agile dynamics. Robust controllers have been reported to perform better than LQG controllers since they address robustness measures directly in the controller synthesis process [6] [26] [28] [40] [54] [64], while seminal work in robust control has been reported in [5] [57] [66] and [67]. Robust control offers several techniques for synthesis of robust MIMO controllers with CACSD tools supporting most of these methods [5] including *MATLAB's* robust control toolbox.

The H_2 and H_∞ optimal control tasks consist of finding controllers that minimize specific norms of $F_l(P,K)$, with:

$$\|F\|_2 = \sqrt{\frac{1}{2\pi}\int_{-\infty}^{\infty} \mathrm{tr}\left[F(j\omega)F^*(j\omega)\right]d\omega} \tag{4.28}$$

$$\|F\|_\infty = \sup_{\omega} \bar{\sigma}\left[F(j\omega)\right] \tag{4.29}$$

An H_∞ multivariable robust controller with loop shaping is now designed with the same previously used linearized model as basis for development. The design methodology follows the loop shaping methodology presented in [33] and applied to the Yamaha R-50 [26]. The objective is to design an optimal loop-shaping controller for the inner-loop with the goal of providing robust attitude stabilization and decoupling of the main four channels. The design technique follows three main steps:

- The nominal plant G is augmented with a pre-compensator W_1 and a post-compensator W_2 in order to shape the singular values of the open-loop transfer function. The augmented plant is $G_s = W_2 G W_1$, with W_1 defining the controller behavior at lower frequencies trying to obtain high gain, which in turn yields good tracking behavior and decoupling and disturbance rejection. For these reasons, PI-type frequency shapes are selected. For W_2, the objective is to provide enough damping at higher frequencies guaranteeing robustness. This is accomplished by selecting low pass filters and constant gain terms; as in [26] the following weighting functions are proposed:

$$W_{1,lat} = \frac{1.13s + 3.4}{s} \quad W_{1,lon} = \frac{1.63s + 6.50}{s} \quad W_{1,ped} = \frac{3.74s + 37.4}{s} \quad W_{1,col} = \frac{8.33s + 10}{s}$$

$$W_{2,lat} = 1 \quad W_{2,lon} = 1.14 \quad W_{2,ped} = \frac{2028}{s^2 + 80s + 2500} \quad W_{2,col} = \frac{2795}{s^2 + 80s + 2500}$$

Both, W_1 and W_2 are diagonal matrices, having the terms above in the main diagonal.

- Find an optimal controller K_∞ that minimizes the H_∞ norm of the transfer function from disturbances to errors - γ is between 1 and 3:

$$\min_{K_\infty} \left\| \begin{bmatrix} I \\ K_\infty \end{bmatrix} (I - G_s K_\infty))^{-1} [G_s, I] \right\|_\infty = \gamma = \varepsilon^{-1}$$

using the `ncfsyn` function from the *MATLAB* robust control toolbox, `[K,CL,GAM]=ncfsyn(G,W1,W2)`

- Construct the feedback controller by combining the synthesized K_∞ controller with W_1 and W_2, such that $K = W_1 K_\infty W_2$

After execution, the `ncfsyn` function returns a valid K_∞ controller, the closed-loop system, and the margin $\gamma = 2.68$ that is within the acceptable interval. Figure 4.21 shows the singular values of three systems: plant, augmented plant and complete open-loop transfer function with the controller. The resulting decoupling is illustrated in Figure 4.22.

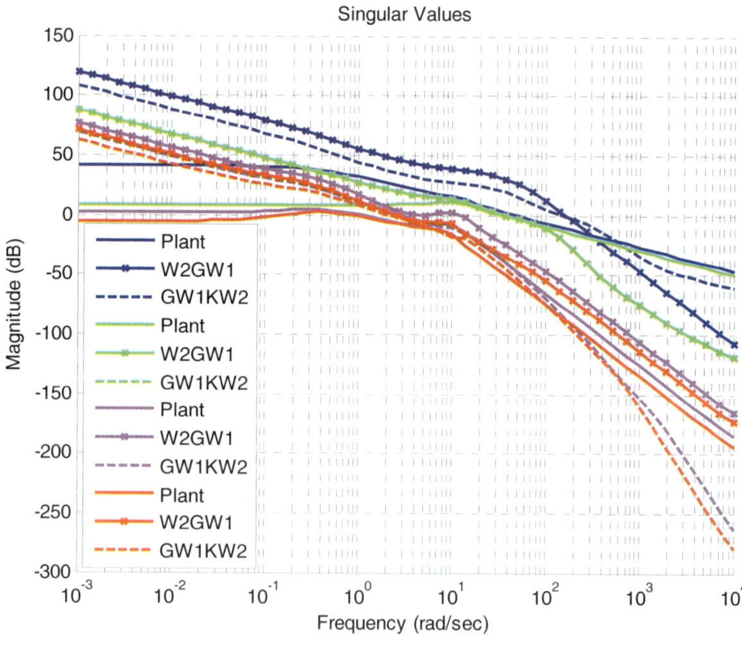

Fig. 4.21 Singular values resulting for loop-shaping design.

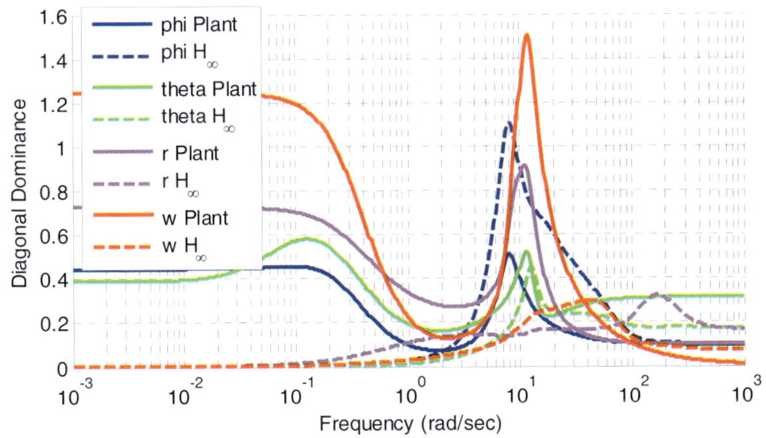

Fig. 4.22. Inner-loop decoupling using H_∞ control with loop-shaping.

Gain at lower frequencies is higher than in the case of the PID controller, but higher than the one achieved in the LQG design. Hence, decoupling and bandwidth, agility, may be just between both previous designs. The behavior at high frequencies is acceptable but not as good as the one achieved with LQG design.

Agility Test: Velocity Tracking for a Specific Trajectory

The three controllers have been used to control the helicopter in a specific trajectory very similar to the one shown in Figure 4.23. The velocities tracked by all three controllers are shown in Figures 4.24 through 4.26. Clearly, the PID design, having the highest gain at lower frequencies, presents the lowest tracking errors. On the other hand, having the H_∞ design the most moderate low frequency gain, it presents relatively larger deviations.

Sensitivity

Figures 4.27 and 4.28 depict the maximum singular values of the sensitivity functions. In the case of the input sensitivity function, the PID controller displays the best behavior, while the LQG controller presents the highest sensitivity to input noise. The output complimentary sensitivity functions indicate that the PID controller presents the worst maximum value, since it presents high sensitivity to sensor noise.

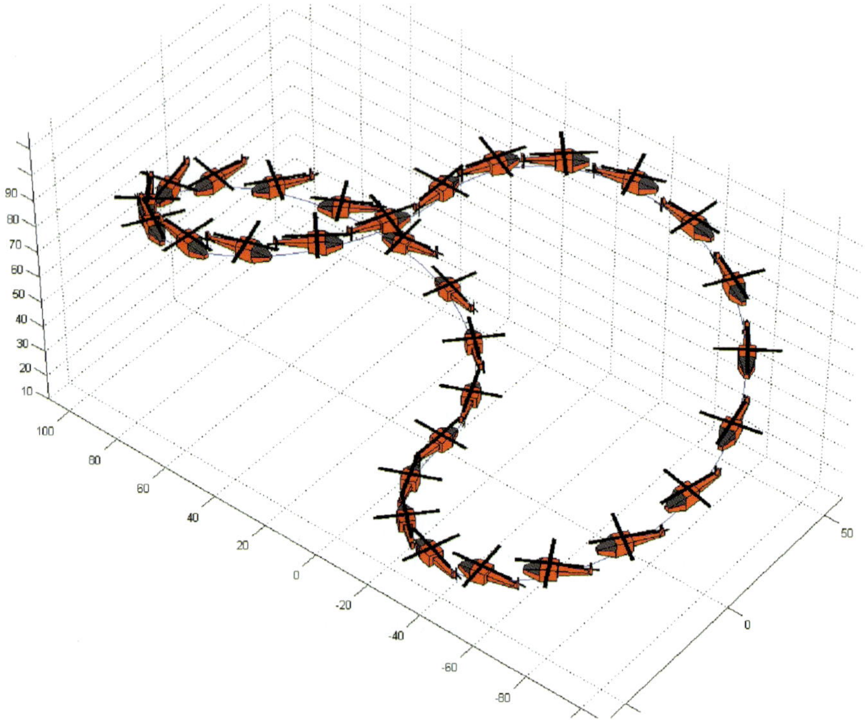

Fig. 4.23. Test trajectory.

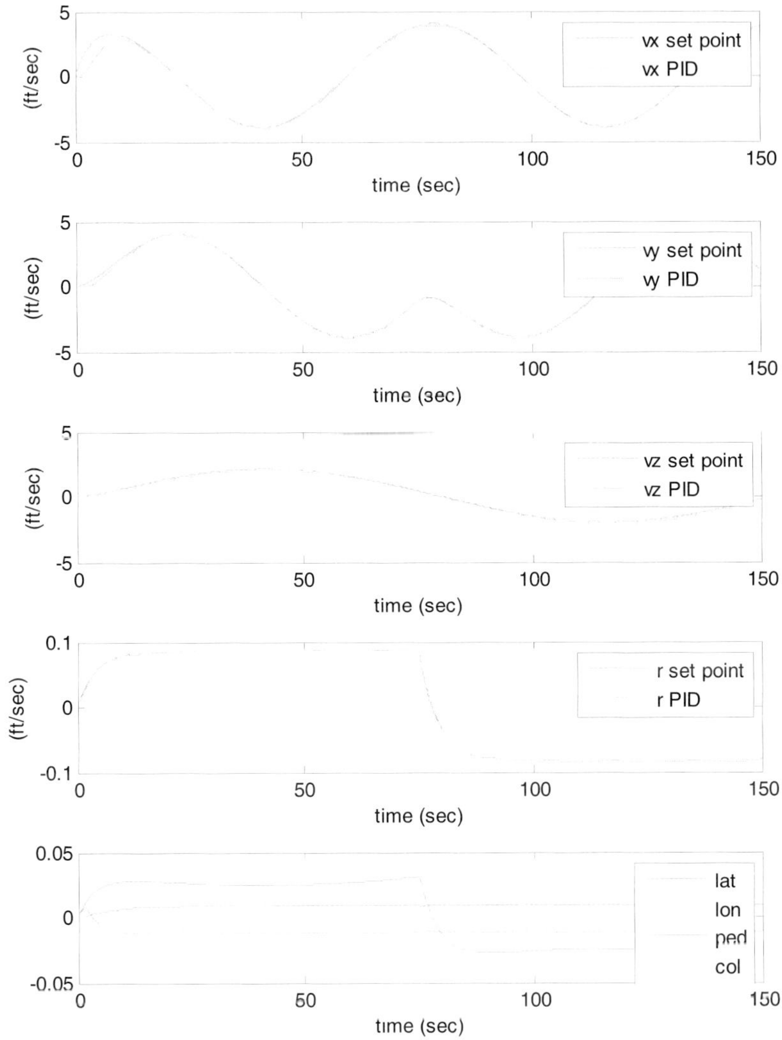

Fig. 4.24. Velocity tracking for PID controller.

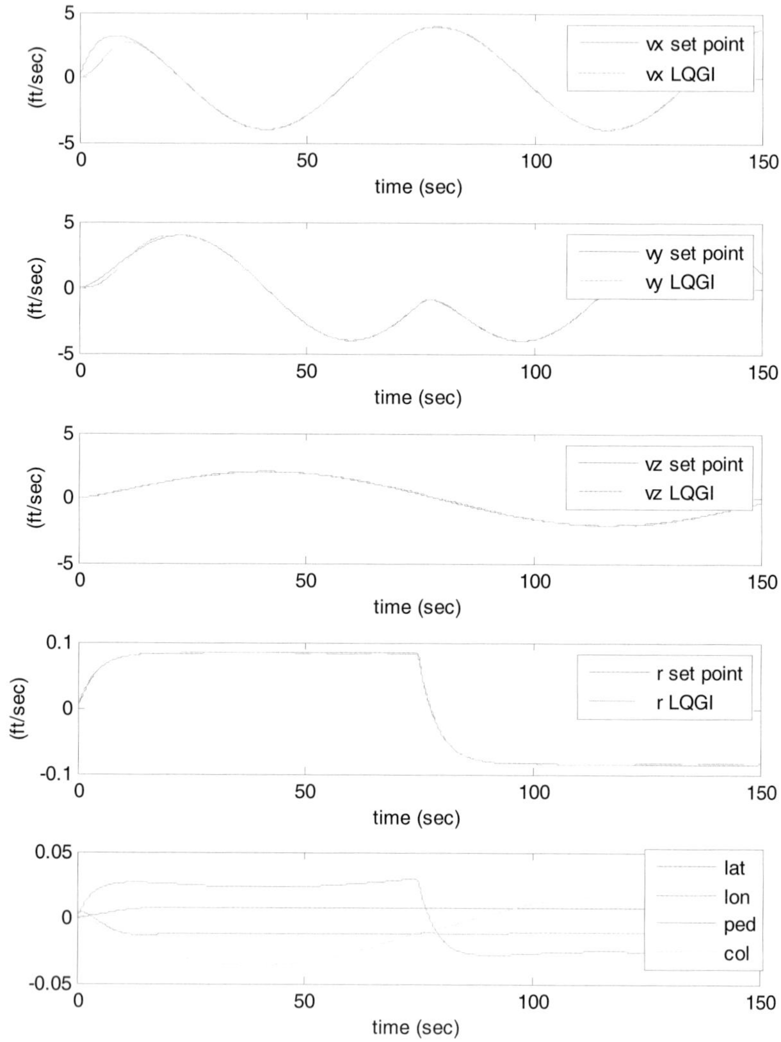

Fig. 4.25. Velocity tracking for LQGI controller.

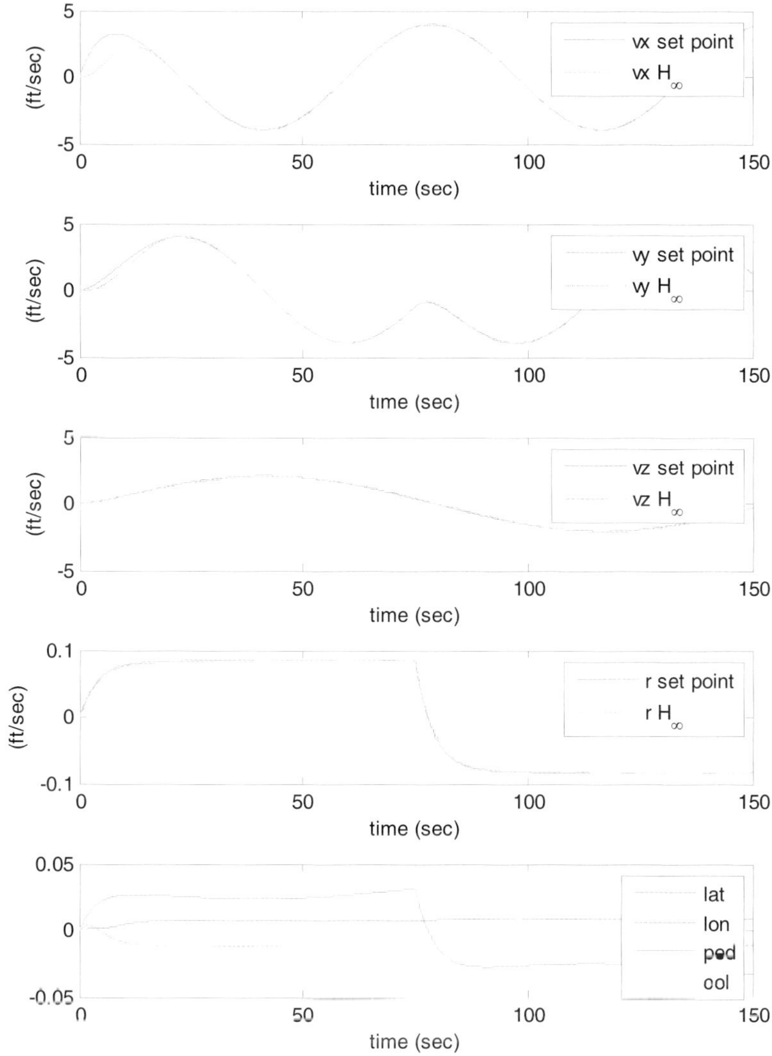

Fig. 4.26. Velocity tracking for H_∞ controller.

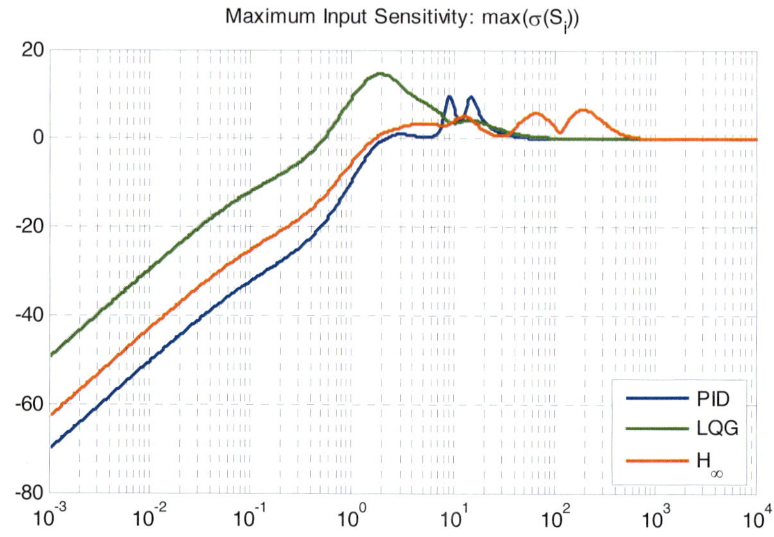

Fig. 4.27. Input sensitivity.

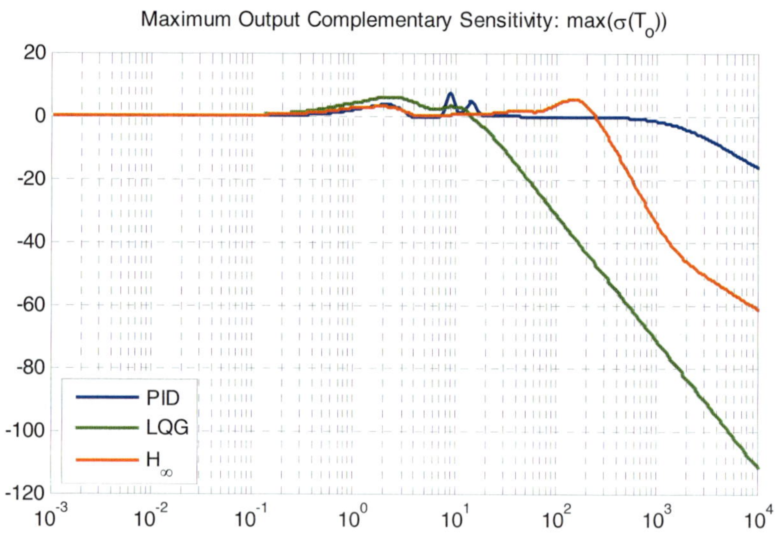

Fig. 4.28. Output complementary sensitivity.

Comparison of the Three Controllers

The three designs have been subject to three evaluation criteria that are extremely relevant for practical implementation of self-governing flights: agility, ability to handle noise and robustness.

A comparative study of the three designs reveals that designs are also subject to trade-offs and compromises. High degree of agility, low steady state errors, robustness against disturbances and robustness against parameter variations cannot be achieved at the same time. The three designs have strengths and weaknesses, and selection of a specific controller depends on the application.

The LQG design exposes a slightly more sluggish dynamic response than the other designs and the worst decoupling behavior of all three. However, it has excellent values for output sensitivity and robustness in general terms. The PID design focused on agility and decoupling; does not provide enough robustness and it is extremely sensitive to output noise. The H_∞ design, thanks to the additional design freedom provided by the loop shaping methodology, presents a good trade-off between agility, robustness and noise sensitivity.

4.6 Conclusions and Future Perspectives

This Chapter presented just a glimpse of controller design for small helicopters. Challenges imposed by unmanned helicopters have contributed positively to advancing theory and applications of control theory, data fusion, flight control systems, and computer vision. New research fields have emerged or gained new relevance as is the case of the SEC program that has helped identifying shortcomings and challenges in computer-based implementation of control systems.

Stronger demands for higher autonomy levels will lead to improving collection and fusion of sensory data and perception provided by diverse suites of sensors, computer vision algorithms and better hardware. Likewise, higher reliability requirements will lead to development of highly intelligent fault-tolerant control systems supporting higher autonomy levels.

However, there are still limitations that prevent deployment of autonomous helicopters in envisioned applications. These limitations not only stem from the constrained payload capabilities and flight endurance of small size helicopters, but also from perspectives of control, autonomy, human robot interaction, and reliability. Most successful autonomous flight control experiments have been carried out under controlled conditions and usually under the supervision of ground operators.

Most nonlinear or hybrid control techniques found in the literature are restricted to moderately successful simulation results. System identification procedures still require lengthy data collection experiments performed with the help of a skilled pilot. The vision of having UAVs deployed by non skilled personnel is yet to be reached. Similarly, the number of human operators per UAV is still represented by a ratio much greater than one.

Stated limitations have not stopped advancement of the unmanned systems field towards new goals as recommended by [43]. Goals in the next years incorporate coordination and cooperation in scenarios where multiple UAVs are required to be deployed. This has generated a profusion of research work in topics such as formation control, pursuit evasion games, swarming, flocking, etc. It is expected that in the near future, all these new theories will blend with intelligent fault-tolerant low-level control systems giving rise to fully autonomous swarms of miniature helicopters.

References

1. Adams R. J. and Banda S. S., "Robust Flight Control Design Using Dynamic Inversion and Structured Singular Value Synthesis", *IEEE Transactions on Control Systems Technology*, 1(2):80–92, 1993.
2. Aerial Robotics Group, Department of Aeronautics and Astronautics, Massachusetts Institute of Technology, "The Development of a Small Autonomous Helicopter Robot for Search and Rescue in Hostile Environments", *Proceedings of the AUVSI Annual Symposium*, Baltimore, MD, July 12, 1999.
3. Amidi O., Kanade T., and Miller J. R., "Vision-Based Autonomous Helicopter Research at Carnegie Mellon Robotics Institute 1991-1997", *American Helicopter Society International Conference,* Paper T7-3, Japan, April 1998.
4. Avila-Vilchis J. C., Brogliato B., Dzul A. and Lozano R., "Nonlinear Modeling and Control of Helicopters", *Automatica*, 39:1583–1596, 2003.
5. Balas G., Chiang R., Packard A. and Safonov M., *Robust Control Toolbox For Use with MATLAB*, The Mathworks, Inc., Natick, MA, September 2006.
6. Bendotti P. and Morris J. C., "Robust Hover Control for a Model Helicopter", *Proceedings, American Control Conference*, pages 682–687, Seattle, Washington, 1995.
7. Brooks R., "A Robust Layered Control System for a Mobile Robot", *IEEE Journal of Robotics and Automation*, 2(1):14–23, March 1986.
8. Brown R. G. and Hwang P. Y. C., *Introduction to Random Signals and Applied Kalman Filtering with MATLAB Exercises and Solutions*, John Wiley & Sons, 3rd Edition, February 2001.

9. Buskey G., Roberts J., Corke P., Ridley P. and Wyeth G., "Sensing and Control for a Small-size Helicopter", *Experimental Robotics*, B. Siciliano and P. Dario, Editors, VIII: 476–487, 2003.
10. Castillo C. L., Alvis W., Castillo-Effen M., Moreno W., and Valavanis K., "Small Scale Helicopter Analysis and Controller Design for Non-aggressive Flights", *Proceedings, IEEE International Conference on Systems, Man and Cybernetics*, Volume 4, pages 3305– 3312, October 2005.
11. Fagg A. H., Lewis M. A., Montgomery J. F. and Bekey G. A., "The USC Autonomous Flying Vehicle: An Experiment in Real-Time Behavior-Based Control", *Proceedings, IEEE/RSJ International Conference on Intelligent Robots and Systems*, pages 1173–1180, Yokohama, Japan, July 1993.
12. Frazzoli E., Dahleh M. A., and Feron E., "Trajectory Tracking Control Design for Autonomous Helicopters Using a Back Stepping Algorithm", *Proceedings, American Control Conference*, pages 4102–4107, Chicago, Illinois, June 2000.
13. Gavrilets V., Martinos I., Mettler B. and Feron E., "Control Logic for Automated Aerobatic Flight of a Miniature Helicopter", *Proceedings, AIAA Guidance, Navigation, and Control Conference and Exhibit*, California, August 2002.
14. Gavrilets V., Mettler B., and Feron E. *Dynamic Model for a Miniature Aerobatic Helicopter*, Technical report, MIT-LIDS report LIDS-P-2580, 2003.
15. Gavrilets V., Shterenberg A., Dahleh M. A. and Feron E., "Avionics System for a Small Unmanned Helicopter Performing Aggressive Maneuvers", *Proceedings, 19th Digital Avionics Systems Conference*, Vol. 1, 2000.
16. Goodwin G. C., Graebe S. F. and Salgado M. E, *Control System Design*, Prentice Hall, September 2000.
17. Huang H.-M., Pavek K., Albus J. and Messina E., "*Autonomy Levels for Unmanned Systems (ALFUS) Framework: An Update*", Proceedings, SPIE Defense and Security Symposium, March 2005.
18. International Aerial Robotics Competition – "*The Ultimate Collegiate Challenge*", http://avdil.gtri.gatech.edu/AUVS/IARCLaunch-Point.html.
19. Isidori A., Marconi L. and Serrani A., *Robust Autonomous Guidance, An Internal Model Approach*, Springer, 2003.
20. Johnson E. N., Debitetto P. A., Trott C. A. and Bosse M. C., "The 1996 MIT/Boston University/Draper Laboratory Autonomous Helicopter System", *Proceedings, 15th Digital Avionics Systems Conference*, pages 381–386, 1996.
21. Johnson E. N. and Kannan S. K., "Adaptive Flight Control for an Autonomous Unmanned Helicopter", *Proceedings, AIAA Guidance, Navigation, and Control Conference and Exhibit*, Monterey, California, August 2002.
22. Kahn A. D. and Foch R. J., "Attitude Command Attitude Hold and Stability Augmentation Systems for a Small-scale Helicopter UAV", *Proceedings, 22nd Digital Avionics Systems Conference*, Vol. 2, 2003.
23. Kim H. J. and Shim D. H., "A Flight Control System for Aerial Robots: Algorithms and Experiments", *Control Engineering Practice*, 11(12):1389–1400, 2003.

24. Kim S. K. and Tilbury D. M., "Mathematical Modeling and Experimental Identification of an Unmanned Helicopter Robot with Flybar Dynamics", *Journal of Robotic Systems*, 21:95–116, March 2004.
25. Koo J. T. and Sastry S., "Output Tracking Control Design of a Helicopter Model Based on Approximate Linearization", *Proceedings, 37th IEEE Conference on Decision & Control*, Tampa, Florida, USA, December 1998.
26. La Civita M., Papageorgiou G., Messner W.C. and Kanade T., "Design and Flight Testing of an H∞ Controller for a Robotic Helicopter", *Journal of Guidance, Control, and Dynamics*, 29(2):485–494, March– April 2006.
27. La Civita M., Papageorgiou G., Messner W. C., and Kanade T., "Integrated Modeling and Robust Control for Full-Envelope Flight of Robotic Helicopters", *Proceedings, IEEE International Conference on Robotics and Automation*, 2003.
28. La Civita M., Papageorgiou G., Messner W. C. and Kanade T., "Design and Flight Testing of a Gain-Scheduled H∞ Loop Shaping Controller for Wide-envelope Flight of a Robotic Helicopter", *Proceedings, American Control Conference*, Vol. 5, page 4195, 4200, June 2003.
29. Ljung L., *System Identification: Theory for the User*, 2nd Edition, Prentice Hall PTR, December 1998.
30. Ljung L., *System Identification Toolbox – For Use with MATLAB*, The Mathworks, Inc., Natick, MA, September 2006.
31. The Mathworks, Inc., Natick, MA, *Control System Toolbox – For Use with MATLAB*, September 2005.
32. Mcconley M. W., Piedmonte M. D., Appleby B. D., Frazzoli E., Feron E. and Dahleh M. A., "Hybrid Control for Aggressive Maneuvering of Autonomous Aerial Vehicles", *Proceedings, 19th Digital Avionics Systems Conference*, Vol.1, 2000.
33. McFarlane D. and Glover Ke. "*A Loop Shaping Design Procedure Using H∞ Synthesis*", IEEE Transactions on Automatic Control, 37:759–769, June, 1992.
34. Mejias L. O., Saripalli S., Cervera P. and Sukhatme G. S., "Visual Servoing of an Autonomous Helicopter in Urban Areas Using Feature Tracking", *Journal of Field Robotics*, 23(3):185–199, 2006.
35. Mettler B., Dever C. and Feron E., *Identification Modeling, Flying Qualities, and Dynamic Scaling of Miniature Rotorcraft*, NATO Systems Concepts & Integration Symposium, Berlin, May 2002.
36. Mettler B., *Identification Modeling and Characteristics of Miniature Rotorcraft*, Springer, 2002.
37. Mettler B., Tischler M. B. and Kanade T., "System Identification of Small-size Unmanned Helicopter Dynamics", *55th Forum of the American Helicopter Society,* Montreal, Quebec, Canada, May 25-27, 1999.
38. Micropilot Inc., http://www.micro-pilot.com, January 2007.
39. Montgomery J. F., Fagg A. H. and Bekey G. A., "The USC AFV-I: A behavior-based entry in the 1994 International Aerial Robotics Competition", *IEEE Expert, (see also IEEE Intelligent Systems and Their Applications)*, 10(2):16–22, 1995.

40. Morris J. C., Van Nieuwstadt M. and Bendotti P., "Identification and Control of a Model Helicopter in Hover", *Proceedings, American Control Conference*, pages 1238–1242, Baltimore, Maryland, June 1994.
41. Nguyen H. T. and Prasad N. R., *Fuzzy Modeling and Control*, CRC, March 1999.
42. Obinata G. and Anderson D.O., *Model Reduction for Control System Design*, Springer, 2001.
43. http://www.fas.org/-irp/program/collect/uav.htm, 2005.
44. Padfeld G. D., *Helicopter Flight Dynamics: The Theory and Application of Flying Qualities and Simulation Modeling*, Co-published by AIAA and Blackwell Science, 1996.
45. Prouty R. W., *Helicopter Performance, Stability, and Control*, Krieger Publishing Company, 2001.
46. Samad T. and Balas G., *Software Enabled Control: Information Technology for Dynamical Systems*, Wiley-IEEE Press, February 2003.
47. Sanders C. P., DeBitetto P. A., Feron E., Vuong H. F. and Leveson N., "Hierarchical Control of Small Autonomous Helicopters", *Proceedings, 37th IEEE Conference on Decision & Control*, pages 3629–3634, Tampa, Florida, USA, December 1998.
48. Saripalli S., Naffin D. J. and Sukhatme G. S., "Autonomous Flying Vehicle Research at the University of Southern California", A. Schultz and L. E. Parker, Editors, *Multi-Robot Systems: From Swarms to Intelligent Automata*, Proceedings of the First International Workshop on Multi-Robot Systems, pages 73–82. Kluwer 2002.
49. Saripalli S., Roberts J. M., Corke P. I., Buskey G. and Sukhatme G. S., "A Tale of two Helicopters", *Proceedings, IEEE/RSJ International Conference on Intelligent Robots and Systems*, Las Vegas, Nevada, October 2003.
50. Saripalli S. and Sukhatme G. S., "Landing a Helicopter on a Moving Target", *Proceedings, International Conference on Robotics and Automation*, Rome, Italy, April 2007.
51. Shim D. H., Hoam C. and Sastry S., "Conflict-Free Navigation in Unknown Urban Environments", *IEEE Robotics & Automation Magazine*, 13(3):27–33, 2006.
52. Shim D. H., Kim H. J. and Sastry S., "Control System Design for Rotorcraft-Based Unmanned Aerial Vehicles Using Time-Domain System Identification", *Proceedings, IEEE International Conference on Control Applications*, Anchorage, Alaska, USA, September 25-27 2000.
53. Shim D. H., Kim H. J. and Sastry S., "Decentralized Nonlinear Model Predictive Control of Multiple Flying Robots in Dynamic Environments", *Proceedings, 42nd IEEE Conference on Decision and Control*, December 2003.
54. Shim D. H., Koo T. J., Hoffman F. and Sastry S., "A Comprehensive Study of Control Design for an Autonomous Helicopter", *Proceedings, 37th IEEE Conference on Decision & Control*, Tampa, Florida, USA, December 1998.
55. Shin J., Fujiwara D., Nonami K. and Hazawa K., "Model-Based Optimal Attitude and Positioning Control of Small-scale Unmanned Helicopter", *Robotica*, 23:51–63, 2005.

56. Shuzhi S. G. and Lewis F. L., Editors, *Autonomous Mobile Robots – Sensing, Control, Decision Making and Applications*, CRC Taylor & Francis Group, May 2006.
57. Skogestad S. and Postlethwaite I., *Multivariable Feedback Control: Analysis and Design*, John Wiley & Sons, July 1996.
58. Sprague K., Gavrilets V., Dugail D., Mettler B., Feron E. and Martinos I., "Design and Applications of an Avionics System for a Miniature Acrobatic Helicopter", *Proceedings, 20th Digital Avionics Systems Conference*, Vol. 1, 2001.
59. Stein G., "Respect the Unstable", *IEEE Control Systems Magazine*, 23(4):12–25, 2003.
60. Stevens B. L. and Lewis F. L., *Aircraft Control and Simulation*, Wiley-Interscience, 2003.
61. Tischler M. and Cauffman M., "Frequency-Response Method for Rotorcraft System Identification: Flight Applications to BO-105 Coupled Fuselage/Rotor Dynamics", *Journal of the American Helicopter Society*, 37:3–17, 1992.
62. http://www.rotomotion.com/prd UAV CTLR.html, January 2007.
63. *UAVs: Payloads and Missions*, Workshop held at the International Conference on Robotics and Automation, May 2006. Retrieved from: http://www.mem.drexel.edu/aerialRobotics/uavWorkshop2006.
64. Weilenmann M. F., Christen U. and Geering H. P., "Robust Helicopter Position Control at Hover", *Proceedings, American Control Conference*, pages 2491–2495, Baltimore, Maryland, June 1994.
65. Zhang P., Gu J., Milios E. E. and Huynh P., "Navigation with IMU/GPS/Digital Compass with Unscented Kalman Filter", *Proceedings, IEEE International Conference on Mechatronics & Automation*, 2005.
66. Zhou K. and Doyle J. C., *Essentials of Robust Control*, Prentice Hall, 1997.
67. Zhou K., Doyle J. C. and Glover K., *Robust and Optimal Control*, Prentice Hall, 1995.

Chapter 5: A Tutorial Approach to Small Unmanned Helicopter Controller Design for Non-aggressive Flights[1]

Helicopters qualify as non-holonomic underactuated systems that are very difficult to control due to gyroscopic effects and inherent instability. Their complex dynamics features, nonlinearities, unstable nature and high degree of coupling among avionics parameters, make controller design a challenge. Therefore, it is of at most importance to design and implement simple but efficient low-level controllers that will guarantee helicopter stable and robust performance within a defined operational profile. This Chapter lays the foundation of a general (tutorial–like) approach to studying, designing, implementing and testing controllers for small unmanned helicopters capable of autonomous non-aggressive flights, putting emphasis on hovering and cruising. It describes simplified, decentralized single input single output, SISO, PID and PID-like Fuzzy Logic controller designs with optimized gains, and a complete design of a multiple inputs multiple outputs, MIMO, linear quadratic regulator (LQR) controller. Testing and comparisons are provided for several flight patterns that reveal valuable information related to the applicability and effectiveness of proposed designs. *Computer Aided Control System Design (CACSD)* and *MATLAB / Simulink* tools are used for controller design. The presented approach is general enough to be applicable to a wide range of small unmanned helicopters.

[1] Written by W. Alvis, C. Castillo, M. Castillo-Effen, W. Moreno, K. P. Valavanis. This work has been supported partially by two Research Grants, ARO W911NF-06-1-0069 and SPAWAR N00039-06-C-0062.

Kimon P. Valavanis (ed.), Advances in Unmanned Aerial Vehicles, 119–170.

5.1 Introduction

Unmanned helicopters are being widely used in a multitude of applications, mostly military, as well as civilian. Their usefulness and applicability results from their ability to fly in very low altitudes, hover, cruise, take off and land almost everywhere, executing missions that require aggressive and non-aggressive flight scenarios.

However, due to their complex dynamics features, nonlinearities, unstable nature and high degree of coupling among avionics parameters, controller design is a challenge, and model simplifications and/or linearization may be possible under certain constraints and around specific operating points.

The families of small / miniature unmanned helicopters considered in this Chapter qualify as non-holonomic underactuated systems that are very difficult to control due to gyroscopic effects and inherent instability [1] [17]. Small / miniature helicopters have also very strict payload capabilities, power supply limitations (batteries) and limited on-board computational power. Therefore, it is of at most importance to design and implement simple but efficient low-level controllers to guarantee helicopter stable and robust performance within a defined operational profile.

This is accomplished by: deriving a simplified helicopter model to control in terms of selecting carefully avionics components; model decoupling given an operational profile; model linearization when applicable; constraining design of low-level control laws to fit in embedded platforms with limited computing power. While following those steps, one should also consider that thrust is the main force produced in a helicopter. For non-aggressive flights, longitudinal and lateral accelerations are small, as are forces involved in producing these accelerations. Longitudinal and lateral forces are the projected components of the thrust in the body-reference frame. Euler angles determine force projections that produce longitudinal and lateral accelerations.

The Chapter focuses on non-aggressive flight scenarios of small / miniature unmanned helicopters like the *Raptor* series, the *R-50* the *Maxi Joker*, or the *Bergen Observer*. It aims at laying the foundation of a rather general method to design controllers for such unmanned vehicles. In essence, it presents in an almost tutorial way a systematic approach to deriving and testing simplified controllers using *CACSD* and *MATLAB / Simulink* tools for achieving autonomous non-aggressive flights, putting emphasis on hovering and cruising.

The starting point is a 13th order state space model derived for an *R-50* helicopter [1], also valid for other helicopters of similar size. The model core is an extension to the stability derivatives model, a linearized form of equations of motion that includes coupled rotor-fuselage dynamics, one of the key characteristics of rotorcraft dynamics. Stated differently, the model is a 'grey box', where equations are first developed through first principles, then given numeric values for specific flight parameters through identification procedures. This combination gives a good balance between simplicity and accuracy, resulting in a parameterized model where parameters are extracted from frequency response experiments. To account for inherent complexities encountered is such systems, a multiple input multiple output (MIMO) controller synthesis procedure is followed; however, under certain conditions, it is possible to use single input single output (SISO) completely decentralized controller design techniques as justified below.

In more details, the linear state-space system model describes the helicopter dynamics with respect to the body-reference frame; however, system behavior is nonlinear because state-space system parameters depend on the flight regime. Transition from one regime to the next depends on the actual inertial speed of the helicopter in the horizontal plane. Thus, parameters used for hovering and cruising represent the linearization of the body-reference frame dynamics around two operating points. The main advantage of this model is the ability to obtain parameters for additional linearization points. This allows for model extension under aggressive flight conditions or for a more accurate description of associated nonlinearities. Model linearization is the result of small angle approximation, followed by a modal decomposition analysis that offers valuable insight information to helicopter instabilities.

Diagonal dominance determines the level of coupling between dynamics model parameters dictating under what conditions it is possible to follow a completely decentralized approach to design (four) SISO controllers, sufficient to control the helicopter. As shown in [13], for non-aggressive flights, using the model in [1] and restricting the magnitude of the Euler angles to less than 2.5^{o}, diagonal dominances have relatively small values allowing neglecting coupling in the helicopter's model.

The same approach is followed for both inner-loop and outer-loop controller design. Three types of controllers are derived and compared for different flight profiles: PID, Fuzzy Logic controller and LQR. Results demonstrate the validity of the approach and the suitability of the aforementioned controllers.

The rationale for this Chapter is to allow researchers with basic control theoretic background to grasp the design process in a conceptual way and apply it to small / miniature unmanned helicopters using available design tools. Further, when a completely decentralized design approach achieves desired performance objectives, advantages when compared with full

MIMO designs are simpler controllers, easier to maintain but also capable of being enhanced in a straightforward way [2].

5.2 Outline of the Design Method

In principle, helicopter controller design follows the well-known system decomposition in inner-loop and outer-loop, regardless of specifics [18] [19]. Figure 5.1 shows the basic cascaded controller block diagram.

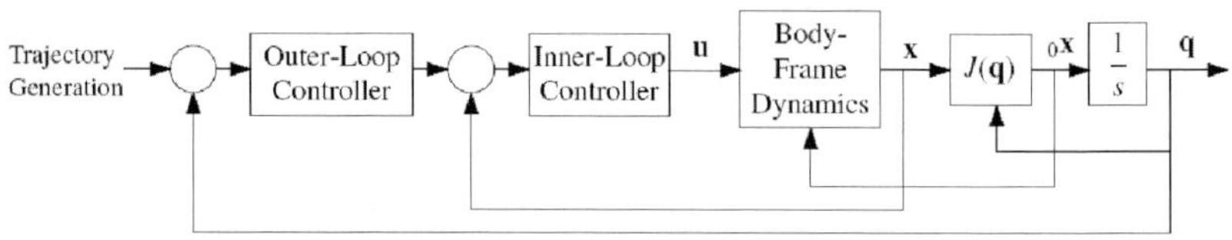

Fig. 5.1. Basic cascaded control structure.

- The *inner-loop* enables stabilization of the unstable plant and partial decoupling of control inputs. It generates the four low level commands, longitudinal and lateral cyclic, collective and pedal.
- The *outer-loop* generates set points for the inner-loop controller. Since the control system is restricted to non-aggressive maneuvers, set points for the outer-loop controller are the four degrees of freedom in the trivial equilibrium point: Cartesian position (x, y, z) and heading (ψ).

The helicopter flight envelope is characterized by parameter changes in the state space configuration. This change of 'operational mode' is analyzed in two ways:

- The plant has hybrid characteristics switching dynamics according to crossing a specified velocity threshold.
- Body-reference frame dynamics are linearized around two operational points, one for hovering, and one for moving forward at a particular speed.

The controller that copes with parameter variations is the inner-loop controller. Parameter variations may be handled by two (proposed) alternatives:

- Design of a single controller with sufficient robustness margins for both flight modes. The disadvantage of this alternative is relatively low performance in both flight modes, not a crucial one because of requirements imposed on non-aggressive flight maneuvering. The advantage relates to simplicity, (guaranteed) stability and robustness. However, the nominal model should average the two available models - obtaining the average model is subject to speculation. A simpler alternative is to use the model for hover flight as the nominal model since the helicopter flies in this mode of operation most of the time when performing non-aggressive flights.

- Design of two controllers based on gain-scheduling. The two controllers result in better performance since each one of them is better tuned. However, the transition from one controller to the other is not smooth because of the existence of two parameter sets - there may be chattering resulting in unexpected behavior. An additional disadvantage is increased complexity.

After the design of the inner-loop through linear techniques, the outer-loop considers nonlinearities due to the transformation from the body-reference to the inertial-reference frame. However, for non-aggressive flights, linearization is possible in the region where the roll and pitch angles are zero.

Detailed design of three controllers is presented. The PID/PD and the PID-like Fuzzy Logic controllers are designed as multiple SISO controllers following the same basic steps. The LQR inner-loop controller with a PID controller used as the outer-loop controller is a MIMO controller and follows a slightly different design procedure.

Figure 5.2 (see next page) illustrates the design approach through a flow chart that assists designers in choosing between SISO and MIMO controller alternatives.

5.3 Fundamentals of Helicopter System Modeling

Figure 5.3 shows a block diagram of Mettler's model [1]. Parameters that characterize the system vary according to the current helicopter speed with respect to the inertial-reference frame. $J(\boldsymbol{q})$ depends on the current helicopter pose.

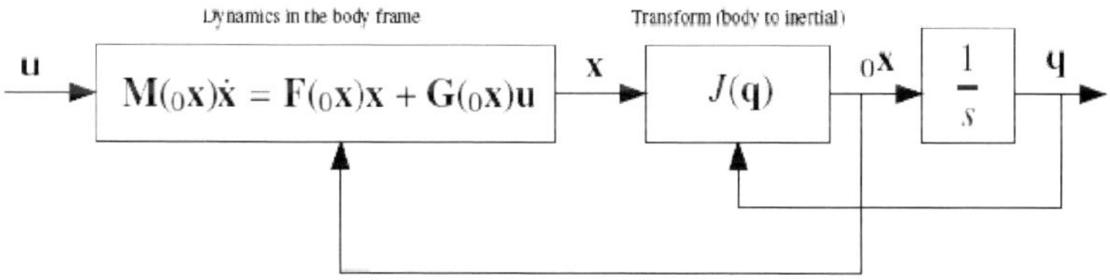

Fig. 5.3. Simplified block diagram of Mettler's model.

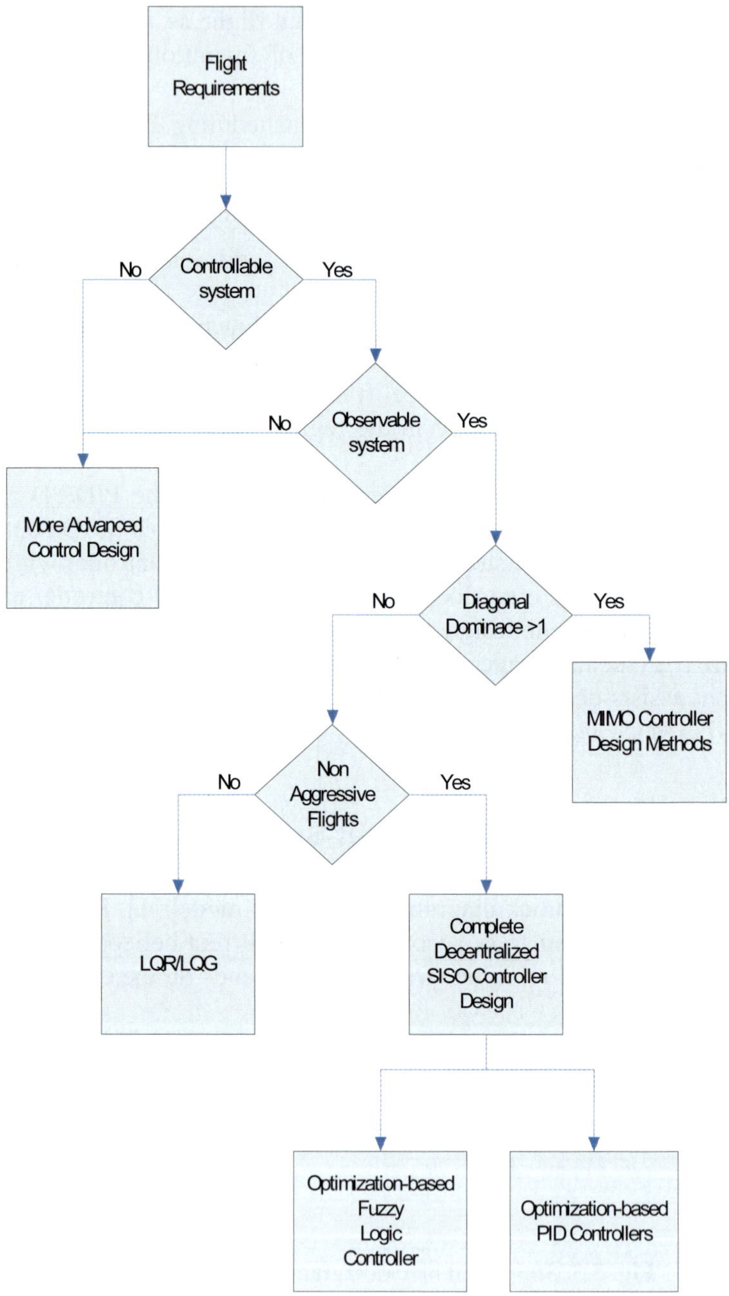

Fig. 5.2. Flow diagram of controller selection.

5.3.1 Dynamics in the Body-Reference Frame

The 13th order model with $\boldsymbol{x}$ and $\boldsymbol{u}$ defined in Tables 5.1 and 5.2 respectively, is:

$$dx/dt = \boldsymbol{Ax} + \boldsymbol{Bu}$$
$$y = \boldsymbol{Cx} \tag{5.1}$$

Part	State	Description
Fuselage Linear Motion	u	x -velocity
	v	y -velocity
	w	z -velocity
Fuselage Angular Motion	p	Roll Angular Rate
	q	Pitch Angular Rate
	r	Yaw Angular Rate
Rotor Tip-Path-Plane	a	Longitudinal Flapping Angle
	b	Lateral Flapping Angle
Stabilizer Bar Tip-Path-Plane	c	Longitudinal Flapping Angle
	d	Lateral Flapping Angle
Augmented Yaw Dynamics	r_{fb}	Yaw Rate Gyro Feedback
Roll	ϕ	Roll Euler Angle
Pitch	θ	Pitch Euler Angle

Table 5.1. Model states.

Control	Description	Units
δ_{lat}	Lateral Cyclic Deflection	Dimensionless [-1,1]
δ_{lon}	Longitudinal Cyclic Deflection	Dimensionless [-1,1]
δ_{ped}	Pedal Control Input	Dimensionless [-1,1]
δ_{col}	Collective Control Input	Dimensionless [0,1]

Table 5.2. Control input variables.

A and ***B*** depend on the flight regime with nominal parameter values for hovering and cruising as defined in [1], and specific form expressed below, while Tables 5.3 to 5.5 present descriptions and values for these constants as also defined in [1].

$$\boldsymbol{A}=\begin{bmatrix} X_u & 0 & 0 & 0 & 0 & -g & X_a & 0 & 0 & 0 & 0 & 0 & 0 \\ 0 & Y_v & 0 & 0 & g & 0 & 0 & Y_b & 0 & 0 & 0 & 0 & 0 \\ L_u & L_v & 0 & 0 & 0 & 0 & 0 & L_b & L_w & 0 & 0 & 0 & 0 \\ M_u & M_v & 0 & 0 & 0 & 0 & M_a & 0 & M_w & 0 & 0 & 0 & 0 \\ 0 & 0 & 1 & 0 & 0 & 0 & 0 & 0 & 0 & 0 & 0 & 0 & 0 \\ 0 & 0 & 0 & 1 & 0 & 0 & 0 & 0 & 0 & 0 & 0 & 0 & 0 \\ 0 & 0 & 0 & -1 & 0 & 0 & -1/\tau_f & A_b/\tau_f & 0 & 0 & 0 & A_c/\tau_f & 0 \\ 0 & 0 & -1 & 0 & 0 & 0 & B_a/\tau_f & -1/\tau_f & 0 & 0 & 0 & 0 & B_d/\tau_f \\ 0 & 0 & 0 & 0 & 0 & 0 & Z_a & Z_b & Z_w & Z_r & 0 & 0 & 0 \\ 0 & N_v & N_p & 0 & 0 & 0 & 0 & 0 & N_w & N_r & N_{rfb} & 0 & 0 \\ 0 & 0 & 0 & 1 & 0 & 0 & 0 & 0 & 0 & K_r & K_{rfb} & 0 & 0 \\ 0 & 0 & 0 & -1 & 0 & 0 & 0 & 0 & 0 & 0 & 0 & -1/\tau_s & 0 \\ 0 & 0 & -1 & 1 & 0 & 0 & 0 & 0 & 0 & 0 & 0 & 0 & -1/\tau_s \end{bmatrix} \tag{5.2}$$

$$\boldsymbol{B}=\begin{bmatrix} 0 & 0 & 0 & 0 \\ 0 & 0 & Y_{\text{ped}} & 0 \\ 0 & 0 & 0 & 0 \\ 0 & 0 & 0 & M_{\text{col}} \\ 0 & 0 & 0 & 0 \\ 0 & 0 & 0 & 0 \\ \dfrac{A_{\text{lat}}}{\tau_f} & \dfrac{A_{\text{lon}}}{\tau_f} & 0 & 0 \\ \dfrac{B_{\text{lat}}}{\tau_f} & \dfrac{B_{\text{lon}}}{\tau_f} & 0 & 0 \\ 0 & 0 & 0 & Z_{\text{col}} \\ 0 & 0 & N_{\text{ped}} & N_{\text{col}} \\ 0 & 0 & 0 & 0 \\ 0 & \dfrac{C_{\text{lon}}}{\tau_s} & 0 & 0 \\ \dfrac{D_{\text{lat}}}{\tau_s} & 0 & 0 & 0 \end{bmatrix} \tag{5.3}$$

Parameter	Description	Hover Value	Cruise Value
X_u	Stability derivative	-0.0505	-0.122
X_a	Stability derivative	-32.2	-32.2
G	force of gravity	-32.2	-32.2
Y_v	Stability derivative	-0.154	-0.155
Y_b	Stability derivative	32.2	32.2
Z_a	Stability derivative	-9.75	--
Z_b	Stability derivative	-131	--
Z_w	Stability derivative	-0.614	-1.01
Z_r	Stability derivative	0.93	--
Y_{ped}	Control derivative	--	11.23
Z_{col}	Control derivative	-45.8	-60.3

Table 5.3. Model constants for fusalage linear motion equations.

Parameter	Description	Hover Value	Cruise Value
A_b	Stability derivative	-0.189	-0.176
A_c	Stability derivative	0.644	0.577
B_a	Stability derivative	0.368	0.419
B_d	Stability derivative	0.71	0.664
τ_f	Main blade time constant	0.046	0.0346
K_r	Stability derivative	2.16	2.18
K_{rfb}	Stability derivative	-8.26	-7.79
τ_s	Stabilizer bar time constant	0.342	0.259
A_{lat}	Cyclic input sensitivity	0.0313	0.0265
A_{lon}	Cyclic input sensitivity	-0.1	-0.0837
B_{lat}	Cyclic input sensitivity	0.14	0.124
B_{lon}	Cyclic input sensitivity	0.0138	0.02
C_{lon}	Cyclic input sensitivity	-0.259	-0.225
D_{lat}	Cyclic input sensitivity	0.273	0.29

Table 5.4. Model constants for tip-path-plane and augmented yaw dynamics.

Parameter	Description	Hover Value	Cruise Value
L_u	Stability derivative	-0.144	--
L_v	Stability derivative	0.143	--
L_b	Stability derivative	166	213
L_w	Stability derivative	--	-0.213
M_v	Stability derivative	-0.0585	--
M_a	Stability derivative	82.6	108
M_u	Stability derivative	-0.0561	--
M_w	Stability derivative	--	0.0728
N_v	Stability derivative	0.0301	0.401
N_p	Stability derivative	-3.53	--
N_w	Stability derivative	0.0857	--
N_r	Stability derivative	-4.13	-3.9
N_{rfb}	Stability derivative	-33.1	-26.4
M_{col}	Control derivative	--	6.98
N_{ped}	Control derivative	33.1	26.4
N_{col}	Control derivative	-3.33	--

Table 5.5. Model constants for angular motion.

5.3.2 Transformation to the Inertial-Reference Frame

The relation between the body-reference and inertial-reference velocities is ${}_0\mathbf{x}=J(\mathbf{q})\mathbf{x}$ with the inertial velocity vector ${}_0\mathbf{x}$ being:

$$
{}_0\boldsymbol{x} = \begin{bmatrix} {}_0\boldsymbol{x}_t \\ {}_0\boldsymbol{x}_r \end{bmatrix} = \begin{bmatrix} v_x \\ v_y \\ v_z \\ \dot{\phi} \\ \dot{\theta} \\ \dot{\psi} \end{bmatrix} \tag{5.4}
$$

${}_0\boldsymbol{x}_t^T$ and ${}_0\boldsymbol{x}_r^T$ are the inertial translational and angular velocities, respectively, composed of the individual Euler angle rates. The velocity vector with respect to the body-reference frame corresponds to a subset of $\boldsymbol{x}$:

$$
\boldsymbol{x} = \begin{bmatrix} \boldsymbol{x}_t \\ \boldsymbol{x}_r \end{bmatrix} = \begin{bmatrix} u \\ v \\ w \\ p \\ q \\ r \end{bmatrix} \tag{5.5}
$$

The Jacobian is decomposed into translational and rotational velocities:

$$\begin{bmatrix} {}_0\boldsymbol{x}_t \\ {}_0\boldsymbol{x}_r \end{bmatrix} = \begin{bmatrix} J_t(\boldsymbol{q}) & 0 \\ 0 & J_r(\boldsymbol{q}) \end{bmatrix} \begin{bmatrix} \boldsymbol{x}_t \\ \boldsymbol{x}_r \end{bmatrix} \tag{5.6}$$

The translational Jacobian for an aerial vehicle is equal to the inverse of the direction cosine matrix (DCM) with *s*, *c* and *t* used for abbreviating *sin(·)*, *cos(·)* and *tan(.)* functions:

$$J_t(\mathbf{q}) = \mathrm{DCM}^{-1} \tag{5.7}$$

$$\mathrm{DCM} = \begin{bmatrix} c\theta c\psi & c\theta s\psi & -s\theta \\ s\phi s\theta c\psi - c\phi s\psi & s\phi s\theta s\psi - c\phi c\psi & s\phi c\theta \\ c\phi s\theta c\psi + s\phi s\psi & c\phi s\theta s\psi - s\phi c\psi & c\phi c\theta \end{bmatrix} \tag{5.8}$$

$\mathrm{DCM}^{-1} = \mathrm{DCM}^T$, therefore:

$$J_t(\boldsymbol{q}) = \begin{bmatrix} c\theta c\psi & s\phi s\theta c\psi - c\phi s\psi & c\phi s\theta c\psi + s\phi s\psi \\ c\theta s\psi & s\phi s\theta s\psi - c\phi c\psi & c\phi s\theta s\psi - s\phi c\psi \\ -s\theta & s\phi c\theta & c\phi c\theta \end{bmatrix} \tag{5.9}$$

The rotational component of the Jacobian is given by:

$$J_r(\boldsymbol{q}) = \begin{bmatrix} 1 & \sin(\phi)\tan(\theta) & \cos(\phi)\tan(\theta) \\ 0 & \cos(\phi) & -\sin(\theta) \\ 0 & \sin(\phi)\sec(\theta) & \cos(\phi)\sec(\theta) \end{bmatrix} \tag{5.10}$$

5.3.3 Model Output

The output of the complete model of Figure 5.3 matches the helicopter degrees of freedom, absolute position (x, y, z) and pose, roll ϕ, pitch θ and yaw ψ:

$$\boldsymbol{q} = \begin{bmatrix} x \\ y \\ z \\ \phi \\ \theta \\ \psi \end{bmatrix} \tag{5.11}$$

5.4 Helicopter Model Analysis Fundamentals

Analysis with respect to the helicopter body-fixed frame results in a thorough understanding of the system itself; analysis relative to the inertial reference frame pertains to, and reveals information about system outputs.

5.4.1 Analysis Relative to the Body-Fixed Reference Frame

This analysis includes open-loop stability (poles and zeros are illustrated in Figure 5.4), mode derivation for hovering and cruising, as well as study of system transmission zeros, controllability and observability.

Modal analysis reveals that there are seven basic modes in the helicopter dynamics for hovering and nine modes for cruise flight. Table 5.6 and Figures 5.5 (modes 1-6), and 5.6 (mode 7) present the seven hovering mode values. Table 5.7 and Figures 5.7 (modes 1-4) and 5.8 (modes 5-9) illustrate similar information for the nine cruising modes. In both cases, modes are listed in descending order with respect to their real part. Unitary eigenvectors are used as initial conditions to excite modes and obtain the dynamic response.

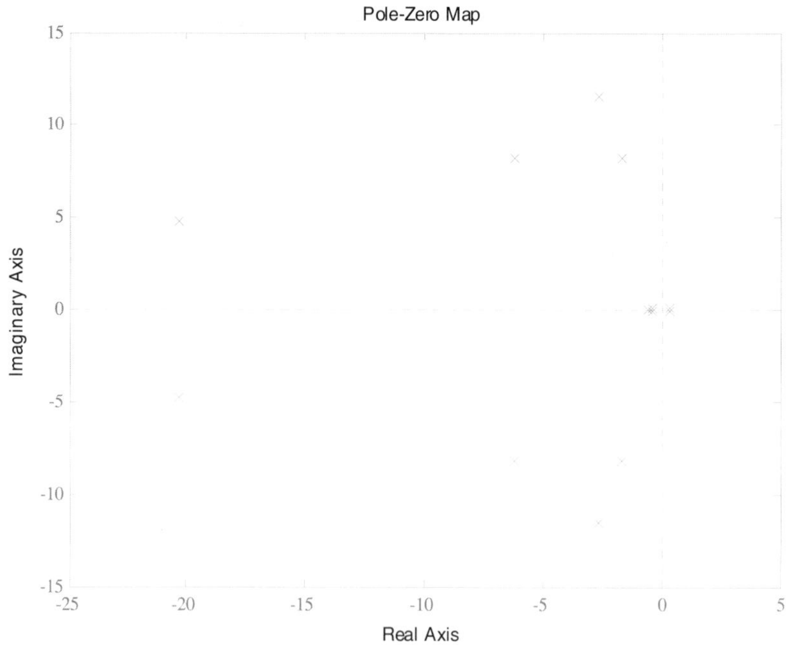

a)

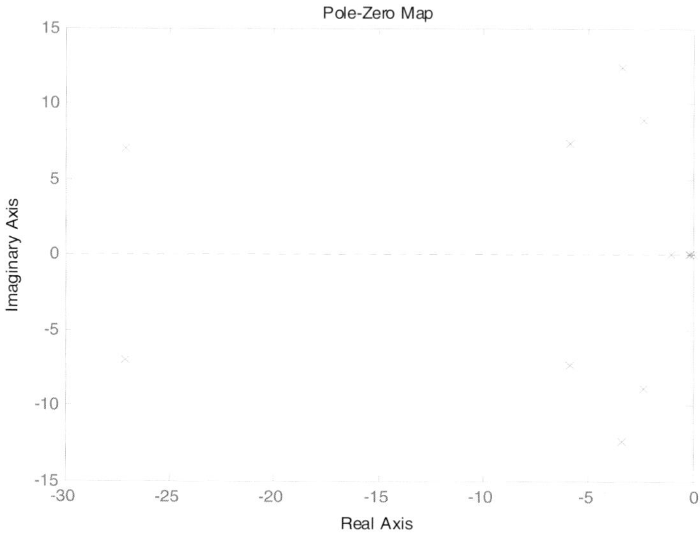

b)

Fig. 5.4. Open-loop poles / zeros: a) hovering, b) cruising modes.

Mode 1	$\lambda_1 = 0.3061 \pm 0.0939i$
Mode 2	$\lambda_2 = -0.4007 \pm 0.0862i$
Mode 3	$\lambda_3 = -0.6078$
Mode 4	$\lambda_4 = -1.6977 \pm 8.1884i$
Mode 5	$\lambda_5 = -2.6605 \pm 11.5571i$
Mode 6	$\lambda_6 = -6.1981 \pm 8.1967i$
Mode 7	$\lambda_7 = -20.3125 \pm 4.7429i$

Table 5.6. Hovering modes.

Mode 1	$\lambda_1 = 0$
Mode 2	$\lambda_2 = 0$
Mode 3	$\lambda_3 = -0.1220$
Mode 4	$\lambda_4 = -0.1550$
Mode 5	$\lambda_5 = -1.0100$
Mode 6	$\lambda 6 = -2.3247 \pm 8.8898i$
Mode 7	$\lambda_7 = -3.3662 \pm 12.3914i$
Mode 8	$\lambda_8 = -5.8500 \pm 7.3314i$
Mode 9	$\lambda_9 = -27.0718 \pm 7.0303i$

Table 5.7. Cruising modes.

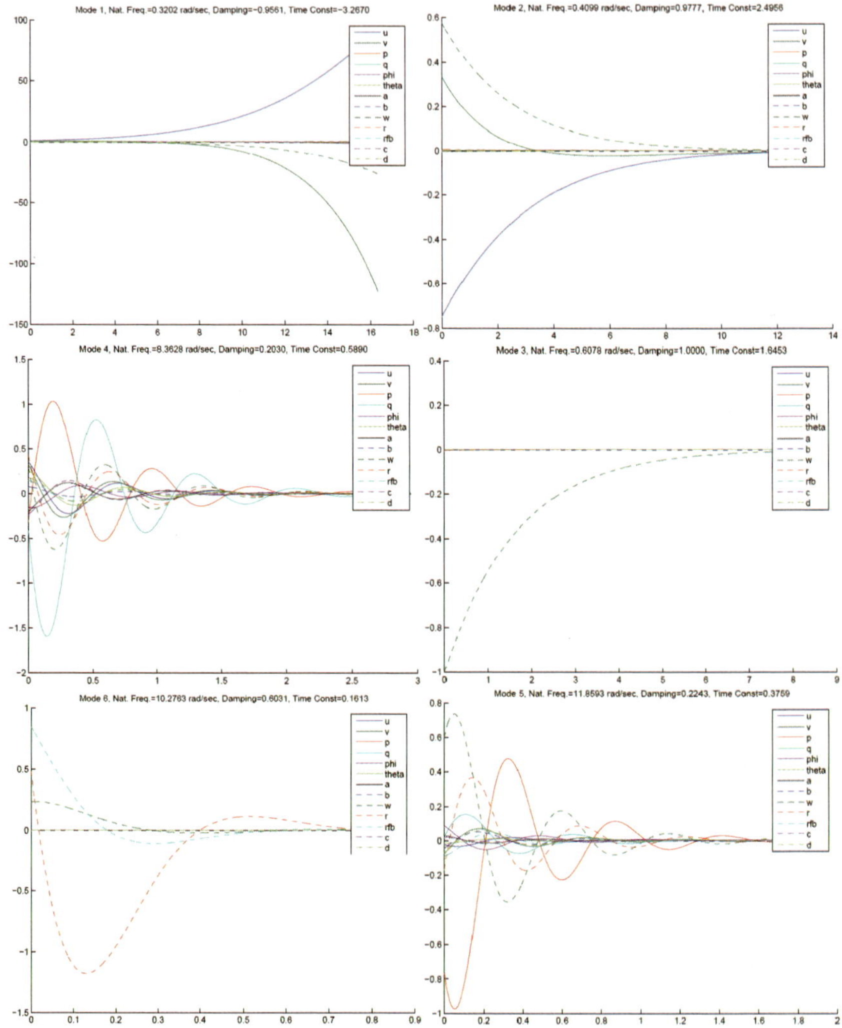

Fig. 5.5. Hovering modes 1-6.

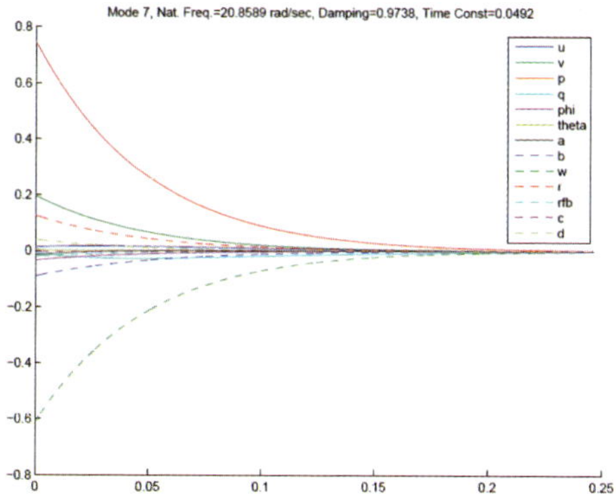

Fig. 5.6. Hovering mode 7.

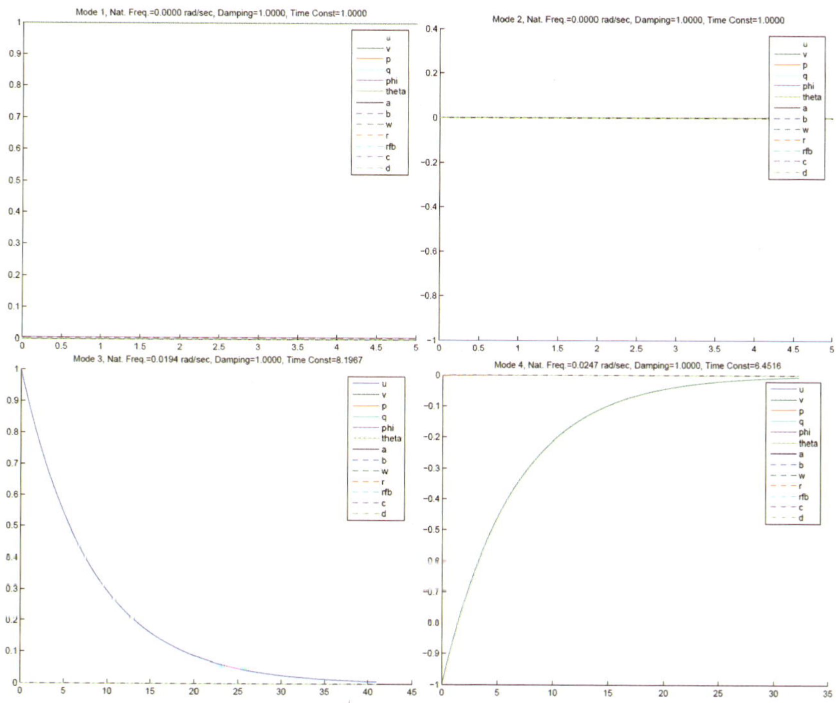

Fig. 5.7. Cruise modes 1-4.

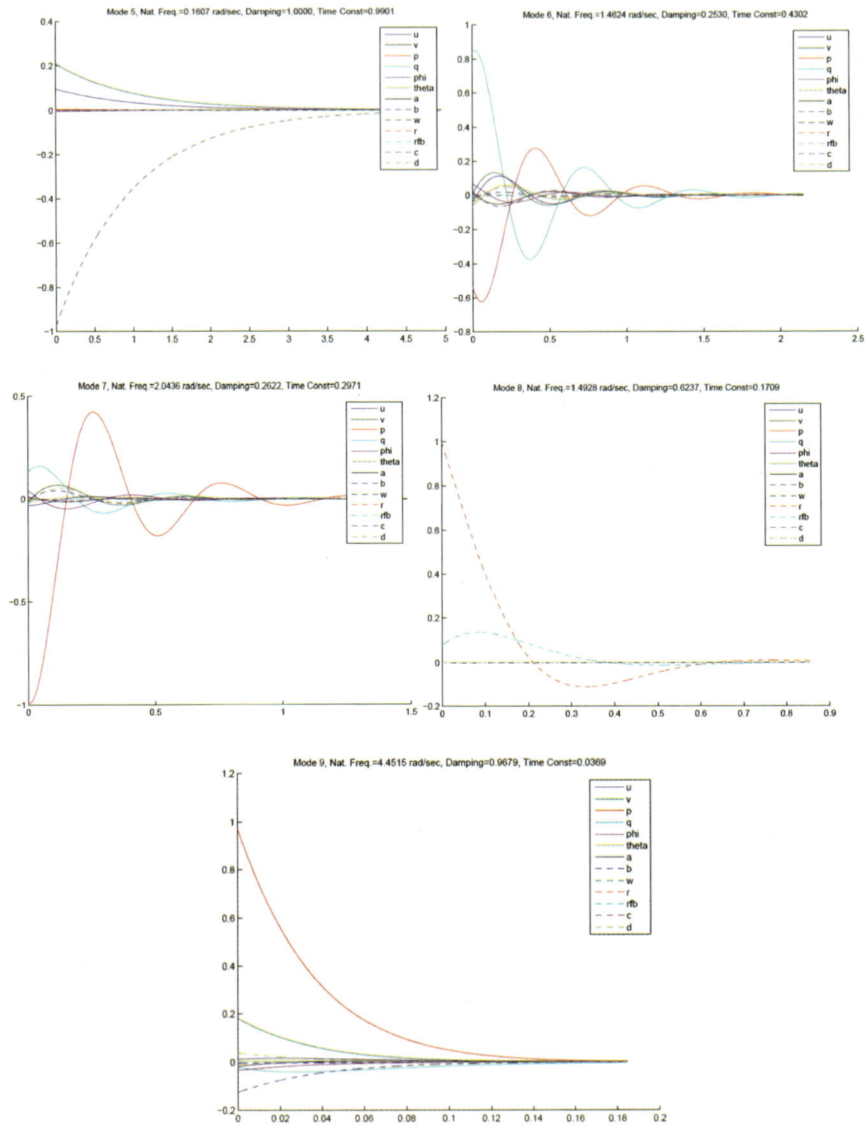

Fig. 5.8. Cruising modes 5-9.

Open-loop Stability: Before system analysis, it is essential to observe existence of any right half-plane (RHP) poles. They affect strongly system behavior and reduce accuracy of analysis based on system frequency response. After defining the state space model, *MATLAB*'s `pzmap` function is used to plot poles and zeros of the MIMO system in the s-plane, allowing for a simple test to determine if the system is open-loop unstable. Figure 5.4 illustrates that there are RHP poles while the helicopter is in hovering mode, but not in cruising mode.

Hovering Modes: Mode 1 is an unstable mode affecting mainly longitudinal and lateral velocities. It does not affect the states rapidly, since the eigenvalue real part is relatively small. Mode 2 is the slowest and most dominant stable mode that shows coupling between all three translational velocities. There is an opposite relationship between the longitudinal velocity and the height. Thus, when the helicopter moves forward it looses height. This mode is almost critically damped showing only small oscillations. Mode 3 affects almost exclusively the velocity in the z-direction. It is a first order mode with a time constant of 1.6s. Mode 4 demonstrates the gyroscopic coupling between rotational velocities. It may be inferred that rotational dynamics is faster than translational dynamics. Besides, the vertical velocity is also coupled in this highly oscillatory mode. Mode 5 reflects the out-of-phase coupling between the pitch rate, p, and the vertical velocity, w. At the same time, the yaw rate, r, is also affected in a similar oscillatory way. Mode 6 demonstrates that the gyro contributes to the dynamics of the plant in a lightly undamped way at relatively high frequencies. The strong coupling between the yaw rate and the yaw rate gyro feedback is only natural. Heave dynamics is also affected. In contrast to Mode 5, Mode 6 affects almost exclusively the three state variables previously mentioned. Mode 7 is the fastest mode and it represents the fastest coupling between the pitch change and the height rate. At the same time, the lateral translational dynamics shows a certain degree of coupling. This mode influences the rest of the state variables to a lesser extent.

Cruising Modes: Mode 1 illustrates the integrating nature of the mode acting on the y-direction. The helicopter behaves as a "free" floating rigid body with respect to the translational movement in the y-direction. Mode 2 is integrating mode acting on the longitudinal direction. The helicopter behaves as a "free" floating rigid body with respect to the translational movement in the x-direction. Mode 3 is the most dominant mode indicating the integrating nature of the helicopter in both directions controlled by the cyclic commands. It acts as a relatively slow mode in the longitudinal direction affecting exclusively the longitudinal velocity. It represents the inertia of the helicopter that prevents it from changing its longitudinal velocity rapidly. Mode 4 is the lateral equivalent to Mode 3, albeit slower, interpreted as the translational inertia of the helicopter that prevents it from changing its lateral velocity swiftly. Mode 5 affects mainly the vertical translational velocity; it relates to Modes 3 and 4, interpreted as inertia acting on the z-direction. It is clear that while cruising, the helicopter responds faster (time constant of 1s) to changes in height than in horizontal direction of travel (time constant around 7s). There is some coupling with the horizontal velocity states. Mode 6 is parallel to Mode 4 in hovering mode. It has a similar time constant and damping. It shows the oscillatory coupling between pitch and roll rate. Hence, it is the most dominant mode that affects the rotational motion of the helicopter in cruise flight. In cruise flight, the yaw rate is less coupled to the other angular rates, while in hover flight all three rotational magnitudes are coupled with the height rate. Mode 6 is parallel to mode 4 in hovering mode. It has a similar time constant and damping. It shows the oscillatory coupling between pitch and roll rate. Hence, it is the most dominant mode that affects the rotational motion of the helicopter in cruise flight. In cruise flight, the yaw rate is less coupled to the other angular rates, while in hover flight all three rotational magnitudes are coupled with the height rate. Mode 7 has a rotational nature and it involves mainly the roll rate. It is undamped and faster than the previous mode where the pitch rate was more dominant. Mode 8 relates to Mode 6 in hovering where the yaw rate dynamics is taken into consideration together with the gyro feedback. However, hover flight is different in the sense that the gyro feedback is more predominant than in cruise flight. Mode 9 is parallel to Mode 7 in hovering flight condition. In both cases, the roll derivative plays a significant role. This mode is the only mode that shows a slight influence of the flapping dynamics.

Several conclusions are drawn from the modal analysis as follows:

- Dynamics in hover flight is unstable; the instability is because of the horizontal translational velocity in x, y directions. This means that special attention must be given to the helicopter lateral and longitudinal controls during the controller design phase.
- Flapping dynamics is almost negligible; its effect is almost not noticeable in any of the modes analyzed. This observation may be used for model order reduction.
- Dynamics in cruise mode shows more damped and slower modes than in hover flight. This result verifies the common assumption that the helicopter behaves more like a fixed-wing aircraft in cruise flight. Furthermore, it means that the helicopter is definitely more difficult to control in hover flight.
- Modes considering the yaw rate have a relatively high damping factor and do not show a large degree of coupling with other state variables. Therefore, a SISO controller is sufficient to tackle heading control.

All features extracted from the modal analysis show a high correlation of the helicopter model with physical insight of a real helicopter verifying its validity (at least qualitatively).

Transmission Zeros: Transmission zeros are MIMO extensions of zeros in SISO systems. Their identification is relevant for application of certain techniques to MIMO systems such as decoupling control or input-output linearization. Their values may be interpreted as frequencies at which the system does not transmit anything given that u_k and the initial condition (ξ_k) have a particular direction that is different from zero. Mathematically, the zeros of a linear system given by ***A, B, C*** and ***D***, can be defined as the values of z_k, such that equations (5.12) through (5.14) hold [14]:

$$\dot{x}(t) = Ax(t) + Bu_k e^{z_k t}, x(0) = \xi_k \tag{5.12}$$

$$y(t) = Cx(t) + Du(t) \tag{5.13}$$

$$y(t) \equiv 0 \quad \forall t > 0 \tag{5.14}$$

Calculations performed in *MATLAB* (using the `tzero` function) reveal that there are no transmission zeros for both modes in the body-fixed reference frame.

Controllability and Observability: As known, the controllability matrix is:

$$C_o = \begin{bmatrix} B & AB & A^2 B \ldots A^{n-1} B \end{bmatrix} \tag{5.15}$$

However, evaluating controllability directly from the controllability matrix may be misleading because it is very sensitive to numerical problems. *MATLAB*'s `ctrbf` function is used to decompose the system into controllable and uncontrollable subspaces and to count the number of controllable states found in each iteration of the algorithm. This is much less prone to numerical problems; therefore, it is the recommended method to determine system controllability. For hovering, there exist 13 controllable states, thus, the system is controllable during hovering:

$$\begin{bmatrix} 4 & 3 & 2 & 2 & 2 & 0 & 0 & 0 & 0 & 0 & 0 & 0 & 0 \end{bmatrix} \tag{5.16}$$

For cruising, there exist 13 controllable states as well:

$$\begin{bmatrix} 4 & 4 & 4 & 1 & 0 & 0 & 0 & 0 & 0 & 0 & 0 & 0 & 0 \end{bmatrix} \tag{5.17}$$

System controllability is also verified by applying variable precision arithmetic to calculating the controllability matrix.

System observability depends on measurable outputs. The assumption is that the complete system state may be observed from measurable outputs.

It is considered that outputs are available for measurement. The output vector is:

$$y = [u, v, w, p, q, r]^T \tag{5.18}$$

The system is decomposed into observable and unobservable subspaces (using the `obsvf` function). For hovering, 13 observable states are found:

$$[4 \;\; 3 \;\; 2 \;\; 2 \;\; 2 \;\; 0 \;\; 0 \;\; 0 \;\; 0 \;\; 0 \;\; 0 \;\; 0 \;\; 0] \tag{5.19}$$

Similarly, for cruising, 13 states are also found:

$$[4 \;\; 4 \;\; 4 \;\; 1 \;\; 0 \;\; 0 \;\; 0 \;\; 0 \;\; 0 \;\; 0 \;\; 0 \;\; 0 \;\; 0] \tag{5.20}$$

DC system analysis: DC-gain analysis reveals information about how inputs affect state variables under steady-state conditions, also indicating how the system ‘behaves’ in the end. It represents an initial step towards choosing appropriate input/output pairs. It provides a rough understanding of the degree of coupling between inputs and state variables, and coupling among state variables. The DC-gain from inputs to outputs is obtained by:

$$\boldsymbol{H}_{\mathrm{DC}} = \lim_{s \to 0} \boldsymbol{H}(s) = \lim_{s \to 0} \left[\boldsymbol{C}(s\boldsymbol{I} - \boldsymbol{A})^{-1} \boldsymbol{B} \right] = \boldsymbol{C}(-\boldsymbol{A})^{-1} \boldsymbol{B} \tag{5.21}$$

Tables 5.8 and 5.9 give the corresponding DC-gains under hovering and cruising. Zero entries correspond to unrelated input/output pairs, and to those establishing a derivative relationship between input/output. Infinity terms reveal an integrating relationship between inputs/outputs.

Observing how inputs/outputs are ‘paired’ leads to applying decentralized control techniques. A measure used to evaluate input/output pairing is the *relative gain array* (RGA) [2], a matrix Λ with ij^{th} elements obtained from:

$$\lambda_{ij} = [\boldsymbol{G}_o(0)]_{ij} \left[\boldsymbol{G}_o^{-1}(0) \right]_{ji} \tag{5.22}$$

$[\mathbf{G}_0(0)]_{ij}$ corresponds to the ij^{th} element of the matrix and $[\mathbf{G}_0^{-1}(0)]_{ij}$ to the ji^{th} element of the inverse of the DC-gain matrix. It is a measure of the sensitiveness between the ith input and the jth output in steady-state. Since this method is applied to square MIMO systems, in this case, different sets of four outputs are chosen and related to corresponding inputs.

The desired outcome is a matrix Λ with ones in the main diagonal.

	δ_{lat}	δ_{lon}	δ_{ped}	δ_{col}
u	169.31	-227.49	0.00	0.00
v	-204.33	-137.50	0.00	0.00
p	0.00	0.00	0.00	0.00
q	0.00	0.00	0.00	0.00
ϕ	-1.30	-0.58	0.00	0.00
θ	-0.24	0.61	0.00	0.00
a	-0.03	-0.25	0.00	0.00
b	0.32	-0.08	0.00	0.00
w	-69.86	20.55	3.96	-75.76
r	-0.95	-0.19	2.62	-0.77
r_{fb}	-0.25	-0.05	0.68	-0.20
c	0.00	-0.26	0.00	0.00
$\boldsymbol{D}$	0.27	0.00	0.00	0.00

Table 5.8. DC-gains in hovering mode.

	δ_{lat}	δ_{lon}	δ_{ped}	δ_{col}
u	∞	∞	0.00	∞
v	∞	∞	72.45	∞
p	1.53	0.10	0.00	0.24
q	0.14	-1.16	0.00	0.19
ϕ	∞	∞	0.00	∞
θ	∞	∞	0.00	∞
a	0.00	0.00	0.00	-0.02
b	0.00	0.00	0.00	-0.06
w	0.00	0.00	0.00	-59.70
r	∞	∞	2.60	∞
r_{fb}	∞	∞	0.73	∞
c	-0.04	0.08	0.00	-0.05
d	-0.11	-0.03	0.00	-0.06

Table 5.9. DC-gains in cruising mode.

The first chosen pair is: $\delta_{lat} \to u$, $\delta_{lon} \to v$, $\delta_{ped} \to w$, $\delta_{col} \to r$ resulting in Λ given by:

$$\Lambda_1 = \begin{bmatrix} 0.3337 & 0.6663 & -0.0000 & 0.0000 \\ 0.6663 & 0.3337 & 0.0000 & 0.0000 \\ 0.0000 & 0.0000 & -0.0156 & 1.0156 \\ 0.0000 & -0.0000 & 1.0156 & -0.0156 \end{bmatrix} \tag{5.23}$$

A second choice is: $\delta_{lat} \to v$, $\delta_{lon} \to u$, $\delta_{ped} \to r$, $\delta_{col} \to w$ resulting in:

$$\Lambda_2 = \begin{bmatrix} 0.6663 & 0.3337 & -0.0000 & 0.0000 \\ 0.3337 & 0.6663 & 0.0000 & 0.0000 \\ 0.0000 & 0.0000 & 1.0156 & -0.0156 \\ 0.0000 & -0.0000 & -0.0156 & 1.0156 \end{bmatrix} \tag{5.24}$$

However, it is known that the longitudinal input is used to change the helicopter pitch moving it in the longitudinal direction; the lateral input is used to change the roll; the pedal is used for changes in the yaw; the collective is used to change the height. This leads to the following pairing: $\delta_{lat} \to \phi$, $\delta_{lon} \to \theta$, $\delta_{ped} \to r$, $\delta_{col} \to w$ that returns the best results used for control design with off diagonal terms very small, close to zero, and diagonal terms close to one:

$$\Lambda_3 = \begin{bmatrix} 0.8529 & 0.1471 & 0.0000 & -0.0000 \\ 0.1471 & 0.8529 & 0.0000 & -0.0000 \\ -0.0000 & 0.0000 & 1.0156 & -0.0156 \\ -0.0000 & 0.0000 & -0.0156 & 1.0156 \end{bmatrix} \tag{5.25}$$

5.4.2 Inertial Reference Frame Analysis

Prerequisite to designing SISO controllers is to consider the level of coupling within the system determining if unwanted effects between inputs/outputs may be controlled realistically by a SISO design.

Zero Angle Approximation: The inner-loop controllers are implemented using roll, pitch, heading (yaw) and height (-z) as feedback variables. The state space model represents the helicopter's dynamics expressed in the body-reference frame. Path planning requires that velocities and positions are known in the inertial (or world) reference frame.

For non-aggressive flights, zero angle approximation for roll and pitch angles (that are relatively small anyway) has little effect on system response [13]. Implementing transformations based on the following equations (5.26), and setting roll and pitch to zero the ***C***-matrix is simplified:

$$\begin{aligned}
\dot{\phi} &= p + q\sin\theta\tan\theta + r\cos\phi\tan\theta \\
\dot{\theta} &= q\cos\phi - r\sin\phi \\
\dot{\psi} &= q\cos\phi - r\sin\phi \\
\dot{z} &= u\sin\theta + v\cos\theta\sin\phi + w\cos\theta\cos\phi
\end{aligned} \tag{5.26}$$

$$\begin{bmatrix} v_x \\ v_y \\ v_z \\ \dot{\phi} \\ \dot{\theta} \\ \dot{\psi} \end{bmatrix} = \begin{bmatrix} \cos\psi & -\sin\psi & 0&0&0&0&0&0&0&0&0&0 \\ \sin\psi & \cos\psi & 0&0&0&0&0&0&0&0&0&0 \\ 0&0&0&0&0&0&0&0&1&0&0&0 \\ 0&0&1&0&0&0&0&0&0&0&0&0 \\ 0&0&0&1&0&0&0&0&0&0&0&0 \\ 0&0&0&0&0&0&0&0&0&1&0&0 \end{bmatrix} \begin{bmatrix} u \\ v \\ p \\ q \\ \phi \\ \theta \\ a \\ b \\ w \\ r \\ r_{fb} \\ c \\ d \end{bmatrix} \tag{5.27}$$

In the case of v_x and v_y heading (yaw angle) normally takes values between 0° and 360° depending on desired helicopter trajectory, which must be taken into consideration when calculating the response for the x and y position. In case of roll rate, pitch rate, heading rate and z-velocity, corresponding terms are reduced to ones and zeros. This simplifies equations and linearizes the output as well.

For the inner-loop, x and y are not involved with feedback controllers, and therefore, zero angle approximation is utilized. For the open-loop analysis, the ***C***-matrix consists of the lower four rows given in equation (5.27). By integrating the system outputs, roll, pitch, yaw and height can be obtained determining the matrix of open-loop transfer functions given by:

$$\boldsymbol{T}(s) = \begin{bmatrix} \frac{\phi}{\delta_{lat}}(s) & \frac{\phi}{\delta_{lon}}(s) & \frac{\phi}{\delta_{ped}}(s) & \frac{\phi}{\delta_{col}}(s) \\ \frac{\theta}{\delta_{lat}}(s) & \frac{\theta}{\delta_{lon}}(s) & \frac{\theta}{\delta_{ped}}(s) & \frac{\theta}{\delta_{col}}(s) \\ \frac{\psi}{\delta_{lat}}(s) & \frac{\psi}{\delta_{lon}}(s) & \frac{\psi}{\delta_{ped}}(s) & \frac{\psi}{\delta_{col}}(s) \\ \frac{z}{\delta_{lat}}(s) & \frac{z}{\delta_{lon}}(s) & \frac{z}{\delta_{ped}}(s) & \frac{z}{\delta_{col}}(s) \end{bmatrix} \tag{5.28}$$

This matrix is calculated in *MATLAB* by setting $\boldsymbol{A}$, $\boldsymbol{B}$, $\boldsymbol{C}$ and $\boldsymbol{D}$, with $\boldsymbol{D}$ equal to zero, using the `ss` command and then using the append and connect commands to include integrations. The `zpk` command is used on the modified state space model to calculate the transfer functions.

Diagonal Dominance: Diagonal dominance determines whether the MIMO system is tightly coupled and whether it may be decomposed into several SISO systems. One method is to place outputs and control inputs along the transfer function matrix diagonal and evaluate the level of influence of the other terms in the row. Diagonal dominance, $\phi_c(j\omega)$, is defined below, where $Z_{rc}(j\omega)$ represents the off diagonal elements of the row, while $Z_{cc}(j\omega)$ represents the diagonal element of the row:

$$\phi_c(j\omega) = \frac{\sum_{\substack{r=1 \\ r \neq c}}^{n} \left| z_{rc}(j\omega) \right|}{\left| z_{cc}(j\omega) \right|} \tag{5.29}$$

In order for SISO methods to be used successfully, diagonal dominance must be less than one [15]. This measure is frequency dependent and it is calculated across the system bandwidth based on a sine wave input to the system that produces a sine wave at the output of the system with a different magnitude; the gain is then obtained from this difference. For systems with a known transfer function, this is done mathematically, allowing for the gain to be calculated even for an open-loop unstable system. In case of existing RHP poles, it should be taken into account that as the RHP poles move further to the right of the s-plane, indicated system behavior will diverge faster from the true system behavior. When this is the case it is better to first design/implement PID/PD controllers using numerical analysis methods and then verify that coupling has been reduced to acceptable limits by applying equation (5.29) to the stabilized inner-loop system (as demonstrated below).

However, in general, diagonal dominance may be determined for an open-loop unstable system as an indication of whether a SISO controller design is likely to be effective. For systems with unstable poles close to the origin, diagonal dominance serves as an acceptable indication of system behavior.

As an example, diagonal dominance of the open-loop system under hovering is depicted in Figure 5.9. Results indicate that maximum diagonal dominance for the height/collective is approximately 1.5. This is to be expected because when the cyclic is used to create forward or sideways motion, it also causes a change in height. However, since the obtained value is still very close to one, it is deduced that a SISO controller still gives an acceptable response.

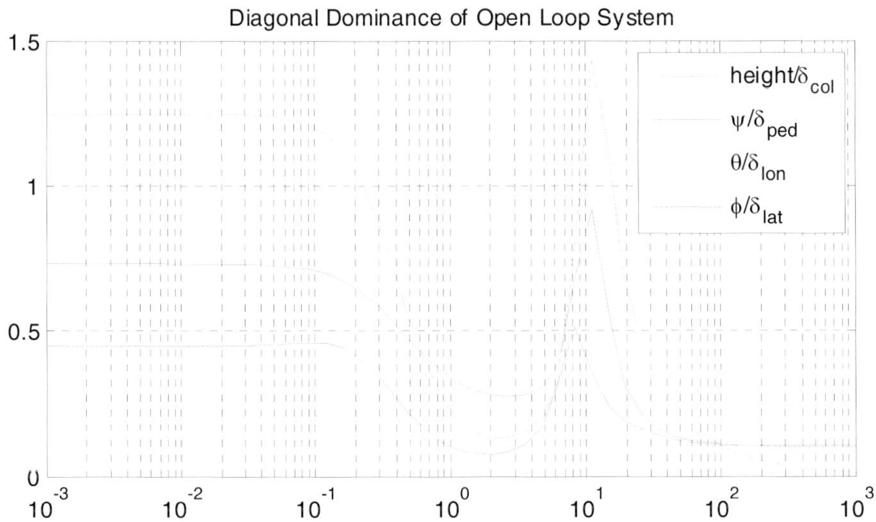

Fig. 5.9. Diagonal dominance of inner-loop transfer functions.

5.5 Completely Decentralized Control

Figure 5.10 shows the fully coupled MIMO helicopter system as well as the multiple input single output (MISO) system that represents the effects of the inputs over the output roll, ϕ with:

$$\phi(s) = \frac{\phi}{\delta_{lat}}(s) * \delta_{lat}(s) + \sum_{\substack{i=lon,\\ ped,\\ col}} \frac{\phi}{\delta_i}(s) * \delta_i(s) \tag{5.30}$$

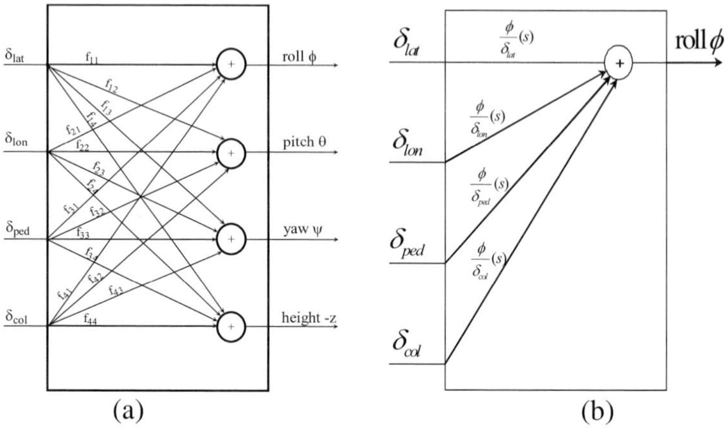

Fig. 5.10. a) Fully coupled MIMO system; b) Effects of inputs on roll.

The key simplifying assumption in completely decentralized control is that a coupling is treated as a form of disturbance [2] with the linear MIMO system treated as multiple SISO systems. Therefore, under the assumption that couplings are considered disturbances and for non-aggressive flights, the helicopter is treated as a set of four SISO subsystems shown in Figure 5.11. The initial MIMO controller design problem has been converted to the design of four SISO controllers.

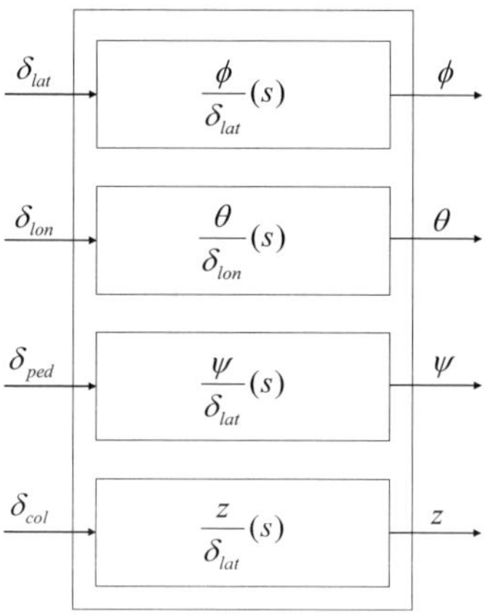

Fig. 5.11. Helicopter control with SISO subsystems.

5.5.1 Designing and Implementing PID Controllers

Figure 5.12 shows a SISO PID controller loop, almost universally used in industrial control [2]. PID controllers are easy to tune if process dynamics are known and a first/second order model can be derived from them. When this is not the case, it is always possible to tune their parameters manually, however, for complex systems like helicopters this is a very tedious and cumbersome task because models are of very high order. The time-domain equation for the PID controller with K_P the proportional gain, K_I the integral gain, and K_D the derivative gain is:

$$u(t) = K_P * e(t) + K_I * \int e(t)dt + K_D * \frac{de(t)}{dt} \tag{5.31}$$

$$e(t) = r(t) - y(t) \tag{5.32}$$

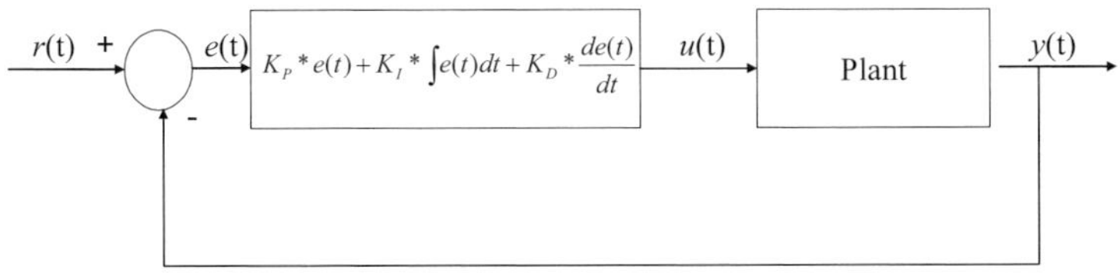

Fig. 5.12. SISO PID closed-loop control configuration.

The derivative term of a PID controller produces an undesirable sharp response to suddenly changing signals (impulse response to step inputs), and to avoid it the derivative term is moved from the closed-loop forward path as in Figure 5.13 resulting in rate feedback [16].

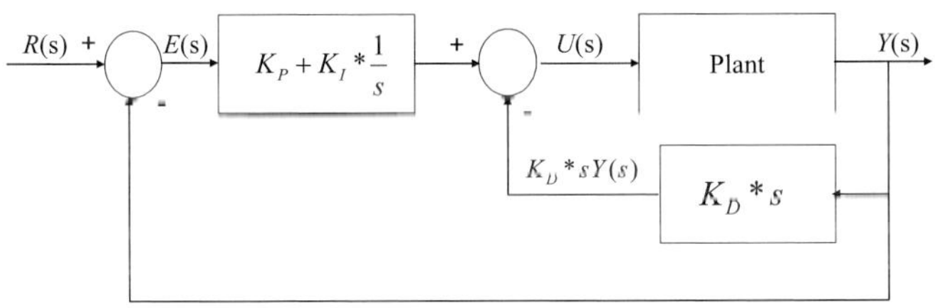

Fig. 5.13. Rate feedback PID configuration.

In cases when plants contain as one of their internal variables the derivative of the output to be controlled, if this derivative is measurable, the rate feedback part of Figure 5.13 may be simplified as in Figure 5.14. This may be beneficial because practical implementation of a differentiator is normally difficult.

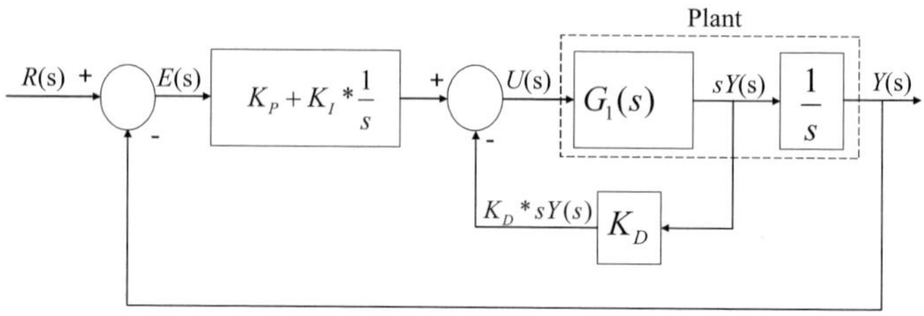

Fig. 5.14. Rate feedback PID with derivative of output measurable.

The helicopter model used has internal variables that are derivatives of some of its controlled outputs; therefore, the inner-loop controller design follows rate feedback PID configuration.

5.5.2 Designing and Implementing Fuzzy Logic Controllers

Fundamentals of a Fuzzy Logic control system are presented in [10] [12], while specific Fuzzy Logic controllers derived to control helicopters are the topic in [3] [4] [6]. Membership functions may be triangular, Gaussian or trapezoidal used for controller input fuzzyfication and output defuzzyfication.

Fuzzy Logic controllers may be tuned by rule modification, membership functions modification (shapes and universe of discourse), and input/output gain or scaling factors modification. Any possible combination of these three basic ways is another tuning procedure. The focus here is in the modification of the inputs/output gain or scaling factors.

A form of Fuzzy IF-THEN rules is [10]:

```
if (antecedent) then (consequent)
```

The antecedent is the set of conditions to be satisfied and the consequent is the set of actions to be realized. The sets of rules used are the General Fuzzy PD and the General Fuzzy PI rule bases [9], presented in Tables 5.10 and 5.11, respectively.

Selection of Fuzzy PID-like controllers is the result of their simplicity and easy tuning. The three types of Fuzzy structural elements of a Fuzzy PID-like controller are described in [17]. Figure 5.15 shows the typical block diagram [9] of a Fuzzy PD-like controller.

		Error, *e(t)*						
		NL	NM	NS	ZR	PS	PM	PL
Change of Error, Δe(t)	PL	ns	ns	ns	ps	pl	pl	Pl
	PM	ns	ns	ns	ps	pl	pl	Pl
	PS	nm	ns	ns	ps	pm	pl	Pl
	ZR	nl	nm	ns	zr	ps	pm	Pl
	NS	nl	nl	mn	ns	ps	ps	Pm
	NM	nl	nl	nl	ns	ps	ps	Ps
	NL	nl	nl	nl	ns	ps	ps	Ps

Table 5.10. Standard rules for a Fuzzy PD-like controller.

		Error, *e(t)*						
		NL	NM	NS	ZR	PS	PM	PL
Change of Error, Δe(t)	PL	ns	ns	ns	ps	pl	pl	Pl
	PM	ns	ns	ns	ps	pl	pl	Pl
	PS	nm	ns	ns	ps	pm	pl	Pl
	ZR	nl	nm	ns	zr	ps	pm	Pl
	NS	nl	nl	mn	ns	ps	ps	Pm
	NM	nl	nl	nl	ns	ps	ps	Ps
	NL	nl	nl	nl	ns	ps	ps	Ps

Table 5.11. Standard rules for a Fuzzy PI-like controller.

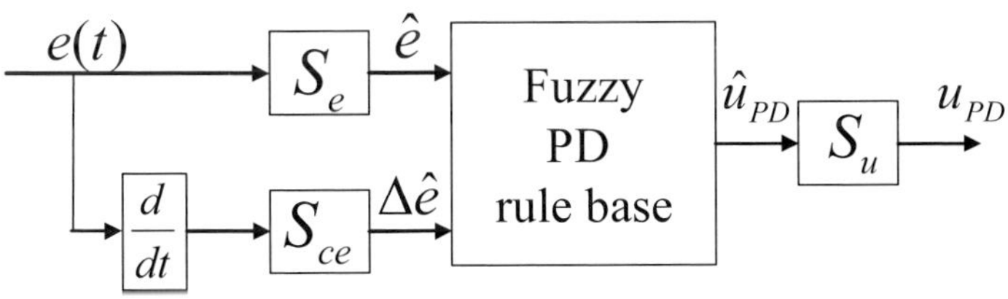

Fig. 5.15. Continuous-time structure of the Fuzzy two-term PD-like controller.

The error e(t) is defined as the difference between the desired signal value (set point) and the real value of the controlled variable; Δe(t) is the error change. S_e is the scaling factor for the error, e(t). S_{ce} is the scaling factor for the change of the error, $\Delta e(t)$. S_u is the scaling factor of the PD-like controller's output. Figure 5.16 shows the typical block diagram of a Fuzzy PID-like controller.

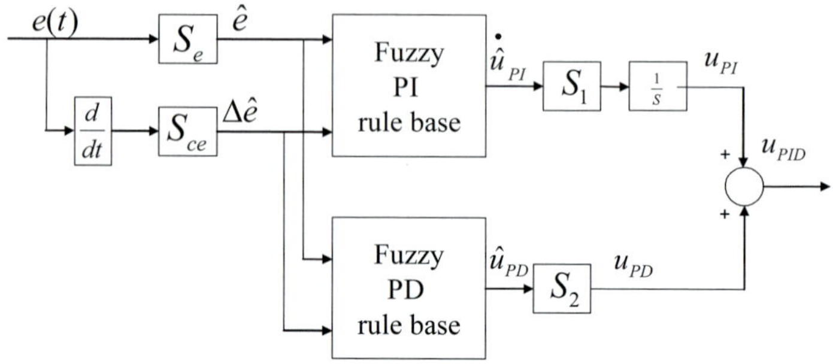

Fig. 5.16. Continuous-time structure of the Fuzzy two-term PID-like controller.

S_1 and S_2 are the output scaling factors of the PI-like and PD-like controller that constitute the Fuzzy PID-like controller.

Using appropriate sets of rules and membership functions mentioned above, Fuzzy PD-like controllers are implemented for height, yaw, *u* and *v* velocity variables, while Fuzzy PID-like controllers for pitch and roll variables.

The output behavior of the controller as a function of the crisp input pair (e(t), Δe(t)) may be represented as an surface, normally called the transfer characteristic of the Fuzzy Logic controller [6], shown in Figure 5.17.

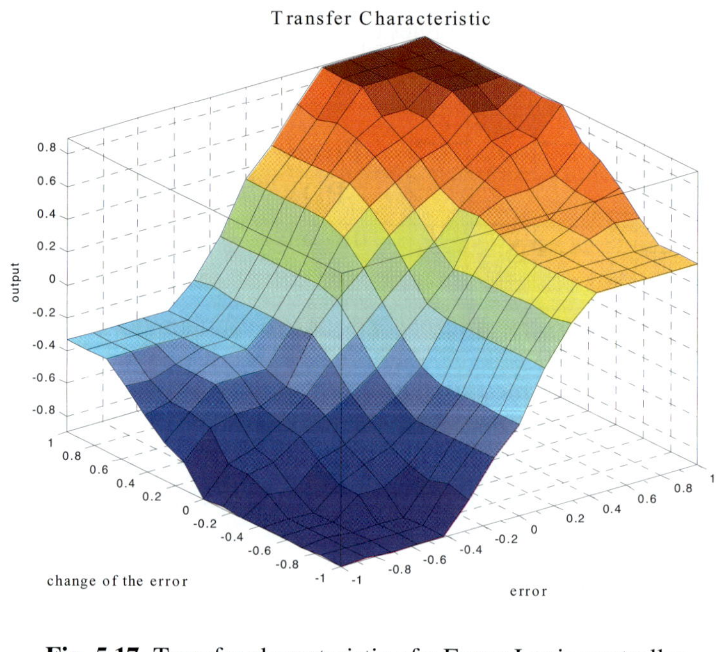

Fig. 5.17. Transfer characteristic of a Fuzzy Logic controller.

5.5.3 Optimization of Controller Parameters

Optimization algorithms help calculate parameter values for best possible performance given a specific objective function. Optimized parameters are the controller gains. Final values are obtained after several runs using the *MATLAB* Optimization Toolbox. The set goal of the objective function under consideration is minimization of the Integral of the Absolute Error (IAE) [11]:

$$\text{IAE}=\int |e(t)| dt \tag{5.33}$$

The IAE defined in [14] is the difference between the step input (set point) and the actual system output, used to obtain the controller's parameters through optimization; however, preliminary optimization results showed that step responses have considerable overshoot. Therefore, the error has been defined as the difference between the desired response of the output and the actual response. The rationale behind this assumption is that if it is possible to obtain controller parameters that produce an exponential output response when the input is excited by a step, then the closed--loop transfer function will be similar to a first order system.

Figure 5.18 shows the configuration of the inner controller built in *MATLAB / Simulink*. The inner controller consists of for four controllers: roll angle controller, pitch angle, yaw angle controller, and the altitude controller. Each block includes parameters that need be optimized as related to the specific type of controller, for example, PID or Fuzzy PID-like.

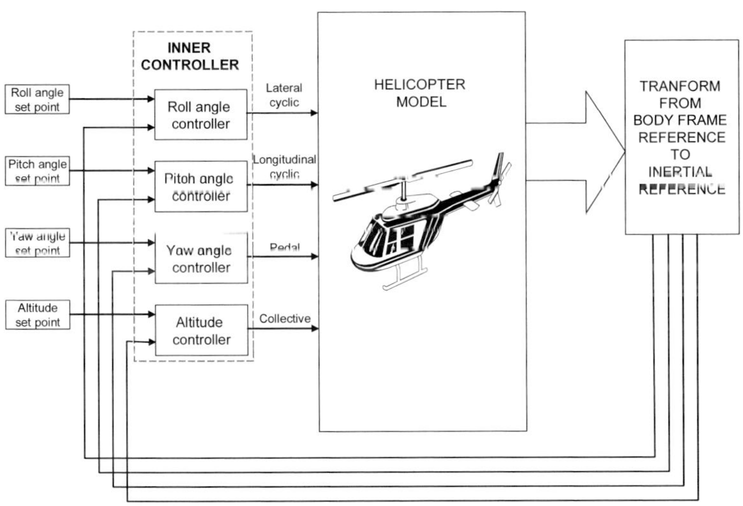

Fig. 5.18. Block diagram used for optimization algorithm for the inner controller.

Figure 5.19 presents the flow chart of the procedure followed to optimize inner-loop controllers. The basic optimization procedure is to apply a step function in the set point of the controller under optimization and keep the set point of the other controllers to a constant value (0° for angles and 10 feet for altitude).

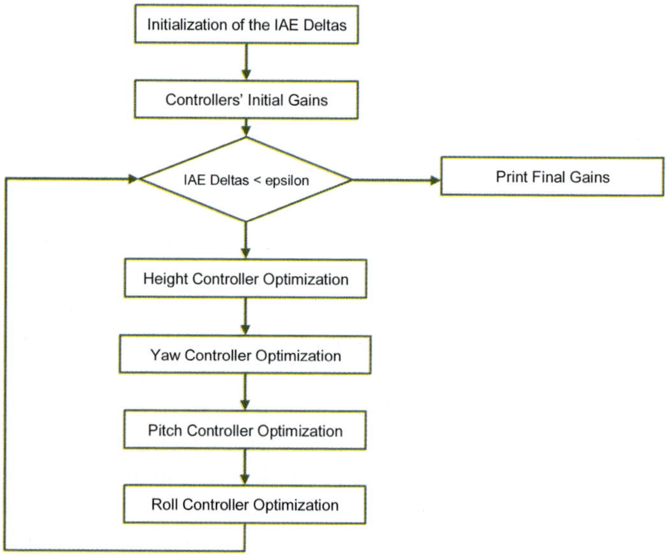

Fig. 5.19. Flow chart of inner-loop controller optimization.

Controller gains have been calculated using the *MATLAB* 7.0 Optimization Toolbox. The IAE Deltas correspond to value changes between two consecutive iterations. The iterative process stops when all IAE Deltas are less than a specified threshold, set to be 0.01. Initial gains are arbitrarily chosen.

The outer-loop controller includes the body frame *u* velocity controller and the body-frame *v* velocity controller with parameters that need be optimized, too. The outer-loop controller configuration used for parameter optimization is depicted in Figure 5.20, with the trajectory generator generating waypoints the helicopter must follow, converting received information to body-reference frame velocities. The yaw angle set point is chosen to be zero, the altitude set point ten meters. Inputs are applied to the set points of the body-reference frame velocities.

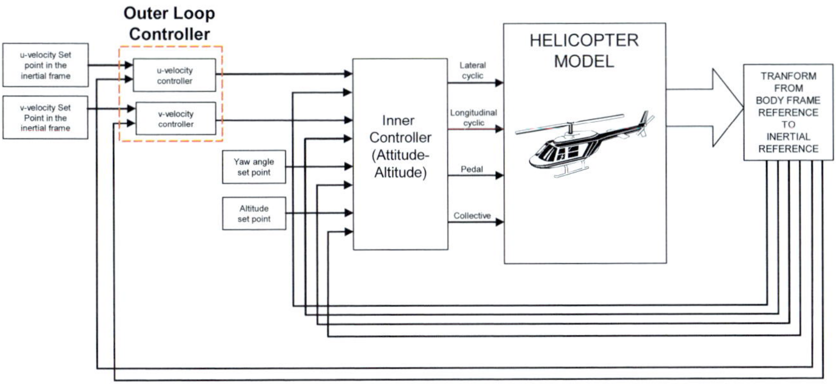

Fig. 5.20. Block diagram for optimization algorithm for the outer controller.

Figure 5.21 shows the flow chart of the procedure followed to optimize the parameters of the *u*- and *v*- velocity controllers. This procedure is very similar to the procedure used for the inner controller.

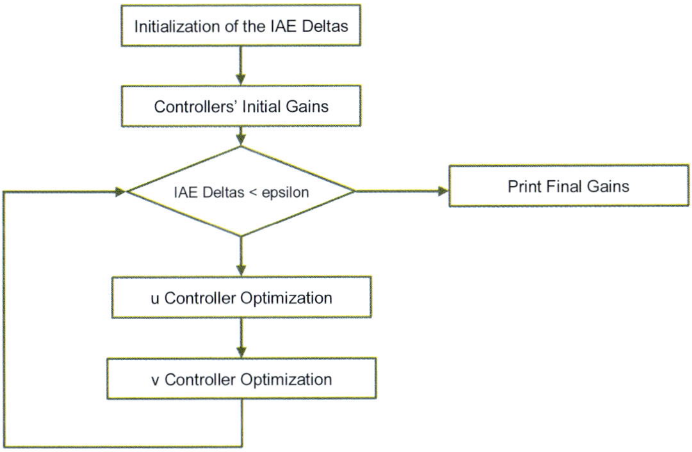

Fig. 5.21. Optimization of outer-loop.

5.5.4 Resulting PID Controllers

Inner-loop Final Parameters: Obtained final values are shown in Table 5.12. Step responses of the different inner-loop PID controllers are shown in Figure 5.22.

	K_P	K_I	K_D
ALTITUDE	4.96983	0	4.81735
YAW	10.514	0	10.1266
PITCH	34.6243	0.0030634	33.8243
ROLL	12.2573	20.068	3.55586

Table 5.12. Proportional, integral, and derivative gains of the PID controllers.

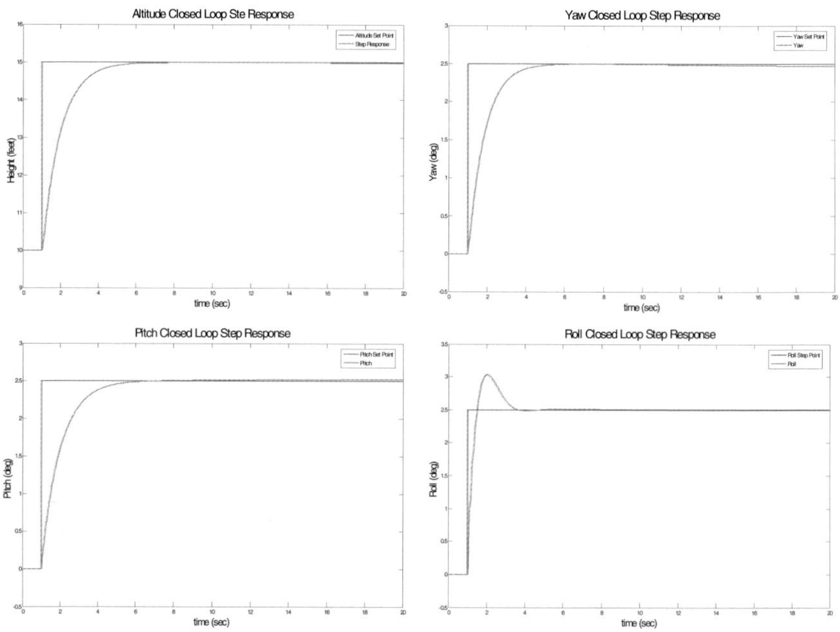

Fig. 5.22. Step responses of the inner-loop controllers.

Inner-loop Analysis: In order to determine diagonal dominance and stability of the inner-loop from a set point input to controlled variable output, a matrix of closed-loop transfer functions is derived:

$$\boldsymbol{T}(s)=\begin{bmatrix} \frac{\phi}{\phi_{sp}}(s) & \frac{\phi}{\theta_{lon}}(s) & \frac{\phi}{\psi_{sp}}(s) & \frac{\phi}{z_{sp}}(s) \\ \frac{\theta}{\phi_{sp}}(s) & \frac{\theta}{\theta_{lon}}(s) & \frac{\theta}{\psi_{sp}}(s) & \frac{\theta}{z_{sp}}(s) \\ \frac{\psi}{\phi_{sp}}(s) & \frac{\psi}{\theta_{lon}}(s) & \frac{\psi}{\psi_{sp}}(s) & \frac{\psi}{z_{sp}}(s) \\ \frac{z}{\phi_{sp}}(s) & \frac{z}{\theta_{lon}}(s) & \frac{z}{\psi_{sp}}(s) & \frac{z}{z_{sp}}(s) \end{bmatrix} \tag{5.34}$$

The block diagram of the closed-loop system for roll and pitch is displayed in Figure 5.23. The block diagram including yaw and height is more complex because the lateral and longitudinal cyclic inputs affect height and yaw, while the collective and pedal does not affect roll and pitch.

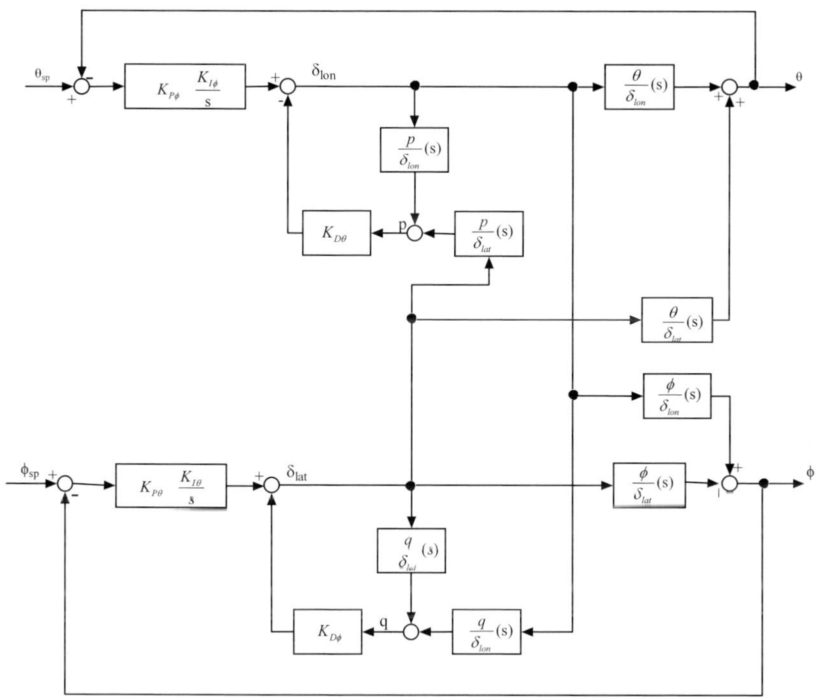

Fig. 5.23. Block diagram of roll and pitch closed-loop system.

Although Mason's rule may be used to obtain each of the closed-loop transfer functions [20], system complexity makes this method difficult and time consuming. *MATLAB* is used to calculate transfer functions.

The integration used to obtain body-reference positions and the controllers are represented in *MATLAB* as transfer functions using `tf`. The open-loop helicopter model is built as a state space model with the roll rate, pitch rate, yaw rate and velocity along the z-axis as outputs using `ss`. Using `append` and `connect` allows integrators to be appended to outputs and controllers to be appended to corresponding inputs, and the feedback loops to be connected. Once the modified state space model is created, the `zpk` command is applied to the modified state space model to obtain the matrix of transfer functions given in equation (5.34).

The system is stable as confirmed by all roots being in the left half s-plane; see Figure 5.24, obtained by using the `zpmap` function on the matrix of closed--loop transfer functions. Diagonal terms correspond to the transfer functions used for the SISO design. This allows diagonal dominance to be calculated over the bandwidth of the system as shown in Figure 5.25. The highest value obtained is for height, approximately 0.32, with the others close to zero. This demonstrates the effectiveness of the optimization in reducing unwanted cross coupling.

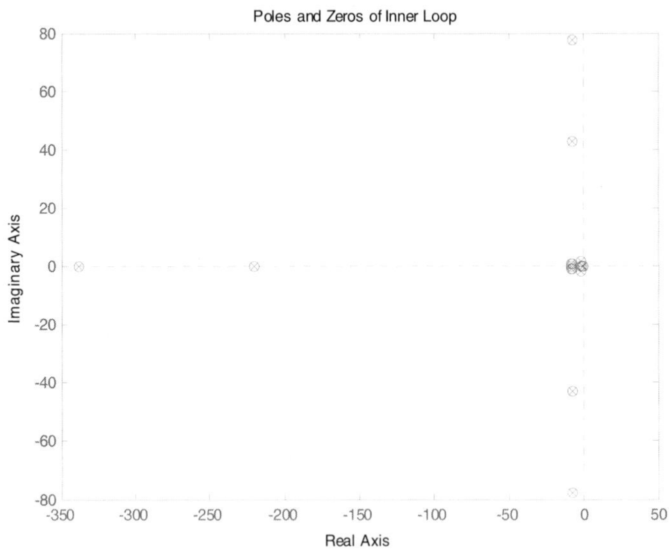

Fig. 5.24. Pole / zero map for inner-loop transfer functions.

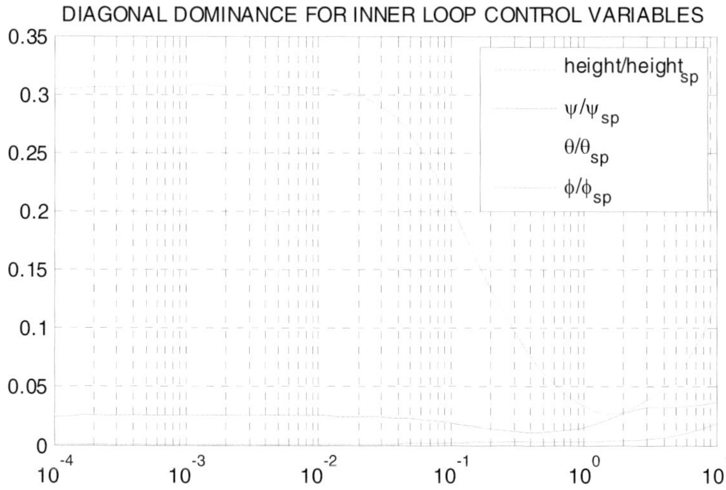

Fig. 5.25. Diagonal dominance for inner-loop control variables.

Outer-loop Final Parameters: Final gain values are given in Table 5.13 while step responses of the outer-loop controller are shown in Figure 5.26.

	K_P	K_I	K_D
u-velocity	0.00982832	0.00156214	0.105471
v-velocity	0.104478	0.00281284	0.179079

Table 5.13. Proportional, integral, and derivative gains of PID controllers.

It can be observed from the step responses that the optimization was able to find the appropriate parameters for the different PID controllers, to obtain the desired response. In this case, the desired response to a step input was a first order response.

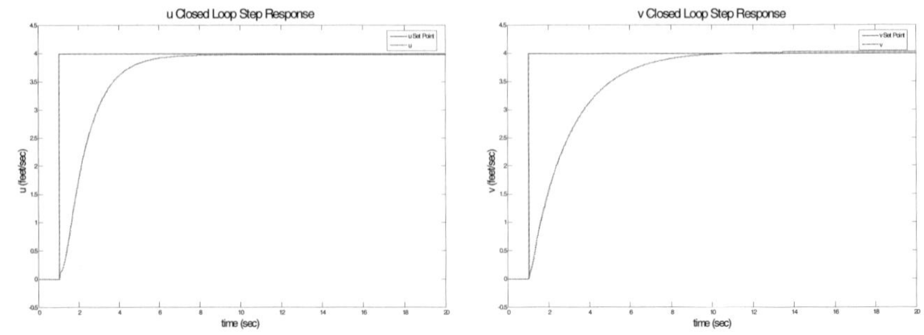

Fig. 5.26. Step responses of the *u*-velocity and *v*-velocity controllers.

5.5.5 Resulting PID-like Fuzzy Controller

A set of PID-like Fuzzy controllers is derived for the inner- outer- loop. Optimized parameters are S_e, S_{ce}, S_1 and S_2, as shown in Figure 5.16. Because of the typical nonlinear behavior of the transfer characteristic of Fuzzy controllers shown in Figure 5.17, the IAE values are calculated based on equation (5.33) [14]. Final parameter values are obtained after several runs using the *MATLAB* Optimization Toolbox, shown in Table 5.14, while step responses of the different inner-loop PID-like Fuzzy controllers are shown in Figure 5.27. For the outer-loop, final values obtained for S_e, S_{ce}, S_1 and S_2 are shown in Table 5.15, while the corresponding step responses of the PID-like Fuzzy controllers are shown in Figure 5.28.

	S_e	S_{ce}	S_1	S_2
ALTITUDE	0.201062	0.167597	0.368128	0.604584
YAW	2.4547	2.46496	2.489	2.53593
PITCH	3.09274	3.12656	3.93068	2.72905
ROLL	2.08249	2.19898	3.09108	0.702223

Table 5.14. S_e, S_{ce}, S_1 and S_2 scaling factors of the PID-like Fuzzy controllers.

	S_e	S_{ce}	S_1	S_2
***u*-velocity**	0.5	1.0	1.0	0.1
***v*-velocity**	10	1.0	1.0	0.0.5

Table 5.15. S_e, S_{ce}, S_1 and S_2 scaling factors of the PID-like Fuzzy controllers.

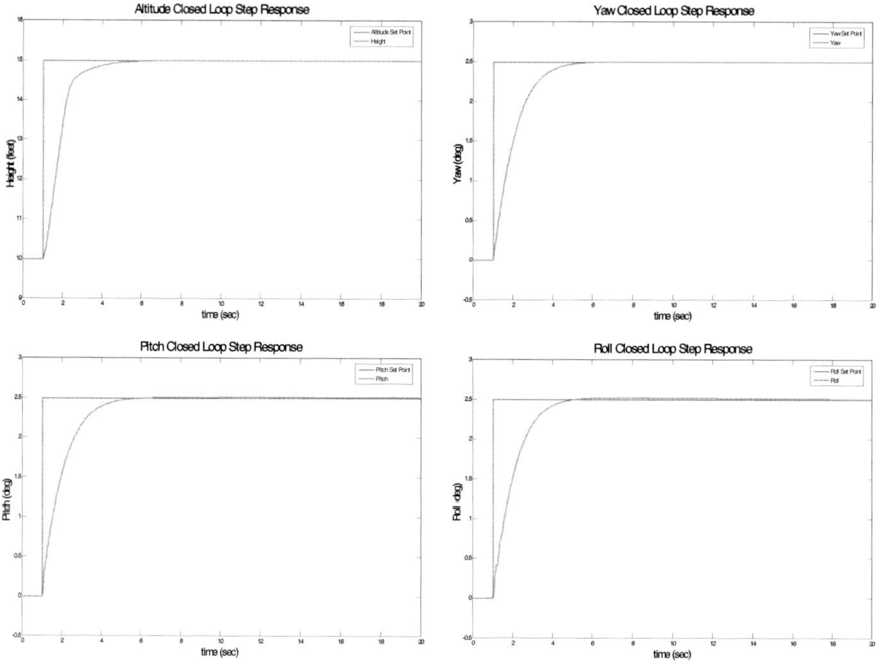

Fig. 5.27. Step responses of the inner-loop controllers.

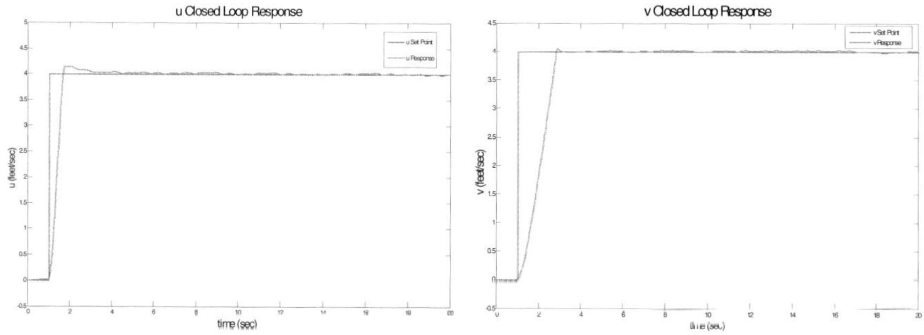

Fig. 5.28. Step responses of the *u*-velocity and *v*-velocity controllers.

It is observed from step responses that through optimization appropriate parameters are found for the different PID-like Fuzzy controllers, to achieve desired response. In this case, the desired response to a step input is a first order response.

5.6 Design Based on Linear Quadratic Techniques

Linear quadratic (LQ) control treats the plant as a full MIMO system. Considering non-aggressive flights, motion planning reduces the vehicle configuration space from *SE(3)* to simply $R^3 x\ S^1$, resulting in control of helicopter position (x, y, z) and heading ψ. Design follows the inner-loop outer-loop decomposition. The inner-loop stabilization, basic tracking and decoupling of interacting variables are handled by the MIMO LQ controller. Since key issues related to the inner-loop are solved by the LQ controller, the outer-loop may be realized by a set of PID controllers where the final dynamic response and steady-state characteristics are defined.

The design of the inner-loop controller follows the separation principle, where initially it is assumed that the state-feedback controller has access to all state variables. In the second step this restriction is lifted because a Kalman filter can be designed coping with the reconstruction of non measurable states and with noisy measurements. The combination of the LQ regulator and Kalman filter constitute the inner-loop Linear Quadratic Gaussian (LQG) controller.

5.6.1 Input-Output Pairing

As previously stated, RGA provides an analytical method for assigning certain outputs to inputs. The best pairing, the one closest to the identity matrix, is: $\delta_{lat}\rightarrow\phi$, $\delta_{lon}\rightarrow\theta$, $\delta_{ped}\rightarrow r$, $\delta_{col}\rightarrow w$. That is, lateral cyclic control changes the roll, longitudinal cyclic control modifies the pitch, pedal control modifies the yaw rate, and collective control changes the heave velocity. In the outer-loop, standard controllers control the four degrees of freedom of the rotorcraft (x, y, z and ψ).

5.6.2 LQR Design

Control of the helicopter body-fixed reference frame dynamics is handled as a steady-state linear quadratic tracking problem. The basic idea is to solve a static optimization problem that minimizes a performance index given by:

$$J = \int_0^T (\boldsymbol{x}'\boldsymbol{Q}\boldsymbol{x} + \boldsymbol{u}'\boldsymbol{R}\boldsymbol{u})dt \tag{5.35}$$

The state feedback control law has the form $\boldsymbol{u}=-\boldsymbol{K}\boldsymbol{x}$, $\boldsymbol{x}' = \boldsymbol{x}^T$, $\boldsymbol{u}' = \boldsymbol{u}^T$. It is well known that the optimization problem is reduced to solving the algebraic Riccati equation (ARE):

$$SA + A'S + C'C - SBR^{-1}B'S = 0 \tag{5.36}$$

S is the solution to the ARE, $A' = A^T$, $C' = C^T$, $B' = B^T$ and the state feedback gain is computed from:

$$K = R^{-1}B'S \tag{5.37}$$

The weighting matrices Q and R are the main design parameters and they are usually chosen by experience and simulation. In this case, the concrete values have been chosen from a manual refinement process where the closed-loop dynamics as well as the avoidance of reaching actuator saturation are the main design constraints. Simulations are run using $Q=20C'C$ and $R=I_{4x4}$. Poles of the closed-loop system are in the left half plane. The calculated feedback gain matrix K is obtained using the *MATLAB* `lqr` function.

The LQR design defines the closed-loop dynamics via pole placement as in any other full state feedback controller. To add basic tracking capabilities to the controller, an input reference matrix may be computed imposing the following requirement:

$$K_{\text{ref}} K_{\text{DC}} \overset{!}{=} I \tag{5.38}$$

where K_{DC} is the steady-state gain of the closed-loop system. Hence:

$$K_{\text{ref}} = K_{\text{DC}}^T \left(K_{\text{DC}} K_{\text{DC}}^T \right)^{-1} \tag{5.39}$$

The input reference matrix provides only a rough open-loop type of unity DC-gain for the closed-loop and it does not eliminate steady-state errors. The controllers in the outer-loop handle steady-state error compensation.

In view of the calculated feedback-gain and the input reference matrix, it is now possible to assess the theoretical behavior of the inner-loop assuming full-state feedback. The frequency response shown as a polar plot using *MATLAB*'s `nyquist` function is depicted in Figure 5.29. The decoupling effect of the LQR regulator is clearly seen. The diagonal terms show unitary DC-gain, whereas the off-diagonal terms have small magnitude. The transfer functions corresponding to the roll and pitch show approximate third-order behavior. The heave velocity and the yaw rate look almost like first-order systems.

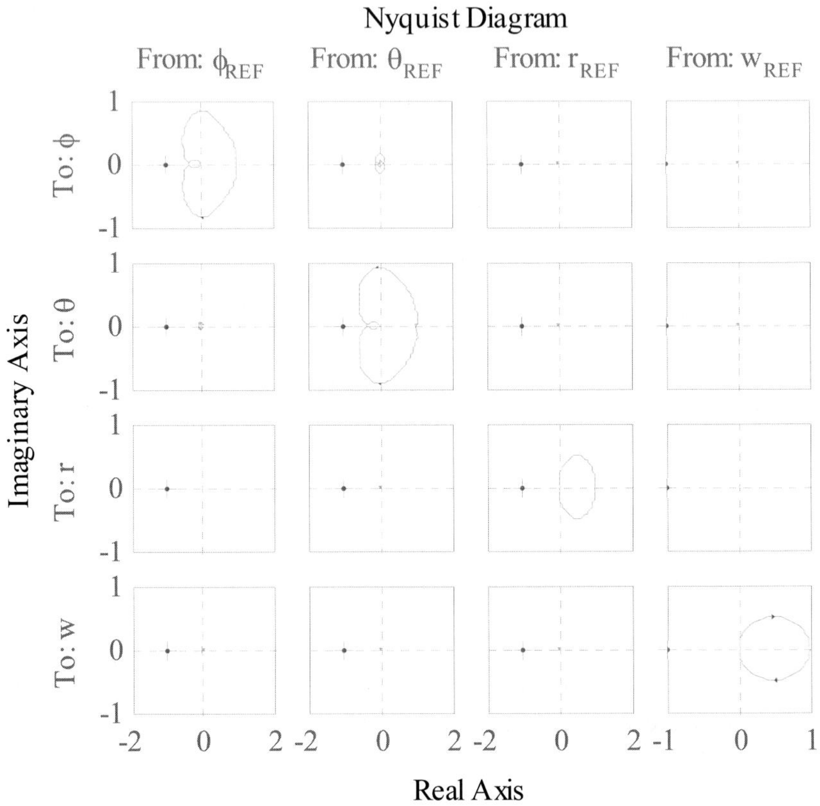

Fig. 5.29. Frequency response of the MIMO system.

5.6.3 Kalman Filter Design

The Kalman estimator is the optimal solution to the problem of estimating the state of a linear stochastic dynamic system described in state-space form, similar to the body-fixed reference frame dynamic model being considered here. The system is subject to noise that can be modeled as an additional input. Measurements are subject to noise as well. The basic state-space model extends to:

$$\dot{x} = Ax + Bu + Gw \tag{5.40}$$

$$y = Cx + Du + Hw + v \tag{5.41}$$

To determine the estimator, it is necessary to have information about the statistical properties of the system noise $\boldsymbol{w}$, and the measurement noise $\boldsymbol{v}$. Correlation matrices $\boldsymbol{Q}_N$ and $\boldsymbol{R}_N$ are defined, where:

$$\boldsymbol{Q}_N = E\left\{\boldsymbol{w}\boldsymbol{w}^T\right\} \qquad \boldsymbol{R}_N = E\left\{\boldsymbol{v}\boldsymbol{v}^T\right\} \tag{5.42}$$

Both can be assumed to be diagonal matrices, due to independence and zero-mean white Gaussian noise distribution.

MATLAB provides a function that solves the ARE associated with the design of the Kalman estimator. For the Kalman estimator design, the Q_N matrix can be used to ponder parameter uncertainties in the model and external disturbances. On the other hand, to determine R_N, it is necessary to have information about the sensors that are going to be used. In an avionics system an inertial measurement unit (IMU) and a GPS provide measurements of various magnitudes closely related to the state vector that needs be obtained.

5.6.4 Complete Solution (LQG)

The LQR controller and the Kalman filter form the LQG controller as in Figure 5.30. For the outer-loop controller, the LQG controller and the rotorcraft model are represented as a system with inputs set points of the LQG controller and outputs signals that are fed back to the controllers. Dynamics must be derived for the analytical design of the outer-loop controllers. The reduction is carried out using *MATLAB*'s model construction commands `append`, and `connect`.

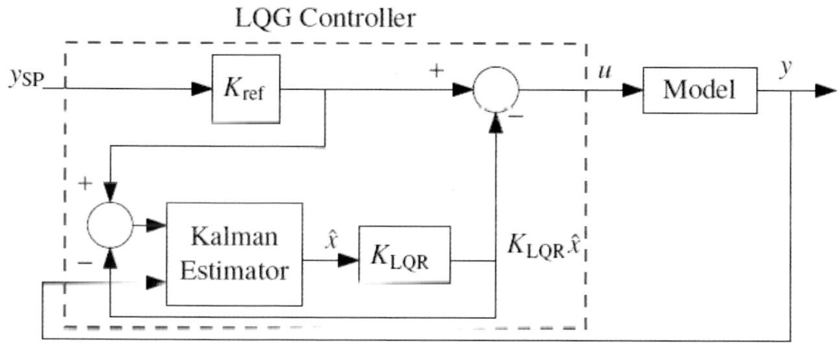

Fig. 5.30. LQG controller.

5.6.5 Design of the Outer-Loop Controller

The outer-loop controller measures the "real outputs", the position (x, y, z), and the heading (ψ). According to the input / output pairing that have been used for the inner-loop design, the outer-loop controller generates the set-points for: roll (ϕ), pitch (θ), heave velocity (r) and yaw rate (w). Considering the stabilizing and decoupling effect of the LQG controller, a reasonable approach to the design of the outer-loop considers the use of SISO PID controllers as shown in Figure 5.31.

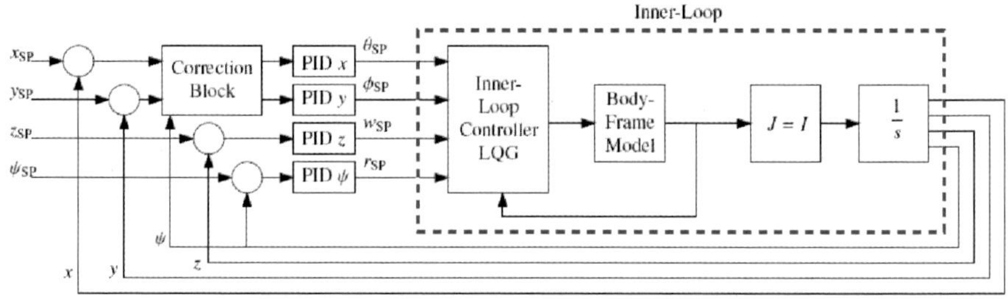

Fig. 5.31. Complete outer- inner- loop controller configuration.

An important assumption for the design of the outer-loop controller is linearization of the system around the equilibrium where Euler angles are close to 0. Thus, controller design becomes much simpler since the Jacobian is the identity matrix. The SISO controllers in the external loop can be tuned using readily available classic control design tools (root locus, frequency response) assuming a stable decoupled inner-loop system.

Commands to control the helicopter in a specified trajectory depend not only on the actual position error, but also on the actual orientation of the helicopter. For this reason, a basic "correction algorithm" (Figure 5.32) generates the "errors" for the PID controllers. This correction block affects mainly the movement of the helicopter in the horizontal plane. Thus, the correction block obtains the actual position error and the heading of the helicopter to compute the inputs for the PID controllers.

The basic setup for testing of the LQG-based design is depicted in Figure 5.33. All blue blocks correspond to the outer-loop control system. The inner-loop blocks are shown in yellow. The blocks that make the model of the helicopter are shown in green. Accessory blocks, such as visualization, trajectory generation and real-time synchronization are shown in magenta.

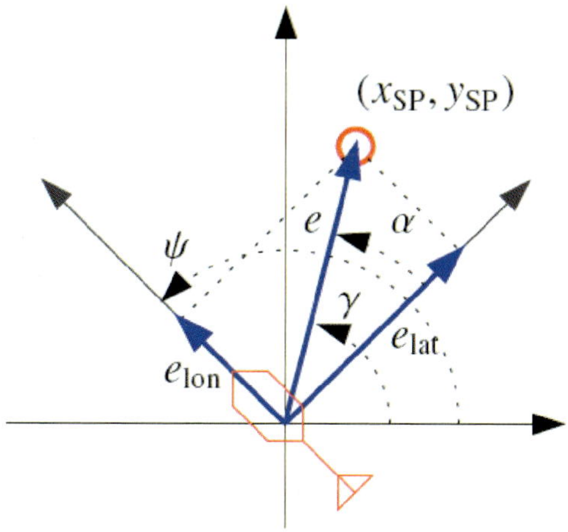

Fig. 5.32. Geometric relationships for the design of the correction block.

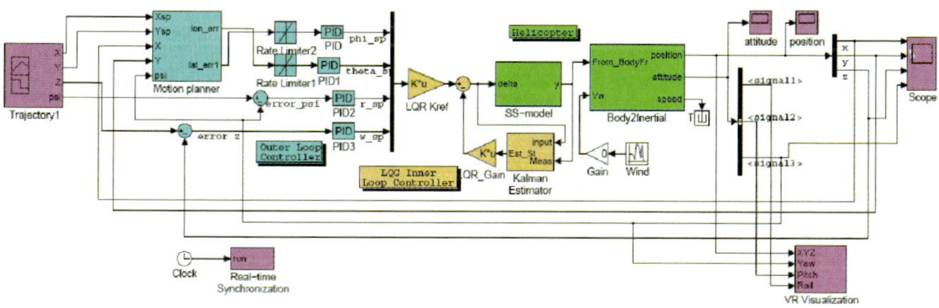

Fig. 5.33. Simulation diagram for the complete system.

5.7 Results

The helicopter model, inner-loop and outer-loop controllers are implemented in *MATLAB / Simulink*. Waypoint trajectories are generated for testing derived controllers. All demonstrated very promising results as shown in Figure 5.34 through Figure 5.41.

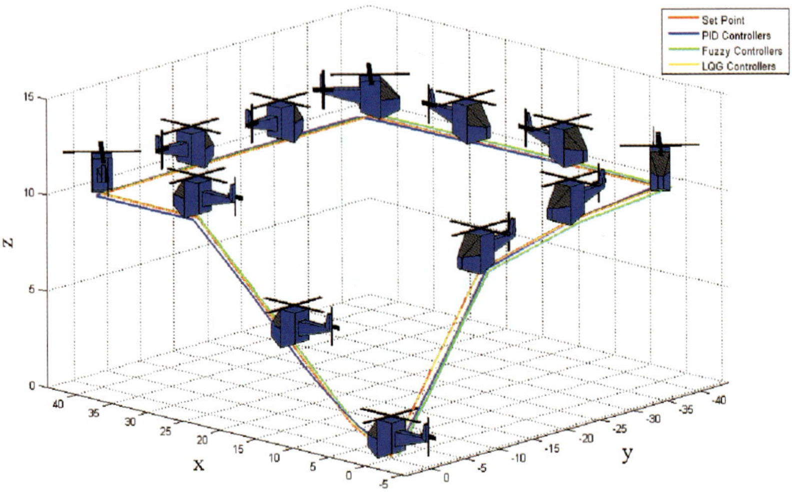

Fig. 5.34. Response to a rectangular trajectory.

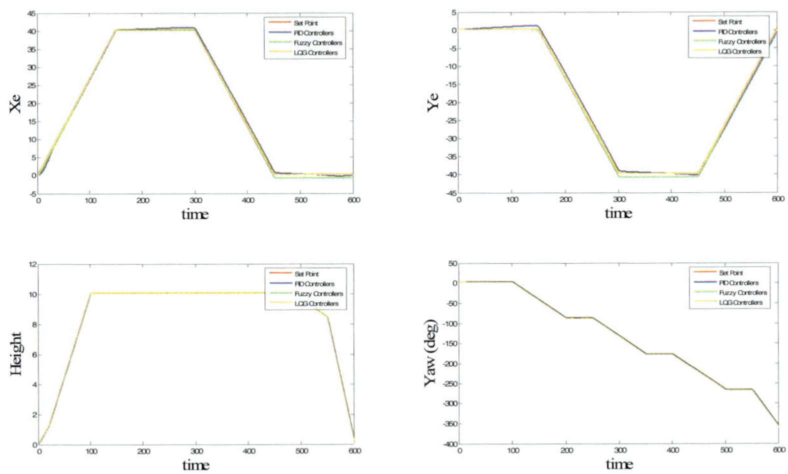

Fig. 5.35. Responses rectangular trajectory. Inertial coordinates and heading.

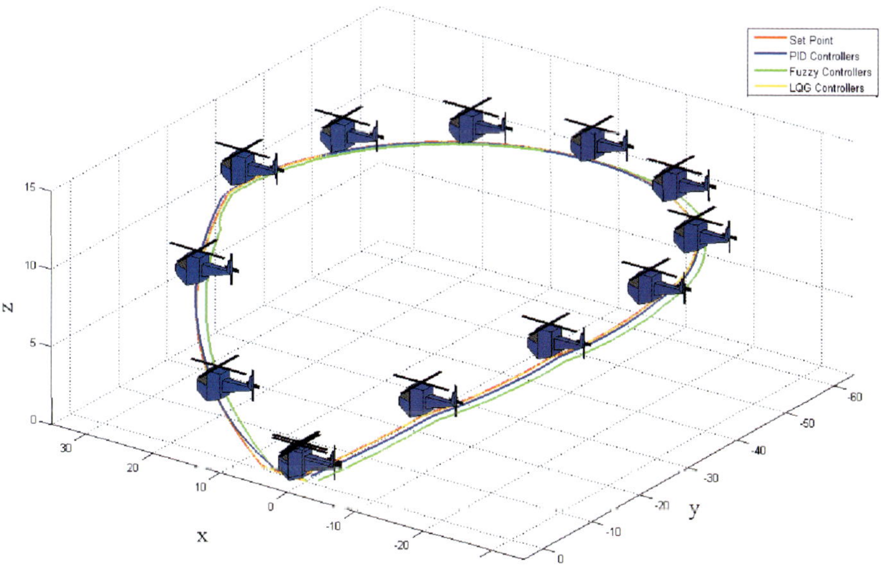

Fig. 5.36. Response to a circular trajectory with zero heading.

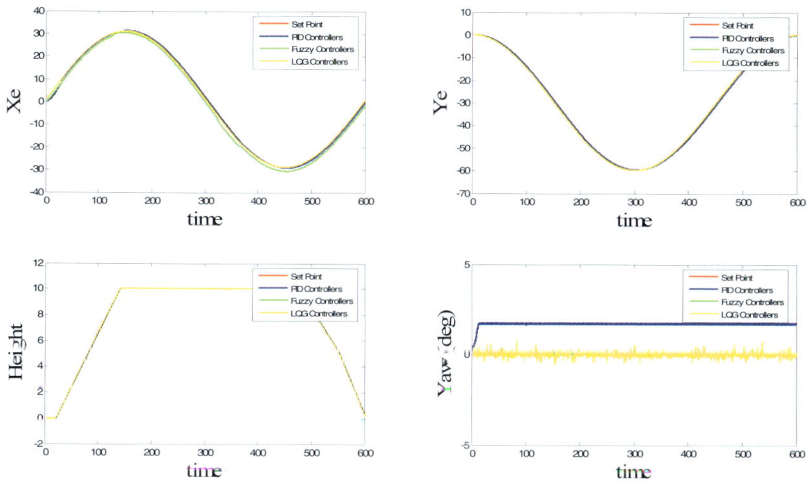

Fig. 5.37. Responses to circular trajectory with zero heading. Inertial coordinates and heading.

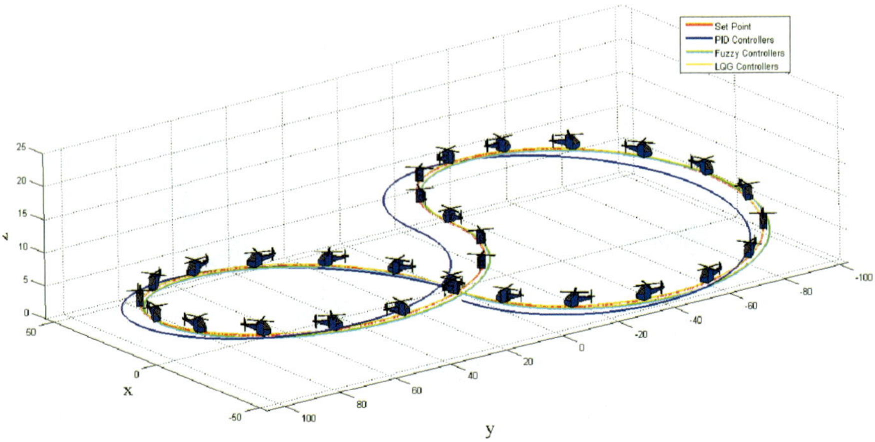

Fig. 5.38. Responses to a double circle trajectory with changing height in 3-D.

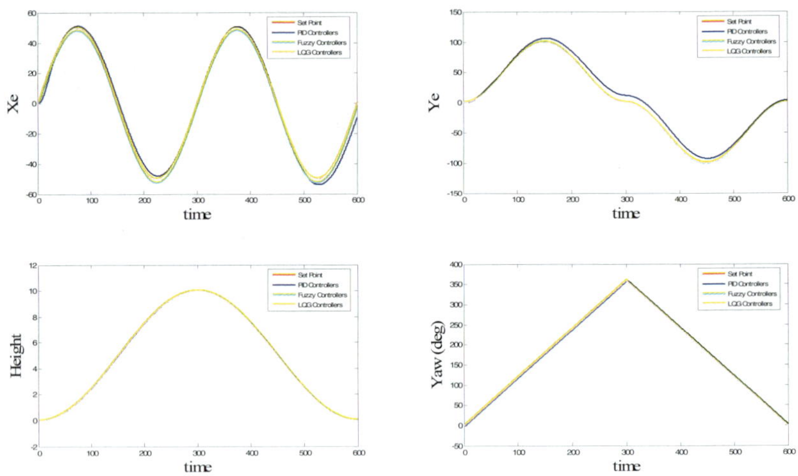

Fig. 5.39. Responses to a double circle trajectory with changing height in 3-D. Inertial coordinates and heading.

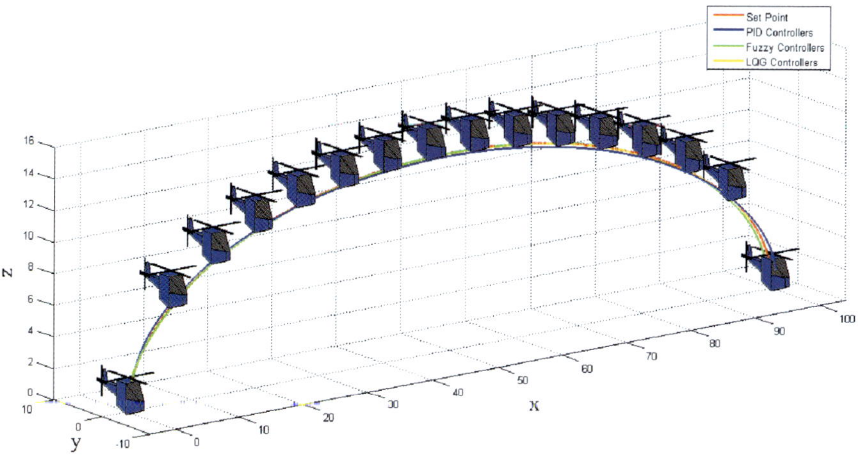

Fig. 5.40. Vertical take-off and landing trajectory in 3-D.

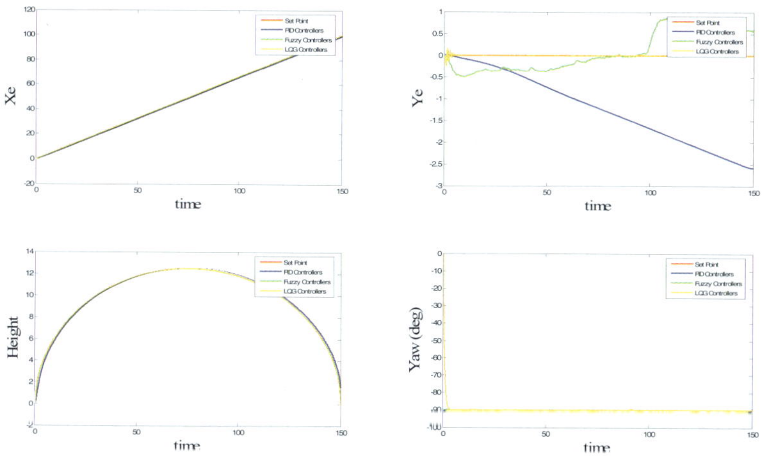

Fig. 5.41. Vertical take-off and landing trajectory in 3-D. Inertial coordinates and heading.

5.8 Conclusions

A tutorial-like approach to designing controllers for small unmanned helicopters has been presented. Derived controllers have been tested by simulation successfully based on real data from non-aggressive flights. The three controllers have different general characteristics, shown in Table 5.16.

Decentralized controllers demonstrate that PID ones are still a simple, yet powerful tool that can be used for tackling difficult to control systems. Furthermore, by making use of available optimization tools it is possible to obtain basic tuning parameters for such controllers fine-tuned from practical flight experiments.

Fuzzy controllers represent an interesting alternative that can be used for obtaining quick results. Due to their nonlinear nature they can adapt better to the nonlinear features of the helicopter. In addition, it can be expected that the Fuzzy controller will handle uncertainty better than all other controllers.

The LQG methodology offers a comprehensive range of tools to handle the stabilization of the rotorcraft as well as an optimal solution for sensor fusion through the Kalman filter.

This research may be extended to include different flight windows and models of helicopters giving way to simplification of design for what once was a difficult procedure requiring a great deal of experience and knowledge of control theory.

	PID	*FUZZY*	*LQG*
Stability	Standard Methods for Testing Available	No Standard Accepted Stability Test Available	Guaranteed by Design (for disturbances and parameter variations within bounds)
Computational Requirements	Low	Moderate	Relatively high (Due to Kalman filter computation)
Model of System Required	Good Approximation	Fair Approximation	Good Approximation

Table 5.16. Comparison of controller characteristics.

References

1. Mettler B., *Identification Modeling and Characterization of Miniature Rotorcraft*, Kluwer Academic Publishers, 2003.
2. Goodwin G., Graebe S., and Salgado M., *Control System Design,* Prentice Hall, Inc, 2001.
3. Sugeno M., Development of an Intelligent Unmanned Helicopter, Chapter 2 of *Fuzzy Modeling and Control: Selected Works of M. Sugeno*, Hung T. Nguyen and Nadipuram R. Prasad (Eds), CRC Press, Boca Raton, 1999.
4. Kadmiry B., and Driankov D., "Fuzzy Control of an Autonomous Helicopter", Joint *9th IFSA World Congress and 20th NAFIS International Conference*, Vol. 5, Canada, July 2001.
5. *Fuzzy Logic Toolbox User's Guide*, Version 2, The MathWorks Inc.
6. Driankov D., Hellendoorn H., and Reinfrank M., *An Introduction to Fuzzy Control*, Springer, Second Edition, 1996.
7. Mann George K. I., Hu B.-G. and Gosine R. G., "Analysis of Direct Action Fuzzy PID Controller Structures", *IEEE Transactions on System, Man, and Cybernetics*, Vol. 29, No 3, June, 1999.
8. Hu B.-G., Mann G. K. I., and Gosine R. G., "A Systematic Study of Fuzzy PID Controllers-Function-Based Evaluation Approach", *IEEE Transactions on Fuzzy Systems*, Vol. 9, No. 5, October 2001.
9. Li H.-X., and Gatland H. B., "Conventional Fuzzy Control and Its Enhancement", *IEEE Transactions on System, Man, and Cybernetics*, Vol. 26, No. 5, October 1996.
10. Wang L.-X., *A Course in Fuzzy System and Control*, Prentice Hall, 1997.
11. Smith C., and Corripio A., *Principles and Practice of Automatic Process Control,* Second Edition, J. Wiley, 1997.
12. Lee C.-C., "Fuzzy Logic in Control Systems: Fuzzy Logic Controller-Part I", *IEEE Transactions on System, Man, and Cybernetics*, Vol. 20, No. 2, pp. 404-418, 1990.
13. Castillo C. L., Alvis W., Castillo-Effen M., Moreno W., and Valavanis K., "Small Scale Helicopter Analysis and Controller Design for Non-Aggressive Flights", *Proceedings, IEEE International Conference on Systems, Man and Cybernetics*, October 2005.
14. Levine, W. (Ed.), *The Control Handbook*, CRC Press, 1995.
15. Munro N., *Symbolic Methods in Control System Analysis and Design*, The Institute of Electrical Engineers, London, 1999.
16. Franklin G., Powel J. D. and Emani-Naeni A., *Feedback Control of Dynamic Systems*, Fourth Edition, Prentice Hall, 2002.
17. Isidori A., Marconi L., and Serrani A., *Robust Autonomous Guidance – An Internal Model Approach*, Springer, 2003.
18. Johnson E.N., and Kannan S. K., "Adaptive Flight Control for an Autonomous Unmanned Helicopter", *Proceedings of AIAA Guidance, Navigation and Control Conference*, AIAA-2002-4439, August 2002.

19. Samad T., and Balas G. (Eds.), *Software-Enabled Control: Information Technology for Dynamical Systems*, IEEE Press, 2003.
20. Dorf R. C., and Bishop R. H. *Modern Control Systems*, Addison-Wesley, 1998.

Chapter 6: Design and Control of a Miniature Quadrotor[1]

This Chapter presents a method for designing and controlling a miniature quadrotor. It starts by presenting an accurate quadrotor simulation model that is used for initial testing and validation via a simulator built using *MATLAB*. A design methodology is then discussed along with an iterative algorithm derived to choose quadrotor components based on design constraints. Testing using several control techniques follows the design phase; the result is adoption of the integral backstepping techniques that achieves superior quadrotor performance. Stability analysis and obstacle avoidance are also discussed. Simulation and experimental results are included to demonstrate effectiveness of the design.

6.1 Introduction

Miniature aerial vehicles (MAVs) have attracted major research interest during the last decade. Recent advances in low power processors, miniature sensors and control theory have contributed to system miniaturization and creation of new application fields.

Miniature flying robots (MFRs) offer major advantages when used for aerial surveillance in complex or cluttered environments like office buildings and commercial centers. MFRs may assist in search and rescue missions after earthquakes, explosions and other natural disasters.

As such, an MFR that is capable of flying in narrow spaces and fit through small openings and collapsed buildings, maneuver around pillars and destroyed wall structures could quickly and systematically search for accident victims or inspect disaster areas without risking human lives. An MFR could, after localization, provide coordinates of potential victims to rescuers guiding and assisting them to complete their mission.

In such hazardous environments, wireless communication is difficult if not impossible because of natural obstacles. Aerial relay could be possible

[1] Written by S. Bouabdallah, R. Siegwart

Kimon P. Valavanis (ed.), Advances in Unmanned Aerial Vehicles, 171–210.

using unmanned aerial vehicles equipped with transceivers, ensuring constant communication.

Reviewing closely related research, the development of a miniature autonomous flight control system and the creation of a multi-vehicle platform for experimentation and validation of multi-agent control algorithms are discussed in [11], while [14] presents results in centimeter-scale quadrotor design and analysis. A recent result reported in [18] is a 13.6 cm micro-helicopter capable of hovering for 3 minutes.

The Autonomous Systems Lab (ASL) of the Swiss Federal Institute of Technology is participating in similar research efforts [1] concentrating in MFRs. The major focus is on fully autonomous MFRs that may be developed through coordinated efforts in system-level optimization, design and control.

The approach that is followed is to design vehicles with optimized mechanics and innovative control techniques. The goal is to keep on miniaturizing the MFR in every redesign step following the latest technological advancements.

The objective of the ASL MFR project is to optimally design and control aerial systems for navigation in cluttered environments. A quadrotor (OS4) and a coaxial (CoaX) vertical take-off and landing (VTOL) systems are considered because of their challenging control problems and their broad field of applications. This is accomplished via a detailed design methodology presented below, based on which the best components are chosen following imposed constraints.

As widely known, when compared with other aerial vehicles, VTOL vehicle systems have specific characteristics like flying in very low altitudes and being able to hover that make them suitable for applications that may be impossible to complete using fixed-wing vehicles.

Different configurations of commonly used MAVs for research purposes and in industry are shown in Table 6.1 along with related advantages and drawbacks. Table 6.1 offers a pictorial comparison that may be used when a new design is proposed.

Configuration	Picture	Advantages	Drawbacks
Fixed-wing (AeroVironment)	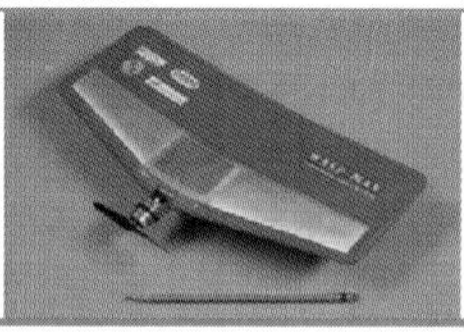	- Simple mechanics - Silent operation	- No hovering

Single (A.V de Rostyne)		- Good controllability and maneuverability	- Complex mechanics - Large rotor - Long tail boom
Axial rotor (Maryland Univ.)	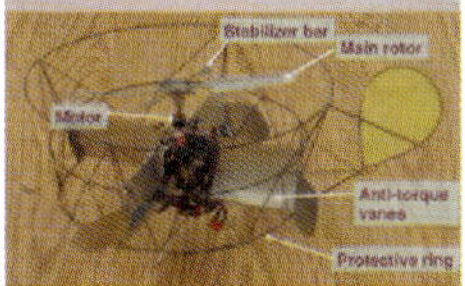	- Compactness - Simple mechanics	- Complex control - Weak maneuvera-bility
Coaxial rotors (ETHZ)		- Compactness - Simple mechanics	- Complex aerodynamics
Tandem rotors (Heudiasyc)		- Good controllability and maneuverability - No aerodynamics Interference	- Complex mechanics - Large size
Quadrotors (ETHZ)		- Good maneuverability - Simple mechanics - Increased payload	- High energy consumption - Large size
Blimp (EPFL)		- Low power consumption - Auto-lift	- Large size - Weak maneuverability
Hybrid (MIT)	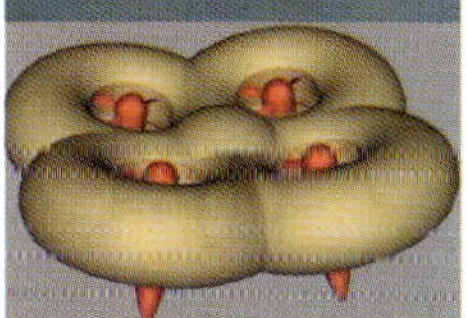	- Good maneuverability - Good survivability	- Large size - Complex design
Bird-like (Caltech)		- Good maneuverability - Low power Consumption	- Complex mechanics - Complex control
Insect-like (UC Berkeley)	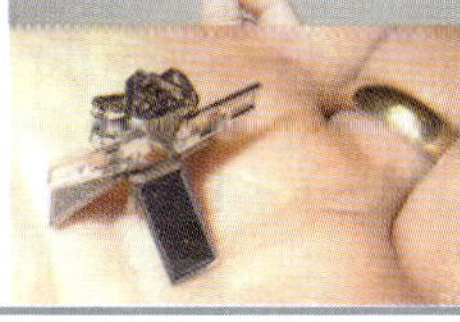	- Good maneuverability - Compactness	- Complex mechanics - Complex control

Fish-like (US Naval Lab)	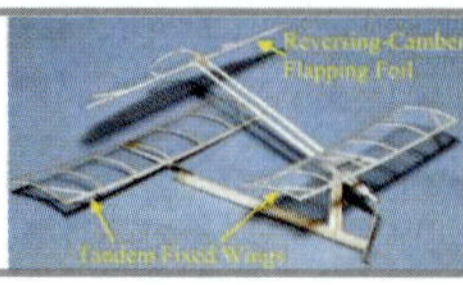	- Multimode mobility - Efficient Aerodynamics	- Complex control - Weak maneuverability

Table 6.1. Common UAV-MAV configurations.

Further, Table 6.2 presents a short and not exhaustive comparison between different VTOL vehicle concepts. It may be observed from Table 6.2 that the quadrotor and the coaxial helicopter are among the best configurations if used as MFRs.

	A	B	C	D	E	F	G	H
Power cost	2	2	2	2	1	4	3	3
Control cost	1	1	4	2	3	3	2	1
Payload/volume	2	2	4	3	3	1	2	1
Maneuverability	4	2	2	3	3	1	3	3
Mechanics simplicity	1	3	3	1	4	4	1	1
Aerodynamics complexity	1	1	1	1	4	3	1	1
Low speed flight	4	3	4	3	4	4	2	2
High speed flight	2	4	1	2	3	1	3	3
Miniaturization	2	3	4	2	3	1	2	4
Survivability	1	3	3	1	1	3	2	3
Stationary flight	4	4	4	4	4	3	1	2
Total	**24**	**28**	**32**	**24**	**33**	**28**	**22**	**24**

A=Single rotor, B=Axial rotor, C=Coaxial rotors, D=Tandem rotors, E=Quadrotor, F=Blimp, G=Bird-like, H=Insect-like.

Table 6.2. VTOL concept comparison (1=Bad, 4=Very good).

6.2 Modeling for Simulation

The OS4 quadrotor simulation model was developed through several successive steps as presented in [2]. Major improvements of the most recent version include hub forces (H), rolling moments (R_m) and variable aerodynamic coefficients. This makes the model more realistic, particularly in forward flight. Previous model versions required often to slightly adjust control parameters for successful experimentation with the OS4 testbed [3]. The current version of the model used in the OS4 simulator incorporates a newly designed Integral Backstepping (IB) controller where simulated control parameters may be directly used on the real OS4 helicopter to demonstrate successful autonomous flights.

The dynamics of a rigid body subject to external forces applied to the center of mass and expressed in the body-fixed reference frame [17] in Newton-Euler formalism are:

$$\begin{bmatrix} mI_{3\times3} & 0 \\ 0 & I \end{bmatrix}\begin{bmatrix} \dot{V} \\ \dot{\omega} \end{bmatrix}+\begin{bmatrix} \omega\times mV \\ \omega\times I\omega \end{bmatrix}=\begin{bmatrix} F \\ \tau \end{bmatrix} \tag{6.1}$$

where, $I\in\Re^{(3\times3)}$ is the inertia matrix, V is the body linear velocity and ω is the body angular velocity. F and τ are, respectively, the body force and torque, while m is the system mass. Consider the earth-fixed frame E and the body-fixed frame B as seen in Figure 6.1. Using Euler angle parameterization, the airframe orientation in space is given by a rotation R from B to E, where R in $SO3$ is the rotation matrix. The frame system is slightly different compared to previous versions in order to conform with the N,E,D (North, East, Down) standard.

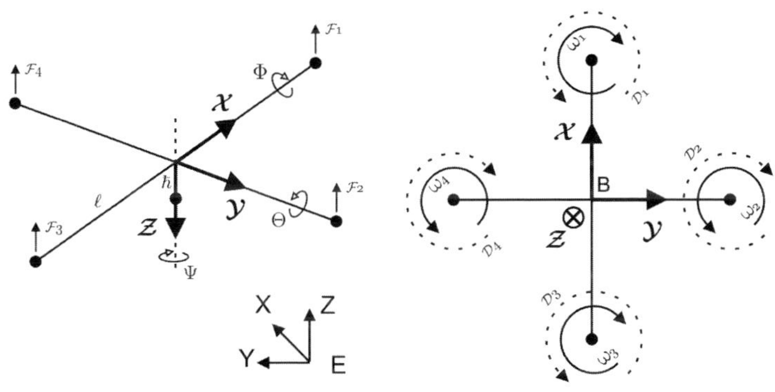

Fig. 6.1. The OS4 coordinate system.

6.2.1 Aerodynamic Forces and Moments

The aerodynamic forces and moments are derived using a combination of momentum and blade element theory [15] based on previously reported research [8]. Equations use, among others, the abbreviated symbols shown next:

Air density	ρ	Rotor radius	R
Solidity ratio	σ	Rotor speed	Ω
Lift slope	a	Rotor area	A
Rotor advance ratio	μ	Pitch of incidence	θ_0
Inflow ratio	λ	Twist pitch	θ_{tw}
Induced velocity	υ	Average drag coefficient	$\overline{C}_d$

Thrust Force

Thrust force is the result of vertical forces acting on all blade elements of one propeller calculated as:

$$\begin{cases} T = C_T \rho A(\Omega R_{rad})^2 \\ \dfrac{C_T}{\sigma a} = (\dfrac{1}{6}+\dfrac{1}{4}\mu^2)\theta_0 - (1+\mu^2)\dfrac{\theta_{tw}}{8} - \dfrac{1}{4}\lambda \end{cases} \tag{6.2}$$

Hub Force

The hub force is the result of horizontal forces acting on all blade elements defined as:

$$\begin{cases} H = C_H \rho A(\Omega R_{rad})^2 \\ \dfrac{C_H}{\sigma a} = \dfrac{1}{4a}\mu\overline{C_d} + \dfrac{1}{4}\lambda\mu(\theta_0 - \dfrac{\theta_{tw}}{2}) \end{cases} \tag{6.3}$$

Drag Moment

Drag moment about the rotor shaft is caused by aerodynamic forces acting on blade elements. Horizontal forces act on the rotor, multiplied by the moment arm and integrated over the rotor. Drag moment is important because it determines the power required to spin the rotor calculated as:

$$\begin{cases} Q = C_Q \rho A(\Omega R_{rad})^2 R_{rad} \\ \dfrac{C_Q}{\sigma a} = \dfrac{1}{8a}(1+\mu^2)\overline{C_d} + \lambda(\dfrac{1}{6}\theta_0 - \dfrac{1}{8}\theta_{tw} - \dfrac{1}{4}\lambda) \end{cases} \tag{6.4}$$

Rolling Moment

The rolling moment of a propeller is present in forward flight when the advancing blade produces more lift than the retreating one. It is defined in terms of integration over the entire rotor of the lift of each section acting at a given radius – it is not to be confused with the overall rolling moment that is caused by a number of other effects. The rolling moment is:

$$\begin{cases} R_m = C_{R_m} \rho A(\Omega R_{rad})^2 R_{rad} \\ \dfrac{C_{R_m}}{\sigma a} = -\mu(\dfrac{1}{6}\theta_0 - \dfrac{1}{8}\theta_{tw} - \dfrac{1}{8}\lambda) \end{cases} \tag{6.5}$$

Ground Effect

Helicopters operating near the ground are subject to thrust augmentation due to better rotor efficiency. This is related to reduction of the induced airflow velocity called In Ground Effect (IGE). One may find in the literature different approaches to deal with this effect, for instance by using adaptive techniques [10]. In this research, the goal is to find a model for IGE that improves autonomous take-off and landing. As such, the idea is to obtain a simple model capturing mainly the variation of the induced inflow velocity. The images method discussed in [15] is used in [5] to state that at constant power $T_{OGE}\, \upsilon_{i.OGE} = T_{IGE}\, \upsilon_{i.IGE}$. The velocity induced at the rotor center by its image is $\delta\upsilon_i = A\upsilon_i/16\pi z^2$. The simple equation shown below was obtained in [5] by assuming that υ and $\delta\upsilon_i$ are constant over the disk allowing for $\upsilon_{i,IGE} = \upsilon_i - \delta\upsilon_i$, with z being the altitude:

$$\frac{T_{IGE}}{T_{OGE}} = \frac{1}{1 - \dfrac{R_{rad}^2}{16z^2}} \tag{6.6}$$

Alternatively, one may consider that the inflow ratio IGE is $\lambda_{IGE} = (\upsilon_{i,OGE} - \delta\upsilon_i - \dot{z}) / \Omega R_{rad}$, where the variation of the induced velocity is $\delta\upsilon_i = \upsilon_i / (4z / R_{rad})^2$. Then, from the previous equations, the thrust coefficient IGE may be re-written as follows:

$$\begin{cases} T_{IGE} = C_T^{IGE} \rho A(\Omega R_{rad})^2 \\ \dfrac{C_T^{IGE}}{\sigma a} = \dfrac{C_T^{OGE}}{\sigma a} + \dfrac{\delta\upsilon_i}{4\Omega R_{rad}} \end{cases} \tag{6.7}$$

The variation of the inflow velocity in and out of the ground effect was also calculated using the OS4 simulator. As such, Figure 6.2 illustrates the plot of the ratio ($\delta \upsilon_i / \upsilon_i$) obtained by simulation and analytically. The influence is minimal, very close to zero when $z / R_{rad} \approx 2$ but it becomes important for values of $z / R_{rad} \leq 1$. It looks like in the case of a quadrotor the ground effect influence is already present at one rotor diameter and becomes important at one rotor radius.

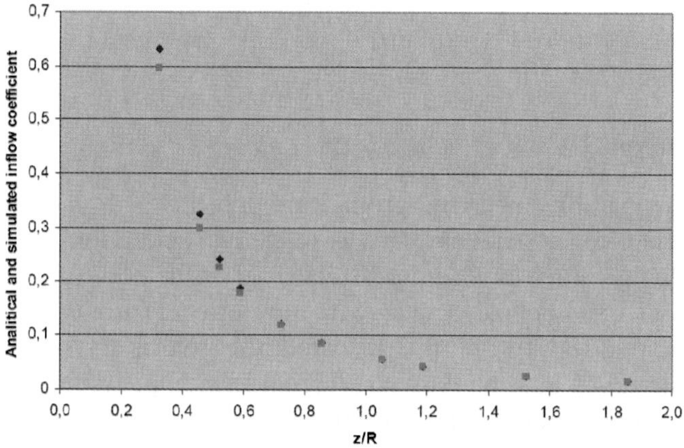

Fig. 6.2. Ground effect influence on the inflow velocity.

In order to empirically verify this assumption, a simple experiment was conducted, which proved that a quadrotor deprived of altitude control can hover at a constant altitude at nearly one rotor diameter from the ground. It should be clear that this result is only an indication of validity and it does not constitute a formal proof.

6.2.2 Moments and Forces

Quadrotor motion is caused by a series of forces and moments. Defining the roll angle as φ, the pitch angle as θ and the yaw angle as ψ, the rotor inertia as J_r, the vertical distance prop./center of gravity as h and the horizontal distance prop./center of gravity as l, the following are defined with respect to rolling, pitching and yawing moments:

Rolling Moments

Body gyro effect: $\dot{\theta}\dot{\psi}(I_{yy} - I_{zz})$

Propeller gyro effect: $J_r \dot{\theta}\Omega_r$

Roll actuators action: $l(-T_2 + T_4)$

Hub moment due to sideward flight: $h\left(\sum_{i=1}^{4} H_{yi}\right)$

Rolling moment due to forward flight: $(-1)^{i+1}\sum_{i=1}^{4} R_{mxi}$

Pitching Moments

Body gyro effect: $\dot{\theta}\dot{\psi}(I_{zz} - I_{xx})$

Propeller gyro effect: $J_r \dot{\theta}\Omega_r$

Roll actuators action: $l(T_1 - T_3)$

Hub moment due to forward flight: $h\left(\sum_{i=1}^{4} H_{xi}\right)$

Rolling moment due to sideward flight: $(-1)^{i}\sum_{i=1}^{4} R_{myi}$

Yawing Moments

Body gyro effect: $\dot{\theta}\dot{\phi}(I_{xx} - I_{yy})$

Inertial counter-torque: $J_r \dot{\Omega}_r$

Counter-torque unbalance: $(-1)^{i}\sum_{i=1}^{4} Q_i$

Hub force unbalance in forward flight: $l(H_{x2} - H_{x4})$

Hub force unbalance in sideward flight: $l(-H_{y1} + H_{y3})$

The forces along the x, y and z axes are defined, respectively as:

Forces along the z- axis

Body gyro effect: $\cos\psi\cos\phi\left(\sum_{i=1}^{4}T_i\right)$

Weight: mg

Forces along the x- axis

Actuators action: $(\sin\psi\sin\phi+\cos\psi\sin\theta\cos\phi)\left(\sum_{i=1}^{4}T_i\right)$

Hub force in y axis: $\left(-\sum_{i=1}^{4}H_{xi}\right)$

Friction: $\frac{1}{2}C_xA_c\rho\dot{x}\,|\,\dot{x}\,|$

Forces along the y- axis

Actuators action: $(-\cos\psi\sin\phi+\sin\psi\sin\theta\cos\phi)\left(\sum_{i=1}^{4}T_i\right)$

Hub force in y axis: $\left(-\sum_{i=1}^{4}H_{yi}\right)$

Friction: $\frac{1}{2}C_yA_c\rho\dot{y}\,|\,\dot{y}\,|$

6.2.3 Equations of Motion

The equations of motion may be derived from (6.1) and the previously stated equations of forces and moments acting on the quadrotor as:

$$\begin{cases} \ddot{\phi} = \sum \tau_x / I_{xx} \\ \ddot{\theta} = \sum \tau_y / I_{yy} \\ \ddot{\psi} = \sum \tau_z / I_{zz} \\ \ddot{z} = g - \sum F_z / m \\ \ddot{x} = \sum F_x / m \\ \ddot{y} = \sum F_y / m \end{cases} \tag{6.8}$$

6.2.4 Rotor Dynamics

The OS4 helicopter is equipped with four fixed-pitch rotors (no swash plate) each including a BrushLess Direct Current (BLDC) motor, a one-stage gearbox and a propeller. The entire rotor dynamics have been identified and validated using the *MATLAB* Identification Toolbox. A first-order transfer function is reasonable to reproduce the dynamics between the propeller's speed set-point and its true speed. It was set to be:

$$G(s) = \frac{0.936}{0.178s + 1} \tag{6.9}$$

It is essential to mention the non-unity gain in (6.9) that is shown in Figure 6.3. Figure 6.3 shows the model output (actuator's output) in red and the sensor data (measured propeller speed) in blue to a step input in green. Since sensorless BLDC motors require a minimum speed to run, the set point does not start from zero.

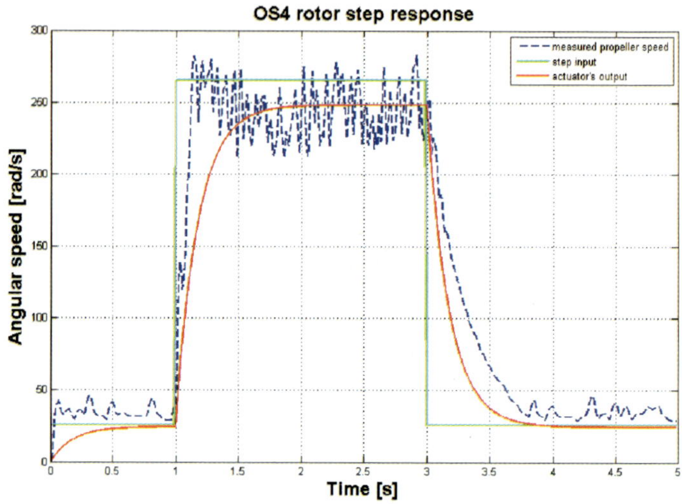

Fig. 6.3. Rotor and model step response, measured at the propeller shaft.

6.3 The OS4 Simulator

The most recent version of the OS4 simulator includes the identified actuator's dynamics, the aerodynamics block, obstacle avoidance controller (OAC) and a high level planner for waypoint definition. Each block is described by one or more *MATLAB* files; the OS4 simulator may be easily incorporated into other simulators. A block diagram of the simulator is depicted in Figure 6.4. Simulation starts by defining the initial state from the dedicated block Initial Conditions. Data is subject to potential delay, white noise filtering before used in the Control block and sent to the Motor Dynamics block. The estimated rotors' speed feeds the Aerodynamics block, which outputs the forces and moments of each propeller, sent to the System Dynamics block, along with the state and the actual rotors' speed to process the new state.

The Control block refers to attitude, altitude and position controllers as detailed in Figure 6.5. Each control loop is simulated at the sampling time of its respective sensor.

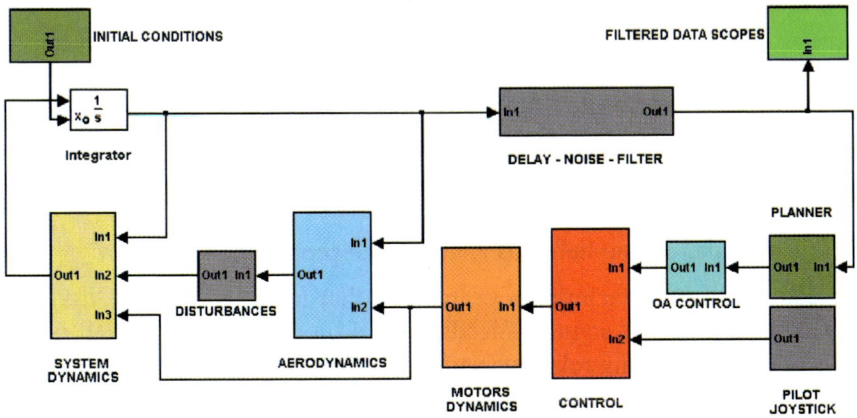

Fig. 6.4. The OS4 simulator block diagram.

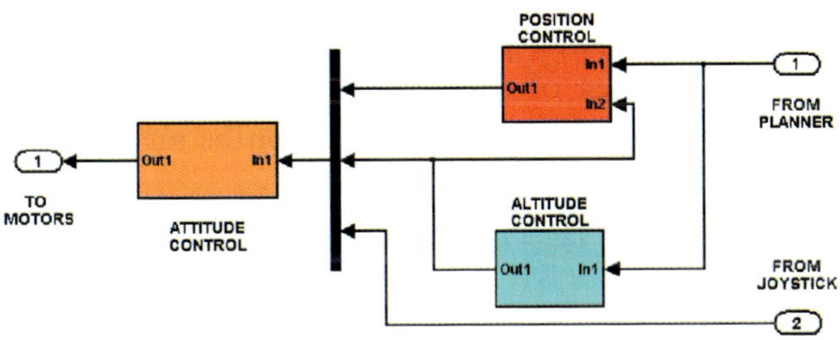

Fig. 6.5. Individual controller blocks in the OS4 simulator.

6.4 Design

In principle, interdependency of components during any design phase makes each component choice to depend on choice of all others and vice-versa. In order to best decide about the OS4 quadrotor design variables and components, a practical method was developed suitable for a small scale rotorcraft. The method described below combines models and databases of relevant components resulting in the best possible selection including battery selection to comply with the total mass constraint.

6.4.1 The General Method

The design process starts by defining three important design constraints for the small scale helicopter: Maximum mass $m_{\max}$, maximum span $s_{\max}$ and target thrust/weight ratio T_w. The three constraints give a good idea about the desired propeller diameter d_{prop}. In practice, the propeller span defines the overall span of the helicopter. Using the propeller diameter d_{prop}, one can estimate the characteristics of the propeller in terms of thrust, drag and power for a range of angular speeds. Based on the analysis in [19], it follows that the following relations should hold:

$$T \propto \Omega^2 L^4,\ D \propto \Omega^2 L^5 \text{ and } P \propto \Omega^3 L^5$$

L is a reference dimension, for example, the center of the blade. So, the mass $m_{\max}$, the drag moment D and the thrust/weight ratio T_w are sufficient to fully define the motor power requirements. This allows for selecting from a database a list of candidate actuators that offer the required power. Then, a rough estimation of the mass of the airframe m_{af} and avionics m_{av} is required to calculate a first estimate of the total mass without considering the battery as shown in Figure 6.6. Considering the mass of the battery, the maximum mass $m_{\max}$ may be defined as:

$$m_{\max} = m_{af} + m_{av} + m_{pg} + m_{bat}$$

where m_{pg} is the propulsion group (propeller, gearbox and motor) mass and m_{bat} is the mass of the battery. The iterative algorithm to calculate m_{pg} and m_{bat} is described next.

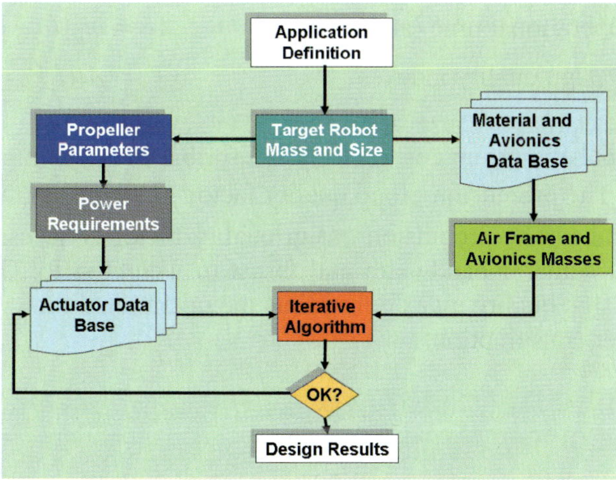

Fig. 6.6. The flowchart of the design method. The user has to define a target mass and size of the system in addition to airframe and avionics mass.

The Iterative Algorithm

A block diagram of the iterative algorithm is shown in Figure 6.7. The algorithm starts by choosing one candidate actuator from the database, its mass being $m_{pg}(i)$. Next, an initial value $m_{bat}(j_0)$ is given to the m_{bat} variable in order to calculate $m_{\max}(i,j) = m_{af} + m_{av} + m_{pg}(i) + m_{bat}(j)$. This 'temporary' total mass of the helicopter $m_{\max}(i,j)$ is used to estimate other variables at the two operational points of hovering and maximum thrust as shown in Figure 6.7: For each candidate actuator, the variable $m_{bat}(j)$ is incremented until $m_{\max}$ is reached. This process allows for estimating under hovering and at maximum thrust, for each candidate actuator and for each increment of $m_{bat}(j)$ the following variables:

- Total mass: m_{bat}
- Total power consumption: P_{tot}
- Propulsion group efficiency: η_{gb}
- Propulsion group cost factor: $C = P_{el}/(T - m_{pg})$
- Propulsion group quality factor: $Q = B_W T_w / \Omega C$

(B_W: PG bandwidth, T_w: thrust/weight ratio)

- Operational time (autonomy): $Au = m_{bat} C_{bat} / P_{el}$
- Design quality index: $Q_{in} = Au / P_{tot}$

The propulsion group cost factor C describes the cost in power of the lifted mass. The propulsion group quality factor Q describes the quality of the mass lifting. The propulsion group quality factor is necessary to take into account actuator bandwidth and thrust to weight ratio. On the other hand, the design quality index constraints the operational time with regards to total power consumption.

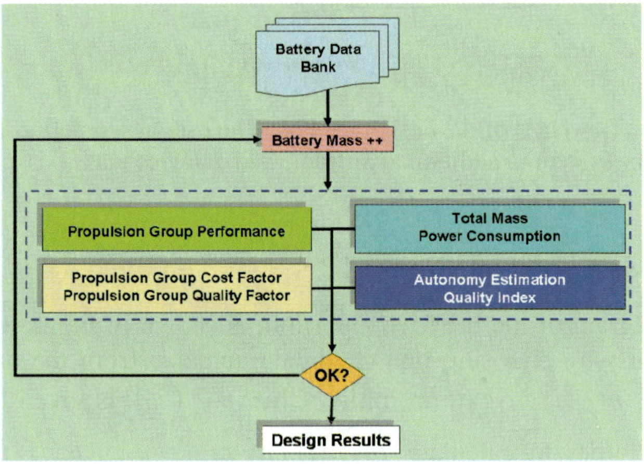

Fig. 6.7. The iterative algorithm flowchart.

6.4.2 The OS4 Quadrotor

The design of the OS4 quadrotor has been based on two main constraints: maximum mass of 500 gr and a maximum span of 800 mm. A propeller with a 300 mm diameter was selected that satisfied the span constraint. Table 6.3 shows the main design variables of a propulsion group that have been used in the models shown in Table 6.4.

Propeller		OS4	Unit
Mass	m_p	5.2	gr
Thrust coefficient	b	3.13e-5	N s^2
Drag Coefficient	d	7.5e-7	Nm s^2
Inertia	J_r	6e-5	kg.m^2
Gearbox		**OS4**	**unit**
Efficiency	η	90	%
Mass	m_{gb}	7	gr
Max torque		0.15	Nm
Max speed		1000	rad/s
Reduction ratio	r	4:1	
Motor		**OS4**	**unit**
Efficiency at hover	η_m	64	%
Mass	m_m	12	gr
Max power	P_{el}	35	W
Internal resistance	R_{mot}	0.6	Ω
Inertia	J_m	4e-7	kgm^2
Torque coefficient	k	5.2	mNm/A

Table 6.3. The OS4 propulsion group design variables.

The choice of the propulsion group is determined by the iterative algorithm that returns a classification based on the cost and quality factors. It returns a battery mass m_{bat}=230 gr (11 V, 3.3 Ah) of Lithium-Polymer. On the rotor side, the tests revealed that a gearbox is necessary. In fact, a direct-drive propulsion group would allow only a thrust to weight ratio of T_w=0.75, which is obviously not enough for lifting purposes. The selected motor is a brushless DC motor (12 gr, 35 W) with a high power to weight ratio, which justifies the choice even including its control electronics. A 6 gr I^2C controller was specially designed for the sensorless out runner LRK195.03 motor as shown in Figure 6.8. Obviously, BLDC motors offer high life time and low electromagnetic noise. The ready to plug propulsion group weighs 40 gr and lifts more than 260 gr.

Component	Model
Propeller	$(b,d)\Omega^2 = (T,D)$
Gearbox	$P_{in}\eta = P_{out}$
DC motor	$-\frac{k^2}{R_{mot}}\omega - D + \frac{k}{R_{mot}}u = J\frac{d\omega}{dt}$
PG cost	$P_{el}/(T - m_{pg}) = C$
PG quality	$B_W T_w / \Omega C = Q$

Table 6.4. Propulsion group (motor, gearbox, propeller) component models.

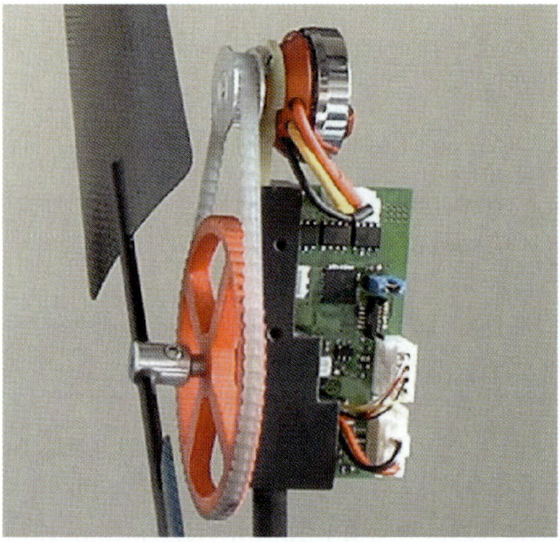

Fig. 6.8. OS4 propulsion group. The module is interfaced through I²C bus and has a local PI speed controller.

Figure 6.9 illustrates the OS4 component block diagram. Embedding the controller is definitely advisable because it avoids delays and interruptions with wireless connections. A miniature computer module, based on a Geode 1200 processor running at 266MHz with 128M of RAM and flash memory was used. The computer module is x86 compatible and offers all standard PC interfaces (see Figure 6.10). The whole computer has a weight of 44 gr, it is 56 mm by 71 mm in size and runs Linux.

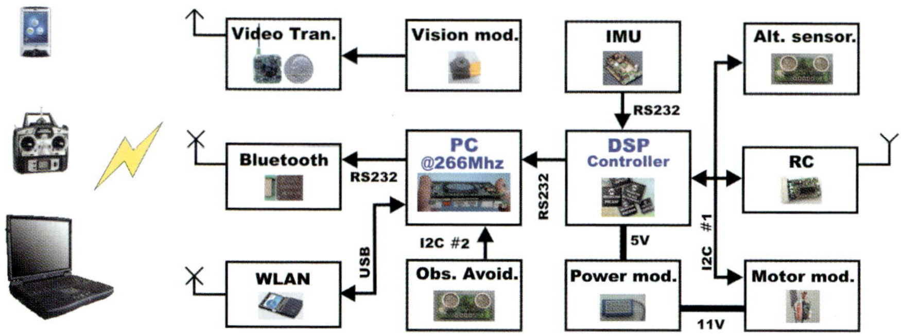

Fig. 6.9. The OS4 block diagram. A DSP processor handles attitude and altitude control. A miniature PC (x-board) handles obstacle avoidance control and communication tasks. The helicopter communicates through a WiFi interface and accepts standard remote control signals.

Fig.6.10. The x-board based, 40 gr, 56x71 mm computer module.

The controller includes an MCU for interfacing Bluetooth with the computer module. The same chip is used to decode the PPM signal picked from a 1.6 gr, 5 channel commercially available RC receiver. This decoding on the MCU makes it possible to interface the RC receiver to an I^{2}C bus and at the same time detects any problem in the channels. It is also possible to control the helicopter using a standard remote control.

Position Sensor

The OS4 position sensor is based on an on board down-looking CCD camera and a simple pattern on the ground. The camera provides a motion blur free image of 320x240 at up to 25 fps. The algorithm detects the pattern, estimates the pose and provides the camera position (x, y) and heading angle (ψ). The image is primarily sent to an off-board computer for processing and then the position data is sent back to the helicopter for control purposes as shown in Figure 6.11.

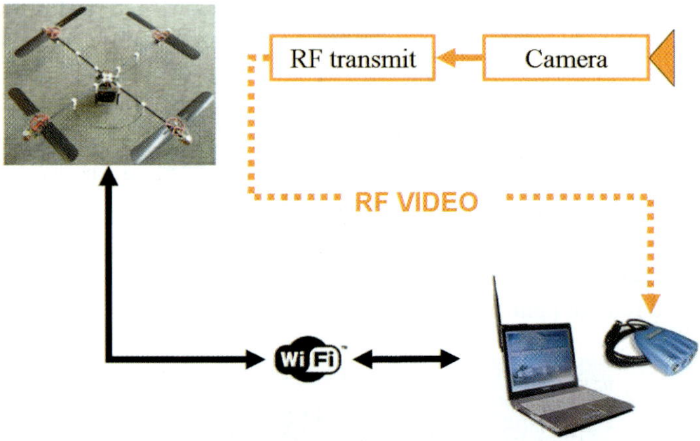

Fig. 6.11. Position sensing setup on the OS4.

Several alternatives have been considered for pattern detection. The first method that was tested refers to detecting five red dots on an A4 paper; this method suffers from sensitivity to lighting conditions. For the second test four LEDs were considered with different colors on an A4 size board; it was hard to tune LED intensity for the overall working volume. Then, a red A4 paper was used with a white spot shifted from the pattern center; this time the pattern was robustly detected using the Canny edge detector and Douglas-Peucker algorithms already implemented in OpenCV. In addition, a least-square based linear regression was used to refine detection. Pose estimation was performed using PnP algorithm [6]. The sensor algorithm was then enhanced to incorporate different situations where the pattern was not or it was partially detected.

All processing takes about 7ms. Image capture takes 1ms with a PCI acquisition card and almost 20ms with a USB 1.1 device on a Pentium 4, 2.4GHz. The algorithm is limited to 25Hz by the camera frame rate (25 fps). The errors obtained in x and y position sensing were about 2 cm at 0.5 m/s. The error on the yaw was about 3° at 180°/s.

Obstacle Detection Setup

Four ultrasound range finders are mounted on the OS4 quadrotor for obstacle detection, one under each propeller as shown in Figure 6.12. Two short plastic tubes are mounted on each sensor in order to reduce the beam width.

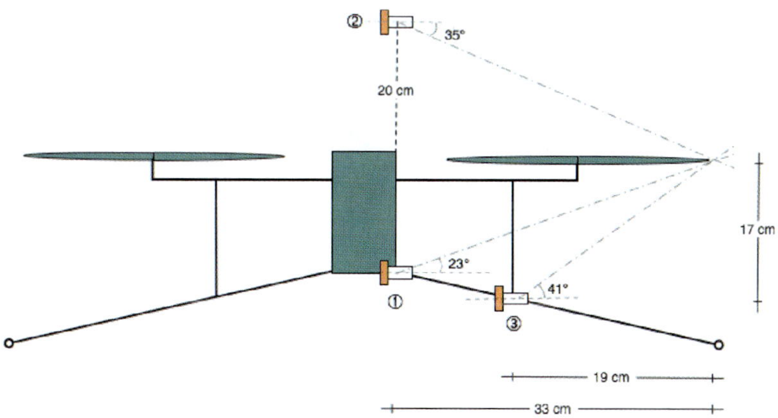

Fig. 6.12. Possible ultrasound sensor arrangement on the OS4. Position (3) was adopted after testing.

Design Results

Figure 6.13 shows the real quadrotor configuration. The quadrotor mass and power distribution are shown in Figure 6.14. The total mass in the initial design was about 520gr, with the battery accounting for almost one-half of the total mass and the actuators accounting for only one-third thanks to BLDC technology. All actuators account for 60W of 66W average power consumption. However, the latter depends on flight conditions and represents a weighted average between the equilibrium (40W) and the worst possible inclination state (120W) without loosing altitude.

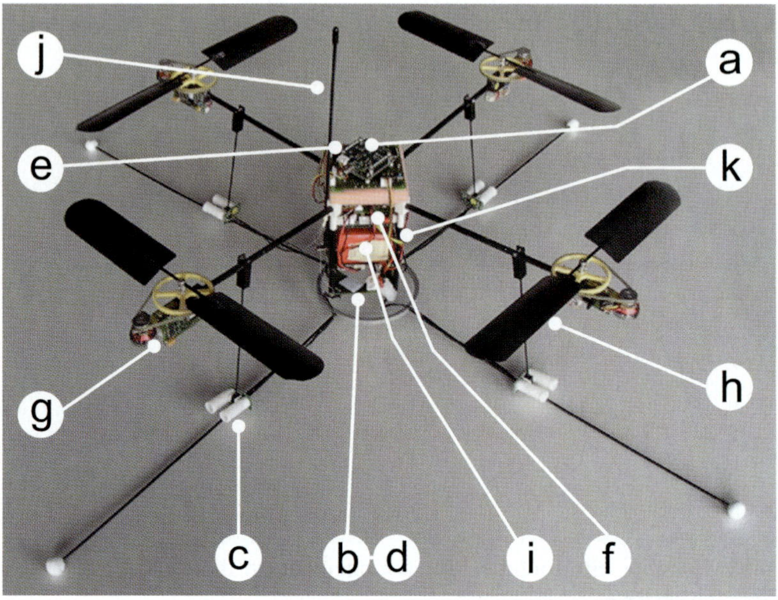

Fig. 6.13. Sensors, actuators and electronics of the OS4 quadrotor. *(a)* inertial measurement unit, *(b)* altitude sensor below the OS4, *(c)* obstacle avoidance sensor with tubes, *(d)* mini camera below the OS4, *(e)* DSP, *(f)* mother board, *(g)* motor module, *(h)* propeller, *(i)* battery, *(j)* RC antenna, *(k)* WiFi dongle.

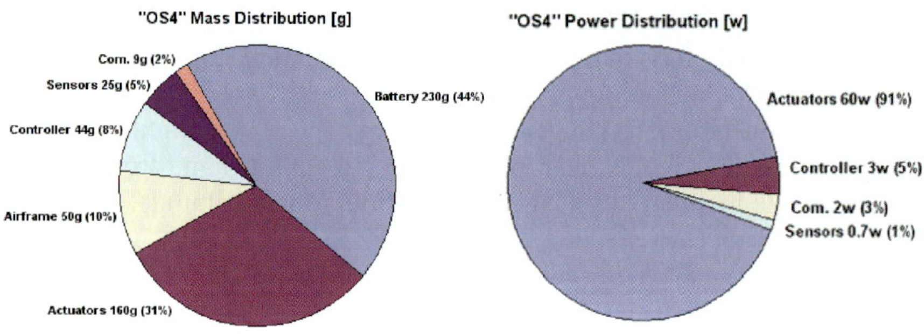

Fig.6.14. Mass and power distribution of the OS4 quadrotor. The battery mass accounts for almost one-half of the total mass while the actuators sink about 90% of the total power.

6.5 Simulation and Control

Improved performance expected from the new generation of MFRs is possible through derivation and implementation of specific control techniques incorporating limitations related to sensors and actuators. For the project under consideration, several control approaches were explored, from theoretical developments to final experimentation.

As a first attempt, two linear controllers were tested, a PID and an LQR that was based on a simplified model. The main result was autonomous hover already presented in [2]. However, strong disturbances (strong wind conditions) were poorly rejected.

In the second attempt control was reinforced using backstepping techniques that allowed for strong disturbance rejection; however, stabilization in hover flight was rather delicate [3].

An additional improvement was to introduce integral backstepping [13]. The idea of using integral action in the backstepping design was first proposed in [12] and applied in [21]. This controller design is derived based on [21]. The result is that the OS4 quadrotor is able to perform autonomous hovering with altitude control and autonomous take-off and landing.

The OS4 controller is structured in terms of six different controllers as illustrated in Figure 6.15. The take-off and landing controller outputs the desired altitude (z_d) to the altitude controller, which outputs the desired overall thrust (T_d) based on sonar data. The position controller receives the OS4 position (x, y) and desired thrust and outputs the desired roll (ϕ_d) and pitch (θ_d) while the desired yaw (ψ_d) comes directly from the user. The attitude controller outputs the desired motor speed to the motor controllers. The integral backstepping technique is used for attitude, altitude and position control. This results in a powerful and flexible control structure.

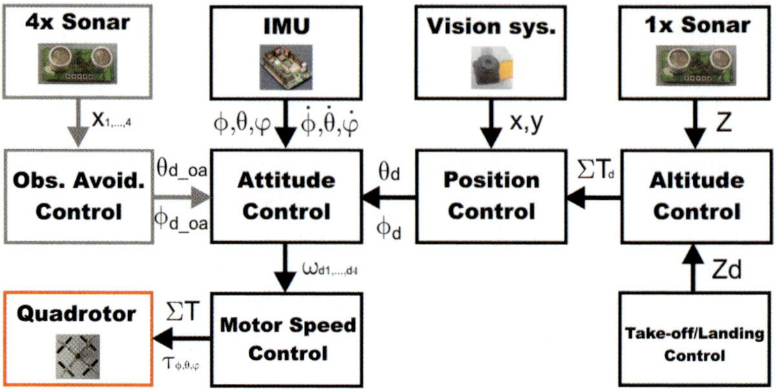

Fig. 6.15. The control structure implemented on the OS4 quadrotor.

6.5.1 Model Simplifications

The differential equations of motion of the quadrotor are shown in (6.8). However, it is advisable to simplify the model in order to comply with real-time constraints of the embedded control loop. Hence, hub forces and rolling moments are neglected, while the thrust and drag coefficients are considered to be constant. The equations of motion may be rewritten in state-space form $\dot{X} = f(X,U)$ with U representing the input vector and X the state vector, chosen as follows:

$$X = [\phi\ \dot{\phi}\ \theta\ \dot{\theta}\ \psi\ \dot{\psi}\ z\ \dot{z}\ x\ \dot{x}\ y\ \dot{y}]^T \tag{6.10}$$

$$U = [U_1\ U_2\ U_3\ U_4]^T \tag{6.11}$$

$$\begin{cases} U_1 = b(\Omega_1^2 + \Omega_2^2 + \Omega_3^2 + \Omega_4^2) \\ U_2 = b(-\Omega_2^2 + \Omega_4^2) \\ U_3 = b(\Omega_1^2 - \Omega_3^2) \\ U_4 = d(-\Omega_1^2 + \Omega_2^2 - \Omega_3^2 + \Omega_4^2) \end{cases} \tag{6.12}$$

The transformation matrix between the rate of change of the orientation angles $(\dot{\phi}\ \dot{\theta}\ \dot{\psi})$ and the body angular velocities $(p\ q\ r)$ can be considered as a unity matrix if perturbations from hover are small. Then, one can write $(\dot{\phi}\ \dot{\theta}\ \dot{\psi}) \approx (p\ q\ r)$. Simulation tests have shown that this assumption is reasonable. From (6.8), (6.10) and (6.11) one obtains:

$$f(X,U) = \begin{pmatrix} \dot{\phi} \\ \dot{\theta}\dot{\psi} a_1 + \dot{\theta} a_2 \Omega_r + b_1 U_2 \\ \dot{\theta} \\ \dot{\phi}\dot{\psi} a_3 + \dot{\phi} a_4 \Omega_r + b_2 U_3 \\ \dot{\psi} \\ \dot{\theta}\dot{\phi} a_5 + b_3 U_4 \\ \dot{z} \\ -g + (\cos\phi\cos\theta)\frac{1}{m} U_1 \\ \dot{x} \\ u_x \frac{1}{m} U_1 \\ \dot{y} \\ u_y \frac{1}{m} U_1 \end{pmatrix} \tag{6.13}$$

$$\begin{pmatrix} a_1 = (I_{yy} - I_{zz})/I_{xx} \\ a_2 = -J_r/I_{xx} \\ a_3 = (I_{zz} - I_{xx})/I_{yy} \\ a_4 = J_r/I_{yy} \\ a_5 = (I_{xx} - I_{yy})/I_{zz} \\ b_1 = l/I_{xx} \\ b_2 = l/I_{yy} \\ b_3 = 1/I_{zz} \end{pmatrix} \tag{6.14}$$

$$\begin{cases} u_x = (\cos\phi\sin\theta\cos\psi + \sin\phi\sin\psi) \\ u_y = (\cos\phi\sin\theta\sin\psi - \sin\phi\cos\psi) \end{cases} \tag{6.15}$$

It is worthwhile to mention that in the latter system the angles and their time derivatives do not depend on translation components. On the other hand, translations depend on the angles. One may ideally imagine the overall system described by (6.13) as being composed of two subsystems, the angular rotations and the linear translations.

6.5.2 Attitude Control

The attitude controller is the heart of the overall control system; it keeps the 3-D orientation of the helicopter at the desired value. Usually roll and pitch angles are forced to zero allowing for hovering. The attitude control loop runs at 76Hz, the update rate of the Microstrain 3DM-GX1 IMU. The latter provides the rates of turn and orientation around (x, y, z) axes with a dynamic accuracy of $\pm 2^\circ$. The first step in the IB controller design is to consider the tracking error $e_1 = \phi_d - \phi$ and its dynamics:

$$\frac{de_1}{dt} = \dot{\phi}_d - \omega_x \tag{6.16}$$

The angular speed ω_x is not considered a control input; it has its own dynamics, so it is set at a desired behavior interpreted as virtual control:

$$\omega_{xd} = c_1 e_1 + \dot{\phi}_d + \lambda_1 \chi_1 \tag{6.17}$$

c_1 and λ_1 are positive constants and $\chi_1 = \int_0^t e_1(\tau) d_\tau$ is the integral of the roll tracking error. So, the integral term is now introduced in (6.17). Since ω_x has its own error e_2, its dynamics are computed using (6.17) as follows:

$$\frac{de_2}{dt} = c_1(\dot{\phi}_d - \omega_x) + \ddot{\phi}_d + \lambda_1 e_1 - \ddot{\phi} \tag{6.18}$$

where e_2, the angular velocity tracking error is defined by:

$$e_2 = \omega_{xd} - \omega_x \tag{6.19}$$

Using (6.17) and (6.19) the roll tracking error dynamics are re-written as:

$$\frac{de_1}{dt} = -c_1 e_1 - \lambda \chi_1 + e_2 \tag{6.20}$$

Replacing $\ddot{\phi}$ in (6.18) by its corresponding expression from (6.13), the real control input U_2 appears in equation:

$$\frac{de_2}{dt} = c_1(\dot{\phi}_d - \omega_x) + \ddot{\phi}_d + \lambda_1 e_1 - \dot{\theta}\dot{\psi} a_1 - \dot{\theta} a_2 \Omega_r - b_1 U_2 \tag{6.21}$$

Therefore, using (6.16), (6.20) and (6.21) tracking errors of the position e_1, angular speed e_2 and integral position tracking error χ_1 are combined to obtain:

$$\frac{de_2}{dt} = c_1(-c_1e_1 - \lambda_1\chi_1 + e_2) + \ddot{\phi}_d + \lambda_1 e_1 - \tau_x/I_{xx} \tag{6.22}$$

where τ_x is the overall rolling torque. The desirable dynamics for the angular speed tracking error is:

$$\frac{de_2}{dt} = -c_2e_2 - e_1 \tag{6.23}$$

This is obtained if one chooses the control input U_2 as:

$$U_2 = \frac{1}{b_1}[(1 - c_1^2 + \lambda_1)e_1 + (c_1 + c_2)e_2 - c_1\lambda_1\chi_1 + \ddot{\phi}_d - \dot{\theta}\dot{\psi}a_1 - \dot{\theta}a_2\Omega_r] \tag{6.24}$$

where c_2 is a positive constant determining the convergence speed of the angular speed loop. Similarly, the pitch and yaw controls are:

$$\begin{cases} U_3 = \dfrac{1}{b_2}[(1 - c_3^2 + \lambda_2)e_3 + (c_3 + c_4)e_4 - c_3\lambda_2\chi_2 + \ddot{\theta}_d - \dot{\phi}\dot{\psi}a_3 + \dot{\phi}a_4\Omega_r] \\ U_4 = \dfrac{1}{b_3}[(1 - c_5^2 + \lambda_3)e_5 + (c_5 + c_6)e_6 - c_5\lambda_3\chi_3] \end{cases} \tag{6.25}$$

with $(c_3, c_4, c_5, c_6, \lambda_2, \lambda_3) > 0$, and (χ_2, χ_3) the integral position tracking error of pitch and yaw angles, respectively.

Stability Analysis

To prove stability, the following candidate Lyapunov function is chosen:

$$V = \lambda\frac{1}{2}\chi_1^2 + \frac{1}{2}e_1^2 + \frac{1}{2}e_2^2 \tag{6.26}$$

It includes the position tracking error e_1, its integration χ_1 and velocity tracking error e_2. From (6.26) and using (6.20) and (6.23) one obtains:

$$\dot{V}(e_1, e_2) < 0, \forall(e_1, e_2) \neq 0 \text{ or } dV/dt = -c_1e^2 - c_2e^2 \tag{6.27}$$

The definition of (6.26) and because of (6.27) guarantee that e_1, χ_1 and e_2 are bounded. The desired position reference ϕ_d is bounded by assumption and $e_1 = \phi_d - \phi$ is also bounded, so, position ϕ is also bounded. This implies that the virtual control ω_x is bounded. Finally, the boundedness of the overall control torque is due to the choice of the control law in (6.24). The system is also globally asymptotically stable from the positive definition of V and the fact that $\dot{V}(e_1,e_2) < 0, \forall(e_1,e_2) \neq 0$ and $\dot{V}(0) = 0$.

Results

Attitude control performance is of crucial importance; it is directly linked to the performance of the actuators. OS4 is equipped with brushless sensorless motors that are powerful enough to avoid amplitude saturation. However, they suffer from low dynamics and, thus, from bandwidth saturation. This was taken into account in control design. Simulation results are shown in Figure 6.16 performed with a model including the actuators' dynamics and amplitude saturation. The simulation takes into account the delay and the noise inherent to sensors. The task was to stabilize roll, pitch and yaw angles at zero. Control parameters used for simulation were $C_1 = 10$, $C_2 = 2$, $C_3 = 10$, $C_4 = 2$, $C_5 = 2$, $C_6 = 2$.

The experimental results shown in Figure 6.17 correspond to a free flight with attitude references set at zero. One may observe in the roll and pitch plots a bounded oscillation of 0.1 rad in amplitude. This oscillation is not felt during flight; nevertheless, it is due to the slow dynamics of the OS4 actuators coupled with the differences between the four propulsion groups. Control parameters used in this experiment were $C_1 = 10.5$, $C_2 = 2$, $C_3 = 10$, $C_4 = 2$, $C_5 = 2$, $C_6 = 2$. These are very close to the parameters used for simulation purposes, which highlights the quality of the model.

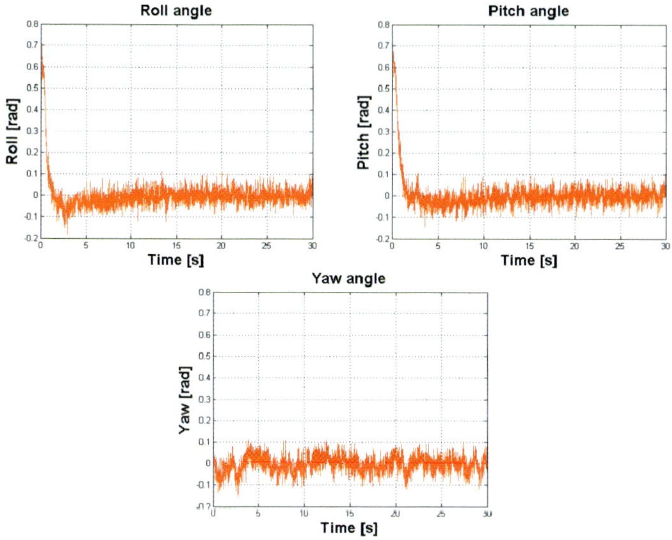

Fig. 6.16. Simulation: Integral backstepping attitude controller has to maintain roll, pitch and yaw angles to zero. Despite of the hard initial conditions and the white noise, the helicopter is quickly brought back to equilibrium.

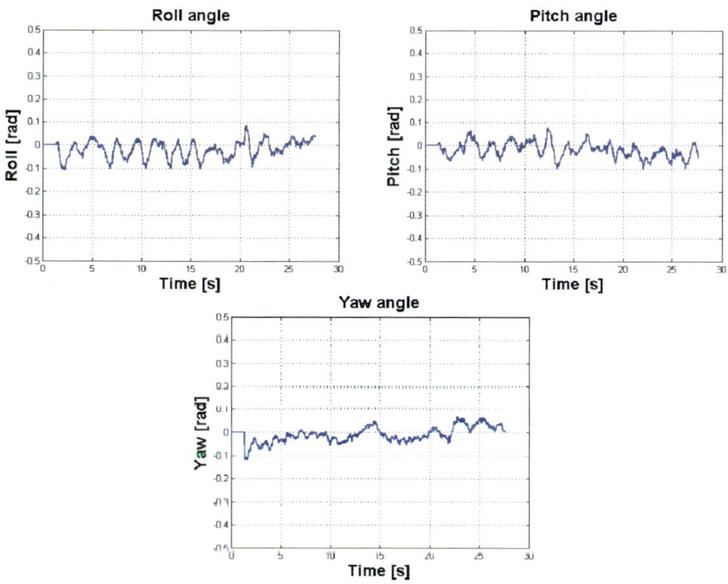

Fig. 6.17. Experiment: Integral backstepping attitude controller maintaining attitude angles to zero during flight. The helicopter is stabilized despite disturbances due to yaw drift, sensor noise and unmodeled effects.

6.5.3 Altitude Control

The altitude controller keeps the helicopter distance from the ground at a desired value. It is accomplished using a sonar (Devantech SRF10) that gives the range to the closest obstacle at 15 Hz. The accuracy depends on the distance, with an error of about 1-2 cm at 1 m. The necessary altitude rate of change is estimated based on the range. For control purposes, the altitude tracking error and the speed tracking error are defined as:

$$e_7 = z_d - z \tag{6.28}$$

$$e_8 = c_7 e_7 + \dot{z}_d + \lambda_4 \chi_4 - \dot{z} \tag{6.29}$$

The control law is then:

$$U_1 = \frac{m}{cos\phi cos\theta}[g + (1 - c_7^2 + \lambda_4)e_7 + (c_7 + c_8)e_8 - c_7\lambda_4\chi_4] \tag{6.30}$$

where (c_7, c_8, λ_4) are positive constants.

Take-off and Landing

The autonomous take-off and landing algorithm adapts the altitude reference z_d to follow the dynamics of the quadrotor for taking-off or landing. One can see in Figure 6.18 that the desired altitude reference is gradually reduced by a fixed step k $(k > 0)$ that depends on the vehicle dynamics and the desired landing speed. Moreover, the fact that the control loop is much faster than the vehicle dynamics makes the landing very smooth. Ground effect was not implemented because the landing skids are long enough to keep the propellers out of ground effect even after touchdown.

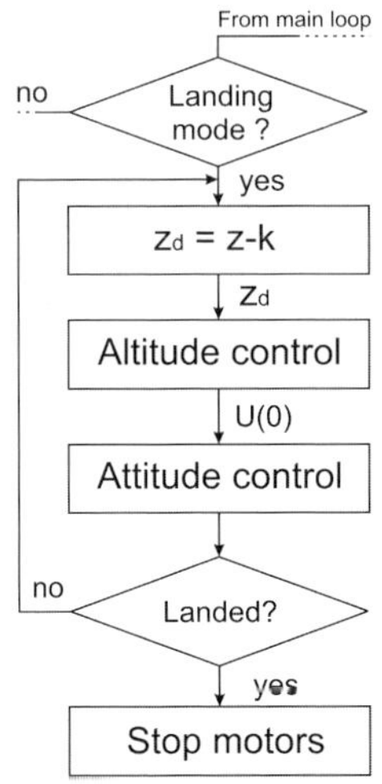

Fig. 6.18. Autonomous landing flowchart. Altitude reference is gradually reduced taking into account the dynamics of the quadrotor.

Results

Altitude control works surprisingly well despite all sonar limitations. Figure 6.19 shows an altitude reference profile (green) followed by the simulated controller (red) and the real controller (blue). The task was to climb to 0.5 m, hover, and then land. Chosen control parameter values were set to $C_7 = 3.5$, $C_8 = 1.5$ for simulation and $C_7 = 4$, $C_8 = 2$ for the experiments. The slight deviation between simulations and experimentation in take-off and landing phases is due to actuators' dynamics where the model was slightly slower, in the raising edge, and slightly faster in the falling one. Take-off is performed in 2 s (0-0.5 m) and landing in 2.8 s (0.5-0 m). Altitude control has a maximum deviation of 3 cm from the reference.

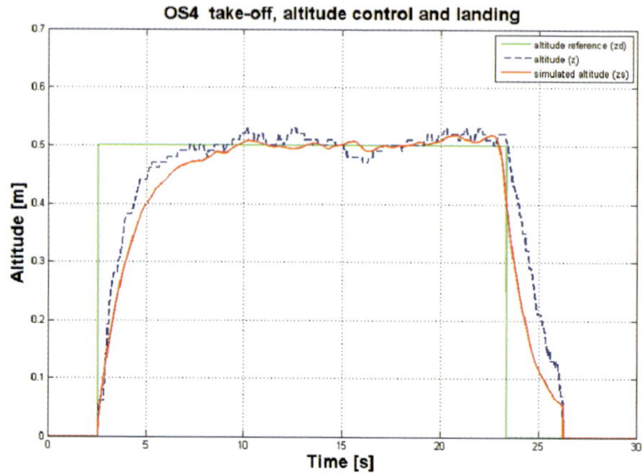

Fig. 6.19. Autonomous take-off, altitude control and landing by simulation and in real flight.

6.5.4 Position Control

Position control keeps the helicopter over the desired point, in this case, the (x, y) horizontal position with respect to a starting point. Horizontal motion is achieved by orienting the thrust vector towards the desired direction of motion. This is done by rotating the vehicle itself in the case of a quadrotor. In practice, one performs position control by rolling or pitching the helicopter in response to a deviation from y_d or x_d references, respectively. Thus, the position controller outputs the attitude references ϕ_d and θ_d, which are tracked by the attitude controller (see Figure 6.15). The thrust vector orientation in the earth-fixed frame is given by R, the rotation matrix. Applying small angle approximation to R gives:

$$R = \begin{bmatrix} 1 & \psi & \theta \\ \psi & 1 & -\phi \\ -\theta & \phi & 1 \end{bmatrix} \tag{6.31}$$

From (6.13) and using (6.31) one can simplify horizontal motion dynamics to:

$$\begin{bmatrix} m\ddot{x} \\ m\ddot{y} \end{bmatrix} = \begin{bmatrix} -\theta U_1 \\ \phi U_1 \end{bmatrix} \tag{6.32}$$

The control law is then derived using the IB technique. Position tracking errors for x and y are defined as:

$$\begin{cases} e_9 = x_d - x \\ e_{11} = y_d - y \end{cases} \tag{6.33}$$

Accordingly, speed tracking errors are:

$$\begin{cases} e_{10} = c_9 e_9 + \dot{x}_d + \lambda_5 \chi_5 - \dot{x} \\ e_{12} = c_{11} e_{11} + \dot{y}_d + \lambda_6 \chi_6 - \dot{y} \end{cases} \tag{6.34}$$

The control laws are then:

$$\begin{cases} U_x = \dfrac{m}{U_1}[(1 - c_9^2 + \lambda_5)e_9 + (c_9 + c_{10})e_{10} - c_9 \lambda_5 \chi_5] \\ U_y = -\dfrac{m}{U_1}[(1 - c_{11}^2 + \lambda_6)e_{11} + (c_{11} + c_{12})e_{12} - c_{11} \lambda_6 \chi_6] \end{cases} \tag{6.35}$$

where $(c_9, c_{10}, c_{11}, c_{12}, \lambda_5, \lambda_6)$ are positive constants.

Results

The main result in position control was obtained by simulation. Observing again Figure 6.15 one sees how the different controllers are cascaded. In fact, only the attitude is driven by the position, the altitude controller is simply feeding signal U_1. Attitude and position loops run at 76 Hz and 25 Hz, respectively. This spectral separation is necessary to avoid a conflict between the two loops; it is often accompanied with gain reductions in the driving loop. Control parameters chosen for simulation purposes were $C_9 = 2, C_{10} = 0.5, C_{11} = 2, C_{12} = 0.5$; results are shown in Figure 6.20.

Waypoint Following

The planner block in Figure 6.4 defines the waypoints, and hence, the trajectories OS4 has to follow. The position of the next waypoint is sent to the position controller that directs the helicopter towards the goal. A waypoint is declared reached when the helicopter enters a sphere around this point. The radius of this sphere (defined to be 0.1 m) is the maximum allowable error. Figure 6.21 shows a square trajectory defined by four waypoints. The task was to climb to 1 m from the ground and then follow the four waypoints of a square of 2 m side.

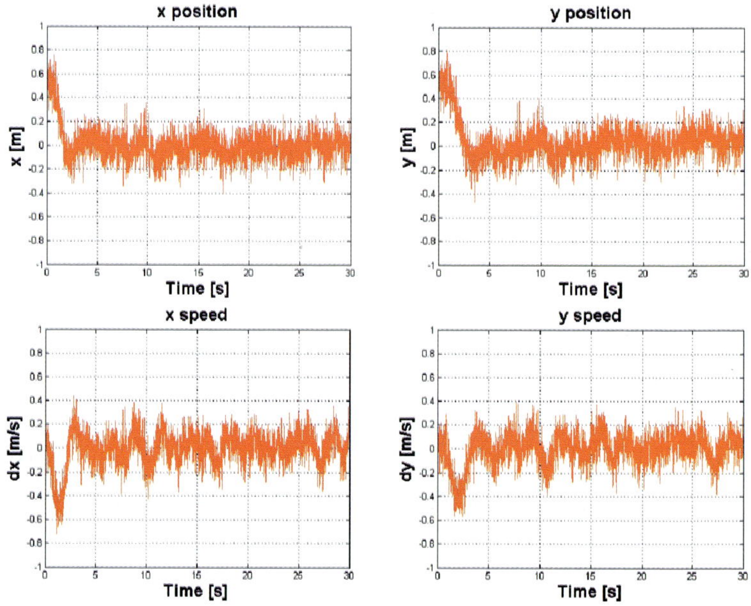

Fig. 6.20. Simulation: Integral backstepping position controller drives attitude controller in order to keep the helicopter over a given point.

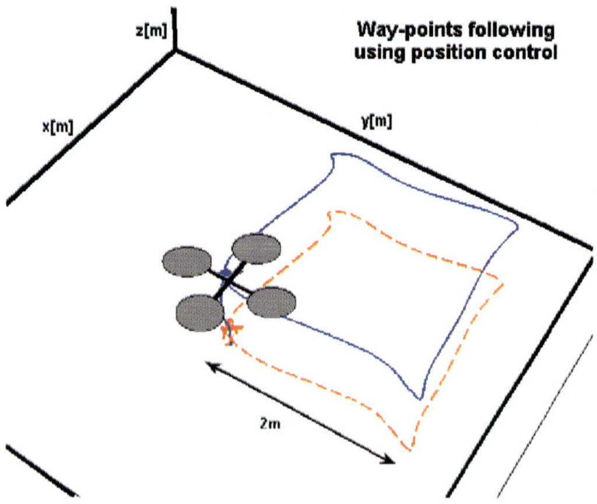

Fig. 6.21. Four waypoints for a square trajectory tracked by OS4.

In order to track the square trajectory, the planner generates the (x_d, y_d) position references, and consequently the position controller generates the (ϕ_d, θ_d) attitude references. Figure 6.22 depicts these signals and shows that the 2 m side square is tracked with about 10% overshoot (20cm), while the trajectory is completed in 20s.

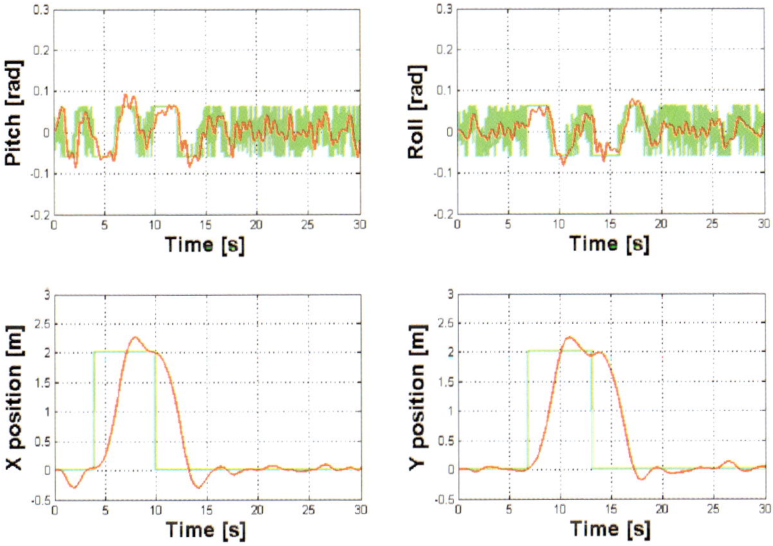

Fig. 6.22. Simulation: The position and attitude signals generated to track the square trajectory.

6.5.5 Obstacle Avoidance

OS4 is equipped with a sonar based obstacle avoidance system composed of four miniature ultrasound range finders in cross configuration. The obstacle avoidance controller is incorporated into the *Simulink* model along with the environment model and sensor libraries. Aiming at simplifying the procedure, it was decided to keep the altitude constant during evasive maneuvers. This would reduce the path planning complexity to a 2-D problem. Direction of flight was also restricted; the OS4 could move only on four directions where the ultrasound sensors were placed. To increase the flight safety, a 90 cm radius security zone was constantly maintained between the helicopter and the environment (see Figure 6.23). This zone assured a 50 cm distance between the helicopter rotors and any obstacle. If an obstacle was detected inside the security zone, a safety loop (that ran in parallel to the obstacle avoidance controller) intervenes with the helicopter flight controller and generates an evasive maneuver. This maneuver is obtained by selecting pre-defined pitch and/or roll angles that would avoid a collision between helicopter and obstacles.

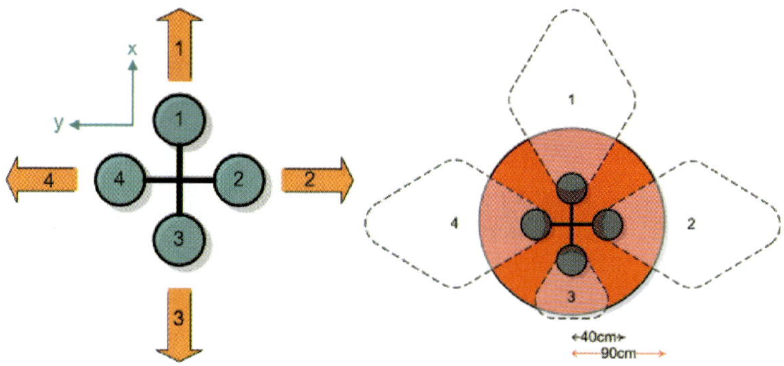

Fig. 6.23. The 4 flight directions (left) and the security zone (right).

Several algorithms were simulated with various results in a 100 m^2 environment, with obstacles represented as columns of 20 cm in diameter and 3 m in height. The approaches followed can be classified as relative position and speed-based approaches. The first obstacle avoidance controller uses the position controller to act on the relative position of the helicopter with respect to the closest obstacle (X_{oa}, Y_{oa}). The second one uses the speed controller to act on the speed of the vehicle $(\dot{X}, \dot{Y})$ if an obstacle is detected. The latter approach was used to develop the main obstacle avoidance controller algorithm. The idea was to act on $\dot{X}$ and $\dot{Y}$ while keeping the heading and altitude constant. When an obstacle is detected its distance from the helicopter is classified based on a given threshold as "far", "close" or "too close". If the obstacle is "far", no avoidance action is needed and there is no intervention with the helicopter normal flight. On the other hand, if the obstacle distance is "close" or "too close" then the obstacle avoidance controller informs the helicopter flight control, reduces its speed, and generates evasive maneuvers using pre-defined flight directions as shown in Figure 6.24. The selection of the direction of the evasive flight depends on the stimulated sensor and the desired flight direction previously selected by the user. However, if the quadrotor is surrounded by obstacles that are "too close", it reduces the speed and keeps a hovering behavior.

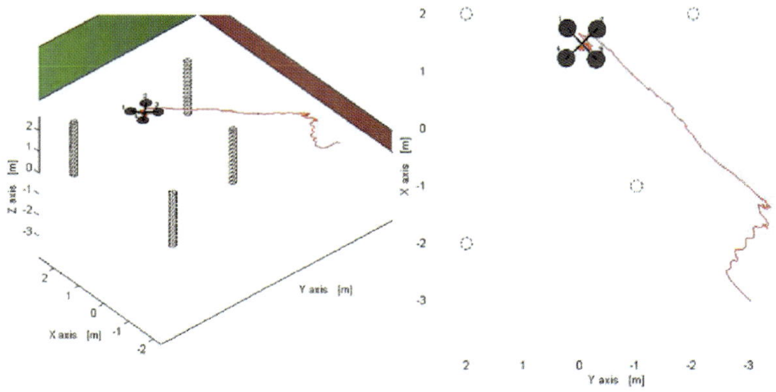

Fig. 6.24. Simulation: OS4 avoiding static obstacles.

The lack of precise sensors for linear speed made the implementation of this approach difficult. A simple collision avoidance algorithm was then developed. The idea was to avoid collision with walls or persons present in the flight area. The inherent noise of the sonar especially in absence of obstacles was affecting the OS4 stability. This was mainly due to interference between the five sonars and the effect of the propellers on the ultrasound waves. Figure 6.25 shows detection of an obstacle with and without the filter. The latter is based on the variation of successive samples and gives a reliable detection signal usable in flight.

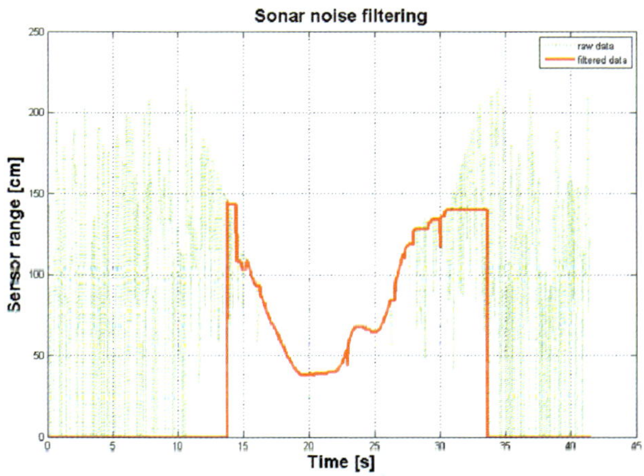

Fig. 6.25. Obstacle detection with and without the filter.

Results

A collision avoidance behavior was practically obtained after numerous tests and tuning. Once the obstacle is detected, a pitch reference is given to fly away the helicopter from the obstacle. Figure 6.26 shows the reaction of OS4 to an obstacle at 40 cm; one can see the distance to the obstacle increasing until the latter disappears; then OS4 recovers a normal flight.

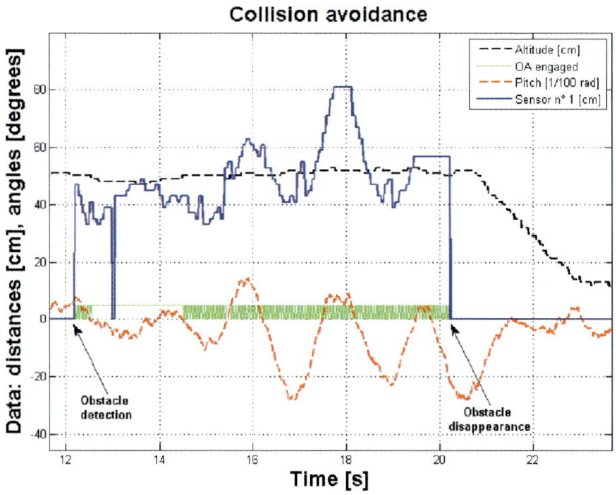

Fig. 6.26. Experiment: Collision avoidance with OS4. The helicopter flies back until the obstacle disappears

6.6 Conclusion and Future Work

This Chapter presented the latest developments of the ASL MFR project. A quadrotor simulation model was introduced. It included major aerodynamic effects modeled with blade element and momentum theory. In addition, the actuator's model was identified and all sensor delays and noise were taken into account. Moreover, a *MATLAB / Simulink* based simulator was developed for testing the controllers. Real experiments were conducted with the same control parameters tuned during simulation. A practical methodology was introduced for miniature rotorcraft design. It was the only tool used to design the OS4 helicopter achieving 100% thrust margin for 30 min autonomy (in the initial design). The last version of OS4 embeds all necessary avionics and energy devices for a fully autonomous flight. This comprises a low cost IMU, a vision based position sensor specifically developed for this project and an obstacle detection setup. The control approach was then presented to design the main controllers: Attitude, altitude, position and the auxiliary ones like take-off and landing, obstacle avoidance and waypoint following. The latter was demonstrated using simulation. Experiments have shown that the OS4 is currently able to take-off, hover, land and avoid collisions automatically thanks to model-based control. Future work is to enhance the propulsion group towards more reliability and high bandwidth, improve the vision sensor in order to get rid of the external pattern and to provide stable vision-based yaw information. It is also possible to reduce the mass of the helicopter by using the tiny single board computers available now. It will be interesting to practically test a waypoint following maneuver with obstacle avoidance capabilities. The OS4 is undoubtedly one of the most integrated quadrotor ever designed. It is the first one practically capable of a collision avoidance maneuver.

References

1. AERO-EPFL, http://aero.epfl.ch/.
2. Bouabdallah S. et al, "PID vs LQ Control Techniques Applied to an Indoor Micro Quadrotor", *Proceedings, IEEE/RSJ International Conference on Intelligent Robots and Systems,* Sendai, Japan, 2004.
3. Bouabdallah S., Siegwart R., "Backstepping and Sliding-Mode Techniques Applied to an Indoor Micro Quadrotor", *Proceedings, IEEE International Conference on Robotics and Automation*, Barcelona, Spain 2005.
4. Bouabdallah S. et al., "Autonomous Miniature Flying Robots: Coming Soon", *IEEE Robotics and Automation Magazine*, to appear.
5. Cheeseman I., Bennett W., "*The Effect of the Ground on a Helicopter Rotor in Forward Flight*", Aeronautical Research Council, No. 3021, 1957.
6. DeMenthon D., Davis L., "Model-Based Object Pose in 25 lines of Code", *International Journal of Computer Vision*, Vol. 15, 2005.
7. Deng X., et al., "Attitude Control for a Micromechanical Flying Insect Including Thorax and Sensor Models", *Proceedings, International Conference on Robotics and Automation*, Teipei, Taiwan, 2003.
8. Fay G., *Derivation of the Aerodynamic Forces for the Mesicopter Simulation*, Stanford University, 2001.
9. Griffiths D., Leishman J., "A Study of Dual-Rotor Interference and Ground Effect Using a Free-Vortex Wake Model", *Proceedings, 58th Annual Forum of the American Helicopter Society*, Montréal, Canada, 2002.
10. Guenard N., et al., "Control Laws for the Teleoperation of an Unmanned Aerial Vehicle Known as an x4-Flyer", *Proceedings, IEEE/RSJ International Conference on Intelligent Robots and Systems*, Beijing, China, 2006.

11. Hoffmann G., et al., "The Stanford Testbed of Autonomous Rotorcraft for Multi Agent Control", *Proceedings, 23rd Digital Avionics Systems Conference*, Salt Lake City, Utah, 2004.
12. Kanellakopoulos I., Krein P., "Integral-Action Nonlinear Control of Induction Motors", *Proceedings, 12th IFAC World Congress*, Sydney, Australia, 1993.
13. Krstic M., et al., *Nonlinear and Adaptive Control Design*, Wiley Interscience, 1995.
14. Kroo I., et al., *The Mesicopter: A Miniature Rotorcraft Concept - Phase 2*, Interim Report, 2000.
15. Leishman J. G., *Principles of Helicopter Aerodynamics*, Cambridge University Press.
16. MARV, http://www.enae.umd.edu/AGRC/Design00/MARV.html.
17. Murray R., et al., *A Mathematical Introduction to Robotic Manipulation*, CRC, Boca Raton, FL 1994.
18. Wang W., et al., "Autonomous Control for Micro-Flying Robot and Small Wireless Helicopter", *Proceedings, IEEE/RSJ International Conference on Intelligent Robots and Systems*, Beijing, China, 2006.
19. Nicoud J. D., Zufferey J. C., "Toward Indoor Flying Robots", *Proceedings, IEEE/RSJ International Conference on Intelligent Robots and Systems,* Lausanne, Switzerland, 2002.
20. OpenCV, http://www.intel.com/research/mrl/research/opencv/.
21. Tan Y., et al., "Advanced Nonlinear Control Strategy for Motion Control Systems", *Proceedings, IEEE Power Electronics and Motion Control Conference*, Beijing, China, 2000.

PART III

Navigation Aspects

Chapter 7: Obstacle and Terrain Avoidance for Miniature Aerial Vehicles[1]

This Chapter presents an approach to miniature aerial vehicle (MAV) obstacle and terrain avoidance that has been developed at Brigham Young University (BYU). The research builds on the notion of utilizing useful but imperfect map information to plan nominal paths through city or mountain terrain. Because maps may be limited in resolution, out of date, or offset in location, MAVs must also utilize sensory information to detect and avoid obstacles unknown to the path planner. In this Chapter, a laser range finder and optic flow sensors are utilized to detect obstacles and terrain. Avoidance algorithms using this sensor information are discussed briefly and flight test results using the BYU MAVs are presented.

7.1 Introduction

Unmanned aerial vehicles (UAVs) are playing increasingly prominent roles in defense programs and strategy around the world. Technology advancements have enabled the development of large UAVs (e.g., Global Hawk, Predator) and the creation of smaller, increasingly capable UAVs. The focus of this Chapter is on smaller fixed-wing miniature aerial vehicles (MAVs), which range in size from ¼ to 2 m in wingspan. As recent conflicts have demonstrated, there are numerous military applications for MAVs including reconnaissance, surveillance, battle damage assessment, and communications relays.

Civil and commercial applications are not as well developed, although potential applications are extremely broad in scope. Possible applications for MAV technology include environmental monitoring (e.g., pollution,

[1] Written by S. Griffiths, J. Saunders, A. Curtis, B. Barber, T. McLain, R. Beard. This work was funded by AFOSR award numbers FA9550-04-1-0209 and FA9550-04-C-0032. This Chapter is the outgrowth of a previously published paper by the authors in the IEEE Robotics and Automation Magazine, Vol. 13, No. 3, 34-43, September 2006.

Kimon P. Valavanis (ed.), Advances in Unmanned Aerial Vehicles, 213–244.

weather, and scientific applications), forest fire monitoring, homeland security, border patrol, drug interdiction, aerial surveillance and mapping, traffic monitoring, precision agriculture, disaster relief, ad-hoc communications networks, and rural search and rescue. For many of these applications to develop to maturity, the reliability of MAVs will need to increase. In addition, their capabilities will need to be expanded and their ease of use will need to be improved. In addition to these technical challenges, the regulatory challenge of integrating UAVs into the national and international air space needs to be overcome.

Critical to the more widespread use of MAVs is making them easy to use by non-pilots including scientists, forest fire fighters, law enforcement officers, or military ground troops. One key capability for facilitating ease of use is the ability to sense and avoid obstacles, both natural and man made. Many potential applications require MAVs to fly at low altitudes in close proximity to structures or terrain. For example, the ability to fly through city canyons and around high-rise buildings is envisioned for future homeland security operations. For MAVs to be effective tools, the challenge of operating in complex environments must be automated, allowing the operator to concentrate on the task at hand.

Performing obstacle and terrain avoidance from a fixed-wing MAV platform is challenging for several reasons. The limited payload and power available on MAV platforms place significant restrictions on the size, weight, and power requirements of potential sensors. Sensors such as scanning LADAR and RADAR are typically too large and heavy for MAVs. Related to limits on sensor payload are those on computing resources. For most MAVs, the primary computational resource is the autopilot microcontroller. Additional computational hardware can be integrated in the MAV, but computers such as PC104-based systems generally exceed the payload capacity and so smaller microcontrollers are typically used.

Another challenge posed by fixed-wing MAVs is that they move fast: ground speeds are often in the range of 10 to 20 m/s (22 to 44 mph). Contrary to the computational limits imposed, obstacle avoidance algorithms must execute and act quickly. Unlike ground robots and unmanned rotorcraft, fixed-wing MAVs cannot stop or slow down while avoidance algorithms process sensor information or plan maneuvers. Reactions must be immediate. Obstacle sensing is further complicated by the fact that sensor readings are altered by changes in aircraft attitude, especially the rolling motions that occur during turns. Attitude changes affect not only the pointing direction of the sensor, but also cause motion of fixed objects in the field-of-view. Obstacle and terrain detection must account for the effects of aircraft attitude changes for avoidance maneuvers to be successful. All of the challenges associated with MAV obstacle and terrain avoidance are compounded by the reality that for MAVs, mistakes are costly or even catastrophic, as crashes can result in damage to or loss of the MAV and failure to complete the objectives of the flight.

As evidenced by the recent DARPA Grand Challenge, capable obstacle avoidance and terrain navigation systems have been developed for ground vehicles. Obstacle avoidance and path planning have been active areas of research for many years and the associated robotics literature is immense. While providing a guiding influence, most of the proposed methods fail to deal with the sensing and computational challenges imposed by the limited payload capabilities of MAVs.

As autonomous MAVs and feasible obstacle sensors are recent technological developments, the body of experimental research directed specifically toward MAV obstacle and terrain avoidance is small. Related to terrain avoidance is work focused on utilizing vision processing techniques to estimate height above ground. In [1], it was demonstrated that mimicking the landing behavior of bees by maintaining constant optic flow during a landing maneuver, could be used to successfully control the descent of a MAV. Development of lightweight sensors for measurement of optic flow has enabled their use in MAVs [2] [3] [4]. In [5], it was demonstrated that these sensors can be used to follow undulations in terrain with low-flying MAVs.

The next Section describes the BYU aerial platforms.

7.2 BYU Miniature Aerial Vehicle Platforms

Over the past five years, BYU has been involved in the development of MAV airframes, autopilots, user interfaces, sensors, and control algorithms. This Section describes the experimental platform developed specifically for the obstacle avoidance research described next.

7.2.1 Airframe

Figure 7.1 shows the airframe used for obstacle avoidance experiments. The airframe has a 1.5 m wingspan and has been constructed with an EPP foam core covered with Kevlar. This design was selected for its durability, useable payload, ease of component installation, and flight characteristics. The airframe can carry a 0.4 kg payload and can remain in flight for over 45 minutes at a time. The collision avoidance sensors that are embedded in the airframe include three optic-flow sensors, one laser ranger, and two electro-optical cameras as shown in Figure 7.2. Additional payload includes the Kestrel autopilot, batteries, a 1000 mW, 900 MHz radio modem, a 12-channel GPS receiver, and a video transmitter.

Fig. 7.1. Airframe used for collision avoidance experiments.

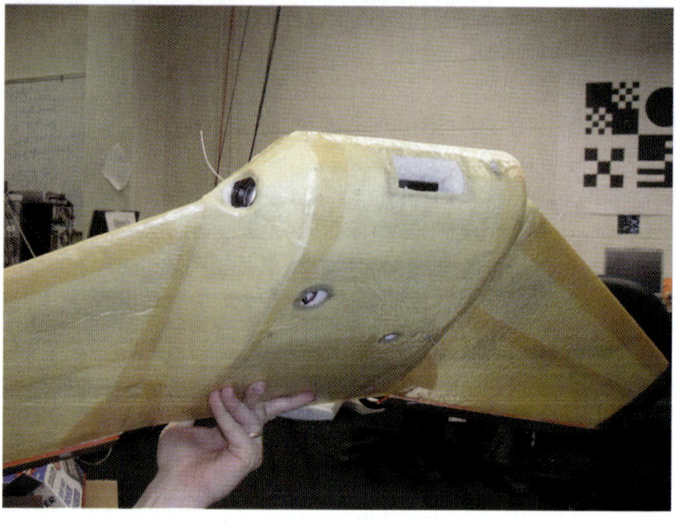

Fig. 7.2. Sensors used for collision avoidance. The round hole on the right and the large hole on the belly are the optic flow sensors. The square hole in the center is the laser ranger, and the other two round holes are for electro-optical cameras.

7.2.2 Kestrel Autopilot

The collision avoidance algorithms have been implemented on Procerus Technologies' Kestrel Autopilot version 2.2 [6], which is shown in Figure 7.3. The autopilot is equipped with a Rabbit 3400 29 MHz processor, three-axis rate gyros, three-axis accelerometers, absolute and differential pressure sensors, and a variety of interface ports. The autopilot measures $3.8 \times 5.1 \times 1.9$ cm and weighs 18 grams. The autopilot also serves as a data acquisition device and is able to log 175 kbytes of user-selectable telemetry at rates up to 60 Hz. The optic flow sensors and the laser ranger are connected directly to the autopilot and the collision avoidance algorithms are executed on-board the Rabbit processor.

Fig. 7.3. The Kestrel autopilot version 2.2. Figure courtesy of Procerus Technologies.

7.2.3 Ground Station

There are two key components of the ground station. The first is the virtual cockpit software that runs on a laptop. A screen shot of the virtual cockpit is shown in Figure 7.4. The virtual cockpit allows the user to configure every parameter on the autopilot during flight. In particular, autopilot and sensor gains can be configured in-flight. In addition, the virtual cockpit was designed to allow TCP/IP connection to other software modules. The TCP/IP interface allows external programs to read telemetry data from the autopilot, and to access internal autopilot parameters. This architecture facilitates a variety of user interfaces. In particular, PDA and voice interfaces have been implemented using the virtual cockpit [18] [19].

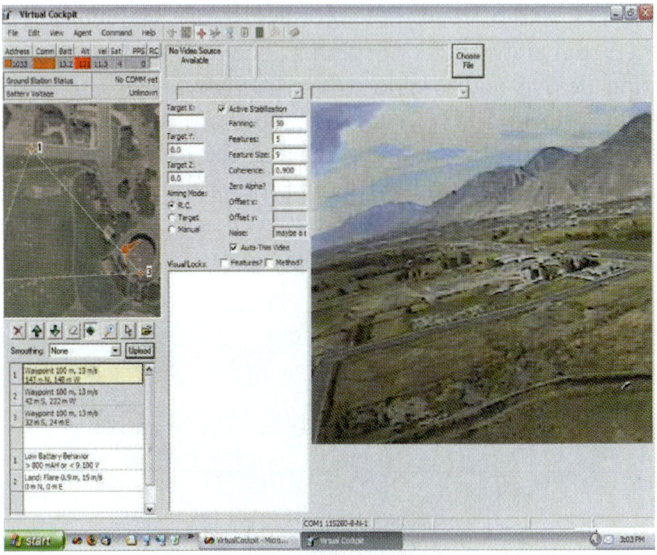

Fig. 7.4. Screen shot of the virtual cockpit used with the Kestrel autopilot. The window on the left is the mission planning window that allows waypoints to be dynamically specified. The window on the right displays the video stream from the MAV.

The second key component to the ground station is the communication box shown in Figure 7.5. The communication box contains a 900 MHz transmitter, a GPS unit, and an interface to an RC transmitter that can be used to maneuver the airframe manually. In addition to standard telemetry the video feed from the cameras is connected to an Imperx VCE-PRO PCMCIA frame grabber [20] hosted on the laptop. The frame grabber provides 640 X480 RGB images at 30 frames-per-second. The image can be displayed in the virtual cockpit and processed for image in the loop applications.

Fig. 7.5. Communication box connected to the laptop. The communication box contains a 900 MHz transceiver, a GPS unit, and an interface to an RC transmitter which can be used as a stand-by fail-safe mechanism to facilitate safe operations. Figure courtesy of Procerus Technologies.

7.2.4 Optic Flow Sensors

The MAV is equipped with three optic-flow sensors. Two of the optic-flow sensors are forward looking but swept back from the nose by $\alpha = 60$ degrees. The third optic flow sensor points down to determine the height above ground. The optic-flow sensors, shown in Figure 7.6, are constructed by attaching a lens to an Agilent ADNS-2610 optical mouse sensor. The ADNS-2610 has a small form factor, measuring only 10 mm by 12.5 mm and runs at 1500 frames per second. It requires a light intensity of at least 80 mW/m^2 at a wavelength of 639 nm or 100 mW/m^2 at a wavelength of 875 nm. The ADNS-2610 measures the flow of features across an 18 by 18 pixel CMOS imager. It outputs two values, δp_x and δp_y, representing the total optic flow across the sensor's field-of-view in both the x and y directions. The flow data in the camera y direction corresponds to lateral motion of the MAV and is ignored.

Fig. 7.6. Optic flow sensors with three different lens configurations: 1.2, 2.5, and 6.5 degree field-of-view. The optic flow sensors are constructed by attaching a lens to an optical mouse chip.

Figure 7.7 indicates how distance is computed using the optic flow sensor. The optical mouse chip outputs an optic flow displacement $(\delta p_x, \delta p_y)^T$ at its internal sample rate of 1500 Hz. Since the collision avoidance loop is executed at $T_s = 20$ Hz, the total optical displacement is integrated over T_s to produce $(\Delta p_x, \Delta p_y)$. The distance to the object D is related to the measured distance d by the expression:

$$D = d\cos\phi\sin\alpha, \tag{7.1}$$

where ϕ is the roll angle of the MAV. From geometry, the measured distance to the object is given by:

$$d = \frac{V_{\text{gps}} T_s}{\tan\left(\frac{\lambda_{\text{eff}}}{2}\right)}, \tag{7.2}$$

where λ_{eff} is the effective field-of-view. The effective field-of-view is given by:

$$\lambda_{\text{eff}} = \lambda_{\text{cam}} \frac{\Delta p_x}{P_x} - \dot{\chi} T_s, \tag{7.3}$$

where λ_{cam} is the field-of-view of the camera, P_x is the size of the pixel array along the direction of motion, and $\dot{\chi}$ is the yaw rate with respect to the ground. Using similar reasoning for left-looking and down-looking optic flow sensors we can derive the following expression for the right, left, and AGL distances:

$$D_{\text{right}} = \frac{V_{\text{gps}} T_s}{\tan\left(\frac{\lambda_{\text{cam}} D_{\text{right}} p_x}{2P_x} - \frac{\dot{\chi} T_s}{2}\right)} \cos\phi \sin\alpha$$

$$D_{\text{left}} = \frac{V_{\text{gps}} T_s}{\tan\left(\frac{\lambda_{\text{cam}} D_{\text{left}} p_x}{2P_x} + \frac{\dot{\chi} T_s}{2}\right)} \cos\phi \sin\alpha \qquad (7.4)$$

$$D_{\text{down}} = \frac{V_{\text{gps}} T_s}{\tan\left(\frac{\lambda_{\text{cam}} D_{\text{down}} p_x}{2P_x} - \frac{\dot{\theta} T_s}{2}\right)} \cos\theta \cos\phi.$$

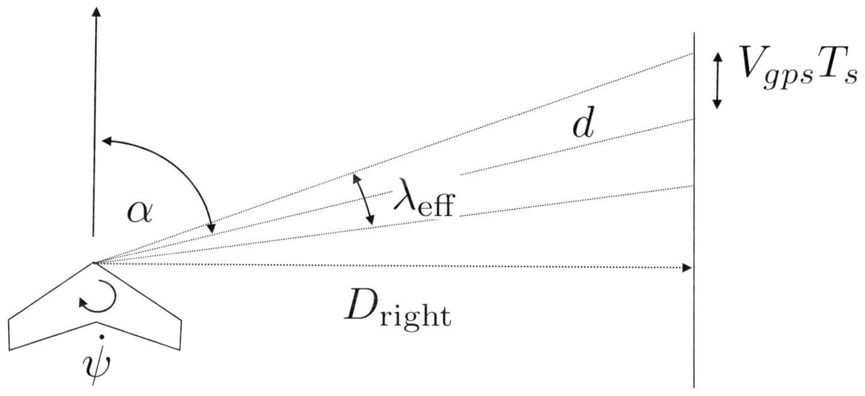

Fig. 7.7. The optic flow sensor is used to compute the distance to an obstacle based on the distance traveled between samples ($V_{\text{gps}} T_s$) and the effective field-of-view λ.

7.2.5 Laser Ranger

For the experiments discussed in this Chapter, the Opti-Logic RS400 Laser rangefinder shown in Figure 7.8 is used. The laser has a sensing range of 400 m with an update rate of 3 Hz. It weights 170 grams and consumes 1.8 W of power. Figure 7.2 shows the laser ranger mounted on the airframe. It is important to note that the RS400 is not a scanning laser rangefinder. Scanning laser range finders are currently too heavy and consume too much power for MAV applications. The RS400 returns a single distance measurement and must be steered by maneuvering the airframe.

Fig. 7.8. The Opti-Logic RS400 laser range finder. The laser range finder was mounted as a strapdown unit on the airframe shown in Figures 7.1 and 7.2.

7.2.6 Signal Flow Diagram

The signal flow diagram for the system is shown in Figure 7.9. The Kestrel autopilot processes GPS, accelerometers, rate gyroscopes, and pressure sensors at a rate of approximately 80 Hz. The laser ranger is processed at a rate of about 3 Hz, and the optic flow sensors are processed at a rate of about 80 Hz. The autopilot implements standard autopilot functions, as well as the reactive path planner and the vector field guidance algorithm. Telemetry data is transmitted to the ground station via a 900 MHz digital transceiver. In addition, an on-board camera transmits video signals to the ground via a 2.4 GHz analog transmitter. The deliberative path planner described in the next section is implemented on the ground station.

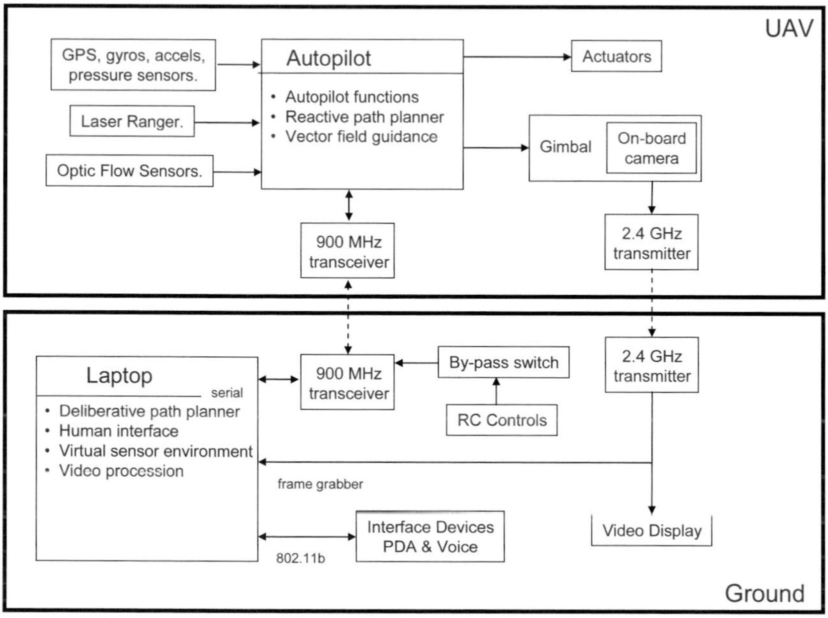

Fig. 7.9. The signal flow diagram for collision avoidance. The collision detection sensors are processed on-board the MAV and provide feedback to the reactive path planner.

7.3 Path Planning and Following

The first step for navigating through complex environments is to plan a nominal path based on known information about the environment, usually in the form of a street map or a topographic map. The MAV must be able to accurately follow the nominal path to avoid known obstacles. This section discusses the methods for planning and following the nominal path. Subsequent sections will discuss reactive, sensor-based obstacle avoidance strategies for maneuvering around obstacles that are unknown during the planning process.

7.3.1 Planning the Nominal Path

When planning paths through complex environments, the computational requirements for finding an optimal path can be significant and unrealistic for near-real-time execution [7]. Because of this, recent research has focused on randomized techniques to quickly find acceptable, though not necessarily optimal, paths [8] [9]. Path planning for MAVs is also difficult because of the dynamic constraints of flight. Many common path planning algorithms are inadequate for fixed-wing MAV systems because they do not handle turn-radius limitations and airspeed constraints effectively.

One randomized method that addresses these limitations is the Rapidly-exploring Random Tree (RRT) algorithm [7] [10]. RRTs use a dynamic model of the system to build a tree of traversable paths. The search space is quickly explored by applying control inputs to configurations already in the tree. Working with the precise control inputs ensures that the dynamic constraints are not violated; however, it also results in an open-loop solution. This would be adequate if we had a perfect model of the system without disturbances, but this method is not satisfactory for an actual MAV because of model inaccuracies and environmental disturbances such as wind.

The basic idea of the Rapidly Exploring Random Tree (RRT) algorithm is to build a tree that uniformly explores the search space. The uniformity is achieved by randomly sampling from a uniform probability distribution. Figure 7.10 gives a graphical depiction the basic RRT algorithm. As shown in Figure 7.10 (a), the input to the RRT algorithm is a start waypoint w_{start}, an end waypoint w_{end}, and the terrain map. The first step of the algorithm is to randomly select a point p in the workspace. As shown in Figure 7.10 (b) a new waypoint w_1 selected a fixed distance D from w_{start} along the line $\overline{pw_{start}}$, and inserted into the tree. At each subsequent step, a random point p is generated in the workspace, and the tree is searched to find the node that is closest to p. As shown in Figure 7.10 (c) a new waypoint is generated that is a distance D from the closest node in the tree, along the line connecting p to the closest node. Before a waypoint path is added to the tree, it needs to be checked for collisions with the terrain. If a collision is detected, as shown in Figure 7.10 (d), then the waypoint is deleted and the process is repeated. When a new node is added, its distance from the end node w_{end} is checked. If it is less than D, then a waypoint path from w_{end} is added to the tree, and as shown in Fig. 7.10 (f), a complete path through the terrain has been found.

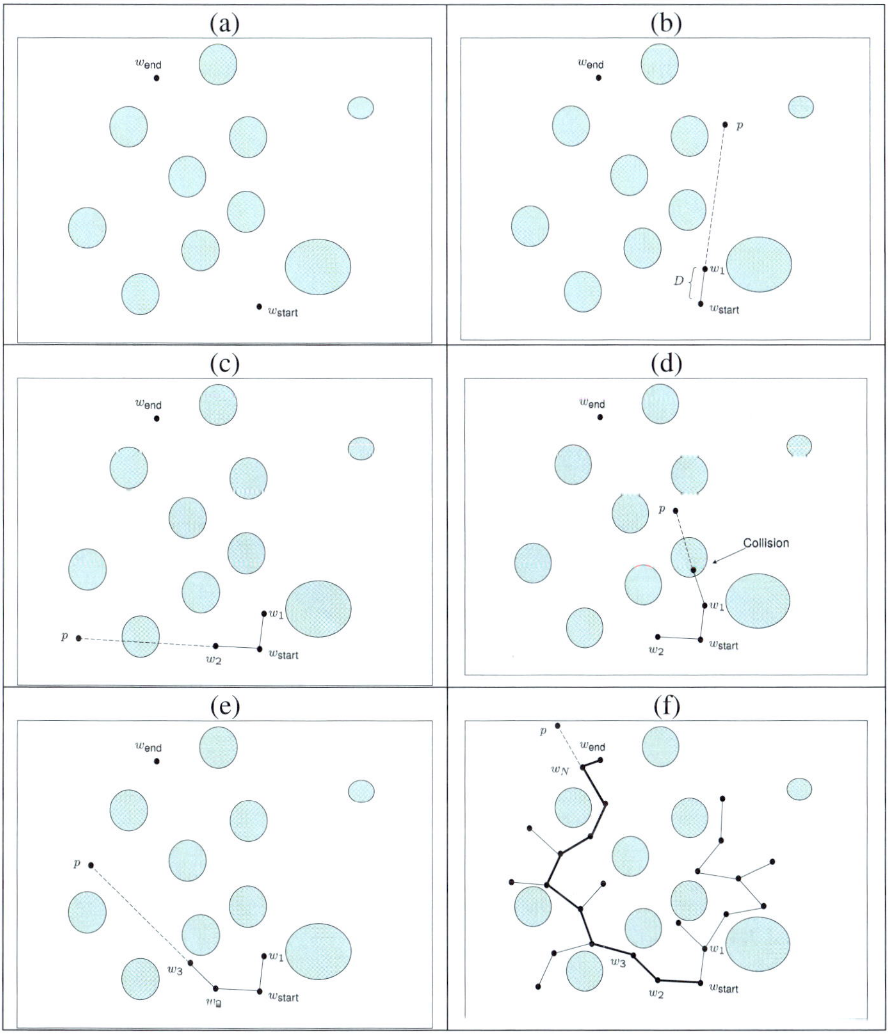

Fig. 7.10. This figure gives a graphical description of the basic RRT algorithm.

The basic algorithm described above is similar to the one presented in [11] in that paths are planned in the output space. The method is easily extended to three dimensions and provides a computationally efficient a priori path planner for the MAVs [12]. Since the MAV will not exactly follow the waypoint path, it is essential to bound the tracking error of the controlled MAV from the waypoint path. For a given waypoint path, one can determine the expected trajectory of the MAV [13] and ensure that only traversable paths are built into the search tree. Branches in the tree are checked to ensure that they pass tests on turn radius and climb rate, and are collision-free. Figure 7.11 depicts the growth of a 3-D RRT path through a simulated urban environment, while real-time results through a canyon are presented in Section 7.5 (Figure 7.27).

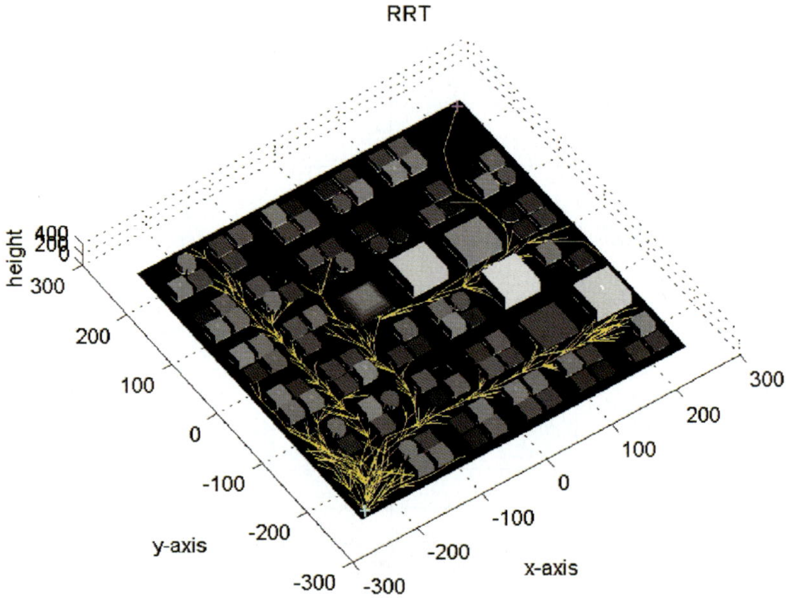

Fig. 7.11. This figure shows the growth of an RRT path tree through a simulated urban environment. The algorithm is terminated once a feasible path to the destination (red X) is found.

This algorithm is used to find nominal paths through different types of terrain. Planned paths through a simulated urban terrain and a canyon represented by USGS terrain data are shown in Figures 7.12 and 7.13, respectively.

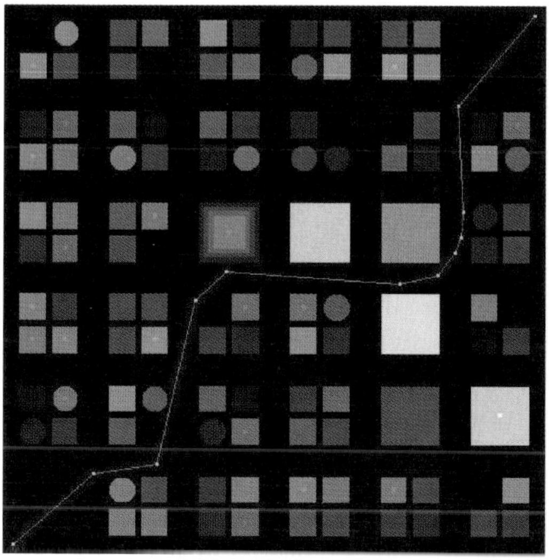

Fig. 7.12. This figure shows the result of the RRT algorithm applied to simulated urban terrain. The minimum turning radius of the MAV is 50 m, and the average street width is 30 m.

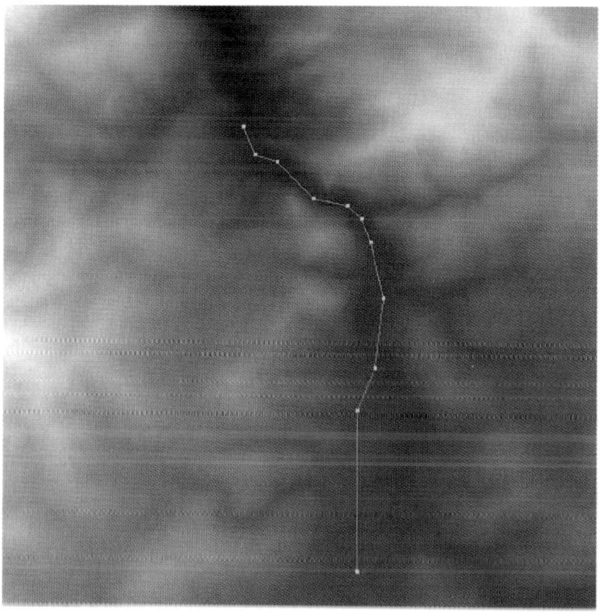

Fig. 7.13. This figure shows the result of the RRT algorithm applied to canyon data downloaded from a publicly available USGS database.

7.3.2. Vector Field Path Following

Given a nominal waypoint path, it is essential for the MAV to have the ability to track the path with precision. MAVs must track these paths despite dynamic limitations, imprecise sensors and controls, and wind disturbances, which are often 20% - 60% of airspeed [14]. Trajectory tracking, which requires the MAV to be at a specific location at a specific time, is difficult in such wind conditions. As an alternative, a path following approach is developed where the focus is simply to be on the path, instead of at a specific point that evolves in time. Similar research in [15] describes a maneuvering method focused on converging to the path then matching a desired speed along the path. The path following method is based on the creation of course vector fields that direct the MAV onto the desired path.

The vector field method produces a field of desired course commands that drive the MAV toward the current path segment. At any point in space, the desired course can be easily calculated. This desired course is used to command heading and roll control loops to guide the MAV onto the desired path. The vector field method uses only the current path segment to find the desired course, avoiding possible singularities and sinks resulting from sums of vectors. Many paths planned for MAVs can be approximated by combinations of straight-line segments and circular arcs [16]. Figure 7.14 shows examples of vector fields for linear and circular paths.

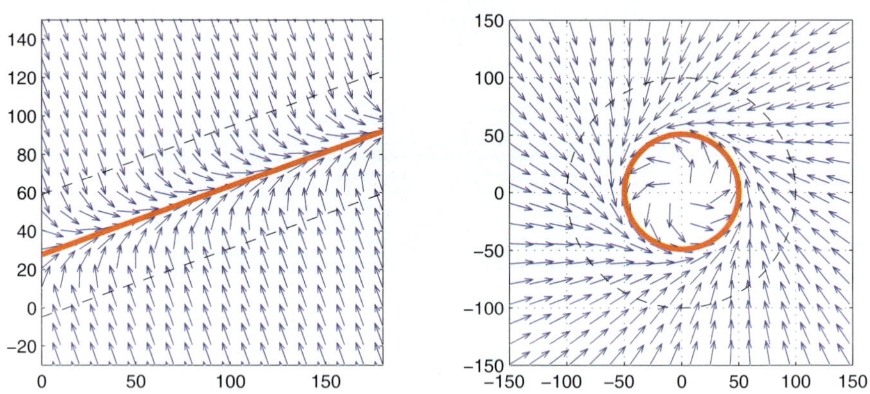

Fig. 7.14. Path following in wind is accomplished by creating a vector field of desired course commands based on the lateral deviation from the path. The Figure on the left shows a possible vector field for a straight-line waypoint path segment. The Figure on the right shows a possible vector field for orbit following.

To account for wind, the course and groundspeed instead of heading and airspeed are used to control the MAV. Ground track motion is the vector sum of the MAV motion relative to the surrounding air mass and the motion of the air mass relative to the ground. Since course direction includes the effects of wind, control based on course is much more effective at rejecting wind disturbances. In implementing the vector field approach, course measurements from GPS are compared with the desired course from the vector field to determine the appropriate control inputs to keep the MAV on the path.

For a given path, the vector field is divided into a transition region and an outer region. This is similar in some respects to the belt zone technique developed in [17]. Outside the transition region, the vector field drives the MAV toward the transition region along a constant course. Once inside, the vector field changes linearly from the entry course direction to the desired course along the path. The effect is to smoothly drive the MAV to follow the path, with larger effort as the error from the path increases. In [14] it is shown that for any initial condition, the MAV will enter the transition region in finite time, and then converge to the desired course asymptotically.

Flight tests have demonstrated the effectiveness of the vector field path following method, even in windy conditions. Figure 7.15 demonstrates path following for straight line segments with acute angles. Wind speeds were approximately 20% of the airspeed during these tests. The vector field method has been shown to be effective in tracking paths of lines and orbits with wind speeds of up to 50% of the airspeed of the MAV. Figure 7.16 shows flight test data for circular orbits and synthetic urban terrain.

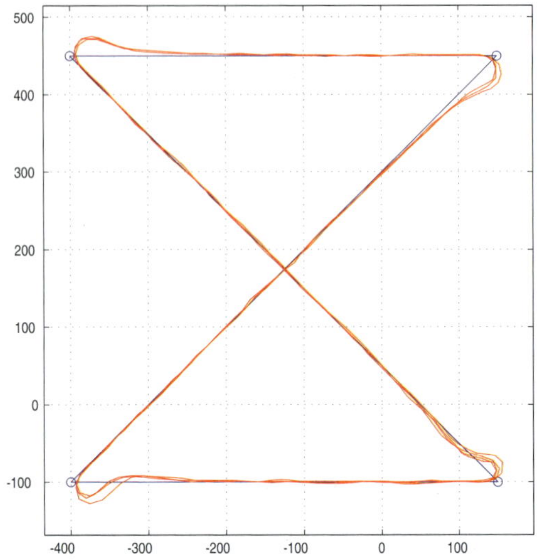

Fig. 7.15. This Figure shows telemetry data for four consecutive traversals of a waypoint path. Wind speeds during the flight were 20% of the MAV airspeed. Note the repeatability of the trajectories even in significant wind.

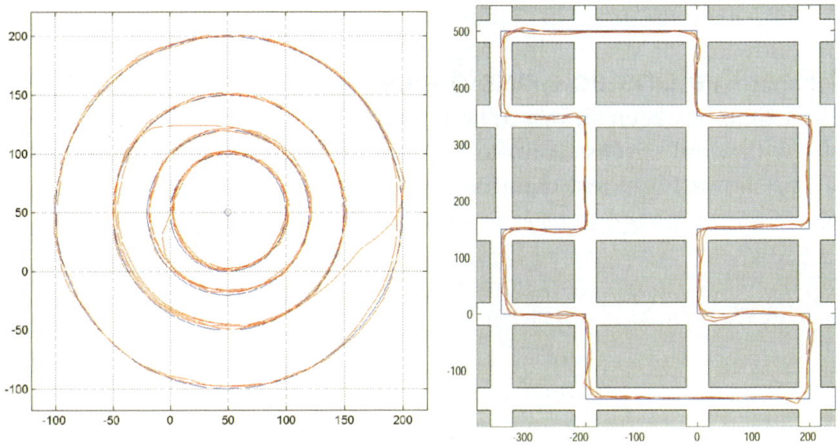

Fig. 7.16. The Figure on the right shows telemetry data for multiple traversals of orbits with radius ranging from 100-300 meters. Wind speeds during the flight were 20% of the MAV airspeed. The Figure on the right shows telemetry data for multiple flights through synthetic urban terrain. While the terrain is synthetic, the data represents actual flight tests in winds that were approximately 30% of the MAV airspeed.

7.3.3 Reactive Obstacle and Terrain Avoidance

Despite having an effective a priori path planner, one cannot guarantee that the flight path will be free of obstacles. This path planner assumes a perfect model of the terrain, but this assumption is not realistic. If an urban terrain model is missing a newly constructed building or a large antenna or tree, a path leading to a collision could result. In fact, the used canyon models are based on 10 m USGS data, which cannot represent small obstacles like trees and power lines. In addition, the GPS sensor used on the MAV has a constant bias that can be as large as 10 m. Path planners can produce a nominal path prior to flight, but the MAV must also have the ability to sense and reactively avoid unanticipated obstacles and terrain in real time.

The following Sections present reactive planners for producing deviations from a nominal path to enable obstacle and terrain avoidance. Section 7.4 presents a method for sensing and avoiding obstacles directly in the flight path and shows results for reactive avoidance of a building. Section 7.5 presents an approach for staying centered between obstacles as might be required for flying through a corridor. Flight test results are presented that demonstrate autonomous navigation of a winding canyon.

7.4 Reactive Obstacle Avoidance

Reactive obstacle avoidance from a MAV platform is challenging because of the size and weight limitations for sensing and computation hardware imposed by the platform. The speed with which avoidance decisions must be made and carried out also causes difficulties. For obstacle avoidance in urban environments, a heuristic algorithm is developed that utilizes a laser ranger to detect and avoid obstacles. The laser ranger points directly out the front of the MAV, and returns range data for objects directly in front of the MAV with a 3 Hz update. For preliminary flight tests, a simple scenario was considered: a single unknown obstacle placed directly in the flight path.

7.4.1 Algorithm

Consider the scenario shown in Figure 7.17 where obstacle avoidance is required. The MAV has a forward ground speed of V and a minimum turn radius R and it is assumed to be tracking the given waypoint path at the time the obstacle is detected by the laser, which has a look ahead distance L. Figure 7.17 (a) shows the instant when the obstacle is detected by the laser ranger. The basic idea is to construct an internal map of obstacles detected by the laser and to modify the waypoint path to maneuver around the obstacles in the internal map. Refer to the internal representation of obstacles as 'map obstacles'. When the laser detects the location of an obstacle, there is uncertainty about the size and height of the obstacle. In this Chapter map obstacles are represented as cylinders with radius R equal to the minimum turn radius of the MAV, and height equal to the current altitude of the MAV. As shown in Figure 7.17 (b), there are two alternate waypoint paths that maneuver around the map obstacle. The endpoints of the waypoint paths are selected so that the new waypoint paths are tangent to the obstacles in the internal map. As shown in Figure 7.18, the new waypoints are located at a distance $Rd/\sqrt{d^2 - R^2}$ from the original waypoint path, where d is the turn away distance from the obstacle. If both waypoint paths are collision free, then the algorithm randomly selects between the two paths as shown in Figure 7.17 (c). Since the map obstacle may be smaller than the actual obstacle, the laser may again detect the obstacle as it maneuvers on the modified path. If that is the case, a new map obstacle is added to the internal map as shown in Figure 7.17 (d). This process is repeated until the MAV maneuvers around the obstacle as shown in Figure 7.17 (e) and 7.17 (f).

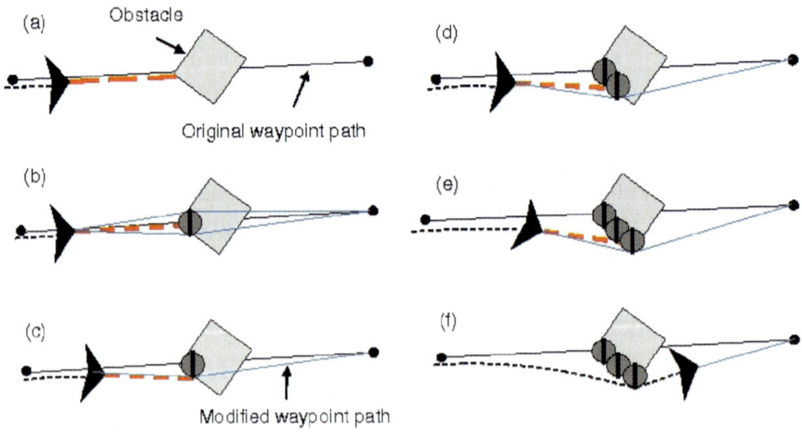

Fig. 7.17. Obstacle avoidance algorithm. (a) The laser detects the obstacle. (b) A map obstacle of radius R is inserted into the map, and two candidate waypoint paths are constructed. (c) A modified waypoint path is randomly selected. (d) The obstacle is again detected by the laser and another map obstacle is constructed. (e-f) The process repeats until the MAV is able to maneuver around the obstacle.

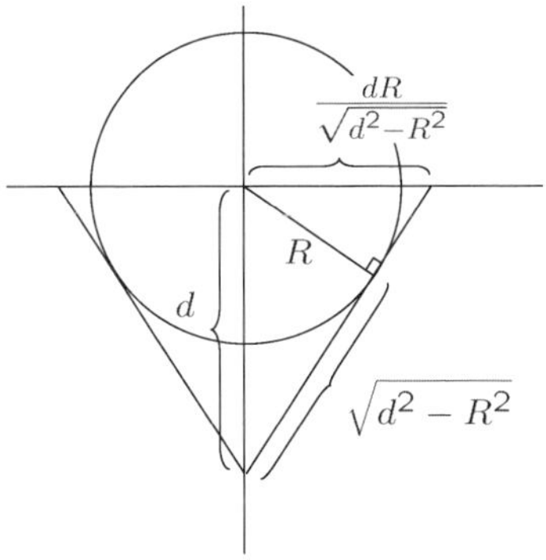

Fig. 7.18. The waypoint path is constructed so that it is perpendicular to the map obstacle. The radius R ensures collision free passage around the map obstacle.

Assuming zero wind, then the 2-D navigation for the MAV is given by:

$$\dot{n} = V \cos \chi \tag{7.5}$$
$$\dot{e} = V \sin \chi$$
$$\dot{\chi} = \frac{g}{V} \tan \phi,$$

where g is the gravitational constant, and ϕ is the roll angle of the MAV.

On most MAVs, the roll angle is limited between $-\bar{\phi} \le \phi \le \bar{\phi}$. Considering that the roll dynamics of the MAV are sufficiently fast to assume near instantaneous transitions between $\pm\bar{\phi}$. Therefore, the minimum turn radius is given by $R = \frac{V^2}{g \tan \bar{\phi}}$.

It is essential to establish a minimum turn away distance D so that there is a guarantee to avoid collision with a single rectangular obstacle. The first step is to determine the bounds on the forward and lateral motion of the MAV when it transitions from one waypoint path to the next.

Proposition 7.1: After the insertion of a map obstacle, the MAV requires at most a forward distance of $\frac{2}{\sqrt{3}} R$ and a lateral distance of $\sqrt{\frac{2}{3}} R$ to transition onto the new waypoint path while avoiding the map obstacle.

Proof: Assuming the ability to roll instantaneously between $\pm\bar{\phi}$, the motion of the MAV during the transition can be constrained to lie on circles of radius R. As shown in [13], the path length of the transition increases monotonically with the angle between the old and new waypoint paths. Therefore, the forward and lateral distances are maximized when the angular separation is maximized, which occurs when instantaneous motion of the MAV follows a circle of radius R that just touches the map obstacle, as shown in Figure 7.19. The proposition follows directly from standard geometrical arguments. Note that the maximum angular separation is therefore given by $\theta = \tan^{-1}\frac{1}{\sqrt{2}} \approx 36^{\circ}$.

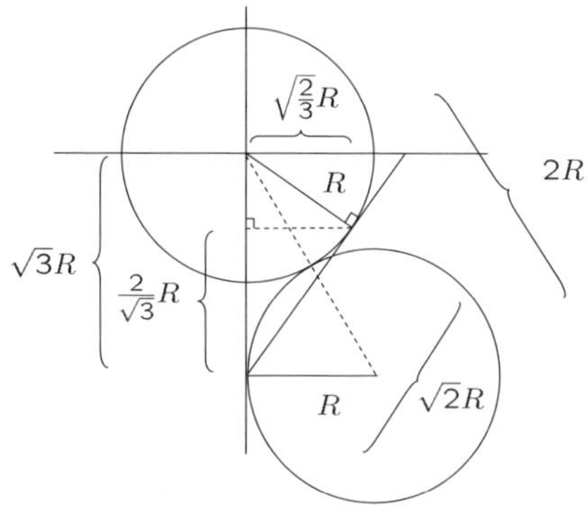

Fig. 7.19. The maximum heading change in waypoint paths is when the MAV must make a full bank to maneuver around the obstacle.

***Proposition** 7.2*: Avoidance of a collision with a flat wall is guaranteed if the turn away distance D satisfies:

$$D > \left(\frac{8 + 2\sqrt{6}}{2\sqrt{3}} \right) R. \tag{7.6}$$

Proof: Consider the worst-case scenario, shown in Figure 7.20, of a MAV that is initially traveling perpendicular to a flat wall. The MAV detects an obstacle and inserts a waypoint at maximum angle $\tan^{-1}\frac{1}{\sqrt{2}}$. After aligning its heading with the waypoint path, the wall is again detected, a map obstacle is inserted, and a new waypoint with maximum angle $\tan^{-1}\frac{1}{\sqrt{2}}$ is planned. This scenario will repeat itself at most three times since $3\tan^{-1}\frac{1}{\sqrt{2}} > \frac{\pi}{2}$. Therefore, the maximum forward direction is bounded by:

$$\sqrt{2}R\left(\left(\sqrt{\tfrac{2}{3}}\right)^{1} + \left(\sqrt{\tfrac{2}{3}}\right)^{2} + \left(\sqrt{\tfrac{2}{3}}\right)^{3} \right) = \left(\tfrac{8+2\sqrt{6}}{2\sqrt{3}}\right)R. \tag{7.7}$$

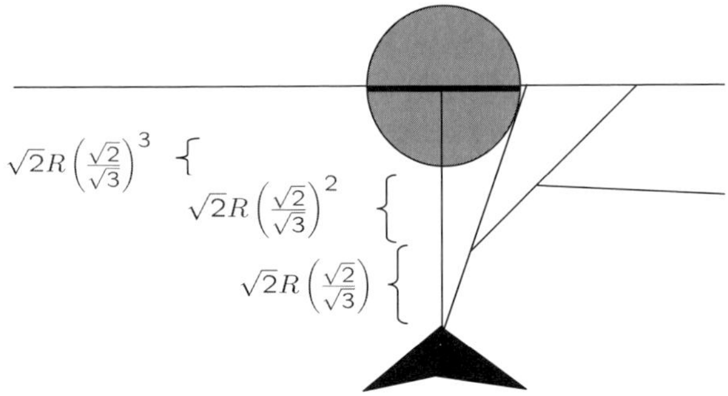

Fig. 7.20. An approximation of the minimum distance required to avoid a straight wall if the laser is only sampled when the MAV is on the waypoint path.

The algorithm described above requires that the laser detect points on the obstacle that are outside of the map obstacles as soon as they become visible. Is this feasible given the update rate of the laser? Let T_s be the time between laser updates.

Proposition 7.3: The maximum distance between laser updates at a range of $d \le L$ is given by:

$$f(d) = 2\sqrt{R^2 + d^2}\,\sin\left(\frac{VT_s}{2R}\right) \tag{7.8}$$

Proof: Assuming the vehicle is turning at its maximum rate, the change in heading between updates is $\frac{VT_s}{R}$. Utilizing the geometry depicted in Figure 7.21, the calculation of $f(d)$ is straightforward. To ensure overlap of map obstacles between samples we require that $f(D) < R$ which implies that:

$$T_s < \frac{2R}{V}\sin^{-1}\left(\frac{R}{2\sqrt{R^2 + D^2}}\right). \tag{7.9}$$

For the BYU airframes, typical values are $V = 13$ m/s, $R = 25$ m, which implies that $D = 93$ m and $T_s < 0.5$ s. The laser ranger sample period of 0.33 s satisfies this constraint, thus ensuring that map obstacles overlap between samples.

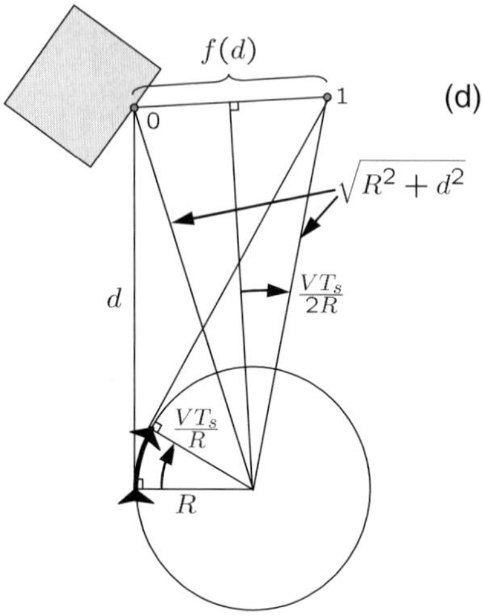

Fig. 7.21. The geometry used to calculate the distance between two consecutive laser updates.

7.4.2 Results

For initial testing of the reactive avoidance algorithm, the choice was to deal with a single obstacle only. It was important that the obstacle be tall enough to allow the MAV to fly at a safe altitude. Flying at an altitude of 40 m also prevented the laser ranger from detecting points on the ground that might be mistakenly interpreted as obstacles, and allowed for losses of altitude that can occur during aggressive maneuvers.

For flight tests, the tallest building on the BYU campus (the Kimball Tower) was used, which is 50 m high and 35 m square. An in-flight image of the building is shown in Figure 7.22. The surrounding buildings are only about 20 m in height. The MAV was directed to fly at 40 m altitude from the south side of the building to the north along a waypoint path that passed directly through the building. Information about the location or the dimensions of the building was not provided to the MAV. A GPS telemetry plot of the results is shown in Figure 7.23.

Fig. 7.22. In-flight image of the Kimball Tower on BYU campus during the collision avoidance maneuver using the laser ranger.

As the MAV approached the building, the laser ranger detected the building and calculated its position. When the MAV came within 93 m of the building, the reactive planner generated a path around the building and the MAV began to track the path. Notice that as the MAV began to pass the building, it turned towards the original waypoint path and detected the building a second time. This caused the MAV to execute a second avoidance maneuver before rejoining the original waypoint path. The MAV successfully avoided the building without human intervention. Figure 7.22 shows images of the MAV and its camera view as it executed the avoidance maneuver.

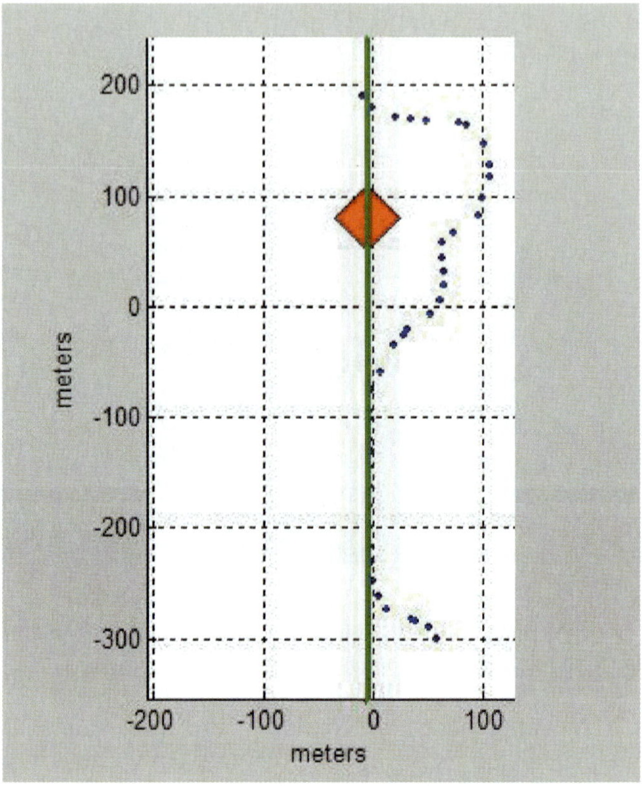

Fig. 7.23. Flight results for collision avoidance using a laser ranger. The green line indicates the planned waypoint path, and the dotted line indicates the GPS track of the MAV.

7.5 Remote Environment Terrain Avoidance

As small MAVs become more reliable and maneuverable, their missions will involve navigating through complex terrain, such as mountainous canyons and urban environments. In this Section, the focus is on terrain avoidance for flying in corridors and canyons. The developed algorithms enable the MAV to center itself within a corridor or canyon, or to fly near walls with a specified offset. The algorithms utilize optic flow sensors like those shown in Figure 7.7. To validate the algorithms, canyon navigation flight experiments were carried out in a mountain canyon.

7.5.1 Canyon Navigation Algorithm

The first step in navigating through a canyon or urban corridor is to select a suitable path through the terrain. This can be done using the RRT algorithm discussed earlier or the operator can utilize maps to define waypoints for the MAV to follow. Preplanned paths will rarely be perfect and some paths could lead the MAV near or even into uncharted obstacles. Reasons for this include inaccurate or biased terrain data, GPS error, and the existence of obstacles that have been added since the terrain was mapped. Therefore, it is important that the MAV be able to make adjustments to its path to center itself between walls and other potential hazards.

In this approach, the MAV follows its preplanned path using the vector field following method. At each time step along the path the MAV computes its lateral distance from objects to the left and right using the optic flow ranging sensors. Using this information, the MAV computes an offset δ from its planned path:

$$\delta = \frac{1}{2}(D_{right} - D_{left}), \qquad (7.10)$$

where D_{left} and D_{right} are distances to the walls on the left and right measured by the optic flow sensors. Shifting the desired path by this offset centers the desired path between the detected walls as shown in Figure 7.24. As Figure 7.25 illustrates, shifting the desired path also shifts the vector field accordingly. To improve the performance of this method the optic ranging sensors are pointed forward at a 30 degree angle. This reduces lag caused by filtering the sensor readings and allows the MAV to detect obstacles ahead of its current position.

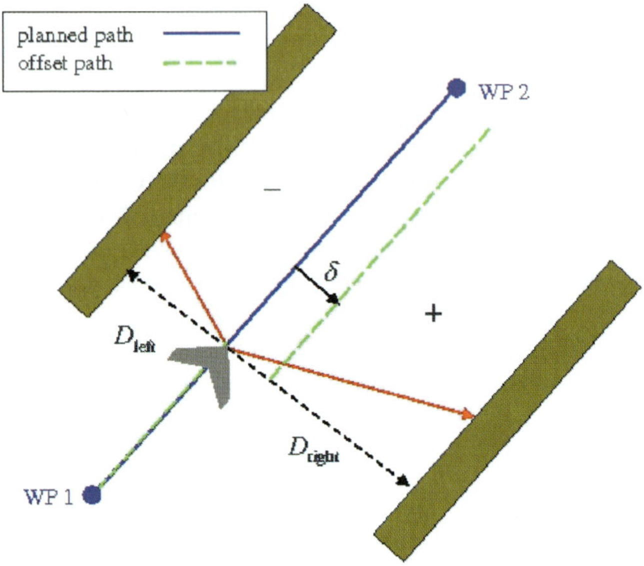

Fig. 7.24. Using the measurements from the optic flow sensors, the planned path (solid blue) is shifted by δ to create a new desired path (dashed green) that is centered between the canyon walls.

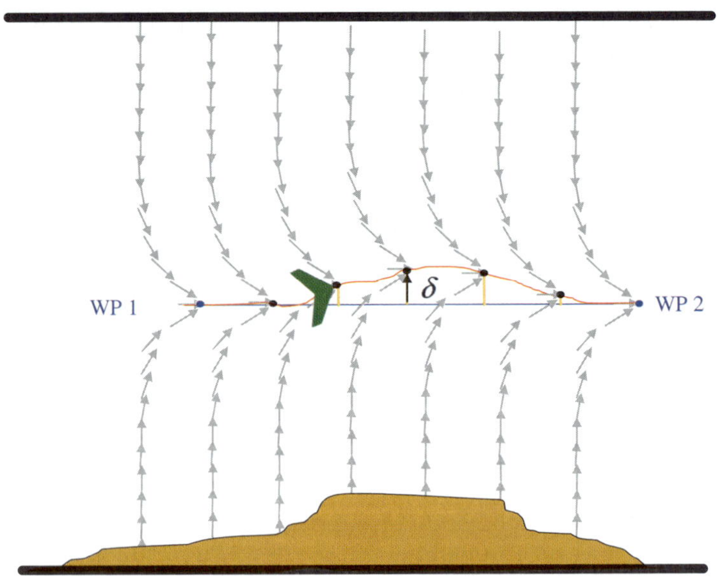

Fig. 7.25. The adjusted path (red) is offset from the preplanned path (blue) by the calculated offset (δ) at each time step to center the desired path between the canyon walls, thus shifting the vector field along with it.

7.5.2 Flight Test Results

Goshen Canyon in central Utah was chosen as a flight test site. This canyon was selected for its steep winding canyon walls that reach over 75 m in height, as well as its proximity to BYU and low utilization by the general population. Flight tests through Goshen Canyon were conducted using the fixed-wing MAV discussed in Section 7.2. Photographs of the flight tests taken by observers and the onboard camera are shown in Figure 7.26. In the first flight through the canyon, the planned path was selected to follow the road. The MAV navigated the canyon with only minor adjustments to its path. For the second flight, the planned path was intentionally biased into the east canyon wall to verify that the navigation algorithms would correct the planned path toward the center of the canyon, enabling the MAV to avoid the canyon walls.

Fig. 7.26. This figure shows the MAV as it enters Goshen Canyon. The inset is an image from the camera on-board the MAV.

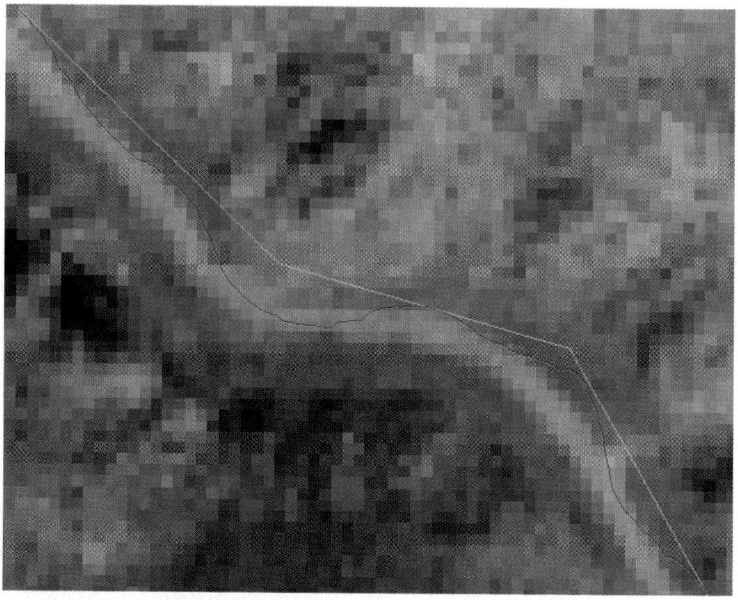

Fig. 7.27 Results from the second flight through Goshen Canyon. Flight test results show the planned path (green) and the actual path (blue). The planned path was intentionally biased to the east forcing the MAV to offset from its planned path to center itself through the canyon.

Telemetry data from the second flight has demonstrated that the MAV biased its desired path up to 10 m to the right to avoid the canyon walls. If the MAV had not biased its path it would have crashed into the east canyon wall.

7.6 Summary

Miniature aerial vehicles have demonstrated their potential in numerous applications. Even so, they are currently limited to operations in open air space, far away from obstacles and terrain. To broaden the range of applications for MAVs, methods to enable operation in environments of increased complexity must be developed. In this Chapter, two strategies have been presented for obstacle and terrain avoidance that provide a means for avoiding obstacles in the flight path and for staying centered in a winding corridor. Flight tests have validated the feasibility of these approaches and demonstrated promise for further refinement.

References

1. Chahl J., Srinivasan M., and Zhang S., "Landing Strategies in Honeybees and Applications to Uninhabited Airborne Vehicles", *International Journal of Robotics Research*, Vol. 23, No. 2, 101–110, 2004.
2. Barrows G., and Neely C., "Mixed-Mode VLSI Optic Flow Sensors for In-flight Control of a Micro Air Vehicle", *Proceedings, SPIE*, 52-63, San Diego, August 2000.
3. Ruffier F., and Franceschini N., "Visually Guided Micro-Aerial Vehicle: Automatic Take off, Terrain Following, Landing and Wind Reaction", *Proceedings, IEEE International Conference on Robotics and Automation*, 2339-2346, New Orleans, April-May 2004.
4. Zufferey J.-C., and Floreano D., "Toward 30-gram Autonomous Indoor Aircraft: Vision-Based Obstacle Avoidance and Altitude Control", *Proceedings, IEEE International Conference on Robotics & Automation*, Barcelona, April 2005.
5. Barrows G. L., Chahl J. S., and Srinivasan M. V., "Biomimetic Visual Sensing and Flight Control", *The Aeronautical Journal, London, The Royal Aeronautical Society*, Vol. 107, 159–168, 2003.
6. Procerus Technologies, http://procerusuav.com/.
7. LaValle S. M., and Kuffner J. J., "Randomized Kinodynamic Planning", *International Journal of Robotics Research*, Vol. 20, No. 5, 378–400, May 2001.
8. Kavraki L. E., Švestka P., Latombe J.-C., and Overmars M., "Probabilistic Roadmaps for Path Planning in High-Dimensional Configuration Spaces", *IEEE Transactions on Robotics and Automation*, Vol. 12, No. 4, 566–580, 1996.
9. Amato N. M., and Wu Y., "A Randomized Roadmap Method for Path and Manipulation Planning", *Proceedings, IEEE International Conference on Robotics and Automation*, Minneapolis, MN, 1996.
10. LaValle S. M., *Rapidly-Exploring Random Trees: A New Tool for Path Planning*, TR 98-11, Computer Science Department, Iowa State University, October 1998.
11. Frazzoli E., Dahleh M. A., and Feron E., "Real-Time Motion Planning for Agile Autonomous Vehicles", *AIAA Journal of Guidance, Control and Dynamics*, Vol. 25, No. 1, 116–129, 2002.
12. Saunders J. B., Call B., Curtis A., Beard R. W., and McLain T. W., "Static and Dynamic Obstacle Avoidance in Miniature Air vehicles", *Proceedings, Infotech@Aerospace Conference*, September 2005.
13. Anderson E. P., *Constrained Extremal Trajectories and Unmanned Air Vehicle Trajectory Generation*, Master's Thesis, Brigham Young University, Provo, Utah, April 2002.
14. Nelson D. R., *Cooperative Control of Miniature Air Vehicles*, Master's Thesis, Brigham Young University, Provo, Utah, December 2005.

15. Skjetne R., Fossen T., and Kokotović P., "Robust Output Maneuvering for a Class of Nonlinear Systems," *Automatica*, Vol. 40, 373–383, 2004.
16. Anderson E. P., Beard R. W., and McLain T. W., "Real Time Dynamic Trajectory Smoothing for Uninhabited Aerial Vehicles", *IEEE Transactions on Control Systems Technology*, Vol. 13, No. 3, 471-477, May 2005.
17. Loizou S., Tanner H., Kumar V., and Kyriakopoulos K., "Closed Loop Motion Planning and Control for Mobile Robots in Uncertain Environments", *Proceedings, 42nd IEEE Conference on Decision and Control*, 2926–2931, 2003.
18. Quigley M. , Goodrich M. A. , and Beard R. W. , "Semi-Autonomous Human-UAV Interfaces for Fixed-Wing Mini-UAVs", *Proceedings, IEEE/RSJ Conference on Intelligent Robots and Systems,* Sendai, Japan, September 2004.
19. Quigley M., Goodrich M. A., Griffiths S., Eldredge A., and Beard R. W., "Target Acquisition, Localization, and Surveillance Using a Fixed-Wing Mini-UAV and Gimbaled Camera", *Proceedings, IEEE International Conference on Robotics and Automation*, Barcelona, Spain, April 2005.
20. Imperx Incorporated, http://www.imperx.co.

Chapter 8: Vision Based Navigation and Target Tracking for Unmanned Aerial Vehicles[1]

The addition of a camera to an unmanned aerial vehicle (UAV) allows the vehicle to perform a variety of tasks autonomously. This Chapter presents UAV vision systems developed and tested at the Georgia Institute of Technology to perform different tasks on the GTMax unmanned research helicopter. The first system demonstrates vision-based navigation. The second system demonstrates an automated search routine for stationary ground targets. The third system demonstrates a ground target tracking architecture for mobile targets.

8.1 Introduction and Motivation

The human pilot in manned aircraft acts as a closed-loop control system by taking in visual data in addition to sensor measurements and then responding to received data. For example, the pilot may observe the changing surroundings to better understand the movements of his aircraft, or the pilot can direct the aircraft to a region of interest after taking ground observations. In unmanned aircraft, the absence of a pilot requires development of automated control systems.

Visual sensors, such as electro-optic or infrared cameras, have historically been included in the on-board sensor suite of many UAVs to increase their value as a surveillance tool [16]. However, images obtained from these sensors are rarely processed as part of an automated control system. Instead, images are returned to a ground operator, who interprets them and reacts accordingly. Recent advances in miniaturization of powerful computing resources have created the potential to add to the autonomy of UAVs by including visual data as an input to an automatic control system. This Chapter discusses the development of UAV visual control systems that contribute to the formation of a navigation solution, identify regions of

[1] Written by B. Ludington, E. N. Johnson, G. J. Vachtsevanos

Kimon P. Valavanis (ed.), Advances in Unmanned Aerial Vehicles, 245–266.

interest in an urban environment, and automatically track a mobile ground target.

8.2 GTMax Unmanned Helicopter

The three developed vision systems were all tested using the GTMax research UAV shown in Figure 8.1.The GTMax consists of four major elements: the basic Yamaha R-Max airframe, a modular avionics system, the baseline software, and a set of simulation tools.

Fig. 8.1. The GTMax research UAV.

The GTMax utilizes the Yamaha R-Max industrial helicopter airframe. The characteristics of the airframe are shown in Table 8.1.

Rotor Diameter	10.2 feet
Length	10.9 feet (including rotor)
Max Weight	205 pounds
Payload Including Avionics	> 66 pounds
Endurance	60 minutes (hover)

Table 8.1. Characteristics of the Yamaha R-Max airframe

The hardware components that make up the baseline flight avionics include general purpose processing capabilities and sensing, and add approximately 35 pounds to the basic airframe, resulting in a total weight of approximately 157 pounds. Research objectives can dictate additional equipment beyond this baseline. The current baseline research avionics configuration includes the following components:

- 266 MHz Pentium II embedded PC, 2 Gb Flash Drive
- 850 MHz Pentium III embedded PC, 2 Gb Flash Drive
- Inertial Science ISIS-IMU Inertial Measurement Unit
- NovAtel OEM-4 Differential GPS
- Honeywell HMR-2300 3-Axis Magnetometer
- Custom Made Ultra-sonic Sonar Altimeter
- Custom Made Optical RPM Sensor
- Actuator Control Interface
- 11 Mbps Ethernet Data Link and Ethernet Switch
- FreeWave 900MHz Spread Spectrum Serial Data Link
- Axis 213R Pan, Tilt, Zoom Network Camera or
- Custom Made Pan, Tilt, Roll Camera Mount
- GMS Video Transmitter

These components are packaged into exchangeable modules: two computer modules, the GPS module, the data link module (wireless Ethernet, wireless serial, Ethernet switch), and the IMU module. These modules are placed in a vibration-isolated rack below the main body of the helicopter, shown in Figure 8.2. Each module has its own self-contained power regulation, air-cooling, and electro-magnetic interference shielding. There is also a sonar/magnetometer assembly at the tail, a power distribution system including circuit breakers near the module rack, and mounting points for camera systems and other components under the nose. The power distribution system utilizes the on-board generator, which outputs 12V DC. It includes a hot-swappable connection to use external power. Each component has a dedicated circuit breaker.

Fig. 8.2. Vibration isolated avionics module rack, from left to right: data link module, GPS module, IMU module, computer module 1 (flight computer), computer module 2 (auxiliary).

The baseline software for both of the on-board computers and the ground station is written in ANSI C/C++ to maintain operating system independence. No special compilers are required. Normally, Microsoft Visual Studio is used for Windows and the GNU C-compiler is used for Linux and QNX.

The on-board software runs on the two on-board computers, referred to as the primary flight computer and the secondary computer. The Ground Control Station (GCS) software runs on the ground, normally on one or more laptop computers, and allows system operators to interact with the on-board systems. Simulation-specific software refers to any software that is not used in the flight configuration. All of the software is included in the GCS or simulation-tool builds. Typically only the on-board software is included in a primary flight computer or secondary computer build in addition to any test-specific software.

The baseline navigation system running on the primary flight computer is a 17 state extended Kalman filter (EKF). The states include: vehicle position, velocity, attitude (quaternion), accelerometer biases, gyro biases, and terrain height error. The system is all-attitude capable and updates at 100 Hz [8]. The baseline flight controller is an adaptive neural network trajectory following controller with 18 neural network inputs, six input layer neurons, five hidden layer neurons, and seven outputs for each of the seven degrees of freedom [9]. These seven degrees of freedom include the usual six rigid-body degrees of freedom plus a degree of freedom for rotor RPM. The controller can also be configured as a more conventional dynamics-inverting controller.

The baseline flight controller and navigation system, when coupled with the simple baseline trajectory generator, is capable of automatic take off, landing, hover, flight up to the maximum attainable by the helicopter (90 feet/sec) and aggressive maneuvering.

Generic and highly-capable data communication software has been developed to support a large number of potential flight and simulator test configurations. First, these routines support serial data reading and writing as necessary for the commercial off the shelf (COTS) sensors and other custom components used. These same routines can also be used to re-route any data through Ethernet or as memory within a single executable. These data routings can be modified in real-time, by a software switch. Almost all operating system specific software is limited to these routines. These data communication routines are used to interface with all sensors, send data over the wireless serial data link, and to repeat all wireless serial data over the wireless Ethernet (for redundancy). Also, any data received over a link can be stored to a binary file. This recorded data can then be played back to simulate selected components. All data received from the helicopter over either data link is stored in this manner during flight.

The simulation tools normally run on high-end personal computers or laptops that use the Windows XP/2000/NT operating system or Linux. The tools include: an aircraft model, the aircraft interface model (YACS), and sensor models (IMU, GPS, sonar, magnetometer, YACS, and camera). The aircraft model has six rigid-body degrees of freedom plus engine, fuel, landing gear, and rotor dynamics. The helicopter interface (YACS) model simulates the servo interface unit functionality and RS-232 serial interface. The sensor models include errors, mounting location and orientation, time delays, and digital interfaces. The scene generator includes a 3-D graphics window, Figure 8.3, showing the aircraft and the terrain, and has additional functionality to aid in data visualization or use in the ground control station

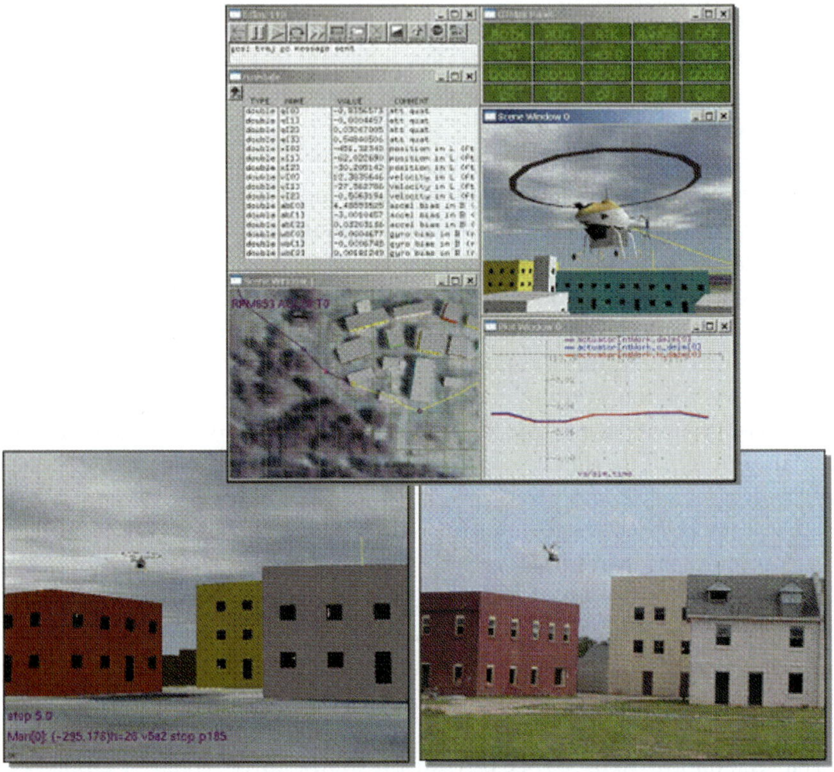

Fig. 8.3. Simulation scene generation capability provides necessary functions for ground control station interface (top), and includes a McKenna Military Operations in Urban Terrain (MOUT) obstacle database (bottom-left synthetic image, bottom-right true photograph of operations at McKenna).

Between the generic communication routines and models for all system components, an extremely large number of simulator configurations are possible. The configurations include an all software configuration, where only the real flight software is utilized. Many combinations of real hardware and simulated components are possible, including: testing only the primary flight computer hardware, only the secondary computer hardware, only the ground station, only a single sensor, or any combination of one or more of these.

8.3 Vision-Based Navigation

UAVs such as the GTMax typically use a combination of GPS and inertial sensors to form a navigation solution, or state estimate. The GPS data bounds the errors that accumulate from integrating acceleration and angular rate measurements. However, in some circumstances, such as urban or low altitude operations, this sensor combination is not reliable enough for sustained autonomous flight since the GPS receiver antenna is prone to losing line-of-sight with the satellites. In these situations, visual data can be used as an alternative to GPS measurements to aid in the formulation of a navigation solution. This Section describes the use of vision measurements along with inertial (accelerometers and gyros) measurements within an EKF to estimate the vehicle state.

Visual data has been previously used to aid in UAV navigation [1] [3] [22] [25]. There has even been some success in controlling aircraft with vision-based sensors as the sole source of feedback [19]. However, there are fewer examples of investigating the application of vision sensors in inertial navigation [11] [20] [21]. In one case, vision sensors were used in the inertial navigation of a ground robot [7]. The work described in this Section builds upon previous work in the field of vision-based navigation by demonstrating the feasibility of vision-aided inertial navigation in the closed-loop control of an airborne system.

In the experiment, it is assumed that a target is within view of the camera on-board the air vehicle. The target is a dark square within a much lighter background, and its position, orientation, and area are part of the *a priori* knowledge. The constraints on the target reduce the processing requirements on the image processor.

In the vision-aided EKF formulation, the state is estimated iteratively in two steps. In the first step the process model is used to form a state estimate using both the helicopter model and the inertial measurements. The estimate is formed using the following set of differential equations:

$$\dot{\hat{\mathbf{x}}}(t) = f\left(\hat{\mathbf{x}}(t), t\right), \tag{8.1}$$

$$\dot{\mathbf{P}}(t) = \mathbf{AP} + \mathbf{PA}^T + \mathbf{Q}, \tag{8.2}$$

where $\dot{\hat{\mathbf{x}}}$ is the state estimate, $f\left(\hat{\mathbf{x}}(t), t\right)$ is the nonlinear helicopter model, $\mathbf{P}$ is the error covariance matrix, $\mathbf{A}$ is a matrix representing the linearized helicopter dynamics, and $\mathbf{Q}$ is a diagonal matrix representing the process noise. The resulting state estimate is used to generate the expected position and size of the target in the image frame using a pinhole camera model. This process is subject to inaccuracies do to model error and the accumulated inertial measurement error. The visual data is used to account for these errors and form an accurate navigation solution.

In the second step, the expected measurements are compared to actual target measurements, and the gain in the EKF is adjusted to optimally fuse the camera measurements with the predicted measurements using the following equations:

$$\mathbf{K} = \mathbf{P}^{-}\mathbf{C}^{T}\left(\mathbf{C}\mathbf{P}^{-}\mathbf{C}^{T} + \mathbf{V}\right)^{-1}, \tag{8.3}$$

$$\hat{\mathbf{x}} = \hat{\mathbf{x}}^{-} + \mathbf{K}\left[\mathbf{z} - h\left(\hat{\mathbf{x}}^{-}\right)\right], \tag{8.4}$$

$$\mathbf{P} = \left(\mathbf{I} - \mathbf{K}\mathbf{C}\right)\mathbf{P}^{-}, \tag{8.5}$$

where **K** is the Kalman gain, **V** is a diagonal matrix representing the measurement noise, **C** is the Jacobian of the measurement vector with respect to the state vector, and $h\left(\hat{\mathbf{x}}^{-}\right)$ is the predicted measurement vector. Minus superscripts in the above equations denote prior values obtained from the process model equations. This innovation step allows the visual data to correct for inaccuracies in both the system model and inertial measurements.

Figure 8.4 illustrates a time history for one test of the vision-based navigation system, showing longitudinal position (forward/backward). The raw differential GPS measurement is compared to the vision-aided inertial navigation system (INS) information. In this test, the vision-aided INS solution is used by the flight control system flying the helicopter. Therefore, the vehicle followed the command without the use of GPS data. Figure 8.5 is an external picture of one of these tests showing the helicopter in a hover, and the black optical target in the lower-right.

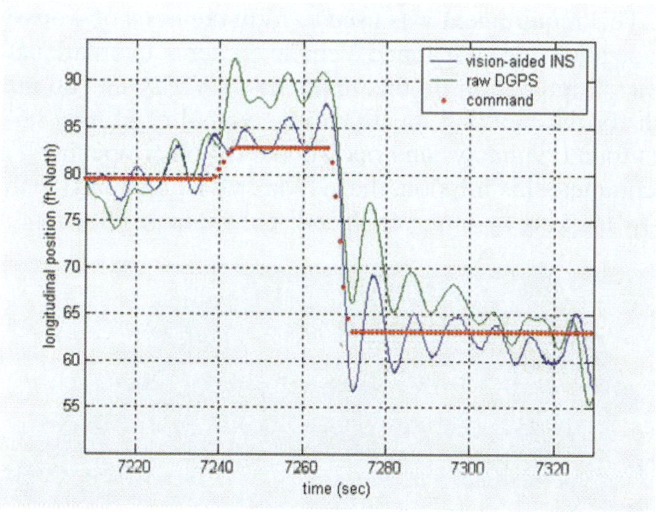

Fig. 8.4. Vision-aided INS position estimation results compared with a Differential GPS measurement.

Fig. 8.5. Photograph of a vision-aided INS flight test, showing the helicopter hovering in the upper left and the black optical target in the lower right.

8.4 Automated Search

When using a UAV to perform a reconnaissance mission above urban terrain, the absence of a human pilot requires the use of an automated system to identify buildings of interest and then gather information about these buildings. This requirement was used to form the basis of a mission as part of the Association for Unmanned Vehicle Systems International (AUVSI) Air Robotics Competition. In this mission, UAVs fly into an urban region and search for the building identification symbol of Figure 8.6. Once the building is found, windows and open doors (or other openings) are identified. To complete this mission, the GTMax was augmented with a symbol and window tracking system, which is discussed in this Section.

Fig. 8.6 During the benchmark reference reconnaissance mission the GTMax searches for the building with the symbol on it. The symbol is shown in the lower-right portion of the building. It consists of a black circle with crossing white lines and letters on it and is one meter in diameter.

Figure 8.7 shows a flow chart outlining the complete process of symbol and window detection, classification and tracking. The symbol detection and classification is done in a two step process. In the first step, incoming images are processed using a shape-matching filter. The filter is performed using Halcon, which is a commercially available image processing library developed by MvTec Software GmBH. A template is constructed using previous flight images of the building identification symbol, Figure 8.6, and an edge detector. This template is then used within the Halcon shape-matching filter to identify candidate symbol locations.

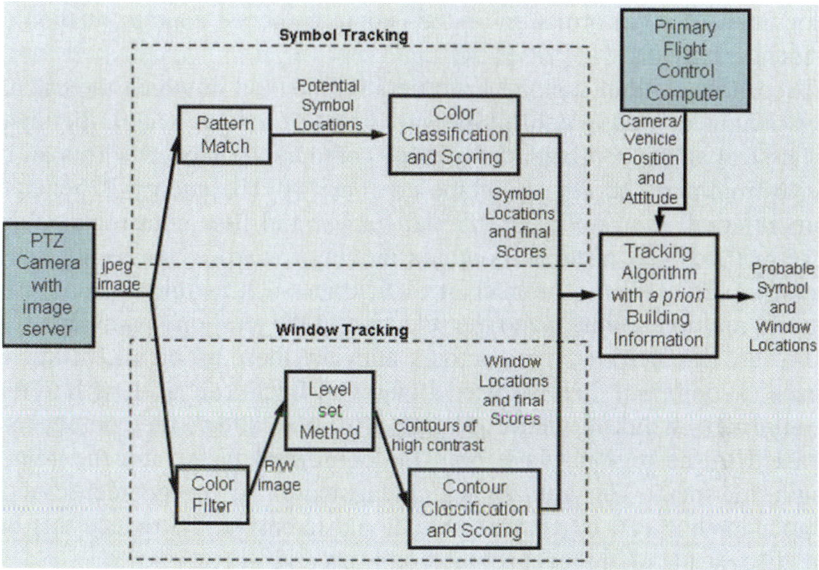

Fig. 8.7. Flow chart indicating the process used to identify the building and openings into it (windows or doors).

In the second step of the symbol detection and classification process, imagery of the candidate symbol location is passed through a color filter. The color filter relies on the symbol being only black and white. The color channels (in the RGB sense) of each pixel within a candidate region are examined. If the difference between the channels is greater than a certain threshold, the pixel is deemed too colorful and is assigned a score of zero. If the channels are similar, then the grayscale value is compared to another threshold and the pixel is assigned a score of one if it is black and two if it is white. The average score is then calculated over the candidate region. If the symbol is within the region the average score should fall between one and two.

The window tracking and classification process is done in three steps. In the first step, the color image is converted into a black and white image as was done in the color filter for the symbol recognition. Each pixel in the image is examined to determine if it has a dominant color. If it does, the pixel is set to white. If it does not, then the grayscale value is examined. If the grayscale value is above a threshold, the pixel is set to white; if it is not, the pixel is set to black.

In the second step of the process, the geometric active contour or snake method is used to identify contours the black and white image from the color filter. More information on the geometric active contour method can be found in [5] [12] [13] [15] [24].

The active contour method is an iterative method in which the calculus of variations is used to control the movement of a curve within the image. In classical active contours, deformable contours are moved within an image to minimize the *energy* of the contour [4]. The energy is generated from external, image forces; internal forces; and user constraints. While these methods can result in contours moving toward edges in the image, they also suffer from a number of difficulties such as initialization, existence of multiple minima, and the selection of the elasticity parameters.

The use geometric active contours alleviate these problems. In this approach, a *conformal factor* is used along with Euclidean arc-length to form a conformal Euclidean metric [5] [10] [12]. This allows the contour to be attracted to the region where both the conformal factor and the contour length are small. The numerical implementation of the geometric active contour method can be very computational intensive. To reduce this burden, the fast-marching level set method is used [23].

In the third step of the process, the contours are extracted from the image and used for classification. First, the contours are examined to see if they are closed. If they are closed, the length of the contour (perimeter) is calculated. The length is used to determine if the contour corresponds to a building opening and to assign the contour a score. If the contour corresponds to a building opening, then location of the corners and center of the window are calculated and sent to the tracking algorithm.

The tracking algorithm is common for both the window and symbol modes. It takes the location and score of each object along with position and attitude information from the primary flight computer, and updates an array containing the position of the objects in the local geographical reference frame. Finally, it calculates a probability for each object based on how many times the object has been seen, the information from the image processor.

Initial flight tests were done with a search area including three of the buildings in the McKenna Military Operations in Urban Terrain Flight Test Site. On all of these tests, the system selected the correct building. In each case, the symbol was placed on a different building. Following this, four attempts were made to search the entire village of 15 buildings. On the first, the symbol was missed (missed detection); the symbol appeared too bright, presumably due to bright sunlight directly on the symbol unlike all previous tests. The nominal distribution range for the symbol score was changed to account for increased brightness. On all three subsequent tests the correct building was selected based on the symbol, appearing for approximately five seconds over a total search flight time of approximately 15 minutes. Typical search results are shown in Figure 8.8.

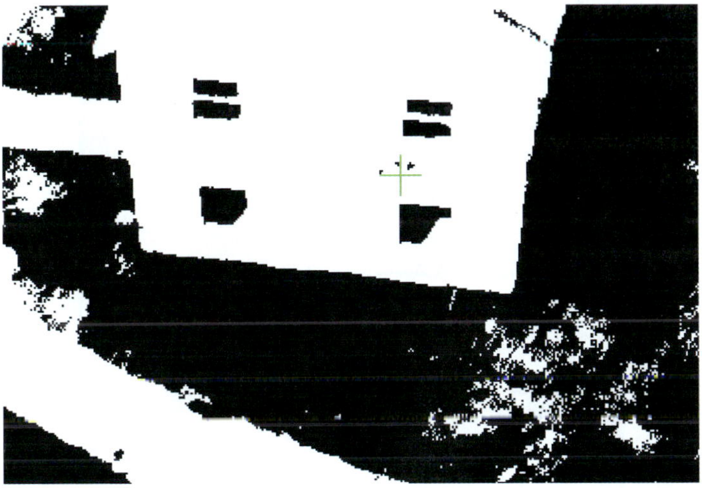

Fig. 8.8. Typical search results. The symbol has been identified as shown by the green crosshairs.

Figure 8.9 shows the results of the image filter with the centroids and corners for the windows that were found using the geometric active contours. Since the initial contour is a circular region, only a portion of the image can be evaluated at one time causing the system to miss openings on the upper floor and some on the lower floor.

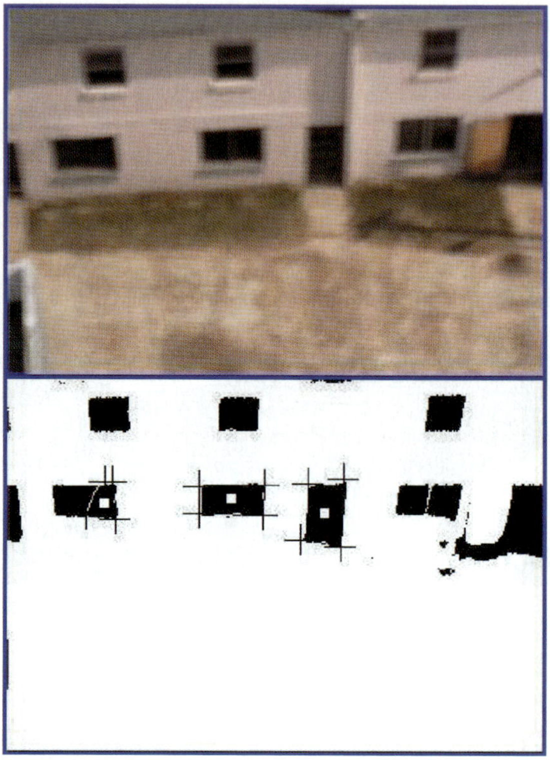

Fig. 8.9. Identification of windows/doors with active contours.

8.5 Ground Target Tracking

UAVs have proven their ability to provide valuable reconnaissance information of maneuvering ground targets [16]. However, this process should be automated to reduce the burden on the vehicle operator. This Section describes an automatic ground target tracking system for the GTMax. The system has three components. The first is a particle filter, which estimates the position of the target within the incoming video frames. The second is a camera controller, which moves the field of view (FOV) to keep the target within the video frames. The third is a trajectory (waypoint) generator, which commands the GTMax to smoothly follow the target.

The visual tracking task is a state estimation problem. The goal is to estimate the probability density of the target position or state within each incoming video frame using information from all past frames. The optimal solution (in a mean squared sense) is given by the Bayesian tracking filter.

When the distribution is Gaussian and the target's dynamics are linear, the Bayesian solution becomes the well-known Kalman filter. However, if the linear, Gaussian assumption is violated, the distribution must be approximated by some other means.

The particle filter is a state estimation tool that is able to approximate a non-Gaussian distribution that evolves according to nonlinear dynamics using a randomly-selected set of weighted samples or particles [2]. It is an effective tool for visual tracking because non-Gaussian distributions are prevalent due clutter in each image frame. As with the Kalman filter or other Bayesian techniques, the estimated distribution is recursively computed using a two step process of prediction and update. As the number of particles increases, the sampled-based distribution converges to the true distribution [6]. When using a particle filter to track targets in images, the state space is typically four dimensional and describes a rectangle. Therefore, the higher weighted particles correspond to the peak of the distribution and are rectangles that are near the target, while the lower weighted particles correspond to the tails of the distribution and are rectangles that are further away from the target.

The system update model used here is similar to [18], where the model is made up of two components:

$$x_k^i = \beta_{RW} N\left(x_{k-1}^i, \sigma_{RW}\right) + \left(1 - \beta_{RW}\right) N\left(z_k^i, \sigma_J\right) \tag{8.6}$$

where x_k^i is the ith particle at time k, x_{k-1}^i is the ith particle at time k-1, $N\left(x, \sigma\right)$ is a Gaussian distribution with mean x and covariance σ, and z_k^i is a randomly selected pixel where motion has been detected. The first component assumes the target moves smoothly from one frame to the next. Therefore, the particles are moved according to a Gaussian random walk model. The second component accounts for the times when the smoothness assumption is violated, which can occur if the target becomes occluded or briefly moves outside of the FOV. This portion of the model is another Gaussian distribution that is centered around a randomly selected pixel where motion has been detected. Motion is detected in a pre-processing step. The two components of the model are combined using a convex combination, and the size of each contribution is changed throughout the tracking process depending on the performance of the system.

After the particles are moved, measurements are taken. Here, color and motion measurements are used. Color cues are generated by first measuring the distance the particle and a reference histogram:

$$D_C\left(\mathbf{h}_k^i, \mathbf{h}_{ref}\right) = \left(1 - \sum_{j=1}^{N_B} \sqrt{h_{k,j}^i h_{ref,j}}\right)^{1/2} \tag{8.7}$$

where $\mathbf{h}_k^i = \left\{h_{k,j}^i, j = 1 \ldots N_B\right\}$ is the histogram for particle x_k^i and $\mathbf{h}_{ref} = \left\{h_{ref,j}, j = 1 \ldots N_B\right\}$ is the reference histogram. This distance is a metric [17]. To keep the measurement robust to lighting changes, the hue, saturation, value (HSV) color space is used.

Once the distance from the reference is measured, an exponential is used to determine the corresponding likelihood:

$$p\left(y_k^C \mid x_k^i\right) \propto \exp\left(-\frac{D_C\left(\mathbf{h}_k^i, \mathbf{h}_{ref}\right)}{\sigma_C^2}\right) \tag{8.8}$$

where σ_C^2 is an adaptable term that controls the slope of the exponential curve.

As in [18] motion cues are used in addition to color cues. Like color, motion cues are formed by first generating a distance:

$$D_M\left(x_k^i\right) = \left(1 - \frac{M_k^i}{A_k^i}\right)^{1/2} \tag{8.9}$$

where M_k^i is the sum of the absolute value of the difference pixels in particle x_k^i and A_k^i is the area of particle x_k^i. The difference pixels are generated by subtracting the values (in the HSV sense) of the pixels within the particle from the corresponding pixels in the previous image. Like the color measurements, the distance is used to form the motion likelihood:

$$p\left(y_k^C \mid x_k^i\right) \propto \exp\left(-\frac{D_M\left(x_k^i\right)}{\sigma_M^2}\right) \tag{8.10}$$

where σ_M^2 is analogous to σ_C^2.

As pointed out in [2], when the particles are re-sampled at each time step, the filter becomes sensitive to outliers. To combat this unwanted sensitivity, a smoothness term is also included in the measurement model. This term rewards the particles that remain close to the previous filter output. As with the color and motion measurements, first a distance measure is created:

$$D_S\left(x_k^i\right)=\left[\left(x_{k,1}^i-x_{k-1,1}\right)^2+\left(x_{k,2}^i-x_{k-1,2}\right)^2\right]^{1/2} \tag{8.11}$$

where $x_{k,1}^i$ and $x_{k,2}^i$ are the horizontal and vertical coordinates of particle x_k^i, respectively and $x_{k-1,1}$, and $x_{k-1,2}$ are the horizontal and vertical coordinates of the filter output at time k-1, respectively. This is a Euclidean distance measure, and is clearly a metric. The distance is used to form the smoothing likelihood:

$$p\left(y_k^S \mid x_k^i\right) \propto \exp\left(-\frac{D_S\left(x_k^i\right)}{\sigma_S^2}\right) \tag{8.12}$$

where σ_S^2 is analogous to σ_C^2 and σ_M^2.

The measurement cues are fused together assuming that they are statistically independent. Therefore, the combined likelihood is given by the product:

$$p\left(y_k \mid x_k^i\right) \propto p\left(y_k^C \mid x_k^i\right)p\left(y_k^M \mid x_k^i\right)p\left(y_k^S \mid x_k^i\right) \tag{8.13}$$

The performance of the filter is estimated by examining properties of the particle distribution via a multi-layer perceptron neural network. The input to the network is a vector containing the following: spread of the ten highest weighted particles, maximum value of the weights, maximum value of the weights generated using only the color cues, standard deviation of the weights, standard deviation of the weights generated using only the color cues, the measurement residual, the color measurement residual, and the expected log likelihood. Each component of the vector is expected to yield information about the shape of the distribution [14]. The network is trained offline using a large set of manually labeled frames.

Once the performance of the filter is estimated the system update and measurement models can be updated to place the particle and fuse the measurements more effectively. Also, the number of particles can be decreased when the filter is performing well. By reducing the number of particles, the inherently large computational burden of the particle filter can be decreased.

Figure 8.10 shows a set of output frames as the particle filter tracked a truck moving across a field. The video was taken from the GTMax and the frames were processed offline after converting the video to a series of frames at 10 frames per second. In this case, the helicopter was hovering and the orientation of the camera was manually controlled. The output of the particle filter is shown as the red and blue rectangles. The ten highest weighted particles are represented by blue rectangles, while the remained of the particle set is represented by red rectangles. The weighted average of the particles is shown as the white crosshairs.

Fig. 8.10. A set of particle filter output frames. The top ten particles are represented by blue rectangles, while the remaining particles are represented by red rectangles. The weight average of the particles is shown with the white crosshairs.

Once the position of the target is estimated, the target can be tracked autonomously by *closing the loop*. First, the camera must be pointed at the target. This is achieved by using a pin-hole camera model (as with the vision-based navigation) and assuming the terrain is flat to estimate the target's ground position. This ground position is used within a linear predictor to estimate the future target position.

The predicted position of the target is also used to generate a new trajectory that will allow the GTMax to follow the target. A new waypoint is assigned that is in the direction of the target. The waypoint will be offset from the target to avoid the GTMax being forced to fly directly over the target. This waypoint is used to generate a trajectory for the GTMax that has reduced acceleration and jerk limits. These limits result in decreased attitude changes for the GTMax. Therefore, the target appears to move smoothly even as the helicopter moves.

A closed loop tracking system was tested in simulation and on-board the GTMax. It was used to track a red sports utility vehicle from a 100 ft altitude. The low altitude required the filter parameters and jerk limits to be set more conservatively. A set of frames from the flight results is shown in Figure 8.11. The particle filter effectively estimated the position of the ground target, and enabled the GTMax to autonomously follow the target.

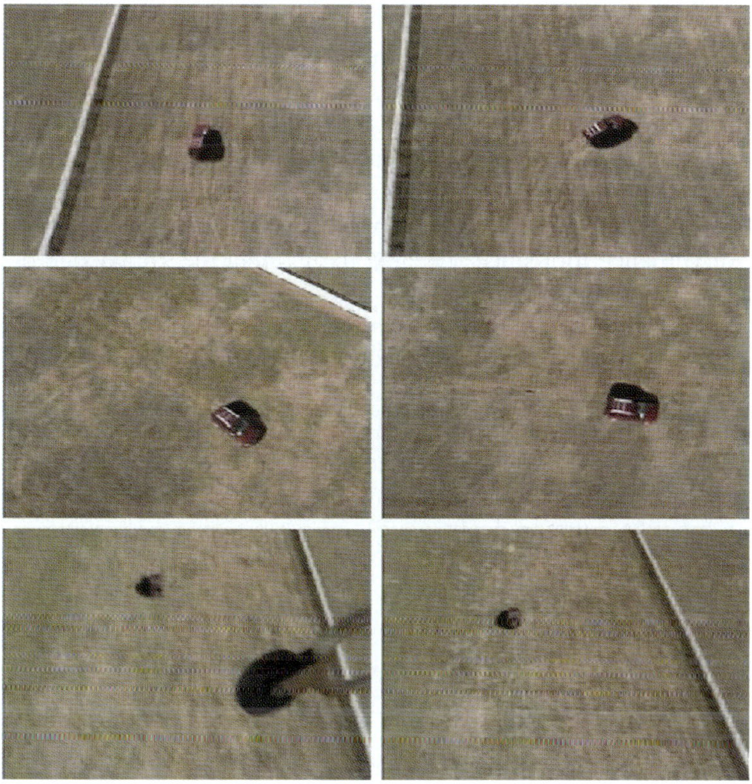

Fig. 8.11. Flight test results using a particle filter for closed loop tracking.

8.6 Conclusion

The use of vision systems on-board UAVs increase the autonomy level and allow unmanned vehicles to effectively perform complex missions in uncertain environments. Three vision systems were presented in this Chapter that allow UAVs to closely mimic piloted aircraft.

The first system allows visual cues to aid in navigation. Through the use of an EKF, vision sensors become a viable alternative to GPS in inertial navigation. This enables a UAV to operate in regions where communication with GPS satellites is unreliable such as low altitude, urban areas, or indoors. In the future, the system described here could be extended to allow multiple targets to be tracked and acquired allowing vehicles to perform a variety of missions without GPS.

The second system presented an automated search system that allows a UAV to search an urban region for a particular building and then identify the openings to the building. Through the use of fast image processing and tracking algorithms, the vehicle was able to autonomously search an urban area with 15 buildings as quickly as a human operator on foot or a remotely piloted vehicle. Such a system increases the value of a UAV as an urban surveillance tool since it is able to augment the knowledge of a city without burdening an operator.

The final system relied upon a particle filter algorithm to track a maneuvering ground target. By estimating the performance of the tracking filter, the system is able to function in real-time. When the particle filter is integrated with a camera controller and waypoint generated, the resulting system is able to track a target autonomously. In the future, the use of an automatic initialization routine will allow create a close-loop tracking system that can function autonomously.

References

1. Amidi O., *An Autonomous Vision-Guided Helicopter*, PhD Thesis, Carnegie Mellon University, 1996.
2. Arulampalam M. S., Maskell S., Gordon N, Clapp T., "A Tutorial on Particle Filters for Online Nonlinear/Non-Gaussian Bayesian Tracking", *IEEE Transactions on Signal Processing*, 50, 174-188, 2002.
3. Bernatz A., Thielecke F., "Navigation of a Low Flying VTOL Aircraft with the Help of a Downward Pointing Camera", *Proceedings of the AIAA Guidance, Navigation, and Control Conference*, 2004.
4. Blake A., Isard M., *Active Contours*, Springer, 1998.

5. Caselles V., Kimmel R., Sapiro G., "Geodesic Active Contours", *International Journal of Computer Vision*, 22, 61-79, 1997.
6. Crisan D., Doucet A., "A Survey of Convergence Results on Particle Filtering", *IEEE Transactions on Signal Processing*, 50, 736-746, 2002.
7. Diel D., DeBitetto P., Teller S., "Epipolar Constraints for Vision-Aided Inertial Navigation", *Proceedings, IEEE Workshop on Motion and Video Computing*, 221-228, 2005.
8. Dittrich J., Johnson E., "Multi-Sensor Navigation System for an Autonomous Helicopter", *Proceedings, 21st Digital Avionics Systems Conference*, 2002.
9. Johnson E., Kannan S., "Adaptive Flight Control for an Autonomous Unmanned Helicopter", *Proceedings of the AIAA Guidance, Navigation, and Control Conference*, 2002.
10. Johnson E. N., Procotor A. A., Ha J., Tannenbaum A., "Visual Search Automation for Unmanned Aerial Vehicles", *IEEE Transactions on Aerospace and Electronic Systems*, 41, 219-232, 2005.
11. Kaminer I., Pascoal A., Kang W., *"Integrated Vision/Inertial Navigation System Design Using Nonlinear Filtering"*, *Proceedings, American Control Conference*, 1910-1914, 1999.
12. Kichenassamy S., Kumar A., Oliver P. J., Tannenbaum A., Yezzi A., "Conformal Curvature Flows: From Phase Transitions to Active Vision", *Archive for Rational Mechanics and Analysis*, 134, 275-301, 1996.
13. Lauziere Y., Siddiqi K., Tannenbaum A., Zucker S., "Area and Length Minimizing Flows for Segmentation", *IEEE Transactions on Image Processing*, 7, 434-444, 1998.
14. Ludington B. T., *Particle Filter Tracking Architecture for Use On-board Unmanned Aerial Vehicles*, PhD Thesis, Georgia Institute of Technology, 2006.
15. Malladi R., Sethian A., Vermuri B., "Shape Modeling with Front Propagation: A Level Set Approach", *IEEE Transactions on Pattern Analysis*, 17, 156-175, 1995.
16. Office of the Secretary of Defense, *Unmanned Aerial Systems Roadmap 2005-2030*, Technical Report, 2005.
17. Perez P., Hue C., Vermaak J., "Color-Based Probabilistic Tracking", *Proceedings, European Conference on Computer Vision*, 661-675, 2002.
18. Perez P., Vermaak J., Blake A., "Data Fusion for Visual Tracking with Particles", *Proceedings of the IEEE*, 92, 495-513, 2004.
19. Proctor A. A., Johnson E. N., "Vision-Only Aircraft Flight Control Methods and Test Results", *Proceedings of the AIAA Guidance, Navigation, and Control Conference*, 2004.
20. Rehbinder H., Ghosh B. K., "Multi-Rate Fusion of Visual and Inertial Data", *Proceedings, IEEE International Conference on Multisensor Fusion and Integration for Intelligent Systems*, 97-102, 2001.
21. Roberts B. A., Vallot L. C., "Vision Aided Inertial Navigation", *Record of the IEEE Position, Location, and Navigation Symposium*, 347-352, 1990.
22. Roberts J., Corke P., Buskey G., "Low-Cost Flight Control System for a Small Autonomous Helicopter", *Proceedings, Australian Conference on Robotics and Automation*, 546-551, 2002.

23. Sethian J. A., *Level Set and Fast Marching Methods*, Cambridge University Press, 1999.
24. Tannenbaum A., "Three Snippets of Curve Evolution Theory in Computer Vision", *Mathematical and Computer Modeling*, 24, 101-119, 2004.
25. Webb T. P., Prazenica R. J., Kurdila A. J., Lind, R., "Vision-Based State Estimation for Autonomous Micro Air Vehicles", *Proceedings of the AIAA Guidance, Navigation, and Control Conference*, 2004.

Chapter 9: Single and Multi-UAV Relative Position Estimation Based on Natural Landmarks[1]

This Chapter presents a vision-based method for unmanned aerial vehicle (UAV) motion estimation that uses as input an image motion field obtained from matches of point-like features. The Chapter enhances vision-based techniques developed for single UAV localization and demonstrates how they can be modified to deal with the problem of multi-UAV relative position estimation. The proposed approach is built upon the assumption that if different UAVs identify, using their cameras, common objects in a scene, the relative pose displacement between the UAVs can be computed from these correspondences under certain assumptions. However, although point-like features are suitable for local UAV motion estimation, finding matches between images collected using different cameras is a difficult task that may be overcome using blob features. Results justify the proposed approach.

---

[1] Written by L. Merino, F. Caballero, P. Forssen, J. Wiklund, J. Ferruz, J. R. Martínez-de-Dios, A. Moe, K. Nordberg, A. Ollero. This Chapter is the outgrowth of the authors' paper "*Multi-UAV Localization Based on Blob Features for Exploration Missions*", published in the IEEE Robotics and Automation Magazine, Vol. 13, No. 3, September 2006. This research has been conducted under the framework of the project "COMETS: "Real-time coordination and control of multiple heterogeneous unmanned aerial vehicles", IST-2001-34304, funded by the IST Programme of the European Commission. Cooperation of all partners and particularly the teams of the Technical University of Berlin (TUB), the Laboratoire d'Architecture et d'Analyse des Systemes (LAAS) and Helivision is acknowledged. The work of the Spanish team has also been also funded by projects AWARE (European Commission IST-2006-3379) and AEROSENS (Spanish National R&D Program DPI2005-02293).

Kimon P. Valavanis (ed.), Advances in Unmanned Aerial Vehicles, 267–307.

9.1 Introduction

The localization problem is usually solved in aerial robotics by means of GPS, Inertial Measurement Units (IMUs) and compass sensors, which are fused to provide UAV position estimation in a global frame. However, GPS estimates are often subject to inaccuracies and errors related to GPS failure or degradation. Also, GPS cannot be directly used for relative positioning of a UAV with respect to other objects, as for example for landing on mobile platforms or for wall following in building inspection.

Therefore, use of exteroceptive sensors for UAV navigation and relative positioning has been considered. For small size UAVs, the most commonly used sensor is vision. Range sensors like laser rangefinders are not very common as yet, and they are mainly used for safe landing.

Many of the developed UAV systems and methods involve only one aerial vehicle. However, in many applications, active cooperation of several UAVs has important advantages. For example, UAVs carrying different sensors can cooperate to exploit sensor synergies [25]. UAV cooperation can be also established at a perception level. Cooperative perception can be defined as the task of creating and maintaining a consistent view of a world containing dynamic objects by a group of agents each equipped with one or more sensors. Cooperative visual perception has become a relevant topic in the multi-robot domain, mainly in structured environments [18] [32] [36], but still has been rarely applied to multi-UAV systems [22] [35].

Cooperative perception techniques integrate results individually obtained by each robot/vehicle, which requires knowing the relative position and orientation of the robots. That is, UAVs should be localized in the same coordinate frame.

This Chapter builds upon previous work to demonstrate how vision-based techniques for single UAV localization extend to solve the problem of multi-UAV relative position estimation. The assumption is that if different UAVs identify common objects in a scene using their cameras, the relative pose displacement between UAVs, under certain assumptions, can be computed from these correspondences, see Figure 9.1. Point-like features are suitable for local UAV motion estimation. Finding matches between images collected by different cameras is based on blob features.

Fig. 9.1. Two UAVs exploring the same region.

9.1.1 Related Research

Vision has been used as a positioning sensor in several UAV projects [26]. It has been argued that some applications requiring very accurate relative position of the UAV with respect to objects, such as close proximity flight, cannot rely on inertial navigation systems or global positioning systems. Thus, vision is used as a method to sense relative position.

The concept of a visual odometer [1] was implemented in the autonomous helicopter developed at CMU. Using this concept the helicopter can visually lock-on to ground objects and sense relative helicopter position and velocity in real-time. Template matching techniques are used to estimate object displacement between consecutive images. This, combined with angular rate sensing, allows for estimating UAV motion. In this case, only a few templates are tracked. In each cycle new targets are selected to avoid losing targets. The same visual tracking techniques, combined with inertial sensors, have been applied to autonomously take off, follow a prescribed trajectory, and landing. The CMU autonomous helicopter has demonstrated autonomous tracking capabilities of moving objects by using only on-board hardware.

The use of vision for autonomous landing has been actively researched. In [5], results on possible use of vision for landing an airplane were presented. In [44], a video based attitude and height sensor was developed for low altitude airborne vehicles using neural networks to compute attitude and height from the measures obtained by a video camera that observes the structured light pattern created by a set of infrared lasers. In [43], the problem of determining position and orientation of an aircraft with respect to a ship from images of several points taken by an infrared camera is considered. All these methods rely on artificial beacons that can be recognized on the images.

In the BEAR project, vision-based pose-estimation of unmanned helicopters relative to a landing target and vision-based landing of an aerial vehicle on a moving deck are researched [33] [42]. In this project, results on multiple view geometry are used to impose constraints over multiple views of a planar landing target with a single camera. These constraints, summarized in the so called multiple view matrix, allow for computing displacement of the UAV given three or more views of the target. The known size of the landing target helps obtaining the scale of the displacement. Using off-the-shelf hardware, the vision system in the BEAR project is able to operate at 30 Hz. Vision is situated in the control loop as a high lever supervisor which sends the current position of the landing target to the navigation controller. Feature extraction and matching are simplified by using an artificial target of known shape.

Computer vision is also used for safe landing in the AVATAR project. In [13], a strategy and an algorithm relying on image processing techniques to search the ground for a safe place to land is presented. A Vision-based technique for landing on a helipad of known shape is presented in [30], where also the case of a slow moving helipad is considered. The system applies target recognition using computer vision (through threshold-based segmentation and using the moments of inertia of the known helipad) and a Kalman filter for target tracking. In [31] they also present a technique for helicopter position estimation using a single CMOS pointing downward camera with a large field of view and a laser pointer to project a signature onto the surface below in such a way that it can be easily distinguished from other features on the ground. In [4] the use of stereo vision is analyzed for height estimation in small size helicopters, while [16] presents a stereo vision for simultaneous localization and mapping (SLAM) using an airship. Recently, work on vision-based position estimation and control of UAVs for inspection in urban-like environments has been presented [21]. Recent work on vision-based control of a fixed-wing aircraft is presented in [28].

9.2 Robust Features for Vision-Based UAV Navigation

Most vision-based methods are based on image motion estimation; most methods estimate image motion from robust feature matching. In this Section, methods are presented for robust feature extraction, which will be used to obtain a sparse motion field on the image plane. The next Section presents how to use these features to compute the UAV displacement.

9.2.1 Point-Like Feature Extraction and Matching

One of the most extended features used for image motion estimation is point-like features, also called interest or salient points. These interest points are pixels on the image plane that present a differencing characteristic respect to the pixels of its neighborhood. The objective is to extract repeatable interest points that are easy to match between images.

In the case of aerial imagery, the feature extraction and matching algorithms should face significant rotational and translational motion, and also scale changes. Therefore, the algorithms should be robust under such conditions.

9.2.1.1 Overview of the Approach

The image matching method presented here is related to the one described in [8], although significant improvements have been made since. As in [37], corner points are selected and used to define fixed-size windows; the initial templates obtained are used to build sequences of matching windows along the stream of video images. The same selection algorithm provides for *direct* candidate windows in each tracking step; correlation-based matching tries, essentially following the Lucas-Kanade approach, are performed between the last known windows in each sequence and a set of selected candidates. A number of competing matching pairs may be thus obtained, so a disambiguation or feature selection process is usually needed.

A detailed description of the feature selection and matching pair generation algorithms is presented in [8], which are similar, except relatively minor details, like the use of normalized correlation in order to achieve greater immunity to change in lighting conditions. Only the most relevant features of the new approach are described next. They stem from a fundamentally different selection approach, which ceases to focus on individual elemental features: clusters are not only built for validation purposes; they are persistent structures which are expected to remain stable for a number of frames, and are searched for as a whole. The 'uncertainty' algorithm changes from a relaxation procedure to a more efficient predictive approach, similar to the one used in [2] for contour matching. Rather than generating an exhaustive database of potential matching pairs as in [37], only selected hypothesis are considered. Each hypothesis, with the help of the persistent cluster database, allows for defining reduced search zones for sequences known to belong to the same cluster as the hypothesis, if a model for motion and deformation of clusters is known.

For this approach to be feasible, the following points should be defined:

- Cluster model;
- Cluster building and validation procedure;
- Cluster based prediction.

Cluster Model

Shape similarity between regions of different images is verified by searching for clusters of windows whose members keep the same relative position, after a scale factor is applied. For a cluster of window sequences, $\Gamma = \{\Phi_1, \Phi_2 \cdots \Phi_n\}$, this constraint is given by the following expression:

$$\left| \frac{\left\| w_{\Phi_k}, w_{\Phi l} \right\|}{\left\| v_{\Phi_k}, v_{\Phi l} \right\|} - \frac{\left\| w_{\Phi_p}, w_{\Phi_q} \right\|}{\left\| v_{\Phi_p}, v_{\Phi_q} \right\|} \right| \leq k_p \ \forall \Phi_k, \Phi_l, \Phi_p, \Phi_q \in \Gamma \tag{9.1}$$

In (9.1) k_p is a tolerance factor, w_i are candidate windows in the next image and v_i are template windows from the preceding image. The constraint is equivalent to verifying that the Euclidean distances between windows in both images are related by a similar scale factor; thus, the ideal cluster may be obtained when Euclidean transformation and scaling can account for the changes in window distribution. In addition, a limit over the maximum difference of rotation angle between pairs of windows is enforced:

$$\left| \alpha_{\Phi_k \Phi_l} - \alpha_{\Phi_p \Phi_q} \right| \leq \gamma_p \ \forall \Phi_k, \Phi_l, \Phi_p, \Phi_q \in \Gamma \tag{9.2}$$

In (9.2) α_{rs} is the rotation angle of the vector that links windows from sequences r and s, if the matching hypothesis is accepted, and γ_p is a tolerance factor.

Although the cluster model is adequately simple and seems to fit the current applications, more realistic local models such as affine or full homography could be integrated in the scheme without much difficulty.

Cluster-Based Prediction for Hypothesis Propagation

It is easy to verify that two hypothesized matching pairs result in predicting the position of the other members of the cluster, if their motion can be modeled approximately by Euclidean motion plus scaling. Using this model, generation of *candidate clusters* for a previously known cluster can start from a primary hypothesis, namely the matching window for one of its window sequences (see Figure 9.2). This assumption restricts the search zone for other sequences of the cluster, which are used to generate at least one secondary hypothesis. Given both hypotheses, the full structure of the cluster can be predicted with the small uncertainty imposed by the tolerance parameters k_p and α_p, and one or several candidate clusters can be added to a database. For a specific cluster, assuming that a primary hypothesis is available, the cluster candidate generation process involves three steps:

- Secondary hypothesis generation;
- Cluster-based prediction;
- Indirect candidate generation.

Secondary hypotheses are generated by single window prediction. As shown in Figure 9.2, a search area can be defined by considering maximum limits of frame-to-frame cluster scaling and rotation, based on knowledge of the environment; thus, search for matching pairs can be restricted to direct candidates contained in the area. In this way, any window sequence linked to the cluster can be used as secondary hypothesis. In Figure 9.2, a primary hypothesis (i.e., a potentially valid matching pair v_2, w_2) is available for sequence 2, a component of cluster Γ_1. The search zone is limited by the minimum and maximum expected distance between the pair of windows $\{w_2, w_5\}$ and the lines obtained by considering maximum clock- and counter-clockwise rotation of the same pair. A rectangular zone that overlaps the described limits is used, for ease of computation.

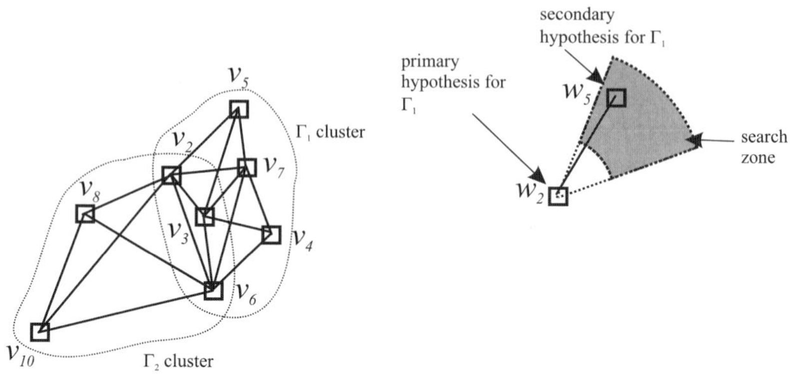

Fig. 9.2. Secondary hypothesis generation.

Thus, several primary/secondary pairs can be obtained. For each one available, a full cluster can be predicted; in this case, tolerance parameters k_p and α_p are used to further constrain the search zone of each component, as Figure 9.3 shows. Matching windows for each sequence can be found by exploring the set of direct or even previously generated indirect candidates located within the computed limits. In Figure 9.3, small search zones are computed for sequences 3, 7, 4 and 6, all components of cluster Γ_1. While several candidates may be found for each sequence in the search zones, this is not frequent for a normal setup of the tolerance parameters where additional cluster candidates should be created when necessary.

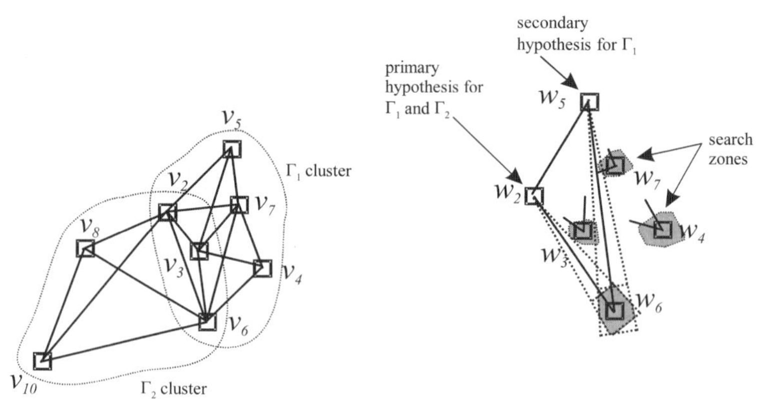

Fig. 9.3. Cluster-based prediction.

If there remain some window sequences without a candidate after the direct candidate expansion, new indirect candidates are generated by prediction of a suitable starting point for the error minimization algorithm.

The creation of any given candidate cluster can trigger creation of others for neighbor clusters, provided that there is some overlap among them; in Figure 9.4, for example, creation of a candidate for cluster Γ_1 can be used immediately to propagate hypothesis and find a candidate for cluster Γ_2 ; a primary hypothesis for sequence 2 is readily available, and a secondary hypothesis for sequence 3 can be found without the need of further error-minimization with the available direct candidates. Direct search of matching windows is thus kept to a minimum by using existing candidates for future hypothesis generation. Indeed, it makes sense to consider a higher likelihood for candidates which already take part in a sufficiently large cluster candidate.

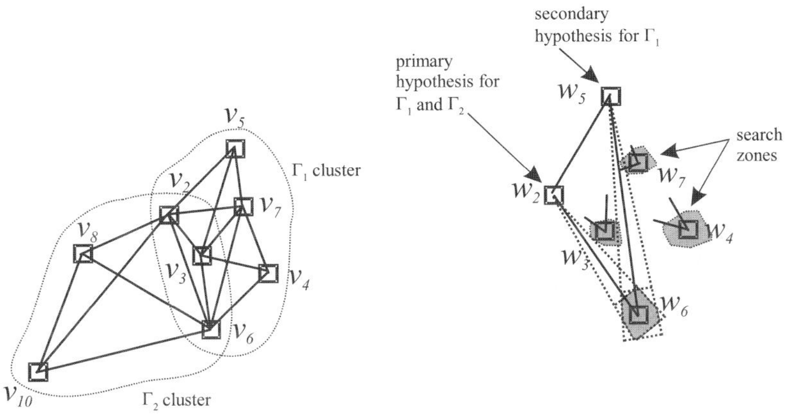

Fig. 9.4. Hypothesis propagation to neighbor clusters.

As Figure 9.5 shows, successful candidate generation can be propagated through neighbor clusters. Two clusters may be said to be *connected* if they can be linked by an arbitrary long chain of overlapping clusters; the set of clusters can be partitioned into subsets of connected elements, or *groups*. The computational efficiency of the process is increased by maximizing propagation, so that candidates are reused whenever possible and direct search and error-minimization is avoided. In addition, window candidates that are already supported by a cluster candidate should be preferred to unsupported candidates.

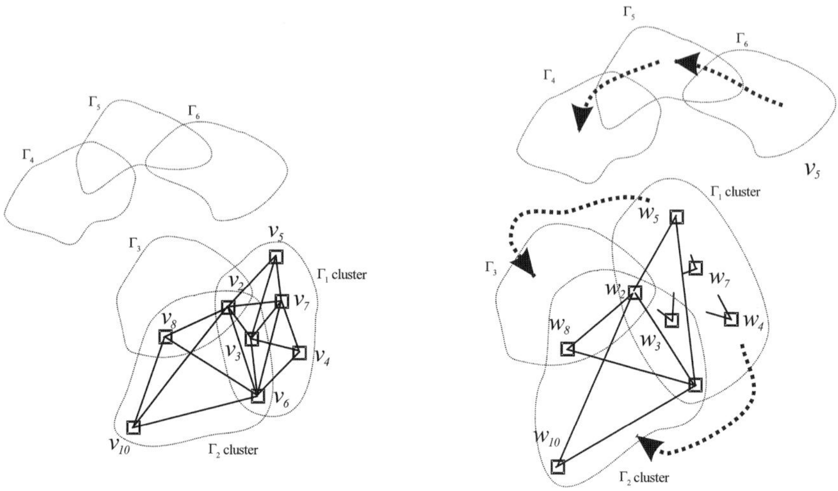

Fig. 9.5. Hypothesis propagation through cluster groups

At the final stage of the method, the best cluster candidates are used to generate clusters in the last image, and determine the matching windows for each sequence. A *dynamic* merit factor is defined for this purpose. The dynamic merit factor is dynamically updated along the candidate generation process using the following expression:

$$m_d = \left(\frac{n_m - n_h}{n_m} \right) m_s \tag{9.3}$$

where m_d is the dynamic merit factor, m_s is the static merit factor, n_m is the number of members in the candidate cluster, and n_h is the number of *inhibited* members. The *inhibited* members of the candidate cluster are those sequence candidates which fit into the cluster, but have failed to be chosen as the best for their specific window sequence, because others have more support from their respective clusters. The static merit factor, n_m, is a fixed parameter that depends only on the characteristics of the candidate cluster - currently, it is just the number of members, but other parameters, such as the residual error of the members, or the fit error in equations (9.1) and (9.2) could be considered as well.

The practical result of the approach described in the preceding paragraphs is to drastically reduce the number of matching tries, which are by far the main component of processing time when a great number of features have to be tracked, and large search zones are needed to account for high speed image plane motion. This is the case in non-stabilized aerial images, particularly if only relatively low frame rate video streams are available.

Some Results

Figure 9.6 shows the matches obtained in two images from an aerial sequence in a low structured environment (during a fire monitoring flight). Figure 9.6, bottom, illustrates the cluster-based tracking algorithm.

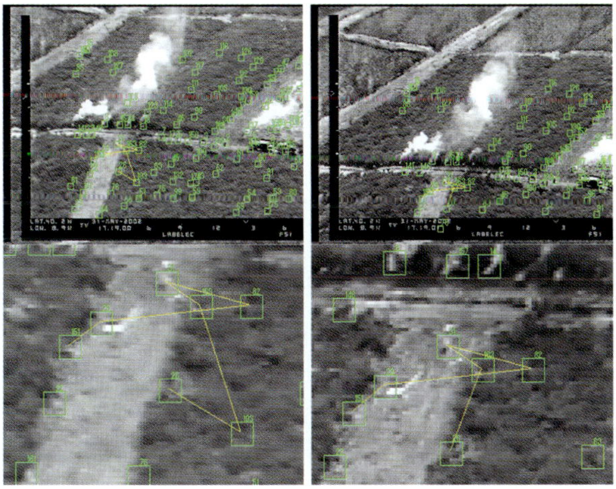

Fig. 9.6. Matches obtained in two images of an aerial sequence. Bottom: details of the cluster formation.

Figure 9.7 shows results obtained during the landing phase of the MARVIN autonomous helicopter (see Section 9.5). In this sequence, the method is able to handle both changes in orientation and scale.

Fig. 9.7. Matches obtained during a landing sequence.

Figure 9.8 illustrates results obtained from a flying helicopter. The images are taken with a nadir point of view. The translation and rotation between images are significant in this sequence.

Fig. 9.8. Matches obtained during a flight with a nadir point of view.

9.2.2 Blob Features

The point features extraction and matching method presented above is suitable to determine the image motion for images collected by one UAV. However, when considering separately the points of view or when the cameras are different (for instance, as in the case of views by different UAVs), other features with a higher degree of invariance are more convenient.

Homogeneity features are called blobs in scale-space theory [19]. In contrast to segmentation, blob detection does not attempt to segment out exact shapes of objects, but to extract robust and repeatable features, discarding exact shapes and thin connections between patches, see Figure 9.9.

Blob features are also related to maximally stable extreme regions (MSER) [23] and to affine invariant neighborhoods [41]. MSER features are defined from regions found by selecting the locally most stable threshold of the intensity function. Regions defined by being above the threshold (MSER+), and regions below it (MSER-), are used to define affine invariant frames [23]. Affine invariant neighborhoods [41] are found by starting at intensity extrema and finding the nearest extrema along rays emanating from the point. These extrema are linked to a closed curve, which is used to define an affine invariant.

In contrast to [23] [41] that use intensity only, blob features are used here that are extracted from color images, using a clustering pyramid built using robust estimation in local image regions [12]. Each extracted blob is represented by its average color $\mathbf{p}_k$, area a_k, centroid $\mathbf{m}_k$, and inertia matrix $\mathbf{I}_k$, each blob is a 4-tuple:

$$B_k = \langle \mathbf{p_k}, a_k, \mathbf{m_k}, \mathbf{I_k} \rangle \quad (9.4)$$

Since an inertia matrix is symmetric, it has *3* degrees of freedom, and total of *3+1+2+3=9* degrees of freedom for each blob. Figure 9.10 gives a brief overview of the blob detection algorithm.

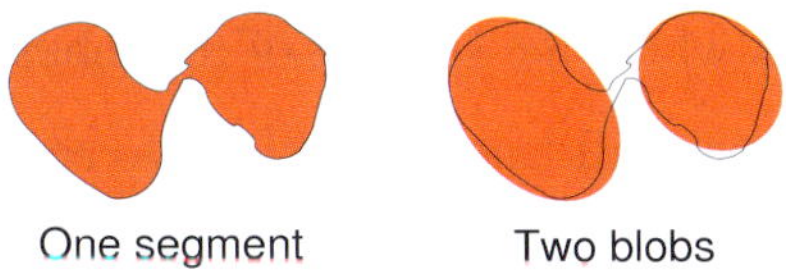

Fig. 9.9. Difference between segmentation and blob detection.

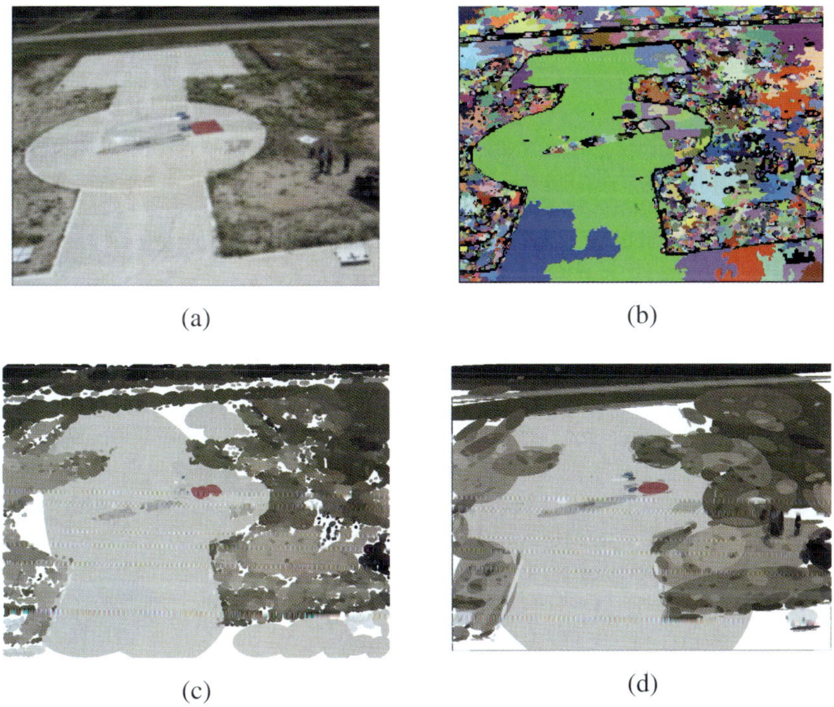

Fig. 9.10. Steps in blob extraction: (a) Input image, (b) label image, (c) raw blobs, and (d) merged blobs.

Starting from an image, the algorithm constructs a clustering pyramid, where a pixel $\mathbf{p}^*$ at a coarser scale is computed as the robust average of *12* pixels $\mathbf{p}_k$ at the lower scale:

$$\arg\min_{\mathbf{p}^*} \sum_k r_k \rho\left(\left\|\mathbf{p}_k - \mathbf{p}^*\right\|\right) \tag{9.5}$$

Figure 9.11, left, shows an illustration. The function $\rho(x)$ is the robust error norm, which in this case is a truncated quadratic, with a cut off parameter $d_{\max}$. Regions where the support of the robust average is below a threshold $c_{\min}$ have their confidence flag r^* set to zero.

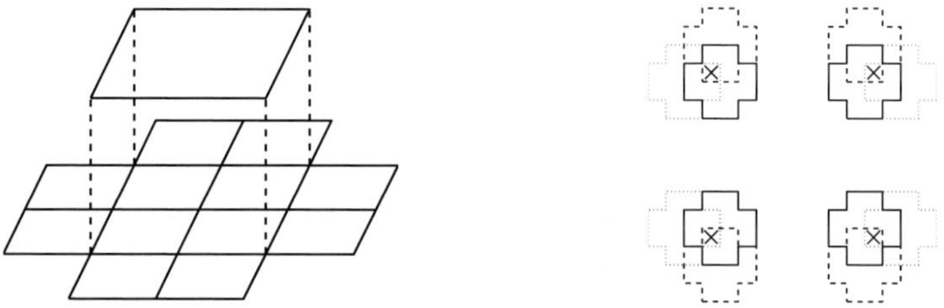

Fig. 9.11. Label propagation. Left: A pixel is computed as the robust average of twelve pixels at the scale below. Right: When traversing the tree top down, there are four possible alignments of a pixel with respect to pixels on the scale above. In each case, the label of a pixel can be determined by comparing its value with three pixels on the scale above.

The algorithm then creates a label image using the pyramid, by traversing it top-down, propagating the label of the most similar pixel on the scale above, see Figure 9.11, right. New labels are assigned to points which have their confidence flag set, but don't contribute to any robust mean on the level above.

The labeling produces a set of compact regions, which are then merged in the final step by agglomerative clustering. Regions with similar color and which fulfill the condition:

$$M_{ij} > m_{thr}\sqrt{\min(a_i, a_j)} \tag{9.6}$$

are merged. Here M_{ij} is the count of pixels along the common border of blobs B_i and B_j, and m_{thr} is a threshold. Typical values for the algorithm parameters are:

$$d_{\max} = 0.16,\ c_{\min} = 0.5,\ m_{thr} = 0.5$$

More information about the feature estimation may be found in [12], and the on-line implementation at http://www.isy.liu.se/~perfo/software/. The current implementation processes $360x288$ RGB images at a rate of 0.22 sec/frame on an AMD Athlon 64, 3800+ CPU 2.4GHz, and produces relatively robust and repeatable features in a wide range of sizes. Analysis of the algorithm robustness is presented in [11].

The blob features can be used to obtain a sparse image motion field by obtaining matches between blobs. Later on in this Chapter it is shown how these matches can be computed for the special case of blobs extracted from images acquired using different cameras.

9.3 Multi-View Motion Estimation for a Single UAV from Planar Homographies

The point-like features presented in the previous Section can be used for image motion estimation. The match of these features along the sequence of images allows one to obtain a sparse image motion field. Under certain circumstances, this local image motion can be transformed into local UAV motion. In the general case, this implies the recovery of the 3D position of the features used for motion estimation. For instance, if artificial landmarks or patterns of known size are used as features, it is possible to obtain metric relations from image motion.

Combining stereo vision with robust feature matching allows obtaining the UAV relative motion, as the range of the features can be computed. The visual odometer of [1] uses a stereo bench for range estimation of the features, and fuses the estimated image displacement with the information provided by angular sensors to estimate the UAV motion. Also, the algorithm presented in [20] combines pixel tracking and stereo vision for motion estimation. In [16] it can be seen the application of this technique for an autonomous blimp. The main problem of using stereo vision with UAVs is the need of wide stereo benches to cope with the associated altitudes.

Another option is to consider approximate models of the scene. If the scene is planar, then it is possible to recover the motion of the UAV without explicit 3-D reconstruction of the scene, as stated in [14] [40]. From the planar *homography* that relates two calibrated views of a planar scene, it is possible to determine the relative motion between the cameras, up to a scale factor. Actually, two potentially valid solutions are obtained. More than two views of the same plane can be used to verify the solutions [33]. The scale factor can be obtained if the distance from the camera to a given point of the plane can be measured (for example, the altitude over the terrain given by a range sensor). This is the approach presented here.

These methods can be used to sense the relative displacement of a UAV. However, in general, the estimated position drifts along the sequence of images. For long-range operation, this error should be compensated. This can be accomplished by using an environment model, comparing the current perceptions with the model. Although SLAM has been widely researched for indoor robotics, only few results for UAVs have been presented. For example, [16] includes a stereo-based motion estimation into a SLAM filter for the localization of a blimp, and [17] presents a SLAM method for airplanes.

An alternative approach is considered here. No mapping is actually done, in the sense that no 3-D information is recovered. Assuming a planar scene, the 3-D information is fully defined by the plane normal and the distance to the vehicle. Instead, a mosaic is built from the images gathered by the UAV using the homographic relation induced by the planar terrain. The mosaic stores all the information gathered by the camera and it contains the information that relates the images and the planar terrain. When a previously registered scene is detected, the algorithm will try to match this view with the mosaic to reduce the accumulated drift.

Figure 9.12 illustrates the steps of the method. By using the information given by the point matching algorithm already described, the local alignment between consecutive images can be obtained. The local transformation is modeled by a homography, computed from the matches using a robust outlier rejection procedure. By composing consecutive homographies, the current image can be transformed into a global frame, in which the mosaic is built. However, if the transformed frame overlaps with the current mosaic, then the common information is used to refine this global alignment. Finally, this global transformation is used to compute the motion between the current camera position and the initial one.

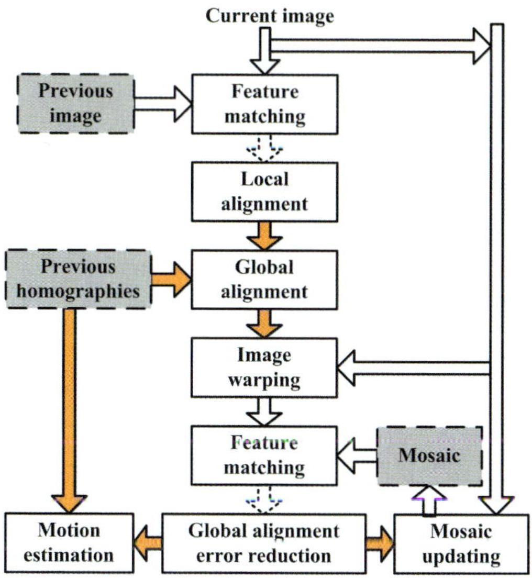

Fig. 9.12. A feature matching algorithm allows for aligning the current image with the previous one. Combining motion, the image is aligned in a global frame. This image motion can be transformed into UAV motion. The drift in the position can be compensated by using a mosaic as a reference map. For images acquired using one camera, the point feature tracking suffices. For images acquired using different cameras, blob features are used.

Moreover, as it will be shown later on, the algorithm of Figure 9.12 can also be applied to sense the relative pose between different UAVs. The main difficulty is to obtain robust matches between features extracted in images gathered from different views. Point-like features did not provide suitable results for this task; hence blob features will be used.

9.3.1 Motion from Homographies: An Overview

The homography that relates two views, 0 and 1, of the same planar scene (see Figure 9.13) may be expressed as:

$$\mathbf{H}_{01} = \mathbf{A} \cdot \mathbf{R}_1 \cdot \left(\mathbf{I} - \omega \cdot \mathrm{t}_1 \cdot \mathbf{n}^T\right) \cdot \mathbf{A}^{-1} \tag{9.7}$$

where $\mathbf{A}$ is the intrinsic calibration matrix of the camera, $\mathbf{t}_1$ is the relative displacement of the second point of view in the first camera coordinate frame, $\mathbf{n}$ is a unitary normal vector to the plane in the first camera coordinate frame (in the outward camera direction), $1/z$, z being the distance from the first camera to the plane, and $\mathbf{R}_1$ is the rotation matrix that transforms a vector in the first camera coordinate frame into a vector expressed in the second camera coordinate frame.

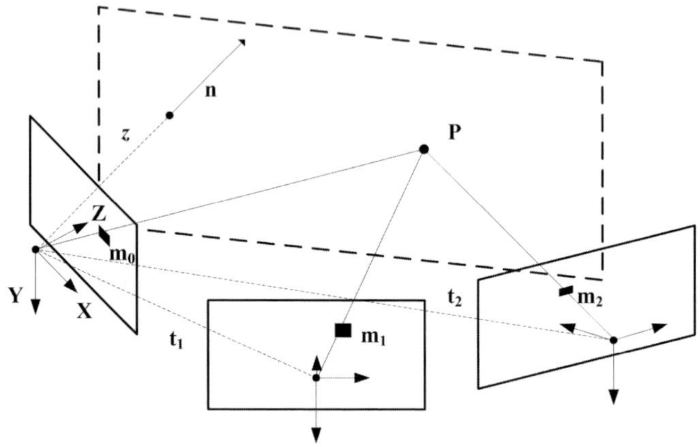

Fig. 9.13. Multi-view geometry for a planar scene.

If the camera is calibrated ($\mathbf{A}$ is known), then it is possible to obtain two valid solutions for $\mathbf{R}_1$, $\mathbf{t}_1$, y $\mathbf{n}$ from $\mathbf{H}_{01}$ up to a scale factor (due to the depth-speed ambiguity). If a third view is considered, knowing $\mathbf{H}_{02}$ it is possible to obtain a unique solution because the plane normal $\mathbf{n}$ must be the same for all relations $\mathbf{H}_{0i}$ [33].

So, given more than two images and the homographies that relate each image with the reference frame, it is possible to compute the two solutions for $\mathbf{R}_i$, $\mathbf{t}_i$, y $\mathbf{n}_i$ associated with each homography $\mathbf{H}_{0i}$, for instance from the *SVD* decomposition of $\mathbf{H}_{0i}$, as described in [39] and [40]. Then, the correct solution should be computed for all images. In [33] a method to compute the correct solution based on the uniqueness of the normal $\mathbf{n}$ is proposed. Given the set of possible unitary normal vectors:

$$S_n = \{\mathbf{n}_1^{\,1}, \mathbf{n}_1^{\,2}, \mathbf{n}_2^{\,1}, \mathbf{n}_3^{\,2}, \mathbf{n}_3^{\,1}, \mathbf{n}_3^{\,2}, \ldots\} \tag{9.8}$$

where the superscript indicates the two possible normal solutions and the subscript the image number sequence. This method defines an empirical threshold ε and considers as the correct solution the normal $\mathbf{n}_1^{\,1}$ or $\mathbf{n}_1^{\,2}$ that allows finding a sequence of normal vectors accomplishing:

$$\left\| \mathbf{n}_1^{\,1,2} - \mathbf{n}_i^{\,1,2} \right\| \leq \varepsilon \qquad \forall i > 1 \tag{9.9}$$

where ε must be carefully tuned to guarantee a proper algorithm operation. If the value of ε is too high then the two possible solutions can accomplish the constraint and it is not possible to detect the correct one. On the other hand, if ε is too low then no solution will accomplish the constraint.

A more reliable method is proposed to detect the correct solution. Thus, if $\mathbf{n}_1^1$ and $\mathbf{n}_1^2$ were correct, there would be two sets of solutions, $S_{\mathbf{n}^1}$ and $S_{\mathbf{n}^2}$. According to the uniqueness of the normal the following constraints are proposed:

$$\left\|\mathbf{n}_1^1 - \mathbf{n}_i^j\right\| \le \varepsilon_1 \qquad \forall \mathbf{n}_i^j \in S_{\mathbf{n}^1} \tag{9.10}$$

$$\left\|\mathbf{n}_1^1 - \mathbf{n}_i^j\right\| \le \varepsilon_1 \qquad \forall \mathbf{n}_i^j \in S_{\mathbf{n}^2} \tag{9.11}$$

where ε_1 and ε_2 are the minimal values that guarantee a unique solution for (9.10) and (9.11), respectively. The pairs $\{S_{\mathbf{n}^1}, \varepsilon_1\}$ and $\{S_{\mathbf{n}^2}, \varepsilon_2\}$ are computed separately by means of the following iterative algorithm:

- *Step 1*: The distances between $\mathbf{n}_1^i$ and the rest of the normal vectors of $S_{\mathbf{n}}$ are computed.
- *Step 2*: ε_i is set to an initial value.
- *Step 3*: For the current value ε_i, check if there exists a unique solution.
- *Step 4*: If no solution is found, increase the value of ε_i and try again using Step 3. If multiple solutions are found decrease ε_i and try again Step 3. If a unique solution is found then finish.

The correct solution is chosen between both options as the one that achieves the minimum ε .

Once $\mathbf{n}$ is known, it is possible to obtain $\mathbf{R}_i$. The translation $\mathbf{t}_i$ is computed up to a scale factor ($\omega|\mathbf{t}_i|$). If an independent sensor is used, such as range sensor, the scale factor can be solved by measuring the distance from the first view to the plane ($1/\omega$).

9.3.2 On-line Mosaicking

In order to compensate for cumulative errors, a mosaic is used as a model, and it is built dynamically while the UAV is flying. The proposed method is based on three steps: *local and global alignment computation*, *global alignment error reduction* and *mosaic updating*. No navigation data is required, and only a minimal image overlap will be assumed with the preceding and the succeeding images.

The first step computes the scene apparent motion between the current image and the previous one. This motion is modeled by a homography and then it is composed with the previous ones in order to align the current image to the reference frame.

The second step is carried out in order to reduce the positioning error of the images inside the mosaic. While the first step computes the relative motion without any consideration about the positioning in the mosaic or previous error in the computation of the relative motion, this stage tries to detect inconsistencies between the estimated position obtained from the composition of the relative motions and the position of the image in the mosaic built until now. If an inconsistence is detected, it will be modeled and compensated.

The third step deals with the functions needed to integrate images into the mosaic.

9.3.2.1 Local and Global Alignment Computation

The goal is to compute the relative motion between two consecutive images, and then, the transformation that aligns the current image with the reference frame.

First, the apparent image motion between consecutive images is extracted by means of the point-like feature tracking function that provides a set of matches between the previous image and the current one. This image motion is modeled by a homography matrix. The model holds when the camera undergoes a pure rotation or when the scene is planar [27], which is a good approximation when an UAV flies at high altitude. The homography computation from point matches is described with detail in [7] and [14].

Given the linear nature of the homographic transformations in homogeneous coordinates, it is possible to express the image motion from image n to the reference frame k, $\mathbf{H}_{kn}$ as a composition of the relative motion between the intermediate images. If T_I is a sequence of n images of a planar scene:

$$T_I = \{I_0, I_1, I_2 \ldots, I_n\} \tag{9.12}$$

and the $n-1$ homography matrices that relate these images are known:

$$\mathbf{H}_{01}, \mathbf{H}_{12}, \mathbf{H}_{23} \ldots \mathbf{H}_{(n-1)n} \tag{9.13}$$

then, the transformation from image n to k will be defined by:

$$\mathbf{H}_{kn} = \prod_{i=k}^{n-1} \mathbf{H}_{i(i+1)} \tag{9.14}$$

Thus, it is possible to compose the complete motion between whatever image of the sequence and the reference frame by means of the relative motion computation. In the rest of the Chapter, the first image (image 0) of the sequence will be taken as reference frame.

The main drawback of this method is the accumulated positioning error of the images inside the mosaic. The homography estimation involves outlier rejection, least squares and robust model estimation over a set of matching candidates. Although by using the techniques presented in [24] the residual error of the computed homography is usually small, equation (9.14) leads to error accumulation. In addition, if the number of outliers is too high (independent object motion or parallax) or the number of matches between images is small (blurred images or small image overlap) the residual error can be considerable.

Figure 9.14 shows a piece of a mosaic that reflects the effects of the accumulative errors in the homography estimation. The errors are more evident when the mosaic is composed by images taken in large loops and the accumulated registration errors generate visible inconsistencies in the global image alignment.

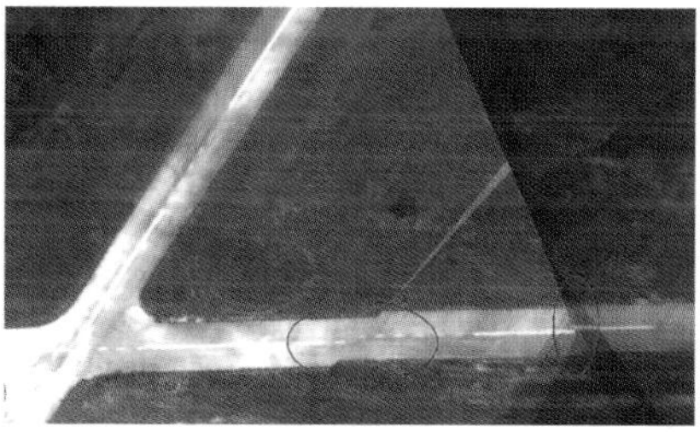

Fig. 9.14. Piece of mosaic with global alignment errors (ellipses).

9.3.2.2 Global Alignment Error Reduction

The idea is to use the information stored in the mosaic to refine the global transformation from the current image to the common frame (transformation that is used later to estimate the motion of the vehicle).

Recently, new iterative methods have been used to minimize global alignment errors in mosaics via topology inference, see [15], [27] and [34]. Unfortunately, real-time constraints led to face this problem by using non-iterative methods. The proposed approach tries to detect global alignment errors each time a new image is introduced in the mosaic.

At time i, the global alignment estimation procedure provides an initial estimation of $\mathbf{H}_{0i}$, that relates the current image I_i and the mosaic M_{i-1}.

Due to the accumulated drift at time i, the warped image $I_i^w = \mathbf{H}_{0i}(I_i)$ will be misaligned respect to its correct position within the mosaic (see Figure 9.15). This alignment error is modeled by a full homography matrix called $\mathbf{H}_i^e$, so that:

$$\mathbf{H}_{0i}^{true} = \mathbf{H}_i^e \cdot \mathbf{H}_{0i} \tag{9.15}$$

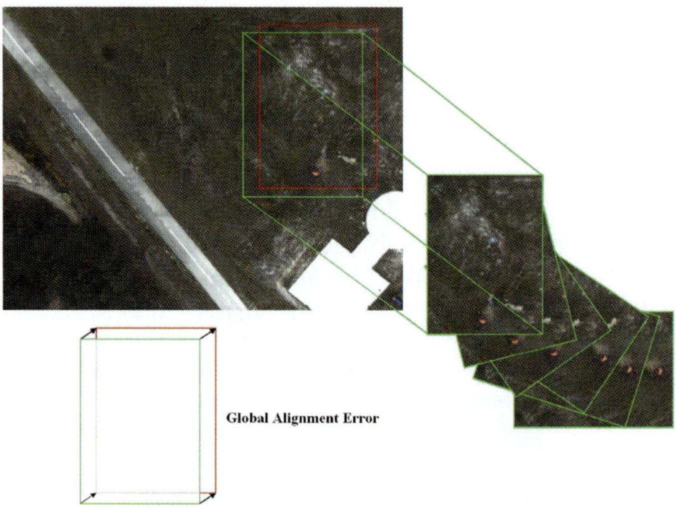

Fig. 9.15. Global alignment errors due to accumulated mis-registration. The red square denotes the correct position inside the mosaic. The green square mark the position computed by using local alignment.

The homography $\mathbf{H}_i^e$ is estimated by matching the warped image I_i^w and the portion of the mosaic M_{i-1} where I_i^w should be placed. This matrix will encapsulate all alignment errors from frame 0 to i. Combining (9.14) and (9.15), the final alignment expression that aligns the image I_i to the mosaic M_{i-1} is given by:

$$\mathbf{H}_{0i}^{true} = \mathbf{H}_i^e \prod_{j=0}^{i} \mathbf{H}_{j(j+1)} \tag{9.16}$$

This relation is then used to compute the motion from the current camera position to the position at the first frame.

The alignment error reduction is carried out only when the UAV visits zones already stored in the mosaic. $\mathbf{H}_i^e$ will be computed if the estimated overlap between the image and the mosaic is greater than the 50% of the image total size. Also, the number of correct matches between the warped image and the mosaic is considered, and if this number is below a threshold, the correction is not used.

9.3.2.3 Mosaic Updating

Finally each pixel of frame i is transformed using the computed homography and inserted into the mosaic. The values of the pixels are interpolated using the most similar pixel approximation. Other interpolation techniques like bilinear or tri-linear were tested but worse results were obtained because these interpolation methods do not take into account the existing information in the mosaic.

Special attention is necessary in the image insertion process in order to increase the mosaic coherence. In outdoors scenarios, the illumination of the images usually changes drastically, mainly due to the camera auto-exposure function. Thus, it is necessary to equalize the illumination of the images introduced in the mosaic. Once the image is warped, it is possible to compare the pixel values between the mosaic M_{i-1} and the image I_i^w. Given the set of overlapped pixels ξ_i and the size of the set $|\xi_i|$, the value:

$$\mu_i = \frac{\sum_x (M_{i-1}(x) - I_i^w(x))/2}{|\xi_i|} \qquad \forall x \in \xi_i \tag{9.17}$$

will define the illumination correction factor for the image i. This factor is an estimation of the illumination error average in the overlapped area. It is important not to use saturated pixels (minimum or maximum gray level) in the computation of μ_i in order to increase the accuracy. This value μ_i is added to the pixel values of the warped image I_i^w, which is finally introduced in the mosaic. Figure 9.16 shows a section of mosaic with illumination factor correction and without.

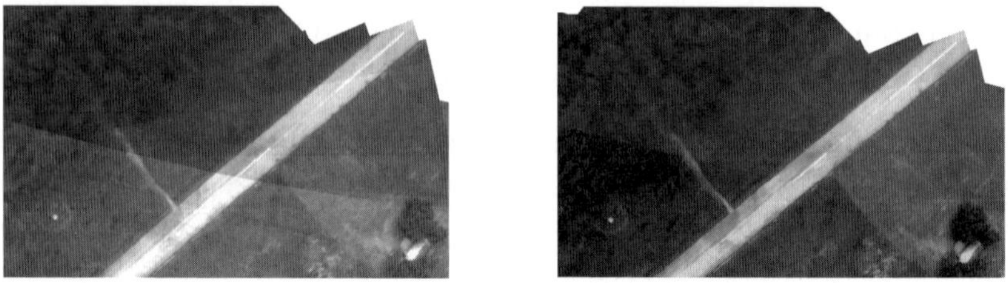

(a) (b)

Fig. 9.16. Illumination correction. (a) Segment of mosaic without illumination correction. (b) Segment of mosaic with illumination correction.

9.3.3 Experimental Results

This method has been applied for simultaneous single UAV mosaic building and motion estimation. Figure 9.17 shows a $3300x3500$ mosaic built with more than 500 images taken by the autonomous airship KARMA, developed by LAAS [16] as part of experiments for the COMETS project. KARMA flew at approximately 150m of altitude in these experiments. The mosaic computation used all previously described techniques: local alignment, global alignment error reduction and illumination error correction.

Figure 9.18 top shows a comparison between some DGPS measures and ego-motion experimental results obtained using this procedure. The effect of the mosaic correction can be observed in Figure 9.18 bottom. It is important to notice that this procedure it is not suitable for big loops because the mosaic building cannot be corrected once an image is fixed within the mosaic. More details and results are presented in [3].

Fig. 9.17. Lousa Airfield (Portugal). On-line mosaic built from more than 500 aerial images from Karma UAV using the proposed technique.

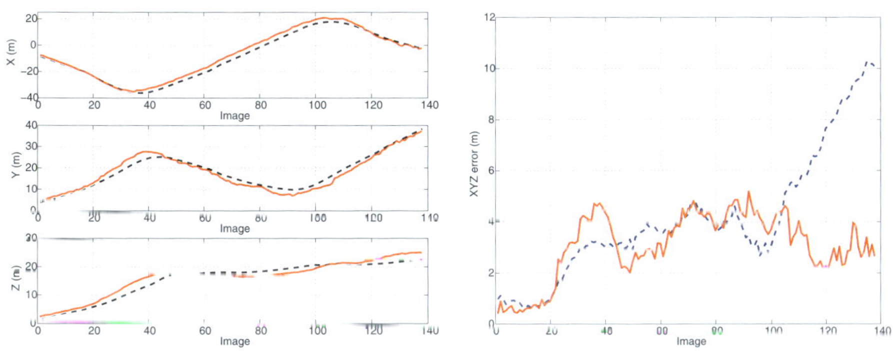

Fig. 9.18. Experiments at Lousa (Portugal). Top. Translation in the X (Easting), Y (Northing) and Z axis (Altitude) in meters. Path of Karma obtained by GPS (dashed) and using the multi-view planar motion method with mosaic correction (solid). Bottom: Absolute error between DGPS and estimated locations for the method with (solid) and without (dashed) mosaic correction.

9.4 Multi-UAV Relative Position Estimation

In the last years, fleets of multiple aerial robots or cooperating aerial-ground robots are being considered. As stated at the beginning of this Chapter, cooperation of several UAVs can involve important advantages in certain applications, such as surveillance or precise monitoring of events. In [42] ground and aerial vehicles cooperate in pursuit-evasion games; in [35] application of multi-UAV SLAM is presented; application of a heterogeneous fleet of UAVs for forest fire detection is presented in [22].

Computation of the relative position between UAVs is needed to share information for surveillance and inspection applications. In some multi-robot systems, the problem is solved if the robots are capable of identifying and localizing each other [18]. However, it is difficult to extrapolate these techniques to fleets of UAVs since they can not detect each other with simple sensors (UAVs fly fast and at different altitude) and in many cases can not carry on board the required complex sensors.

An alternative approach is followed here. If different UAVs identify, using their cameras, common objects in the scene, the relative pose displacement between the UAVs can be computed from these correspondences under certain assumptions. The method described in Section 9.3 obtains the relative displacement between two views of the same planar scene. The relations presented there also hold if the views are taken by different calibrated cameras. Thus, the relative displacement between two UAVs can be obtained by computing the plane-induced homography matrix between their cameras.

The main problem is to obtain matches between images gathered by different UAVs to obtain this homography. The features to be matched should be identified in a robust and repeatable way in images from different cameras at different locations and orientations, and maybe different lighting conditions. It is difficult to find point-like features with these characteristics. Thus, first, it is presented how the blob features described can be used to solve this problem.

9.4.1 Blob-Based Image Matching and Homography Estimation between Images from Different UAVs

This technique uses image matching techniques based on blobs representing natural landmarks. To facilitate image matching a color correction method compensates the color spectrum of the images to be matched. Then, a RANSAC-based [10] approach is used to estimate the homography between two views. The homographies are estimated from matches of sets of so called blob-chains.

9.4.1.1 Color Correction

Color correction techniques transform pixel values in an image I to a new image I', such that they appear as if acquired with a different camera, or under different lighting conditions. The color correction mapping converts an RGB color vector $\mathbf{p}$ from I to a new vector $\mathbf{p}'$, which is how $\mathbf{p}$ should look if it had been acquired using the camera and lighting conditions in I'. The conversion is a plain linear mapping:

$$\mathbf{p} = \mathbf{W}\mathbf{p}' \tag{9.18}$$

When only lighting differs between the two images, the mapping $\mathbf{W}$ is a $3x3$ diagonal matrix, and if correction between two different cameras is sought a full $3x3$ matrix is used [6] [9].

In [38] a method is described for estimating a color correction mapping $\mathbf{W}$ from a pair of images I and I', using their color moments of order two and three. Tran's algorithm assumes that the contents of the two images are near identical. This reduces the problem of finding $\mathbf{W}$ to that of finding a transformation that maps the color covariance matrix in image I to that of image I':

$$\mathbf{C}' = \frac{1}{N}\sum \mathbf{p}\mathbf{p}'^T = \frac{1}{N}\sum \mathbf{W}\mathbf{p}\mathbf{p}^T\mathbf{W}^T = \mathbf{W}\mathbf{C}\mathbf{W}^T \tag{9.19}$$

By also adding constraints on the transformation of third degree moments, it is then possible to define $\mathbf{W}$ uniquely, and to estimate it.

This method is applied to a collection of photos of babies, and finds that is does not work satisfactorily, since the images have too different contents. The method is amended by first segmenting the images into foreground and background, and then applying the method between the foregrounds and the backgrounds separately.

The previous improvement is enhanced further. Consider a pair of images of the same scene, from similar view points, but acquired with two different cameras, see Figure 9.19, left and center. Detect blobs in these images, and find a set of blob correspondences, $\mathrm{B}_i \leftrightarrow \mathrm{B}'_j$. From the color vectors $\mathbf{p}_i$ and $\mathbf{p}'_j$ in the corresponding blobs it is now possible to define a set of constraints on $\mathbf{W}$:

$$\mathbf{p}_i = \mathbf{W}\mathbf{p}'_j \qquad \forall \mathrm{B}_i \leftrightarrow \mathrm{B}'_j \tag{9.20}$$

For three correspondences **W** will be uniquely defined, and for a higher number of correspondences the least-squares fit can be used to estimate **W** . Figure 9.19, right, shows the center image after color-correction.

Fig. 9.19. Left and center: Two images from two different UAVs with a video sensor and a digital photo camera, respectively. Right: The image from the high resolution camera is color corrected to match the color spectrum of the image from the video camera.

Note that the color-correction method has eliminated the need for each color to occupy approximately the same fraction of the images. However, for very large differences in illumination this method will most likely encounter problems, since large illumination differences will reduce the number of blob correspondences that can be found. As in all color correction methods this one also requires that the set of corresponding blobs contain sufficiently many different colors.

Typically, the color correction is only estimated once for a specific set of cameras and lighting conditions. The same mapping can be reused throughout the Session.

9.4.1.2 Finding Blob-Chain Correspondence Hypotheses

Blob matching extracted from images with different points of view is not an easy problem. The approach for solving this problem is to establish relations between blobs as well as between chains of blobs extracted from each image.

The first step is to find initial guesses of correspondences between a blob in one of the images and the blobs in the other, resulting in starting blobs for a set of chains. This is done using only the color of the blobs. Once these correspondence guesses are made, the rest of the blobs are matched by using criteria based on the features of the blobs and also based on normalized distances and relative angles between blobs, see Figure 9.20. This procedure assures that the matching is invariant to scale and rotation.

The matching is done by picking a blob close to the previous blob and then search for blobs with similar relations among the correspondence guesses. If the extracted features for the blobs to match are similar, the blobs are included in the blob chains. If not, another blob to search for is picked. The errors in the blobs of a chain are accumulated. This error is used to estimate the confidence that the two blob chains is the same. If the confidence falls below some threshold, the chain is removed from the list of chains. Also if another chain has a much higher confidence the chain is removed.

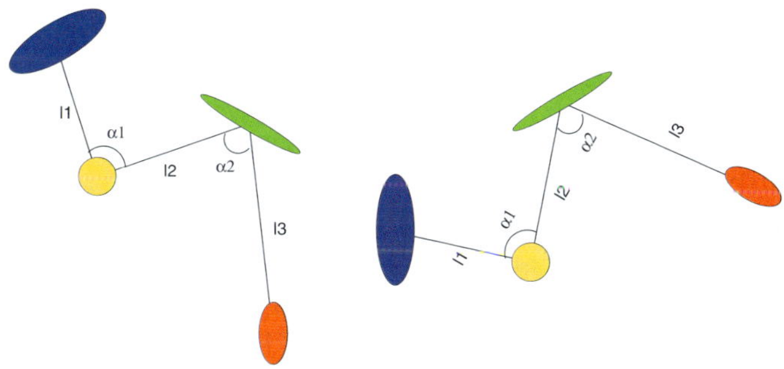

Fig. 9.20. Blob chains contain the ratios between consecutive distances and relative angles in the chain. These inter-blob features make the feature vector of the chain rotation and scale invariant.

The probability of matching correct blobs is high when finding close blobs since the predicted position of the next blob degrades as the distance to the previous blob increases, especially with large differences in perspective between the views to match. The probability of matching correct blobs is further increased by giving higher priority in the selection of blobs to those with unusual color. This also implies a fast reduction of the number of chains. The whole procedure can be summarized as:

- *Step 1*: Select the blob with the most unusual color in the image from one of the views. The blob should not have already been used for matching.
- *Step 2*: Find all similar blobs in the image from the other view. These are the beginnings of the chains.
- *Step 3*: Select a close and unique blob to the latest matched blob which has not yet been picked.
- *Step 4*: Try to find matching blobs for the chains and update the error measures. Remove the chains with too high error.
- *Step 5*: If the chain match between the two views is not unique, or there are too few blobs in the chain, go to Step 3.
- *Step 6*: Store the position of the blobs if a unique chain with at least a certain number of blobs was found.

- *Step 7*: If the total number of stored blobs is less than a threshold, go to Step 1 and pick another start blob.

The result of this step is a set of blob-chain pairs, which can be used as initial hypotheses in the homography estimation process. Typically, there is in the order of 100 blob features found in each view. About one quarter of these are successfully matched. The number of blob-chain pairs is in the order of 10. These numbers indicate that the processing time and memory consumption are low.

9.4.1.3 Homography Estimation from Blob Correspondences

In order to apply the method described in Section 9.3.1 to determine the relative displacement between two cameras, the homography matrix $\mathbf{H}$ that relates the two images must be computed. The matrix $\mathbf{H}$ has 8 unknowns so it requires finding correspondences between at least 4 blobs. A total least squares method with iterative outlier rejection is used to estimate $\mathbf{H}$. The estimation is made using a RANSAC based algorithm:

- *Step 1*: Take two pairs, at random, of corresponding chains and estimate a homography from these correspondences.
- *Step 2*: Transform all blobs with this homography and find new blob correspondences. Estimate a new homography with these correspondences.
- *Step 3*: Iterate Step 2 two-tree times to refine the homography.
- *Step 4*: A confidence measure is calculated as a function of how well the homography was fitted to the blobs, the number of blobs used in the homography estimation and the distribution of the blobs in the image.
- *Step 5*: If the confidence measure is higher than any previous value, save the estimated homography as the best solution found so far.
- *Step 6*: Go to Step 1 as many times as feasible. Typically this implies a limited number of iterations.

If a sequence of images is used, the extrinsic camera parameters estimated using the procedure described in the previous section can be stabilized with a Kalman filter and the predictions of the parameters be used to make the blob matching faster and more accurate. Given a prediction of the extrinsic camera parameters, a predicted homography can be determined.

The matching can then be improved by performing operations such as the following:

- Match only with blobs close to the positions given by the predicted homography.

- Compensate for the perspective distortions of the blobs by transforming the coordinate system with the predicted homography.
- If the confidence for the predicted homography is high, use it as input to the iterative homography estimation directly.

The result of a wide baseline matching algorithm on blob features is shown in Figure 9.21. The left image originates from on of the UAVs presented in Section 5, with a color video sensor. The right image originates from a different UAV with a high resolution digital camera. The matching blobs are overlaid in both views in Figure 9.21. As pointed out earlier, finding the homography assumes that the blob features are located on a planar surface. Figure 9.21 shows that this is not always true. However, in practice the procedure can cope with situations when the planar surface requirement is somewhat relaxed. This means that the distance from the camera to the ground should be fairly large compared to the height differences on the ground. If this is the case, the problem of occlusion in the views is also reduced.

Fig. 9.21. Blobs fulfilling the estimated homography between two images from different UAVs with video sensor and a digital photo camera, respectively.

One major problem is to get a sufficiently large common area in the two views, in order to get enough blob features to be able to calculate the correspondences. Since blobs which are crossing the border of the image cannot be used, there is a requirement that there must be enough distinctive area patches inside the common area of each view. This can be accomplished in different ways; one possibility is to use partial mosaic images to get larger area coverage for the blob matching. The procedure has shown to give accurate estimates of the homography when the confidence measure is high.

9.4.2 Multi-UAV Relative Position Estimation

Assume that multiple networked UAVs are exploring a region and that at a certain time instant (the time of image 0) the same scene is viewed by two or more UAVs, for instance UAV 1, which is taken as the reference UAV, and UAV 2. The blob matching procedure is applied with the aim to match image 0 of the reference UAV with its most recent captured image of UAV 2. If a number of good blob matches are obtained, the algorithm of Section 3 is used to obtain the relative displacement between UAV1 and UAV2 at that time. Otherwise, the images are discarded.

In general exploration missions, the probability of having the same scene in the field of view of two or more UAVs is not very high. In Figure 9.22, $\mathbf{H}^{12}$ is the homography matrix that relates two images gathered by UAVs 1 and 2 for image 0 computed by blob matching. $\mathbf{H}^{j}{}_{0i}$, the homography matrix between image i and image 0 for UAV j is computed by composing the homography matrices between consecutive images. For instance, $\mathbf{H}^{2}{}_{02} = \mathbf{H}^{2}{}_{01} \cdot \mathbf{H}^{2}{}_{12}$. Combining the homographies $\mathbf{H}^{2}{}_{0i}$ with $\mathbf{H}^{12}$ determines the homography matrices of UAV 2 at different time instants along its trajectory with respect to the view of UAV 1 at image 0.

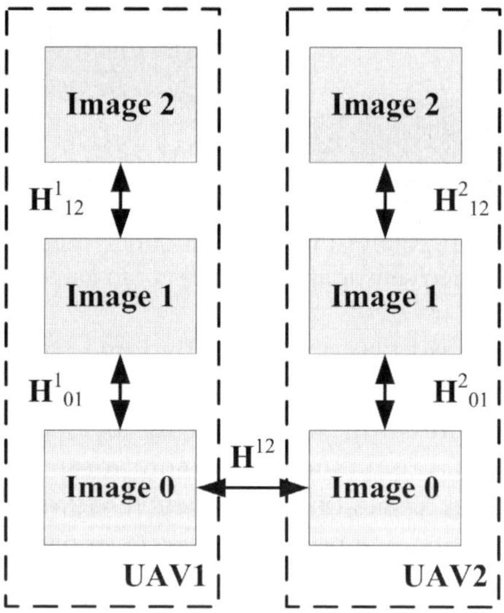

Fig. 9.22. Scheme of relations between homographies.

Figure 9.12 may be extended to multi-UAV systems. The homography matrices obtained from blob matches are used to compute the local alignment between images from one UAV and the reference UAV. This local alignment is refined (global alignment in Figure 9.12) by using the mosaic built by the reference UAV. This refined alignment is used to determine the relative position between UAVs. For illustration, Figure 9.23 shows a multi-UAV mosaic that is built from individual mosaics generated by the Heliv and Marvin UAVs in the experiments.

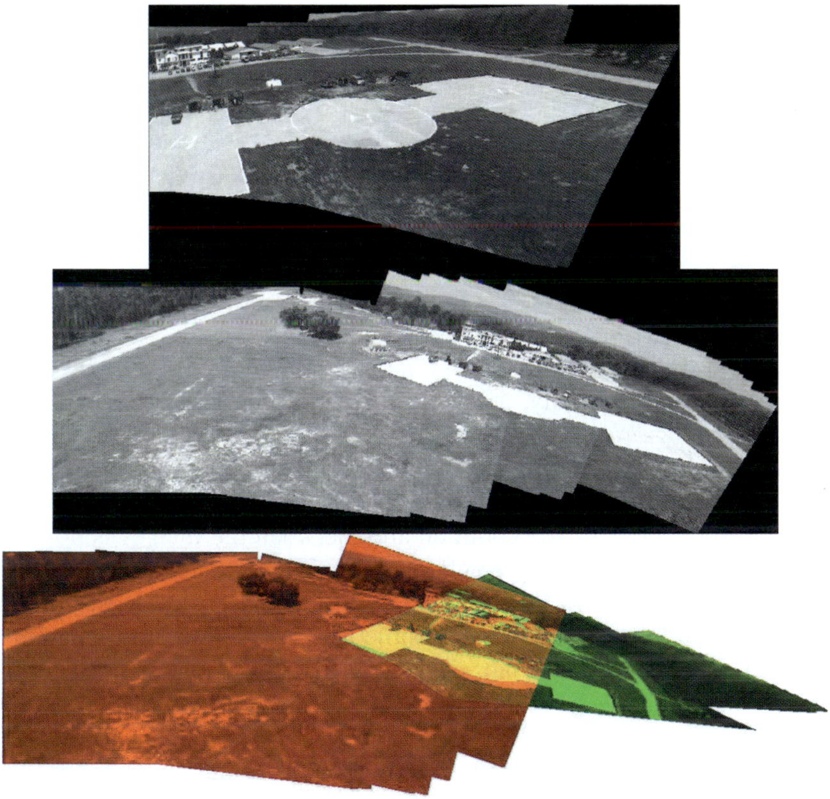

Fig. 9.23. Mosaics built individually by the Heliv and Marvin (top) and mixed using homography $\mathbf{H}^{12}$ (bottom).

Assuming that the position of the reference UAV in a global frame is known, it is easy to obtain the position of all the UAVs in the global frame. For example, locations of UAV 2 at time of image 2 with respect the reference UAV at time of image 0 are given respectively by $\mathbf{H}^{12}\mathbf{H}^{2}_{01}\mathbf{H}^{2}_{12}$. Moreover, as presented in Section 9.3, the algorithm provides the distance to the ground plane. This method could be used as a reliability tool in multi-robot fleets in case of failures of GPS. Figure 9.24 top illustrates a measured with DGPS (dashed line) and the results of the method (solid line).

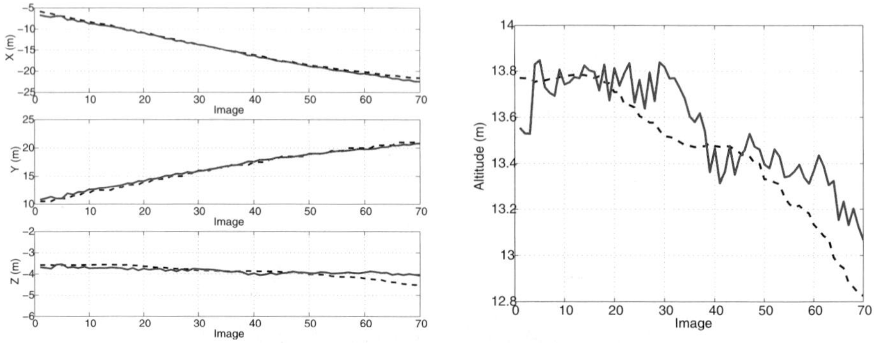

Fig. 9.24. Left: vision-based (solid) and GPS based (dashed) estimations of relative displacement of Marvin with respect to Heliv. Right: distance to the ground using the procedure described (solid) and GPS measures (dashed).

As mentioned previously, it is necessary to have distinctive areas to get a reliable homography computation. If the scene is too homogeneous, the method would not work. Most of the images taken during the experiments, such as those in Figure 9.21, contained enough structure and the method worked satisfactorily. If the common area between the images is not large enough, it would be unlikely to obtain enough blob matches. In this case, mosaics individually created by each UAV can be used, at a cost of memory storage (although it could be possible to store a blob representation of the mosaic, instead of the mosaic itself). The blob matching technique could be applied between an image and the mosaic so-far built. This would allow obtaining homographies even if the UAVs are not seeing the same scene at the same time at a cost of higher computational requirements.

9.5 Applications

In the COMETS project a distributed system for cooperative activities using heterogeneous UAVs was designed, implemented and demonstrated for cooperative fire detection and localization.

The objective is that a group of networked UAVs survey an area looking for fire alarms. If an alarm is detected, the fleet should localize it and confirm or discard it, taking benefit of the diversity of sensors.

The cooperative perception system tested on these experiments is a partially distributed system, with local UAV modules that perform low and mid level processing functions (like feature extraction, matching and segmentation) and a node that fuses the data collected by them. This node only receives the fire alarms and their positions. Thus, the bandwidth requirements are kept low. Also, the architecture can consider local UAV modules on ground depending on the processing power on-board each UAV.

The UAVs used in the experiments are the helicopters Marvin and Heliv and the airship Karma. Marvin is an autonomous helicopter developed by the Real-Time Systems and Robotics Group of the Technische Universität Berlin (Germany) [29]. Karma is an airship developed by LAAS at Toulouse (France) [16]. Heliv is the result of the evolution of a conventional remotely piloted helicopter, adapted by the Robotics, Vision and Control Group at the University of Seville (Spain) by adding sensing, perception, communication, and control functions.

Figure 9.25 shows the three UAVs in an experiment. Figure 9.26 shows some of the on-board electronics. Given the hard vibrations and motions induced by the UAV engine, the camera lenses were screwed to guarantee the same calibration along the flight. These assumptions were valid and sufficient in the experiments.

Fig. 9.25. Marvin, Karma and Heliv in a test in Lousa, Portugal, May 2005.

Fig. 9.26. Electronic box and cameras (left) mounted on board Heliv (right).

In the experiments carried out, Marvin and Heliv were commanded to patrol an area looking for fire alarms. Marvin is equipped with a camera and a fire sensor (photo detector adjusted to respond to ultraviolet radiations characteristic of fire emission). Heliv has an infrared and a visual camera.

The techniques described above have been applied over the images recorded by each UAV to generate an individual mosaic of the area such as those shown in Figure 9.23. Also, the technique in Section 9.4 computes the relative locations between the UAVs when they are looking at the fire.

The application of blob matching techniques in these images allows to compute the homography matrix $\mathbf{H}^{12}$. The composition of $\mathbf{H}^{12}$ with the local homographies is used to estimate the relative locations between the UAVs along the experiment. Figure 9.27 shows two views during a fire experiment with the resulting blob matches. From these matches, it is possible to estimate the location of the Heliv in the local frame of Marvin. Figure 9.28 shows the estimated position of Heliv compared to GPS measures. These results were obtained when the helicopters are hovering looking over the same scene, and therefore the probability of finding matches is high. In this experiment Marvin is equipped with a 2 cm accuracy DGPS system. Heliv carries a GPS with poor accuracy (several meters) and low rate (1 Hz), which results in high localization errors. These techniques can be used to compensate for these errors.

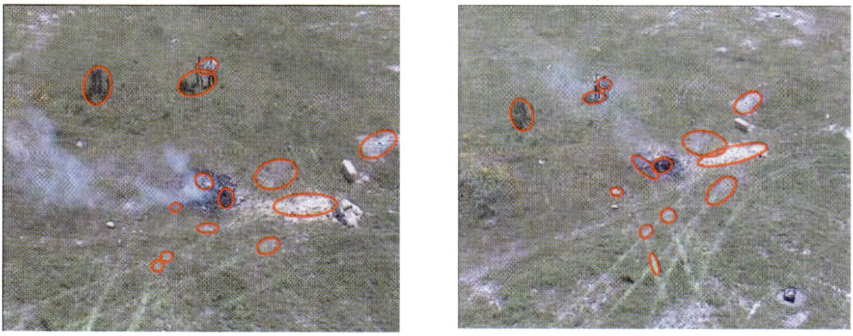

Fig. 9.27. Blob matching of a fire experiment from different points of view.

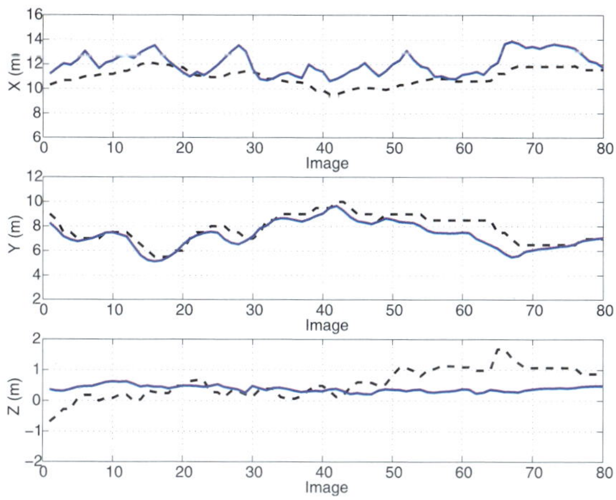

Fig. 9.28. Comparison between the estimated relative position of Heliv with respect to Marvin (solid) and GPS measures (dashed).

9.6 Conclusions

The Chapter has presented a vision-based method for single and multi-UAV localization. First, a method for single-UAV motion estimation from homographies between images of the same planar scene is described. These homographies are computed from matches between features extracted from consecutive images. The drift errors in the motion estimation procedure are compensated by simultaneously building a mosaic that stores past information about the environment, and that allows to discover inconsistencies in the estimated position.

Moreover, the use of blob features to obtain natural landmarks in low-structured scenes is proposed. If these blobs can be matched between images from different UAVs, provided that the scene is planar, the relative displacement between the UAVs can be estimated. The method has been tested field experiments with several UAVs. For a more thorough presentation of performance, stability, and repeatability of the blob feature matching algorithm [11].

The proposed technique increases the reliability in UAV operation. It can be applied to avoid GPS failures as a backup positioning system. The method has some limitations. The scene should contain enough structure, which is a common limitation for vision-based methods. The method worked satisfactorily in the field experiments carried out. Equally good results could be expected in urban scenes. The planar assumption is not a hard limitation since the effect of non-planar terrains is reduced if the altitude of the UAV is high compared to the deviations from the flat surface.

An interesting idea is to use blob representations of mosaics and the blob-chain matching procedure between images and the mosaic to relax the limitation of having same scene in the field of view of the UAVs at the same time. Further work includes the improving in the accuracy of the mosaic computation. It would be very helpful to associate uncertainty measures to the relations between the different patches of the mosaic. These relations would then be updated as new data are available. In this case, it would be possible to compute the accuracy of the different parts of the mosaic and obtain a measure about the mosaic uncertainty.

References

1. Amidi O., Kanade T., and Fujita K., "Visual Odometer for Autonomous Helicopter Flight", *Journal of Robotics and Autonomous Systems*, 28:185-193, August 1999.
2. Ayache N., *Artificial Vision for Mobile Robots*, MIT Press.
3. Caballero F., Merino L., Ferruz J., and Ollero A., "Improving Vision-Based Planar Motion Estimation for Unmanned Aerial Vehicles Through Online Mosaicking", *Proceedings, International Conference on Robotics and Automation*, 2860-2865, Orlando, Florida, May 2006.
4. Corke P. I., Sikka P., and Roberts J. M., "Height Estimation for an Autonomous Helicopter", *Proceedings, ISER*, 101-110, 2000.
5. Dickmanns E.D. and Schell F. R., "*Autonomous* Landing of Airplanes Using Dynamic Machine Vision", *Proceedings, IEEE Workshop on Applications of Computer Vision*, 172-179, 1992.
6. Drew M. S., Wei J., and Li Z.-N., "On Illumination Invariance in Color Object Recognition", *Pattern Recognition*, 31(8):1077-1087, 1998.

7. Faugeras O., and Luong Q-T., *The Geometry of Multiple Images: The Laws that Govern the Formation of Multiple Images of a Scene and Some of Their Applications*, MIT Press, 2001.
8. Ferruz J., and Ollero A., "Real-Time Feature Matching in Image Sequences for Non-Structured Environments: Applications to Vehicle Guidance", *Journal of Intelligent and Robotic Systems*, Vol. 28, Issue 1/2, pp. 85-123, June 2000.
9. Finlayson G. D., Drew M. S., and Funt B. V., "Color Constancy: Generalized Diagonal Transforms Suffice", *Journal of the Optical Society of America, A*, 11(11):3011-3019, 1994.
10. Fischler M. A., and Bolles R. C., "Random Sample Consensus: A Paradigm for Model Fitting with Applications to Image Analysis and Automated Cartography", *Communications of the ACM*, 24:381-395, 1981.
11. Forssén P.-E., and Moe A., "View Matching with Blob Features", *Proceedings, 2nd Canadian Conference on Robot Vision*, 228-235, Victoria, BC, Canada, May 2005.
12. Forssén P.E., *Low and Medium Level Vision Using Channel Representations*, PhD Thesis No. 858, Linköping University, 2004.
13. Garcia-Pardo P. J., Sukhatme G. S., and Montgomery J. F., "Towards Vision-Based Safe Landing for an Autonomous Helicopter", *Robotics and Autonomous Systems*, 38(1):19-29, 2001.
14. Hartley R. I., and Zisserman A., *Multiple View Geometry in Computer Vision*, Cambridge University Press, 2nd Edition, 2004.
15. Hsu S., Sawhney H. S., and Kumar R., "Automated Mosaics via Topology Inference", *IEEE Computation and Graphics Applications*, 22(2):44-54, 2002.
16. Hygounenc E., Jung I-K., Soueres P., and Lacroix S., "The Autonomous Blimp Project of LAAS-CNRS: Achievements in Flight Control and Terrain Mapping", *The International Journal of Robotics Research*, 23(4-5):473-511, 2004.
17. Kim J., and Sukkarieh S., "Autonomous Airborne Navigation in Unknown Terrain Environments", *IEEE Transactions on Aerospace and Electronic Systems*, 40(3):1031-1045, July 2004.
18. Konolige K., Fox D., Limketkai B., Ko J., and Stewart B., "Map Merging for Distributed Robot Navigation", *Proceedings, IEEE/RSJ International Conference on Intelligent Robots and Systems*, 212- 217, 2003.
19. Lindeberg T., *Scale-Space Theory in Computer Vision*, Kluwer, 1994.
20. Mallet A., Lacroix S., and Gallo L., "Position Estimation in Outdoor Environments Using Pixel Tracking and Stereo Vision", *Proceedings, IEEE International Conference on Robotics and Automation*, 3519- 3524, 2000.
21. Mejías L., Saripalli S., Campoy P., and Sukhatme G. S., "Visual Servoing of an Autonomous Helicopter in Urban Areas Using Feature Tracking", *Journal of Field Robotics*, 23(3-4):185-199, 2006.
22. Merino L., Caballero F., Martínez de Dios J.R., Ferruz J., and Ollero A., "A Cooperative Perception System for Multiple UAVs: Application to Automatic Detection of Forest Fires", *Journal of Field Robotics*, 23(3- 4):165-184, 2006.

23. Obdrzálek S., and Matas J., "Object Recognition Using Local Affine Frames on Distinguished Regions", *Proceedings, 13th BMVC*, 113-122, September 2002.
24. Ollero A., Ferruz J., Caballero F., Hurtado S., and Merino L., "Motion Compensation and Object Detection for Autonomous Helicopter Visual Navigation in the COMETS System", *Proceedings, IEEE International Conference on Robotics and Automation*, 19-24, 2004.
25. Ollero A., Lacroix S., Merino L., Gancet J., Wiklund J., Remuss V., Veiga I., Gutierrez L. G., Viegas D. X., Gonzalez M. A., Mallet A., Alami R., Chatila R., Hommel G., Colmenero F. J., Arrue B., Ferruz J., Martinez J. R., and Caballero F., "Multiple Eyes in the Sky: Architecture and Perception Issues in the COMETS Unmanned Air Vehicles Project", *IEEE Robotics and Automation Magazine*, 12(2):46- 57, June 2005.
26. Ollero A., and Merino L., "Control and Perception Techniques for Aerial Robotics", *Annual Reviews in Control*, 28:167-178, 2004.
27. Pizarro O. and Singh H., "Toward Large-Area Mosaicing for Underwater Scientific Applications", *IEEE Journal of Oceanic Engineering*, 28(4):651-672, October 2003.
28. Proctor A. A., Johnson E. N., and Apker T. B., "Vision-Only Control and Guidance for Aircraft", *Journal of Field Robotics*, 23(10):863-890, 2006.
29. Remuss V., Musial M., and Hommel G., "Marvin - An Autonomous Flying Robot-Bases on Mass Market", *Proceedings of the Workshop WS6, Aerial Robotics, IEEE/RSJ International Conference on Intelligent Robots and Systems*, 2002.
31. Saripalli S., Montgomery J. F., and Sukhatme G. S., "Visually Guided Landing of an Unmanned Aerial Vehicle", *IEEE Transactions on Robotics and Automation*, 19(3):371- 380, June 2003.
32. Saripalli S., Naffin D., and Sukhatme G. S., "Autonomous Flying Vehicle Research at the University of Southern California", *Proceedings, First International Workshop on Multi-Robot Systems*, 2002.
33. Schmitt T., Hanek R., Beetz M., Buck S., and Radig B., "Cooperative Probabilistic State Estimation for Vision-Based Autonomous Mobile Robots", *IEEE Transactions on Robotics and Automation*, 18:670- 684, October 2002.
34. Shakernia O., Vidal R., Sharp C., Ma Y., and Sastry S., "Multiple View Motion Estimation and Control for Landing an Aerial Vehicle", *Proceedings, International Conference on Robotics and Automation*, 2793-2798, May 2002.
35. Shum H.-Y., and Szeliski R., "Construction and Refinement of Panoramic Mosaics with Global and Local Alignment", *Proceedings, Sixth International Conference on Computer Vision*, Washington, DC, 1998.
36. Sukkarieh S., Nettleton E., Kim J.-H., Ridley M., Goktogan A., and Durrant-Whyte H., "The ANSER Project: Data Fusion across Multiple Uninhabited Air Vehicles", *The International Journal of Robotics Research*, 22(7-8):505-539, 2003.
37. Thrun S., "A Probabilistic Online Mapping Algorithm for Teams of Mobile Robots", *International Journal of Robotics Research*, 20(5):335-363, 2001.

38. Tomasi C., *Shape and Motion from Image Streams: A Factorization Method*, PhD Thesis, Carnegie Mellon University.
39. Tran L. V., *Efficient Image Retrieval with Statistical Color Descriptors*, PhD Thesis No. 810, Linköping University, 2003.
40. Triggs B., "Autocalibration from Planar Scenes", *Proceedings, 5th European Conference on Computer Vision*, 89-105, London, UK, 1998.
41. Tsai R. Y., Huang T. S., and Zhu W.-L., "Estimating Three-Dimensional Motion Parameters of a Rigid Planar Patch, ii: Singular Value Decomposition", *IEEE Transactions Acoustic, Speech and Signal Processing*, 30(8):525-524, August 1982.
42. Tuytelaars T., and Gool L. V., "Matching Widely Separated Views Based on Affinely Invariant Regions", *International Journal of Computer Vision*, 59:61-85, 2004.
43. Vidal R., Sastry S., Kim J., Shakernia O., and Shim D., "The Berkeley Aerial Robot Project (BEAR)", *Proceedings, IEEE/RSJ International Conference on Intelligent Robots and Systems,* 1-10, 2002.
44. Yakimenko O. A., Kaminer I. I., Lentz W. J., and Ghyzel P. A., "Unmanned Aircraft Navigation for Shipboard Landing Using Infrared Vision", *IEEE Transactions on Aerospace and Electronic Systems*, 38(4):1181-1200, 2002.
45. Zhang Z., and Hintz K. J., "Evolving Neural Networks for Video Attitude and Height Sensor", *Proceedings, SPIE International Symposium on Aerospace/Defense Sensing and Control*, 2484:383-393, 1995.

Chapter 10: Evolutionary Algorithm Based Path Planning for Multiple UAV Cooperation[1]

This Chapter describes an evolutionary based off-line / on-line path planner for cooperating Unmanned Aerial Vehicles (UAVs). It considers the environment characteristics, the flight envelope and mission constraints of cooperating UAVs. The scenario under consideration assumes that several UAVs are launched from the same or different but known initial locations. Then, the main goal is to produce 3-D trajectories that ensure a collision free operation with respect to mission constraints. The path planner produces curved routes that are represented by 3-D B-Spline curves. Two types of path planner are discussed: i) the off-line planner that generates collision free paths in environments with known characteristics and flight restrictions; ii) the on-line planner, based on the off-line one, that generates collision free paths in unknown static environments by using acquired information from the UAV on-board sensors. This information is exchanged between cooperating UAVs in order to maximize knowledge of the environment. For each UAV, the on-line planner generates rapidly a near optimum path that guides the vehicle safely to an intermediate position within the already scanned territory, taking into account mission and cooperation objectives and constraints. The process is repeated for each UAV until the final position is reached by one of them. Then, each remaining UAV uses the acquired information about the environment in order to compute a curved path that connects its current position to the final one. Both off-line and on-line path planning problems are formulated as optimization problems, with a differential evolution algorithm serving as the optimizer.

10.1 Introduction

Path planning refers to the generation of a space path between an initial location and the desired destination, with an optimal or near-optimal per-

[1] Written by I. K. Nikolos, N. C. Tsourveloudis, K. P. Valavanis

Kimon P. Valavanis (ed.), Advances in Unmanned Aerial Vehicles, 309–340.

formance under specific constraints [1]. A path planning algorithm may produce different candidate plans, which should be compared and evaluated based on specific criteria. Such criteria are generally related to *feasibility* and *optimality* of the path generation. The first criterion relates to derivation of a plan that moves safely a UAV (an object) to its final state, without taking into account the quality of the produced plan. The second criterion refers to derivation of optimal, yet feasible, paths, with optimality defined according to the problem under consideration [2]. However, searching for optimal paths is not a trivial task; in most cases this requires excessive computational time; in some cases even computation of just one feasible path is a rather involved task. Therefore, the search focuses mostly on suboptimal or just feasible solutions.

Compared to the path-planning problem in other application areas, path planning for UAVs requires special characteristics that need be considered [3] [4] [5]: i) feasibility, which refers to limitations from using UAVs, such as limited endurance and range, minimum turning angle, minimum and maximum speed, etc; ii) stealth, in order to minimize the probability of detection by hostile sensors; iii) acceptable performance related to assigned mission, which imposes special requirements, including maximum climbing / diving angle, minimum and/or maximum flying altitude, etc; iv) real-time implementation, which asks for computationally efficient algorithms, and, v) cooperation between UAVs in order to maximize the possibility of mission accomplishment.

Cooperation between UAVs has recently gained increased interest as systems of multiple vehicles engaged in cooperative behavior show specific benefits compared to a single vehicle [6]. Such systems of cooperating UAVs may be used for fire fighting applications, military missions, search and rescue scenarios or exploration of unknown environments (space-oriented applications).

In order to establish a robust framework for efficient and reliable cooperation of multiple UAVs, several issues must be addressed: i) the task assignment problem that relates to the number of UAVs required to perform a mission, with predefined order, on a number of targets; ii) the path planning problem to provide feasible, flyable and near optimal trajectories that connect starting to destination points; iii) the exchange of information between cooperating UAVs, and data fusion expected to enhance team effectiveness; iv) cooperative sensing of the targets defined as how to 'use' UAV sensors in terms of their locations to achieve optimal estimation of the state of each target [7], and, v) cooperative sensing of the environment for better situation awareness.

10.1.1 Related Work

UAV path planning algorithms were initially derived for solving the single vehicle case. However, the continuously increasing interest for cooperating UAVs has resulted in development of algorithms that take into account special characteristics of multi-UAV missions.

Path planning problems are computationally demanding multi-objective multi-constraint optimization problems [8]. Problem complexity increases when multiple UAVs are used. Several approaches have been reported for coordinated route planning, such as Voronoi diagrams [9], mixed integer linear programming [10] [11], and dynamic programming [12] formulations.

In [9] a group of UAVs is required to transition through a number of known target locations, with a number of threats present in the region of interest. Some threats were known a priori, while some others 'popped up' or became known only when a UAV flew near them. The motion planning problem was decomposed into a waypoint path planner and a dynamic trajectory generator. The path planning problem was solved via a Voronoi diagram and Epstein's *k*-best paths algorithm. The trajectory generator problem was solved via a real-time nonlinear filter that explicitly accounted for the dynamic constraints of the vehicle and modified the initial path.

In [13], the motion planning problem for a limited resource of Mobile Sensor Agents (MSAs) was investigated in an environment with a number of targets larger than the available MSAs. The problem, being a combination of problems of sensor resource management and robot motion planning, was formulated as an optimization one with the objective to minimize the average time duration between two consecutive observations of each target.

Computational intelligence methods, such as Neural Networks [14], Fuzzy Logic [15] and Evolutionary Algorithms (EAs) [5] [16] have been successfully used to derive trajectories for guiding mobile robots in known, unknown or partially known environments.

Besides their computational cost, EAs are considered as a viable candidate to solve path planning problems effectively, because of their high robustness compared to other existing directed search methods, their ease of implementation in problems with a relatively high number of constraints, and their high adaptability to the special characteristics of the problem under consideration [16].

EAs have been successfully used in the past for the solution of the path finding problem in ground based or sea surface navigation [17]. A common practice was to model the path using straight line segments, connecting successive waypoints. In the case of partially known or dynamic environments a feasible and safe trajectory was planned off-line by the EA and the algorithm was used on-line whenever unexpected obstacles were sensed [18] [19]. EAs have been also used for solving the path finding problem in a 3-D environment for underwater vehicles, assuming that the path is a sequence of cells in a 3-D grid [20] [21].

In [5] a route planner for UAVs was proposed based on evolutionary computation. The generated routes enabled the vehicles to arrive at their destination simultaneously by taking into account the exposure of UAVs to potential threats. The flight route consisted of straight-line segments, connecting the waypoints from the starting to the goal points. A real coded chromosome representation with certain design variables and state variable were used. These variables provided information on the feasibility of the corresponding way point and the route segment composed of these points. The cost function penalized the route length, high altitude flights or routes that come dangerously close to known ground threats. The imposed constraints on route segments were relevant to minimum route leg length, maximum route distance, minimum flying height, maximum turning angle, maximum climbing/diving angle, simultaneous arrival at target location and no collision between vehicles.

In [22] a multi-task assignment problem for cooperating UAVs was formulated as a combinatorial optimization problem; a Genetic Algorithm was utilized for assigning the multiple agents to perform different tasks on multiple targets. Integer encoding was used for the chromosomes, which were composed of two rows; the first row presented the assignment of a vehicle to perform a task on the target appearing on the second row. The proposed algorithm solved the problem efficiently, while taking into account requirements, such as task precedence and coordination, timing constraints, and flyable trajectories.

In [16] an EA based framework was utilized to design an off-line / on-line path planner for UAVs autonomous navigation. The path planner calculated a curved path line, represented using B-Spline curves in a 3-D terrain environment; the coordinates of B-Spline control points served as design variables. The off-line planner produced a single B-Spline curve that connected the starting and target points with a predefined initial direction. The on-line planner gradually produced a smooth 3-D trajectory aiming at reaching a predetermined target in an unknown environment; the produced trajectory consisted of smaller B-Spline curves smoothly connected with each other. For both off-line and on-line planners, the problem is formulated as an optimization one; each objective function is formed as the weighted sum of different terms, which take into account the various objectives and constraints of the corresponding problem. Constraints are formulated using penalty functions.

10.1.2 Problem Definition and Proposed Solution

The main scenario considered in this Chapter is the following: A number of UAVs are launched from the same or different but known initial locations with predefined initial directions. The main objective is to derive 3-D trajectories, represented by 3-D B-Spline curves, which connect the initial locations with a single destination location and ensure a collision free operation with respect to mission constraints. The UAVs are assumed to be equipped with a set of on-board sensors, including radar, GPS (Global Positioning System) / DGPS (Differential GPS), INS (Inertial Navigation System) and gyroscopes, through which they sense their surroundings and position. Each vehicle is assumed to be a point and its actual size is taken into account by equivalent obstacle – ground growing.

Two types of path planner are discussed: The off-line planner that generates collision free paths in environments with known characteristics and flight restrictions. The on-line planner, being an extension of the off-line one based on already reported research in [16]. Knowledge of the environment is gradually acquired through the on-board sensors that scan the area within a certain range from each UAV. This information is exchanged between cooperating UAVs in order to maximize sensor effectiveness. The on-line planner rapidly generates a near optimum path for each UAV that will guide the vehicle safely to an intermediate position within the known territory, taking into account the mission and cooperation objectives and constraints. The process is repeated until the corresponding final position is reached by an UAV. Then, each one of the remaining UAVs uses acquired information about the environment and the off-line planner output to compute a path that connects its current position to the final destination. Both path planning problems are formulated as optimization (minimization) problems, where specially constructed functions take into account mission and cooperation objectives and constraints, with a Differential Evolution algorithm to serve as the optimizer.

The rest of the Chapter is organized as follows: Section 10. 2 discusses B-Spline and differential EA fundamentals; the solid terrain formulation, used for experimental simulations, is also presented. The off-line path planner for a single UAV is briefly discussed in Section 10.3. Section 10.4 deals with the concept of cooperating UAV on line path planning using differential evolution. The problem formulation is described, including assumptions, objectives, constraints, cost function definition and path modeling. Simulation results are presented in Section 10.5, followed by discussion and conclusions in Section 10.6.

10.2 B-Spline and Differential Evolution Fundamentals

10.2.1 Path Modeling Using B-Spline Curves

Straight line segments that connect a number of waypoints have been used in the past to model UAV paths in 2-D or 3-D space [23] [5]. However, these simplified paths cannot be used for accurate simulation of a UAV flight, unless a large number of waypoints are adopted. Furthermore, if an optimization procedure is used, the number of design variables explodes, especially if cooperating flying vehicles are considered. As a result, computation time becomes impractical for real world applications.

In [9], paths from the initial vehicle location to the target location are derived from a graph search of a Voronoi diagram that is constructed from the known threat locations. The resulting paths consist of line segments, which are subsequently smoothed around each way point, in order to provide feasible trajectories within the dynamic constraints of the vehicle.

In [24] car formulation has been proposed as an alternative approach to modeling UAV dynamics. Each UAV is assumed to fly with constant altitude, constant flight speed and to have continuous time kinematics [25]. This approach seems inefficient to model real world scenarios, such as those including 3-D terrain avoidance and following of stealthy routes; however, it seems to be sufficient for task assignment purposes with cooperating UAVs flying at safe altitudes [13] [22] [25].

B-Spline curves have been used for trajectory representation in 2-D [26] or 3-D environments [16] [27]. They fit in an optimization procedure as they need a few variables, the coordinates of their control points, to define complicated curved paths. Each control point has a local effect on the curve's shape and small perturbations in its position produce changes in the curve only in the neighborhood of the repositioned control point if the degree of the curve remains small.

B-Spline curves are parametric curves; their construction is based on blending functions [28] [29]. Their parametric construction produces non-monotonic curves, like the trajectories of moving objects.

If the number of control points of the corresponding curve is $n+1$, with corresponding coordinates $(x_0, y_0, z_0),\ldots, (x_n, y_n, z_n)$, the coordinates of the B-Spline curve may be written as:

$$x(u) = \sum_{i=0}^{n} x_i \cdot N_{i,p}(u), \tag{10.1}$$

$$y(u)=\sum_{i=0}^{n} y_i \cdot N_{i,p}(u), \tag{10.2}$$

$$z(u)=\sum_{i=0}^{n} z_i \cdot N_{i,p}(u), \tag{10.3}$$

where u is the free parameter of the curve, $N_{i,p}(u)$ are the blending functions of the curve and p is its degree, which is associated with the curve's smoothness, $p+1$ being its order. Higher values of p correspond to smoother curves.

The blending functions are defined recursively in terms of a *knot* vector $U=\{u_0,\ldots,u_m\}$, which is a non-decreasing sequence of real numbers, with the most common form being the *uniform non-periodic* one, defined as:

$$u_i = \begin{cases} 0 & if \quad i<p+1 \\ i-p & if \quad p+1 \le i \le n \\ n-p+1 & if \quad n<i. \end{cases} \tag{10.4}$$

The blending functions $N_{i,p}$ are computed using the knot values defined above, as:

$$N_{i,0}(u)=\begin{cases} 1 & if \ u_i \le u < u_{i+1} \\ 0 & otherwise, \end{cases} \tag{10.5}$$

$$N_{i,p}(u)=\frac{u-u_i}{u_{i+p}-u_i} N_{i,p-1}(u)+\frac{u_{i+p+1}-u}{u_{i+p+1}-u_{i+1}} N_{i+1,p-1}(u). \tag{10.6}$$

If the denominator of either of the fractions is zero, that fraction is defined to have zero value. Parameter u varies between 0 and $(n-p+1)$ with a constant step, providing the discrete points of the B-Spline curve. The sum of the values of the blending functions for any value of u is always 1.

The use of B-Spline curves for the determination of a flight path provides the advantage of describing complicated non-monotonic 3-D curves with controlled smoothness with a small number of design parameters, i.e. the coordinates of the control points. Another valuable characteristic of the adopted B-Spline curves is that the curve is tangential to the control polygon at the starting and ending points. This characteristic may be used in order to define the starting or ending direction of the curve, by inserting an extra fixed point after the starting one, or before the ending control point. Figure 10.1 shows a quadratic B-Spline curve ($p=2$) in 2-D with its control points and the corresponding control polygon.

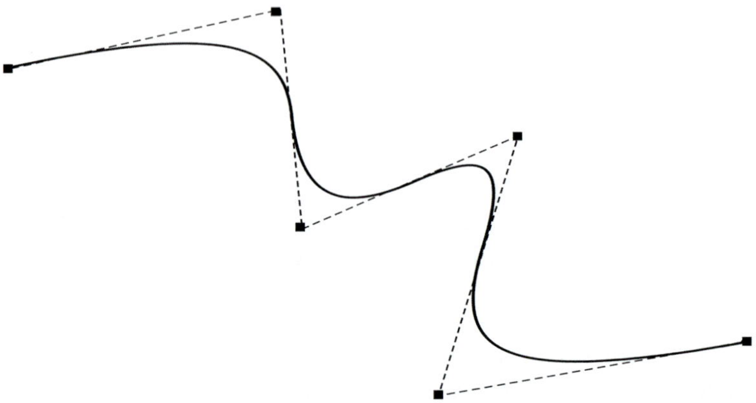

Fig. 10.1. Quadratic 2-D B-Spline curve, produced using a uniform non-periodic knot vector, and its control polygon.

10.2.2 The Solid Boundary Representation

The terrain is represented by a meshed 3-D surface using mathematical functions of the form:

$$z(x,y)=\sin(y+a)+b\cdot\sin(x)+c\cdot\cos\left(d\cdot\sqrt{x^2+y^2}\right)+ \tag{10.7}$$
$$e\cdot\cos(y)+f\cdot\sin\left(f\cdot\sqrt{x^2+y^2}\right)+g\cdot\cos(y),$$

where a, b, c, d, e, f, g are constants experimentally defined, in order to produce a surface with mountains and valleys, as shown in Figure 10.2.

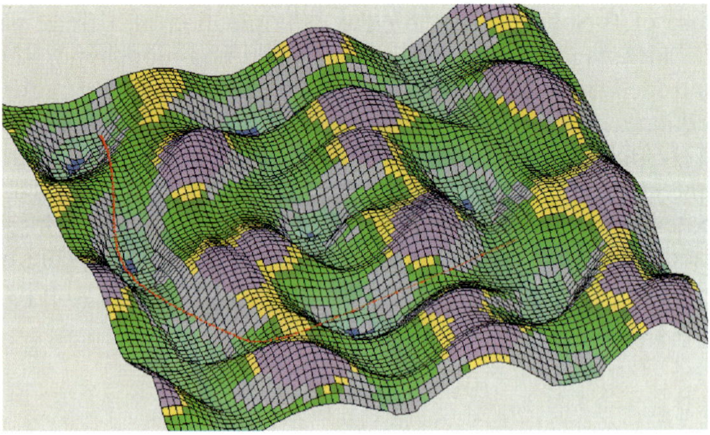

Fig. 10.2. A typical terrain used for simulation studies.

A graphical interface has been developed for visualization of the terrain surface along with the path lines [16]. The corresponding interface deals with different terrains produced either artificially or based on real geographical data, providing an easy verification of the feasibility and the quality of each solution. The path planning algorithm considers the boundary surface as a group of quadratic mesh nodes with known coordinates.

10.2.3 Differential Evolution Algorithm

Differential Evolution (DE) [30] [31] is used as an optimization tool. A DE algorithm, being a recent development in the field of optimization algorithms, represents a type of evolutionary strategy, especially formed so that it can effectively deal with continuous optimization problems. The classic DE algorithm evolves a fixed size population, which is randomly initialized. After initializing the population, an iterative process is started and at each generation G, a new population is produced until a stopping condition is satisfied. At each generation, each element of the population can be replaced with a new generated one. The new element is a linear combination between a randomly selected element and the difference between two other randomly selected elements. In detail, given a cost function f :

$$f(X): R^n \rightarrow R\, ,\, n{=}n_{param} \tag{10.8}$$

the optimization target is to minimize the value of this cost function by optimizing the values of its parameters (design variables):

$$X = \left(x_1, x_2, \ldots, x_{n_{param}}\right),\ x_j \in R \tag{10.9}$$

where X denotes the vector composed of n_{param} cost function parameters (design variables). These parameters take values between specific upper and lower bounds:

$$x_j^{(L)} \le x_j \le x_j^{(U)},\ \ j = 1, \ldots, n_{param} \tag{10.10}$$

The DE algorithm implements real encoding for the values of the objective function design variables. In order to obtain a starting point for the DE algorithm, initialization of the population takes place. Initialization ($G = 1$) is established by randomly assigning values to the parameters j of each i member of the population within the given boundaries, as follows:

$$x_{i,j}^{(1)} = r \cdot \left(x_j^{(U)} - x_j^{(L)}\right) + x_j^{(L)},\ \ i = 1, \ldots, n_{pop},\ \ j = 1, \ldots, n_{param} \tag{10.11}$$

r is a uniformly distributed random value within the range [0, 1]. The DE algorithm mutation operator is based on a triplet of randomly selected different individuals. For each i member of the population a new parameter vector $V_i^{(G)}$ is generated by adding the weighted difference vector between the two members of the triplet to the third one, the donor:

$$V_i^{(G)} = X_{r3_i}^{(G)} + F \cdot \left(X_{r1_i}^{(G)} - X_{r2_i}^{(G)} \right)$$
$$V_i^{(G)} = \left(v_{i,1}^{(G)}, v_{i,2}^{(G)}, \ldots, v_{i,n_{param}}^{(G)} \right) \tag{10.12}$$

$X_{r3_i}^{(G)}$ is called the 'donor', G is the current generation, and:

$$i = 1,\ldots,n_{pop},\ j = 1,\ldots,n_{param}$$
$$r1_i \in \left[1,\ldots,n_{pop}\right],\ r2_i \in \left[1,\ldots,n_{pop}\right], r3_i \in \left[1,\ldots,n_{pop}\right] \tag{10.13}$$
$$r1_i \neq r2_i \neq r3_i \neq i$$
$$F \in \left[0,1+\right],\ r \in \left[0,1\right]$$

In this way a perturbed individual is generated. The perturbed individual $V_i^{(G)}$ and the initial population member $X_i^{(G)}$ are then subjected to a cross-over operation that generates the final candidate solution $U_i^{(G+1)}$:

$$u_{i,j}^{(G+1)} = \begin{cases} v_{i,j}^{(G)} & if\ \left(r \le C_r \vee j = k\right) \forall\ j = 1,\ldots,n_{param} \\ x_{i,j}^{(G)} & otherwise \end{cases} \tag{10.14}$$
$$C_r \in \left[0,1\right]$$

where k is a random integer within $[1, n_{param}]$, chosen once for all members of the population. The random number r is seeded for every gene of each chromosome. F and C_r are DE algorithm control parameters, which remain constant during the search process and affect the convergence behaviour and robustness of the algorithm. Their values also depend on the objective function, the characteristics of the problem and the population size.

The population for the next generation ($G+1$) is selected between the current population and the final candidates. If each candidate vector is better fitted than the corresponding current one, the new vector replaces the vector with which it was compared. The DE selection scheme is described as follows (for a minimization problem):

$$X_i^{(G+1)} = \begin{cases} U_i^{(G+1)} & if\ f\left(U_i^{(G+1)}\right) \le f\left(X_i^{(G)}\right) \\ X_i^{(G)} & otherwise \end{cases} \tag{10.15}$$

A new scheme [32] to determine the donor for the mutation operation has been adopted to accelerate the convergence rate. In this scheme, the donor is randomly selected (with uniform distribution) from the region within the 'hyper-triangle', formed by the three members of the triplet:

$$donor_i^{(G)} = \sum_{k=1}^{3}\left(\lambda_k \Big/ \sum_{m=1}^{3}\lambda_m\right) X_{rk_i}^{(G)}, \quad \lambda_m = rand[0,1] \tag{10.16}$$

where $rand[0,1]$ denotes a uniformly distributed value within the range [0, 1]. With this scheme the donor comprises the local information of all members of the triplet, providing a better starting point for the mutation operation that result in a better distribution of the trial vectors.

The random number generation (with uniform probability) is based on the algorithm presented in [33]. For each different operation inside the DE algorithm that requires a random number generation, a different sequence of random numbers is produced by using a different initial seed for each operation and a separate storage of the corresponding produced seeds.

The off-line path planner is presented next.

10.3 Off-line Path Planner

The off-line path planner is presented to introduce the concept of UAV path planning using EAs. The off-line planner generates collision free paths in environments with known characteristics and flight restrictions. The derived path line for each UAV is a single continuous 3-D B-Spline curve, while the solid boundaries are interpreted as 3-D surfaces. The starting and ending control points of each B-Spline curve are fixed. A third point close to the starting one is also fixed, determining the initial flight direction of the corresponding UAV; this control point is placed in a pre-specified distance from the starting control point. Between the fixed control points, additional free-to-move control points determine the shape of the curve. For each path, the number of the free-to-move control points is user-defined.

10.3.1 Path Modeling Using B-Spline Curves

Each path is constructed using a 3-D B-Spline curve; each B-Spline control point is defined by its three Cartesian coordinates $x_{k,j}$, $y_{k,j}$, $z_{k,j}$, $k=0,\ldots,n$, $j=1,\ldots,N$; N is the number of UAVs, while $n+1$ is the number of control points in each B Spline curve, the same for all curves. The first ($k=0$) and last ($k=n$) control points of the control polygon are the initial and target points of the jth UAV, which are predefined by the user. The second ($k=1$) control point is positioned in a pre-specified distance from the initial control point, in a given altitude, and in a given direction, in order to define the initial direction of the corresponding path.

The control polygon of each B-Spline curve is defined by successive straight line segments as shown in Figure 10.3. For each segment, its length $seg\_length_{k,j}$ and its direction $seg\_angle_{k,j}$ are used as design variables ($k=2,...,n-1, j=1,...,N$). Design variables $seg\_angle_{k,j}$ are defined as the difference in degrees between the direction of the current segment and the previous one. For the first segment ($k=1$) of each control polygon $seg\_angle_{1,j}$ is measured with respect to the x-axis. Additionally, the control points' altitudes $z_{k,j}$ are used as design variables, except for the three predefined fixed points ($k=0$, $k=1$, and $k=n$). In the first segment ($k=1$), $seg\_length_{1,j}$, and $seg\_angle_{1,j}$ are pre-specified in order to define the initial direction of the path, and they are not included in the design variables of the optimization procedure. The lower and upper boundaries of each independent design variable are predefined by the user.

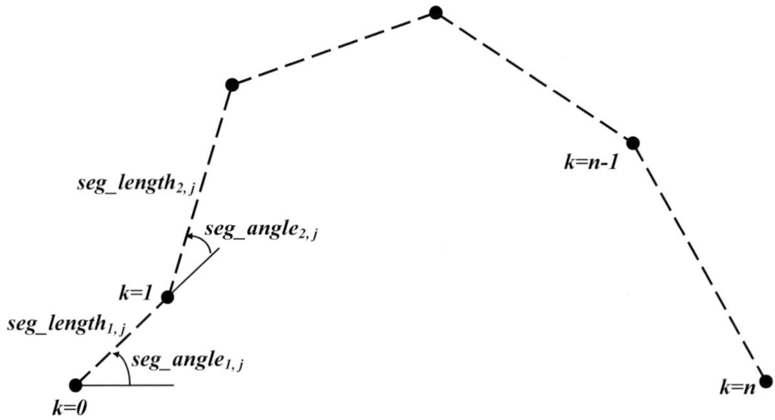

Fig. 10.3. B-Spline control polygon in the horizontal plane.

The coordinates of each B-Spline control point $x_{k,j}$ and $y_{k,j}$ can be easily calculated by using $seg\_length_{k,j}$ and $seg\_angle_{k,j}$ along with the coordinates of the previous control point $x_{k-1,j}$ and $y_{k-1,j}$. The use of $seg\_length_{k,j}$ and $seg\_angle_{k,j}$ as design variables instead of $x_{k,j}$ and $y_{k,j}$ was adopted for three reasons: i) abrupt turns of each flight path can be easily avoided by explicitly imposing short lower and upper bounds for the $seg\_angle_{k,j}$ design variables; ii) by using the proposed design variables, a better convergence rate is achieved compared to the case using the B-Spline control points' coordinates as design variables, a consequence of shortening the search space following the proposed formulation; iii) by using $seg\_length_{k,j}$ as design variables, an easier determination of the upper bound for each curve's length is achieved, along with a smoother variation of the lengths of each curve's segments.

10.3.2 Cost Function Formulation for a Single UAV

For the case of a single UAV the optimization problem to be solved minimizes a set of five terms based on set objectives and constraints associated with the feasibility of the curve, its length and a safety distance from the ground. The cost function to be minimized is defined as:

$$f = \sum_{i=1}^{5} w_i f_i \tag{10.17}$$

Term f_1 penalizes the non-feasible curves that pass through the solid boundary. In order to compute this term, discrete points along each curve are computed, using B-Spline equations (10.1) to (10.6) and a pre-specified step for B-Spline parameter u. The value of f_1 is proportional to the number of discrete curve points located inside the solid boundary; consequently, non-feasible curves with fewer points inside the solid boundary show better cost function than curves with more points inside the solid boundary.

Term f_2 is the length of the curve (non-dimensional with the distance between the starting and destination points) and it is used to provide shorter paths.

Term f_3 is designed to provide flight paths with a safety distance from solid boundaries. For each discrete point i ($i=1,\ldots,n_{line}$, where n_{line} is the number of discrete curve points) of the B-Spline curve its distance from the ground is calculated (recall that the ground is described by a mesh of n_{ground} discrete points). Then the minimum distance of the curve and the ground d_{min} is computed. Term f_3 is defined as:

$$f_3 = \left(d_{safe} / d_{\min}\right)^2 \tag{10.18}$$

while d_{safe} is a safety distance from the solid boundary.

Term f_4 is designed to provide B-Spline curves with control points inside the pre-specified space. If a control point results with an x or y coordinate outside the pre-specified limits, a penalty is added to term f_4 which is proportional to violating the following constraints:

$$\begin{aligned}
&if\ x_{k,j} > x_{\max} \Rightarrow f_4 = f_4 + C_1 * \left|x_{k,j} - x_{\max}\right| \\
&if\ y_{k,j} > y_{\max} \Rightarrow f_4 = f_4 + C_1 * \left|y_{k,j} - y_{\max}\right| \\
&if\ x_{k,j} < x_{\min} \Rightarrow f_4 = f_4 + C_1 * \left|x_{k,j} - x_{\min}\right| \\
&if\ y_{k,j} < y_{\min} \Rightarrow f_4 = f_4 + C_1 * \left|y_{k,j} - y_{\min}\right| \\
&\forall k,\ k = 0,\ldots,n,\ \forall j,\ j = 1,\ldots,N
\end{aligned} \tag{10.19}$$

C_1 is a constant, and x_{min}, x_{max}, y_{min}, y_{max} define the borders of the working space. An additional penalty is added to f_4 in case its value is greater than zero, in order to ensure that curves inside the pre-specified space have a smaller cost function than those having control points outside of it. This can be formally written as:

$$if\ f_4 > 0 \Rightarrow f_4 = f_4 + C_2 \tag{10.20}$$

where C_2 is a constant.

Term f_5 is defined to provide path lines within the known terrain. This characteristic is particularly useful when the off-line path planner is used together with the on-line one, as it will be explained later. Each control point of the B-Spline curve is checked for whether it is placed over a known territory. The ground is modeled as a mesh of discrete points and the algorithm computes the mesh shell (on the x-y plane) that includes each B-Spline control point. If the corresponding mesh shell is characterized as unknown then a constant penalty is added to f_5. A mesh shell is characterized as unknown if all its 4 nodes are unknown (have not been scanned by a sensor).

Weights w_i are experimentally determined, using as criterion the almost uniform effect of the last four terms in the objective function. Term $w_1 f_1$ has a dominant role in (10.17) providing feasible curves in few generations, since path feasibility is the main concern. The minimization of (10.17), through the DE procedure, results in a set of B-Spline control points, which actually represent the desired path.

Initially, the starting and ending path-line points are determined, along with the direction of flight. The limits of the physical space, where the vehicle is allowed to fly (upper and lower limits of their Cartesian coordinates), are also determined, along with the ground surface. The determined initial flight direction is used to compute the third fixed point close to the starting one.

The DE randomly produces a number of chromosomes to form the initial population. Each chromosome contains the z coordinates of the free-to-move B-Spline control points ($k=2, \ldots, n-1$), along with the corresponding seg_length_k and seg_angle_k design variables ($k=2, \ldots, n-1$). For each chromosome the Cartesian coordinates of all B-Spline control points are then computed. Using (10.1) to (10.6), with a constant step of parameter u, a B-Spline curve is calculated for each chromosome of the population in the form of a sequence of discrete points. Subsequently, each B-Spline is evaluated using the aforementioned cost function f. The population of candidate solutions evolves during the generations; at the last generation the population member with the smallest value of the cost function is the solution to the problem and corresponds to the path line with the best characteristics according to the aforementioned criteria.

The simulation runs have been designed in order to search for path lines between ‘mountains’. For this reason, an upper ceiling for flight height has been enforced by providing an upper limit to the z coordinates of the B-Spline control points.

10.4 Cooperating UAVs On-line Path Planning

This Section describes the development and implementation of an on-line path planner for cooperating UAV navigation and collision avoidance in completely unknown static environments. The problem formulation is described, including assumptions, objectives, constraints, cost function definition and path modeling.

Given N UAVs launched from the same or different known initial locations, the objective is to derive N 3-D trajectories, aiming at reaching a predetermined target location while ensuring collision avoidance with the environmental obstacles. Additionally, produced flight paths should satisfy specific route constraints. Each vehicle is assumed to be a point, while its actual size is taken into account by equivalent obstacle – ground growing.

The general problem constraint is collision avoidance between UAVs and the ground. The route constraints are based on:

- Predefined initial and target coordinates for all UAVs;
- Predefined initial directions for all UAVs;
- Predefined minimum and maximum limits of allowed-to-fly space, expressed in terms of minimum and maximum allowed Cartesian coordinates for all path points.

The first two route constraints are explicitly taken into account by the optimization algorithm. The third route constraint is implicitly handled by the algorithm through the definition of the cost function. The cooperation objective is that all UAVs should reach the same target point.

The on-line planner is based on the ideas developed in [16] for a single UAV. It uses acquired information from all UAV on-board sensors (that scan the area within a certain range from the corresponding UAV). The on-line planner rapidly generates a near optimum path, modeled as a 3-D B-Spline curve that will guide each vehicle safely to an intermediate position within the already scanned area. The information about the already scanned area by each UAV is passed to the other cooperating UAVs in order to maximize environment knowledge. The process is repeated until the final position is reached by a single UAV. Then the other UAVs turn to the off-line mode and a single B-Spline path for each UAV is computed to guide it from its current position, through the already scanned territory to the common final destination. As a result, each path line from the corresponding starting point to the final goal is a smooth, continuous 3-D line that consists of successive B-Spline curves smoothly connected to each other.

10.4.1 Path Modeling

As the terrain is completely unknown and radars (or equivalent sensors) gradually scan the area, it is impossible to generate feasible paths that connect each starting point with the target point. Instead, at certain moments, each sensor scans a region around the corresponding moving UAV and this region is added to the already scanned regions by all cooperating UAVs. For the UAV under consideration a path line is generated that connects a temporary starting point with a temporary ending point. Each temporary ending point is also the next curve's starting point for the corresponding vehicle. Therefore, what is finally generated is a group of smooth curve segments connected to each other, eventually connecting the starting point to the final destination for each UAV. This procedure is represented in Figures 10.4 to 10.6 for a single UAV.

In the on-line problem only four control points define each B-Spline curve, the first two of which are fixed and determine the direction of the current UAV path. The remaining two control points are allowed to take any position within the scanned by the radars known space, taking into consideration given constraints.

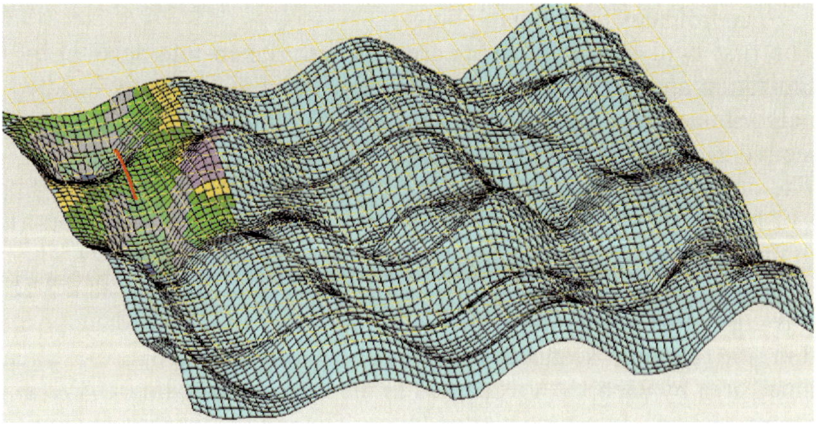

Fig. 10.4. Scanned area (in color), single UAV, movement along first segment.

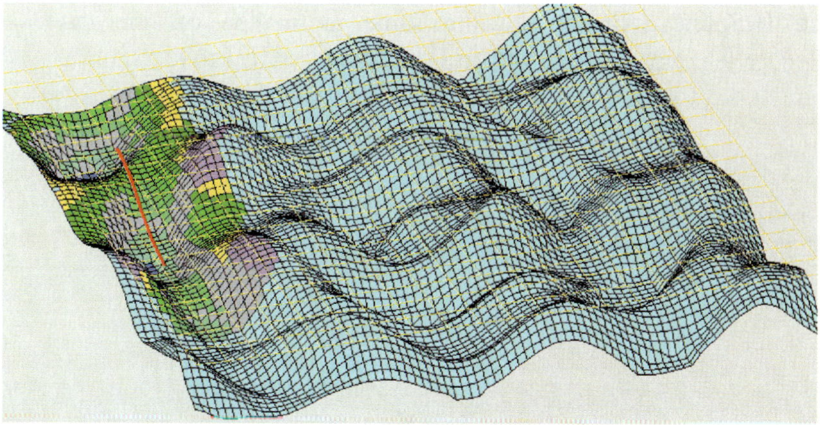

Fig. 10. 5. Scanned area (in color), single UAV, movement along second segment.

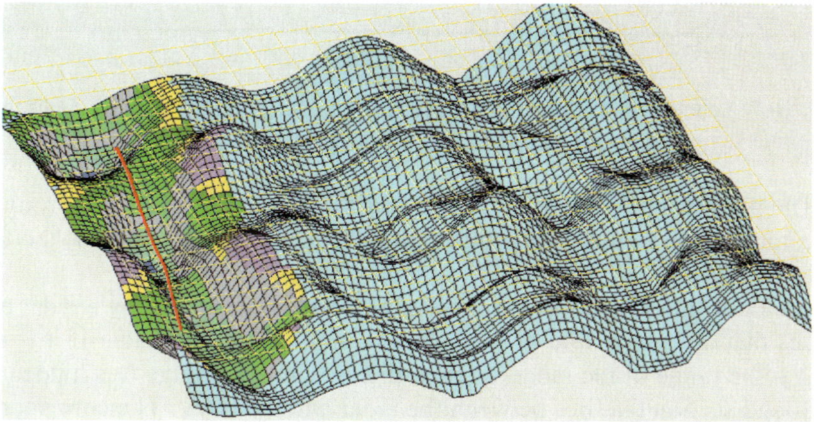

Fig. 10.6. Scanned area (in color), single UAV, movement along third segment.

When the next path segment is generated, only the first control point of the B-Spline curve is known; it is the last control point of the previous B-Spline segment. The second control point is not random, since it is used to guarantee at least first derivative continuity of the two connected curves at their common point. Hence, the second control point of the next curve lies on the line defined by the last two control points of the previous curve as shown in Figure 10.7. It is also desirable that the second control point is near the first one, so that the UAV may easily avoid any obstacle suddenly sensed in front of it. This may happen because the radar scans the environment not continuously, but at intervals. The design variables that define each B-Spline segment are the same as in the off-line case, i.e., $seg\_length_{k,j}$, $seg\_angle_{k,j}$, and $z_{k,j}$ (k=2, 3, and j=1,...,N).

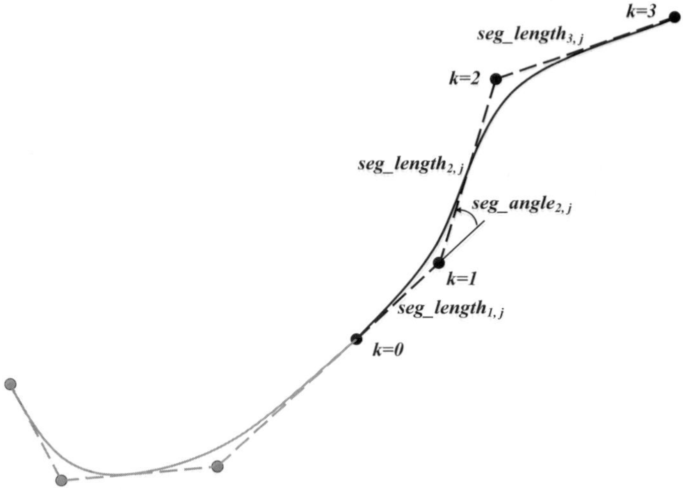

Fig. 10.7. Schematic representation of formation of a complete path by successive B-Spline segments projected on the horizontal plane.

The path-planning algorithm considers the scanned surface as a group of quadratic mesh nodes. All ground nodes are initially assumed to be unknown.

An algorithm is used to distinguish between nodes visible by a radar and nodes that are not visible by it. A node is not visible by a radar if it is not within the range of the radar, or even if it is within its range it is hidden by a ground section that lies between the radar and the UAV. The corresponding algorithm, simulates the radar and checks whether the ground nodes within the radar range are 'visible' or not and consequently 'known' or not. If a newly scanned node is characterized as 'visible', it is added to the set of scanned ground nodes, which is common for all cooperating UAVs.

Radar information is used to produce the first path line segment for the corresponding UAV. As the vehicle moves along its first segment and until it has traveled about 2/3 of its length, its radar scans the surrounding area, returning a new set of visible nodes, which are subsequently added to the common set of scanned nodes. The on-line planner, then, produces a new segment for each UAV, whose first point is the last point of the previous segment and whose last point lies somewhere in the already scanned area. The on-line process is repeated until the ending point of the current path line segment of one UAV lies close to the final destination. Then the other

UAVs turn into the off-line process, in order to reach the target using B-Spline curves that pass through the scanned terrain.

The position at which the algorithm starts to generate the next path line segment for each UAV (here taken as the 2/3 of the segment length) depends on the radar range, the UAV's velocity and the algorithm computational demands.

10.4.2 Cost Function Formulation

The computation of intermediate path segments for each UAV is formulated as a minimization problem. The cost function to be minimized is the weighted sum of eight different terms:

$$f = \sum_{i=1}^{8} w_i f_i \tag{10.21}$$

where w_i are the weights and f_i are the corresponding terms described below.

Terms f_1, f_2, and f_3 are the same with terms f_1, f_3, and f_4 respectively of the off-line procedure. Term f_1 penalizes the non-feasible curves that pass through the solid boundary. Term f_2 is designed to provide flight paths with a safety distance from solid boundaries. Only already scanned ground points are considered for this calculation. Term f_3 is designed to provide B-Spline curves with control points inside the pre-specified working space.

Term f_4 is designed to provide flight segments with their last control point having a safety distance from solid boundaries. This term was introduced to ensure that the next path segment that is going to be computed will not start very close to a solid boundary (which may lead to infeasible paths or paths with abrupt turns). The minimum distance D_{min} from the ground is calculated for the last control point of the current path segment. Only already scanned ground points are considered for this calculation. Term f_4 is then defined as:

$$f_4 = \left(d_{safe} / D_{\min}\right)^2 \tag{10.22}$$

while d_{safe} is a safety distance from the solid boundary.

The value of term f_5 depends on the potential field strength between the initial point of the UAVs path and the final target [16]. This potential field between the two points is the main driving force for the gradual development of each path line in the on-line procedure. The potential is similar to the one between a source and a sink, defined as:

$$\Phi = \ln \frac{r_2 + c \cdot r_0}{r_1 + c \cdot r_0} \tag{10.23}$$

where r_1 is the distance between the last point of the current curve and the initial point for the corresponding UAV, r_2 is the distance between the last point of the current curve and the final destination, r_0 is the distance between the initial point for the corresponding UAV and the final destination and c is a constant. This potential allows for selecting curved paths that bypass obstacles lying between the starting and ending point of each B-Spline curve [16].

Term f_6 is similar to term f_5 but it corresponds to a potential field between the current starting point (of the corresponding path segment) and the final target.

Term f_7 is designed to prevent UAVs from being trapped in small regions and to force them move towards unexplored areas. It may be possible that some segments of the path lines are concentrated in a small area, away from the final target. In order to help the UAV leave this area, term f_7 repels it from the points of the already computed path lines (of all UAVs). Furthermore, if a UAV is wandering around to find a path that will guide it to its target, the UAV will be forced to move towards areas not visited before by this or other UAVs. This term has the form:

$$f_7 = \frac{1}{N_{po}} \sum_{k=1}^{N_{po}} \frac{1}{r_k} \tag{10.24}$$

where N_{po} is the number of the discrete curve points produced so far by all UAVs and r_k is their distance from the last point of the current curve segment.

Term f_8 represents another potential field, which is developed in a small area around the final target. When the UAV is away from the final target, the term is given a constant value. When the UAV is very close to the target the term's value decreases proportionally to the square of the distance between the last point of the current curve and the target. Thus, when the UAV is near its target, the value of this term is quite small and prevents the UAV from moving away.

Weights w_i in (10.21) are experimentally determined, using as criterion the almost uniform effect of all the terms, except the first one. Term $w_1 f_1$ has a dominant role, in order to provide feasible curve segments in a few generations, since path feasibility is the main concern.

10.5 Simulation Results

The same artificial environment was used for all test cases considered, with different starting and target points. The artificial environment is constructed within a rectangle of 20x20 (non-dimensional lengths). The (non-dimensional) radar's range for each UAV was set equal to 4. The safety distance from the ground was set to d_{safe}=0.25. The experimentally optimized settings of the DE algorithm during the on-line procedure were as follows: *population size* = 20, F = 0.6, C_r = 0.45, *number of generations* = 70. For the on-line procedure two free-to-move control points were considered, resulting in 6 design variables.

The corresponding settings during the off-line procedure were as follows: *population size* = 30, F = 0.6, C_r = 0.45, *number of generations* = 70. For the off-line procedure eight control points were used to construct each B-Spline curve, including the initial (k=0) and the final one (k=7). These correspond to five free-to-move control points, resulting in 15 design variables. All B-Spline curves have a degree p equal to 3.

All experiments search for path lines between 'mountains'. For this reason, an upper ceiling for flight height has been enforced in the optimization procedure by explicitly providing an upper bound for the z coordinates of all B-Spline control points.

The first test case corresponds to the on-line path planning for a single UAV over an unknown environment. Results of how the path is formed in terms of successive snapshots of path formulation have already been illustrated through Figures 10.4 to 10.6. Figures 10.8 and 10.9 depict the path that finally succeeds in guiding the UAV towards the target location, although the initial flight direction drives the UAV away from the target. In this case, term f_3 is activated when the path line exceeds the borders of the pre-specified workspace as observed in the lower left corner of Figure 10.9, and enforces the path line to return within the limits. Although the complete path is constructed of 15 successive B-Spline curves, the final curve is smooth enough to be followed by a flying vehicle.

Four additional test cases of on-line path planning for a single UAV are shown in Figures 10.10 to 10.13, respectively. For the second case the starting point and the initial direction is the same with the first case. The third and forth test cases have the same initial point close to the center of the terrain. The fifth test case has its starting point near the right lower corner of the terrain. As observed, although the planner some times produces complicated paths, it succeeds in finding the final destination. However, as shown in Figure 10.13, an abrupt turn occurs, which is the result of the effort to avoid exceeding workspace limits.

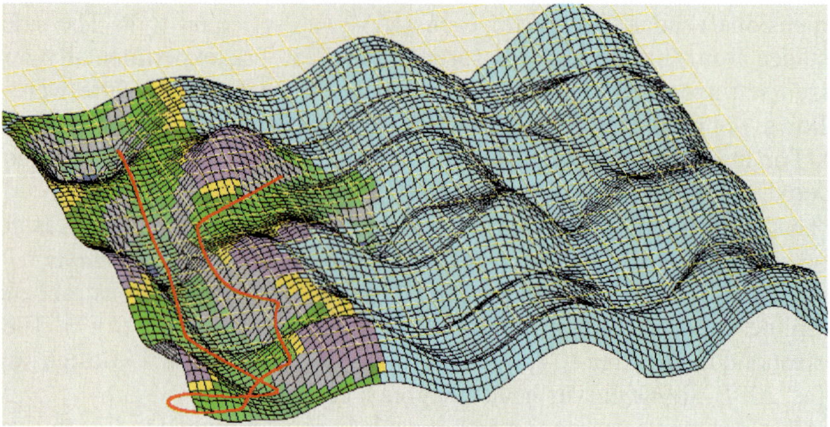

Fig. 10.8. Test case 1, single UAV: On-line path planning with the scanned area shown in color. Path shown in an intermediate position of the flight; consists of 12 B-Spline segments.

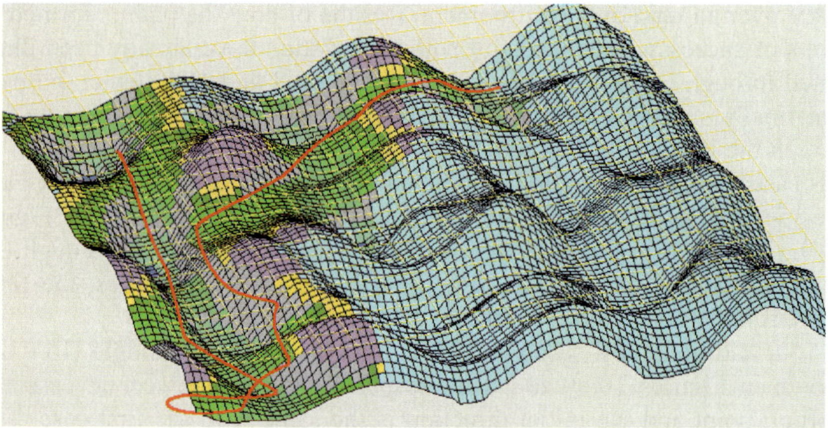

Fig. 10.9. The completed path for test case 1 consisting of 15 B-Spline segments.

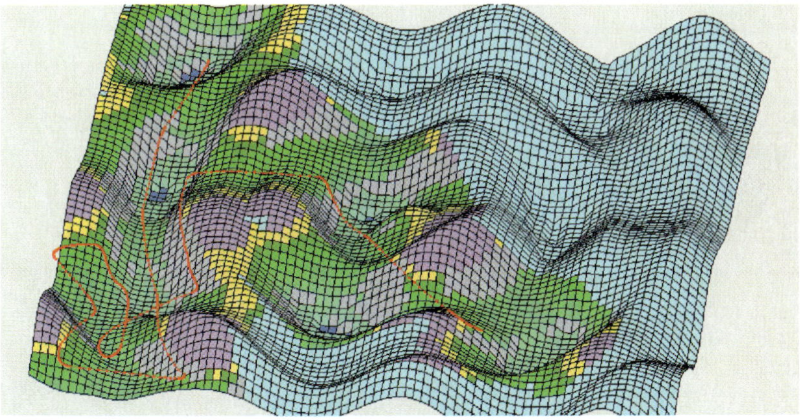

Fig. 10.10. Test case 2, single UAV: On-line path planning; Completed path with the starting point being the same with test case 1, close to the upper left corner.

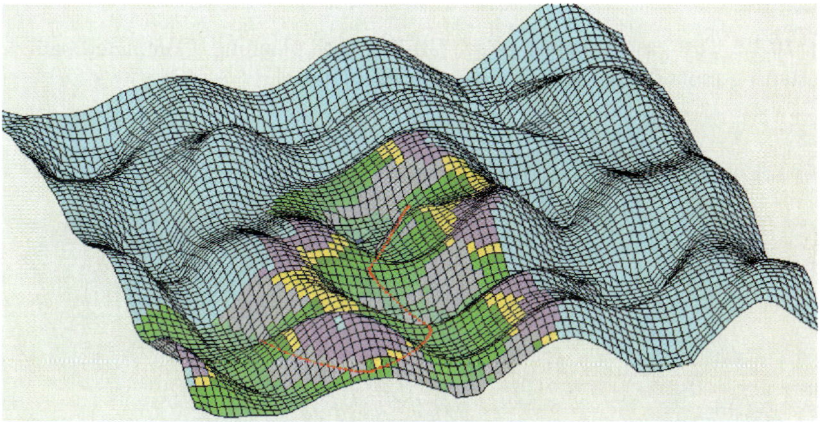

Fig. 10.11. Test case 3, single UAV: On-line path planning; The starting point is close to the center of the terrain.

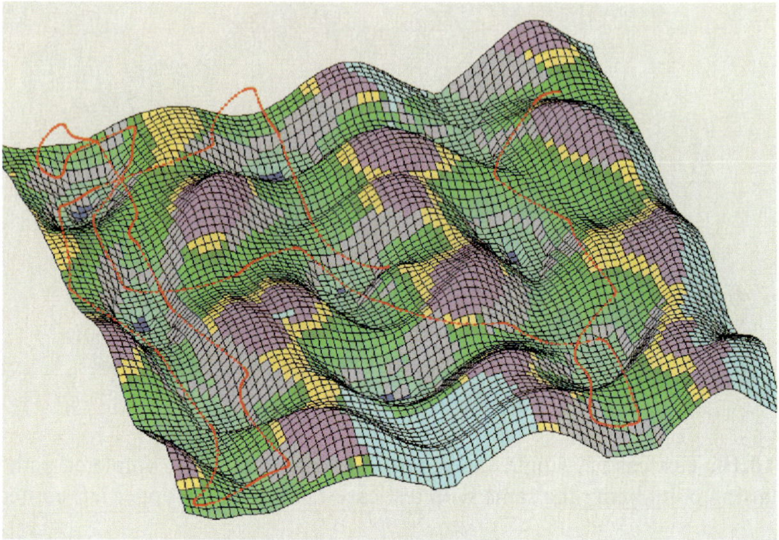

Fig. 10.12. Test case 4, single UAV: On-line path planning; Completed path with the starting point being close to the center of the terrain.

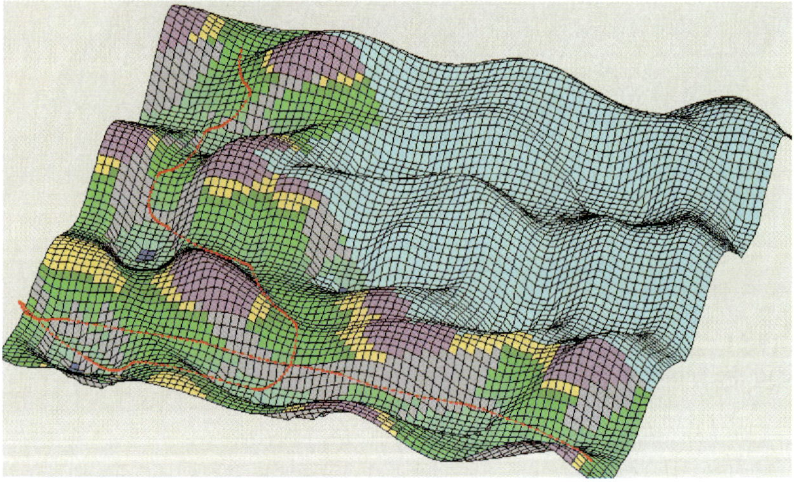

Fig. 10.13. Test case 5, single UAV: On-line path planning with the starting point being close to the right lower corner of the terrain.

Test case 6 corresponds to on-line path planning of 2 UAVs as shown in Figures 10.14 and 10.15. Figure 10.14 shows the path lines when the first UAV (blue line) reaches the target. At that moment the second UAV (red line) turns into the off-line mode, in order to compute a feasible path line that connects its current position with the target through the already scanned area. The final status is demonstrated in Figure 10.15. The starting point for the first UAV is near the lower left corner of the terrain; the starting point of the second UAV is near the upper left corner.

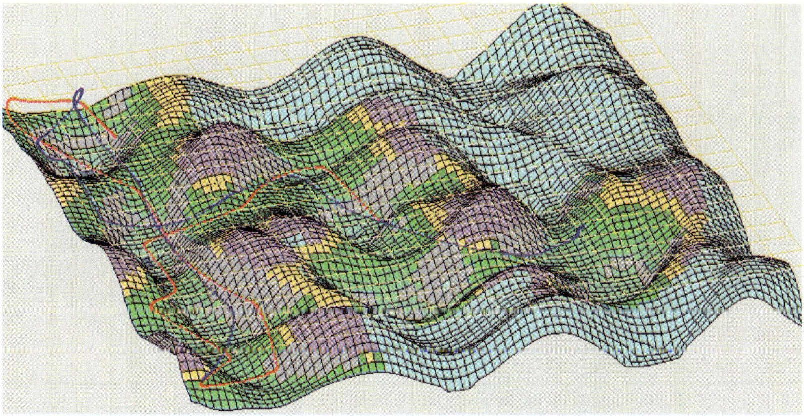

Fig. 10.14. Test case 6, two UAVs. On-line path planning; when the first UAV (blue line) reaches the target the second one turns into the off-line mode.

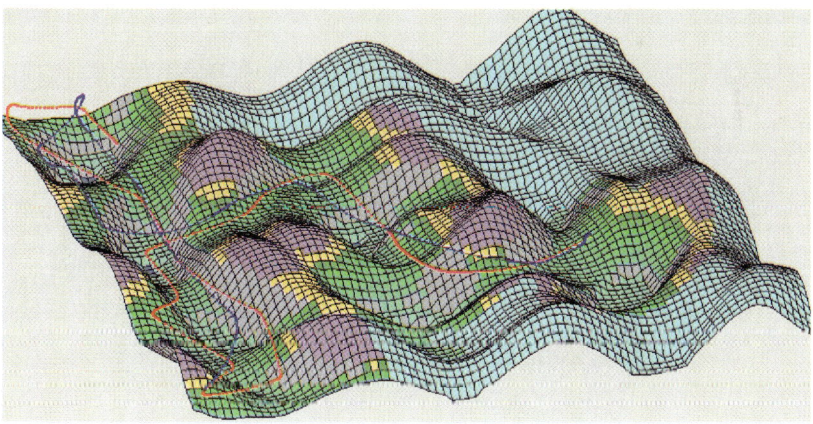

Fig. 10.15. Final status of the path lines for the test case 6.

Test case 7 corresponds to on-line path planning for 3 UAVs as shown in Figures 10.16 and 10.17. Figure 10.16 shows the status of the two path lines when the first UAV (blue line) reaches the target. At that moment the second UAV (red line) and the third one (black line) turn into the off-line mode, in order to compute feasible path lines that connect their positions with the target. The final status is demonstrated in Figure 10.17. The starting point of the first and second UAV are the same as in test case 6, while the starting point of the third UAV is near the middle of the left side of the terrain.

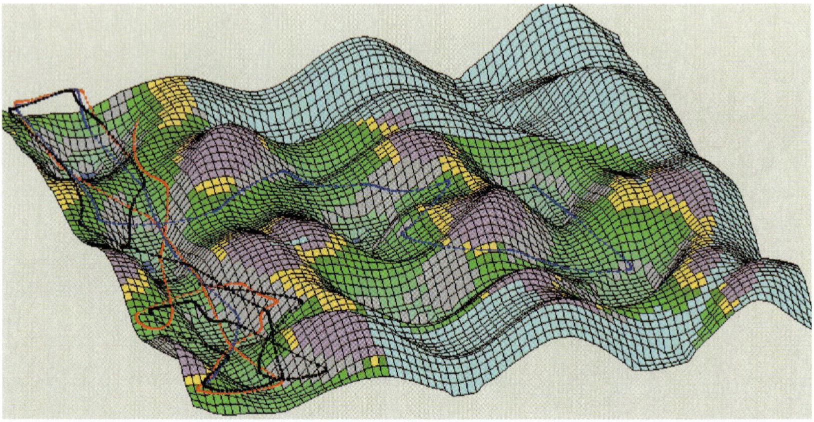

Fig. 10.16. Test case 7, three UAVs. On-line path planning; it shows the paths when the first UAV (blue line) reaches the target.

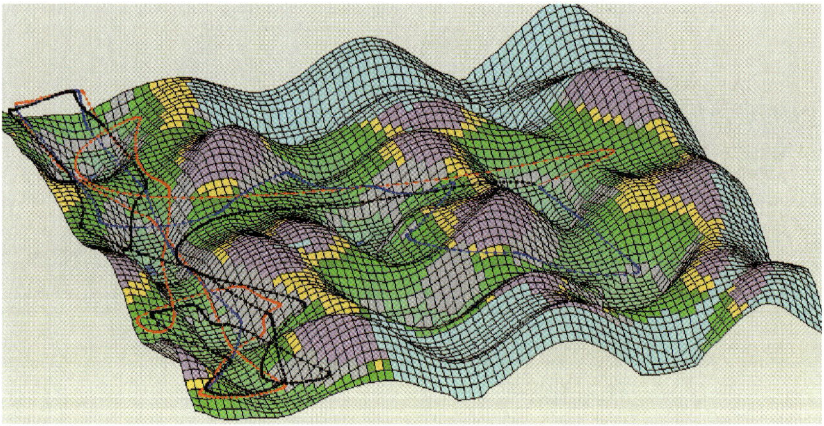

Fig. 10.17. The final status of the path lines of test case 7

Test case 8 refers to another on-line path planning scenario for 3 UAVs as shown in Figures 10.18 and 10.19. Figure 10.18 depicts the status of the two path lines when the first UAV (blue line) reaches the target. The final status is demonstrated in Figure 10.19. As the first UAV (blue line) is close to the target, it succeeds in reaching it using just one B-Spline segment. The other two UAVs turn into off-line mode to reach the target.

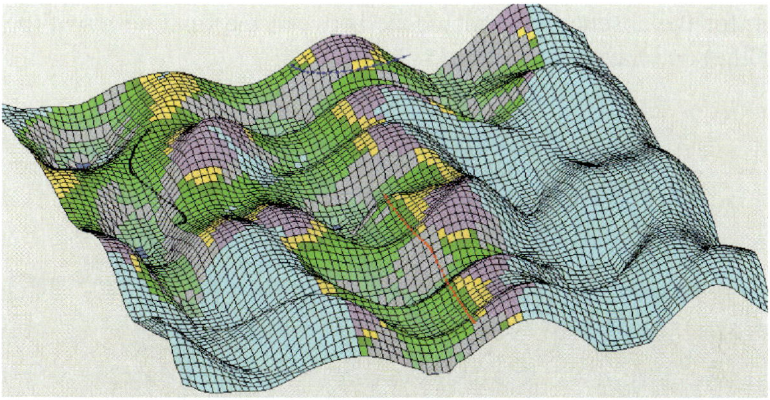

Fig. 10.18. Test case 8, three UAVs. On-line path planning; it shows the path lines when the first UAV (blue line) reaches the target.

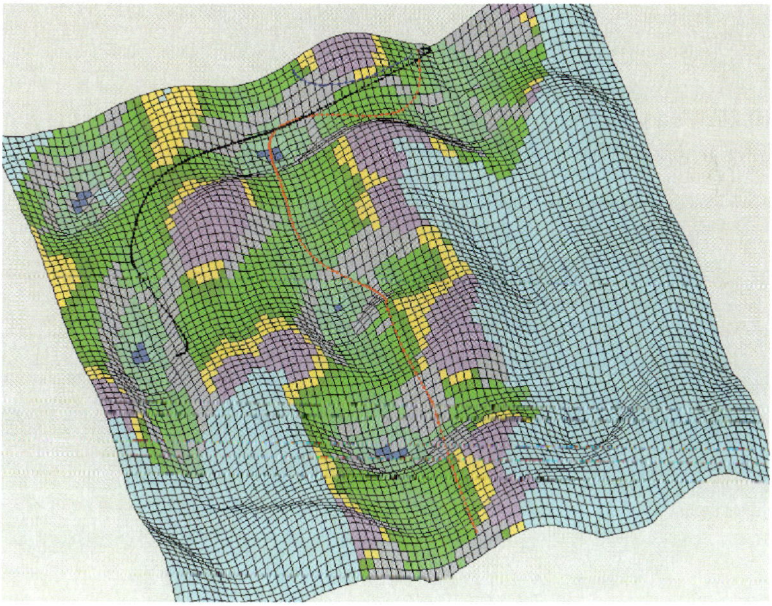

Fig. 10.19. The final status of the path lines of test case 8.

In test case 9 three UAVs are launched from the same point in the center of the working space but with different directions. Figure 10.20 shows the status of the two path lines when the first UAV (blue line) reaches the target. The final status is demonstrated in Figure 10.21. During the on-line procedure, when the final point of a curve segment is within a small range from the final destination the on-line procedure is terminated; this is the reason for the absence of coincidence between the final points of the first (blue line) and the rest path lines.

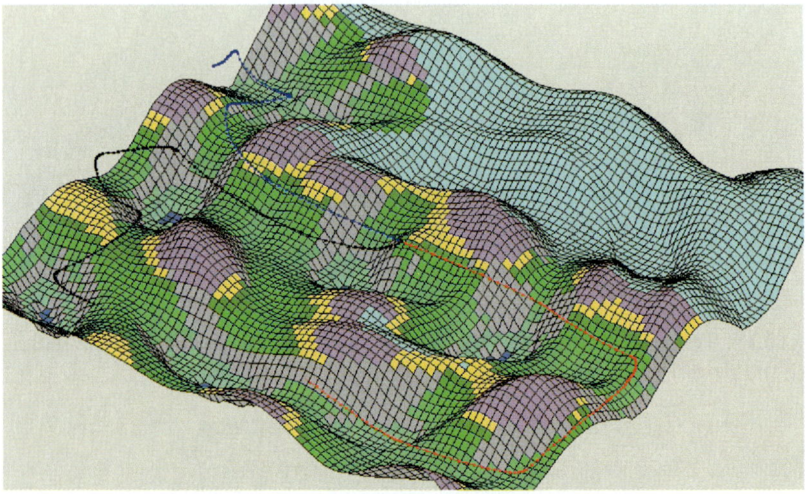

Fig. 10.20. Test case 9, three UAVs. On-line path planning; it shows the status of path lines when the first UAV (blue line) reaches the target.

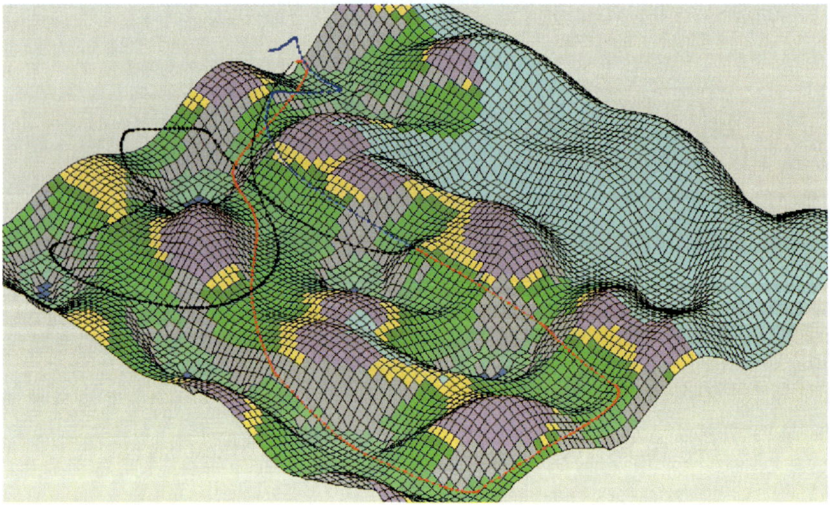

Fig. 10.21. The final status of the path lines of test case 9.

10.6 Conclusions

A path planner has been presented that is suitable for navigating a group of cooperating UAVs avoiding collisions with environment obstacles. The planner is capable of producing smooth path curves in known or unknown static environments.

Two types of path planner were presented. The off-line path planner generates collision free paths in environments with known characteristics and flight restrictions. The on-line planner, which is based on the off-line one, generates collision free paths in unknown environments. The path line is gradually constructed by successive, smoothly connected B-Spline curves, within the gradually scanned environment. The knowledge of the environment is acquired through the UAV on-board sensors that scan the area within a certain range from each UAV. This information is exchanged between the cooperating UAVs; as a result, each UAV utilizes the knowledge of a larger region than the one scanned by its own sensors.

The on-line planner generates for each vehicle a smooth path segment that will guide the vehicle safely to an intermediate position within the known territory. The process is repeated for all UAVs until the corresponding final position is reached by an UAV. Then, the rest vehicles turn into the off-line mode in order to compute path lines consisting of a single B-Spline curve that connect their current positions with the final destination. These path lines are enforced to lie within the already scanned region. Both path planners are based on optimization procedures, and specially constructed functions are used to encounter the mission and cooperation objectives and constraints. A DE algorithm is used as the optimizer for both planners.

The introduced potential fields are the main driving forces for the gradual generation of the path lines in the on-line planner. As demonstrated by the presented test cases, potential fields may be effectively used to generate curves that bypass the solid ground obstacles positioned between the starting and target positions, and when combined with the other terms of the fitness (cost) function, they can produce path lines that escape from concave areas.

The use of a confined workspace for the UAV flight was proven to be another useful characteristic of the on-line procedure. The planner enforces the path lines to be constructed within this confined space; as a result, the search for each path is restricted within a finite area. By utilizing a special term in the cost function, the on-line planner is enforced to explore new areas and prevent each UAV from being trapped in the same region for a long time. As the search space is confined, this term enables the planner to explore this space, and even if a vehicle is initially driven towards the wrong direction, it eventually returns towards its final destination.

The selection of the B-Spline design variables was also proven very effective in providing smooth curves. For all *seg_angle* variables a range of variation between -90 and +90 degrees was explicitly defined, which resulted in smooth turns for both off-line and on-line planners. If the B-Spline control point coordinates were used as design variables, the avoidance of abrupt turns would be a much more difficult procedure that would require additional special terms in the cost function.

Results are very encouraging and support further research to implement the algorithms to small UAVs in real-time.

References

1. Gilmore J. F., "Autonomous vehicle planning analysis methodology", *Proceedings, Association of Unmanned Vehicles Systems Conference*, 503-509, 1991.
2. LaValle S. M., *Planning Algorithms*, Cambridge University Press, 2006.
3. Bortoff S., "Path planning for UAVs", *Proceedings, American Control Conference*, 364–368, 2000.
4. Szczerba R. J., Galkowski P., Glickstein I. S., and Ternullo N., "Robust Algorithm for Real-time Route Planning", *IEEE Transactions on Aerospace Electronic Systems*, 36, 869–878, 2000.
5. Zheng C., Li L., Xu F., Sun F., Ding M., "Evolutionary Route Planner for Unmanned Air Vehicles", *IEEE Transactions on Robotics*, 21, 609-620, 2005.
6. Uny Cao Y., Fukunaga A. S., Kahng A. B., "Cooperative Mobile Robotics: Antecedents and Directions", *Autonomous Robots*, 4, 7-27, 1997.
7. Schumacher C., "Ground Moving Target Engagement by Cooperative UAVs", *Proceedings, American Control Conference*, Oregon, June 2005.
8. Mettler B., Schouwenaars T., How J., Paunicka J., and Feron E., "Autonomous UAV Guidance Build-up: Flight-test Demonstration and Evaluation Plan", *Proceedings of the AIAA Guidance, Navigation, and Control Conference*, 2003.
9. Beard R. W., McLain T. W., Goodrich M. A., Anderson E. P., "Coordinated Target Assignment and Intercept for Unmanned Air Vehicles", *IEEE Transactions on Robotics and Automation*, 18, 911-922, 2002.
10. Richards A., Bellingham J., Tillerson M., and How J., "Coordination and Control of UAVs", *Proceedings of the AIAA Guidance, Navigation and Control Conference,* Monterey, CA, 2002.
11. Schouwenaars T., How J., and Feron E., "Decentralized Cooperative Trajectory Planning of Multiple Aircraft with Hard Safety Guarantees", *Proceedings of the AIAA Guidance, Navigation, and Control Conference and Exhibit*, 2004.

12. Flint M., Polycarpou M., and Fernandez-Gaucherand E., "Cooperative Control for Multiple Autonomous UAV's Searching for Targets", *Proceedings, 41st IEEE Conference on Decision and Control*, 2002.
13. Tang Z., and Ozguner U., "Motion Planning for Multi-Target Surveillance with Mobile Sensor Agents", *IEEE Transactions on Robotics*, 21, 898-908, 2005.
14. Gomez Ortega J., and Camacho E. F., "Mobile Robot Navigation in a Partially Structured Static Environment, Using Neural Predictive Control", *Control Engineering Practice*, 4, 1669-1679, 1996.
15. Kwon Y. D., and Lee J. S., "On-line Evolutionary Optimization of Fuzzy Control System Based on Decentralized Population", *Intelligent Automation and Soft Computing*, 6, 135-146, 2000.
16. Nikolos I. K., Valavanis K. P., Tsourveloudis N. C., Kostaras A., "Evolutionary Algorithm Based Offline / Online Path Planner for UAV Navigation", *IEEE Transactions on Systems, Man, and Cybernetics – Part B*, 33, 898-912, 2003.
17. Michalewicz Z., *Genetic Algorithms + Data Structures = Evolution Programs*, Springer, 1999.
18. Smierzchalski R., "Evolutionary Trajectory Planning of Ships in Navigation Traffic Areas", *Journal of Marine Science and Technology*, 4, 1-6, 1999.
19. Smierzchalski R., and Michalewicz Z., "Modeling of Ship Trajectory in Collision Situations by an Evolutionary Algorithm", *IEEE Transactions on Evolutionary Computation*, 4, 227-241, 2000.
20. Sugihara K., and Smith J., "Genetic Algorithms for Adaptive Motion Planning of an Autonomous Mobile Robot", *Proceedings, IEEE International Symposium on Computational Intelligence in Robotics and Automation*, Monterey, California, 138-143, 1997.
21. Sugihara K., and Yuh J., *GA-Based Motion Planning for Underwater Robotic Vehicles*, UUST-10, Durham, NH, 1997.
22. Shima T., Rasmussen S. J., Sparks A. G., "UAV Cooperative Multiple Task Assignments Using Genetic Algorithms", *Proceedings, American Control Conference*, Oregon, June 2005.
23. Moitra A., Mattheyses R. M., Hoebel L. J., Szczerba R. J., Yamrom B., "Multivehicle Reconnaissance Route and Sensor Planning", *IEEE Transactions on Aerospace and Electronic Systems*, 37, 799-812, 2003.
24. Dubins L., "On Curves of Minimal Length with a Constraint on Average Curvature, and with Prescribed Initial and Terminal Position", *American Journal of Mathematics*, 79, 497–516, 1957.
25. Shima T., Schumacher C., "Assignment of Cooperating UAVs to Simultaneous Tasks Using Genetic Algorithms", *Proceedings, AIAA Guidance, Navigation, and Control Conference and Exhibit*, California, 2005.
26. Martinez-Alfaro H., and Gomez-Garcia S., "Mobile Robot Path Planning and Tracking Using Simulated Annealing and Fuzzy Logic Control", *Expert Systems with Applications*, 15, 421-429, 1988.
27. Nikolos I. K., Tsourveloudis N. C, and Valavanis K. P., "Evolutionary Algorithm Based 3-D Path Planner for UAV Navigation", CD-ROM Proceedings,

9th *Mediterranean Conference on Control and Automation*, Dubrovnik, Croatia, 2001.
28. Piegl L., Tiller W., *The NURBS Book*, Springer, 1997.
29. Farin G., *Curves and Surfaces for Computer Aided Geometric Design, A Practical Guide*, Academic Press, 1988.
30. Storn R., and Price, K., *DE - A Simple and Efficient Adaptive Scheme for Global Optimization over Continuous Space*, ICSI, Technical Report TR-95-012, 1995.
31. Price K. V., Storn R. M., Lampinen J. A., *Differential Evolution, a Practical Approach to Global Optimization*, Springer, 2005.
32. Hui-Yuan F., Lampinen J., Dulikravich G. S., "Improvements to Mutation Donor Formulation of Differential Evolution", *Proceedings of EUROGEN Conference on Evolutionary Methods for Design, Optimization and Control, Applications to Industrial and Societal Problems*, CIMNE, Barcelona, 2003.
33. Marse K. and Roberts S. D., "Implementing a portable FORTRAN Uniform (0, 1) Generator", *Simulation*, 41-135, 1983.

PART IV

Applications

Chapter 11: Robust Nonlinear Observers for Attitude Estimation of Mini UAVs[1]

A key requirement for navigation and control of any autonomous flying vehicle is availability of a robust attitude estimate. Small scale aerial robotic vehicles such as mini or micro aerial vehicles use low cost lightweight inertial measurement units (characterized by high noise levels and time varying additive biases) and embedded avionics systems that make classical stochastic filter techniques unreliable. This Chapter proposes a suite of nonlinear attitude observers that fuse angular velocity and orientation measurements in an analogous manner to that of a complementary filter for a linear system. By exploiting the natural geometry of the group of rotations an attitude observer is derived that: requires only accelerometer and gyro outputs; it is suitable for implementation on embedded hardware; it provides robust attitude estimates as well as estimating the gyro biases on-line. Experimental results from a robotic testbed and a radio controlled unmanned aerial vehicle are provided to verify the filter performance.

11.1 Introduction

The last decade has seen an intense world wide effort in developing mini aerial vehicles (mAVs). Such vehicles are characterized by small scale (dimensions of the order of 60cm), limited payload capacity and embedded avionics systems. A key component of the avionics system in a mAV is the attitude estimation subsystem [2] 12] [30]. Such systems must be highly reliable and have low computational overhead to avoid overloading the limited computational resources available in some applications. Traditional linear and extended Kalman filter (EKF) techniques [14] [3] [20] suffer from issues associated with poor system modeling (in particular characterization of noise within the system necessary for tuning filter parameters) as well as potentially high computational requirements [28] [30]. An alter native is to use deterministic complementary filter and nonlinear observer

[1] Written by R. Mahony, T. Hamel

Kimon P. Valavanis (ed.), Advances in Unmanned Aerial Vehicles, 343–375.

design techniques [35] [2] [1] [34]. Recent work has focused on some of the practical issues encountered when data is obtained from low cost inertial measurement units (IMU) [26] [1] [34] [23] as well as observer design for partial attitude estimation [27] [21] [22]. It is also worth mentioning the related problem of fusing IMU and vision data [16] [25] [13] [7] [6] and the problem of fusing IMU and GPS data [24] [34]. A key issue in attitude observer design for systems with low-cost IMU sensor units is on-line identification of gyro bias terms. This problem is also important in IMU calibration of attitude observers for satellites [14] [8] [4] [32] [17]. An important development that has resulted from early work on estimation and control of satellites was the use of the quaternion representation for the attitude kinematics [29] [9] [32] [31]. The nonlinear observer designs that are based on this work have strong robustness properties and deal well with the bias estimation problem [34] [32]. The quaternion representation for attitude can provide a significant computational advantage in the implementation of observer algorithms. However, it tends to obscure the underlying geometric structure of the algorithms proposed.

In this Chapter the design of nonlinear attitude observers on the group of rotations, the special orthogonal group $SO(3)$ is studied. The proposed observers are called *complementary filters* because of the similarity of the architecture to that of linear complementary filters, although, for the nonlinear case there is no frequency domain interpretation. A general formulation of the error criterion and observer structure is described on the Lie-group structure of $SO(3)$. This formulation leads to defining two nonlinear observers on $SO(3)$, termed the *direct complementary filter* and *passive complementary filter* [19]. The direct complementary filter corresponds (up to some minor technical differences) to nonlinear observers proposed using the quaternion representation [29] [32] [34]. There is no knowledge of a prior reference for the passive complementary filter. The passive complementary filter has several practical advantages associated with implementation and low sensitivity to noise. In particular, it is shown that the filter can be reformulated in terms of direct measurements from the IMU system, a formulation called the *explicit complementary filter* [11]. The explicit complementary filter does not require on-line algebraic reconstruction of attitude, an implicit weakness in prior work on nonlinear attitude observers due to the computational overhead of the calculation and poor error characterization of the constructed attitude. As a result the observer is ideally suited for implementation on embedded hardware platforms. Furthermore, the relative contribution of different data can be preferentially weighted in the observer response, a property that allows the designer to adjust for application specific noise characteristics. Finally, the explicit complementary filter remains well defined even if the data provided is insufficient to algebraically reconstruct the attitude. This is the case, for example, for an IMU with only accelerometer and rate gyro sensors. Although the principal results of the Chapter are developed in the matrix Lie group representation of $SO(3)$, the equivalent quaternion representation of the observers are also derived. It is recommended that the quaternion representations are used for hardware implementation.

This Chapter consists of eight Sections. Section 11.2 provides a short discussion on linear complementary filter design. Section 11.3 provides a quick overview of the sensor model, geometry of $SO(3)$ and introduces the notation used. Section 11.4 details the derivation of the direct and passive complementary passive filters. The development here is deliberately kept simple to be clear. Section 11.5 integrates on-line bias estimation into the observer design and provides a detailed stability analysis. Section 11.6 develops the *explicit complementary filter*, a reformulation of the passive complementary filter directly in terms of error measurements. A suite of experimental results, obtained during flight tests of the Hovereye (see Figure 11.7), are provided in Section 11.7 that demonstrate the performance of the proposed observers. Finally Section 11.8 concludes the Chapter.

11.2 A Review of Complementary Filtering

Complementary filters provide a means to fuse multiple independent noisy measurements of the same signal that have complementary spectral characteristics [1]. For example, consider two measurements $y_1 = x + \mu_1$ and $y_2 = x + \mu_2$ of a signal x where μ_1 is predominantly high frequency noise and μ_2 is a predominantly low frequency disturbance. Choosing a pair of complementary transfer functions $F_1(s) + F_2(s) = 1$ with $F_1(s)$ low pass and $F_2(s)$ high pass, the filtered estimate is given by:

$$\hat{X}(s) = F_1(s)Y_1 + F_2(s)Y_2 = X(s) + F_1(s)\mu_1(s) + F_2(s)\mu_2(s).$$

The signal $X(s)$ is all pass in the filter output while noise components are high and low pass filtered as desired. This type of filter is also known as *distortionless filtering* since the signal $x(t)$ is not distorted by the filter [5].

Complementary filters are particularly well suited to fusing low bandwidth position measurements with high band width rate measurements for first order kinematics systems. Consider the linear kinematics:

$$\dot{x} = u. \tag{11.1}$$

with typical measurement characteristics:

$$y_x = L(s)x + \mu_x, \quad y_u = u + \mu_u + b(t) \tag{11.2}$$

where $L(s)$ is a low pass filter associated with sensor characteristics, μ represents noise in both measurements and $b(t)$ is a deterministic perturbation that is dominated by low frequency content. Normally for the low pass filter it is true that $L(s) \approx 1$ over the frequency range on which the measurement y_x is of interest. The rate measurement is integrated $\frac{y_u}{s}$ to obtain an estimate of the state and the noise and bias characteristics of the integrated signal are dominantly low frequency effects. Choosing:

$$F_1(s) = \frac{C(s)}{C(s)+s}$$

$$F_2(s) = 1 - F_1(s) = \frac{s}{C(s)+s}$$

with $C(s)$ all pass such that $L(s)F_1(s) \approx 1$ over the bandwidth of $L(s)$. Then:

$$\hat{X}(s) \approx X(s) + F_1(s)\mu_x(s) + \frac{\mu_u(s)+b(s)}{C(s)+s}$$

By suitable choice of $C(s)$ it is possible to tune the filters $F_1(s)$ and $1/(C(s)+s)$ to obtain satisfactory noise attenuation.

In practice, the filter structure is implemented by exploiting the complementary sensitivity structure of a linear feedback system subject to load disturbance. Consider the block diagram in Figure 11.1.

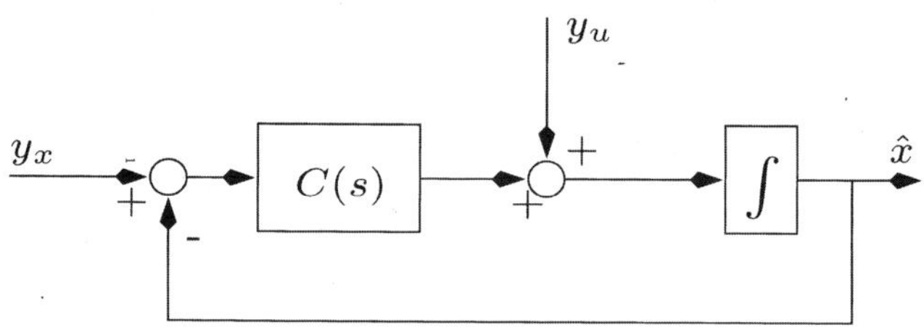

Fig. 11.1. Block diagram of a classical complementary filter.

The output $\hat{x}$ can be written:

$$\hat{x}(s) = \frac{C(s)}{s + C(s)} y_x(s) + \frac{s}{C(s) + s} \frac{y_u(s)}{s}$$
$$= T(s) y_x(s) + S(s) \frac{y_u(s)}{s}$$

where $S(s)$ is the sensitivity function of the closed-loop system and $T(s)$ is the complementary sensitivity. This architecture is easy to implement efficiently and allows one to use classical control design techniques for $C(s)$ in the filter design.

The simplest choice is a proportional feedback $C(s) = k_P$. In this case the closed-loop dynamics of the filter are given by:

$$\dot{\hat{x}} = y_u + k_P(y_x - \hat{x}). \tag{11.3}$$

The frequency domain complementary filters associated with this choice are $F_1(s) = \frac{k_P}{s+k_P}$ and $F_2(s) = \frac{s}{s+k_P}$. Note that the crossover frequency for the filter is at k_P rad/s. The gain k_P is typically chosen based on the low pass characteristics of y_x and the low frequency noise characteristics of y_u to choose the best crossover frequency to tradeoff between the two measurements. If the rate measurement bias, $b(t) = b_0$, is a constant then it is natural to add an integrator into the compensator to make the system of Type I:

$$C(s) = k_P + \frac{k_I}{s}. \tag{11.4}$$

A Type I system will reject the constant load disturbance b_0 from the output. Gain design for k_P and k_I is typically based on classical frequency design methods.

The nonlinear development in the body of the Chapter requires a Lyapunov analysis of the closed-loop system (11.3). Applying the PI compensator of (11.4) one obtains state space filter with dynamics:

$$\dot{\hat{x}} = y_u - \hat{b} + k(y_x - \hat{x}), \quad \dot{\hat{b}} = -k_I(y_x - \hat{x})$$

The negative sign in the integrator state is introduced to indicate that the state $\hat{b}$ will cancel the bias in y_u. Consider the Lyapunov function:

$$L = \frac{1}{2} | x - \hat{x} |^2 + \frac{1}{2k_I} | b_0 - \hat{b} |^2$$

Abusing notation for the noise processes, and using $\tilde{x} = (x - \hat{x})$, and $\tilde{b} = (b_0 - \hat{b})$, one has:

$$\frac{d}{dt} L = -k_P |\tilde{x}|^2 - \mu_u \tilde{x} + \mu_x (\tilde{b} - k\tilde{x})$$

In the absence of noise one may apply the Lyapunov direct method to prove convergence of the state estimate. LaSalle's principal of invariance may be used to show that $\hat{b} \to b_0$. When the underlying system is linear, then the linear form of the feedback and adaptation law ensures that the closed-loop system is linear and stability implies exponential stability.

11.3 Notation and Definitions

11.3.1 Notation and Mathematical Identities

The special orthogonal group is denoted $SO(3)$. The associated Lie-algebra is the set of anti-symmetric matrices:

$$\mathfrak{so}(3) = \{A \in R^{3\times3} \mid A = -A^T\}$$

For any two matrices $A, B \in R^{n\times n}$ then the Lie-bracket (or matrix commutator) is $[A, B] = AB - BA$. Let $\Omega \in R^3$ then define:

$$\Omega_\times = \begin{pmatrix} 0 & -\Omega_3 & \Omega_2 \\ \Omega_3 & 0 & -\Omega_1 \\ -\Omega_2 & \Omega_1 & 0 \end{pmatrix}.$$

For any $v \in R^3$ then $\Omega_\times v = \Omega \times v$ is the vector cross product. The operator $\mathrm{vex}: \mathfrak{so}(3) \to R^3$ denotes the inverse of the $\Omega_\times$ operator:

$$\mathrm{vex}(\Omega_\times) = \Omega, \quad \Omega \in R^3.$$

$$\mathrm{vex}(A)_\times = A, \quad A \in \mathfrak{so}(3)$$

For any two matrices $A, B \in R^{n\times n}$ the Euclidean matrix inner product and Frobenius norm are defined:

$$<< A, B >> = \mathrm{tr}(A^T B) = \sum_{i,j=1}^{n} A_{ij} B_{ij}$$

$$\|A\| = \sqrt{<< A, A >>} = \sqrt{\sum_{i,j=1}^{n} A_{ij}^2}$$

The following identities are used:

$$(Rv)_\times = R v_\times R^T, R \in SO(3), \quad v \in R^3$$

$$(v \times w)_\times = [v_\times, w_\times], \quad v, w \in R^3$$

$$v^T w = < v, w > = \frac{1}{2} << v_\times, w_\times >>, \quad v, w \in R^3$$

$$v^T v = |v|^2 = \frac{1}{2} \| v_\times \|^2, \quad v \in R^3$$

$$<< A, v_\times >> = 0, \quad A = A^T \in R^3, \quad v \in R^3$$

$$\mathrm{tr}([A, B]) = 0, \quad A, B \in R^{3\times 3}$$

The following notation for frames of reference is also used: $\{A\}$ denotes an inertial (fixed) frame of reference; $\{B\}$ denotes a body-fixed-frame of reference; $\{E\}$ denotes the estimator frame of reference.

Let P_a, P_s denote, respectively, the anti-symmetric and symmetric projection operators in square matrix space:

$$P_a(H) = \frac{1}{2}(H - H^T), \qquad P_s(H) = \frac{1}{2}(H + H^T).$$

Let (θ, a) ($|a| = 1$) denote the angle-axis coordinates of $R \in SO(3)$ with:

$$R = \exp(\theta a_\times), \qquad \log(R) = \theta a_\times$$

$$\cos(\theta) = \frac{1}{2}(\mathrm{tr}(R) - 1), \qquad P_a(R) = \sin(\theta) a_\times.$$

For any $R \in SO(3)$ then $3 \geq \mathrm{tr}(R) \geq -1$. If $\mathrm{tr}(R) = 3$ then $\theta = 0$ in angle-axis coordinates and $R = I$. If $\mathrm{tr}(R) = -1$ then $\theta = \pm\pi$, R has real eigenvalues $(1, -1, -1)$.

The unit quaternion representation of rotations is commonly used for the realization of algorithms on $SO(3)$ since it is efficient in code implementation. The quaternions set is denoted $Q = \{q = (s, v) \in R \times R^3 : |q| = 1\}$. The set Q is a group under the operation:

$$q_1 \otimes q_2 = \begin{bmatrix} s_1 s_2 - v_1^T v_2 \\ s_1 v_2 + s_2 v_1 + v_1 \times v_2 \end{bmatrix}$$

with identity element $\mathbf{1} = (1, 0, 0, 0)$. The group of quaternions is homomorphic to $SO(3)$ via the map:

$$F : Q \to SO(3), \quad F(q) := I_3 + 2sv_\times + 2v_\times^2$$

This map is a two to one mapping of Q onto $SO(3)$ with kernel $\{(1, 0, 0, 0), (-1, 0, 0, 0)\}$, thus, Q is locally isomorphic to $SO(3)$ via F. Given that $R \in SO(3)$ such that $R = \exp(\theta a_\times)$, then it follows that $F^{-1}(R) = \{\pm(\cos(\frac{\theta}{2}), \sin(\frac{\theta}{2})a)\}$. Let $\Omega \in \{A\}$ denote a body-fixed frame velocity, then the pure quaternion $\mathbf{p}(\Omega) = (0, \Omega)$ is associated with a quaternion velocity.

11.3.2 Measurements

The measurements available from a typical IMU unit are 3-axis rate gyros, 3-axis accelerometers and 3-axis magnetometers measurements. The reference frame of the strap down IMU is termed the body-fixed-frame $\{B\}$. The inertial frame is denoted $\{A\}$. The rotation $R = R_B^A$ denotes the relative orientation of $\{B\}$ with respect to $\{A\}$.

Rate Gyros: The rate gyro measures angular velocity of $\{B\}$ relative to $\{A\}$ expressed in the body-fixed-frame of reference $\{B\}$. The error model used is:

$$\Omega^y = \Omega + b + \mu \in R^3$$

where $\Omega \in \{B\}$ denotes the true value, μ denotes additive measurement noise and b denotes a constant (or slowly time-varying) gyro bias.

Accelerometer: Denote the instantaneous linear acceleration of $\{B\}$ relative to $\{A\}$, expressed in $\{A\}$, by $\dot{v}$. An ideal accelerometer, 'strapped down' to the body-fixed frame $\{B\}$, measures the instantaneous linear acceleration of $\{B\}$ minus the (conservative) gravitational acceleration field g_0 (where g_0 is expressed in the inertial frame $\{A\}$), and provides a measurement expressed in the body-fixed frame $\{B\}$. In practice, the output a from a MEMS component accelerometer has added bias and noise:

$$a = R^T(\dot{v} - g_0) + b_a + \mu_a,$$

where b_a is a bias term and μ_a denotes additive measurement noise. Normally, the gravitational field $g_0 = |g_0| e_3$ where $|g_0| \approx 9.8$ dominates the value of a for low frequency response. Thus, it is common to use:

$$v_a = \frac{a}{|a|} \approx -R^T e_3$$

as a low-frequency estimate of the inertial z-axis expressed in the body-fixed-frame.

***Magnetometer*:** The magnetometers provide measurements of the magnetic field:

$$m = R^T m^A + B_m + \mu_b$$

where m^A is the Earths magnetic field (expressed in the inertial frame), B_m is a body-fixed-frame expression for the local magnetic disturbance and μ_b denotes measurement noise. The noise μ_b is usually quite low for magnetometer readings; however, the local magnetic disturbance can be very significant, especially if the IMU is strapped down to a mAV with electric motors. Only the direction of the magnetometer output is relevant for attitude estimation. A vector measurement is used for subsequent derivations:

$$v_m = \frac{m}{|m|}$$

The measured vectors v_a and v_m can be used to construct an instantaneous algebraic measurement of the rotation $R_B^A : \{B\} \to \{A\}$:

$$R_y = \arg \min_{R \in SO(3)} \left(\lambda_1 |e_3 - Rv_a|^2 + \lambda_2 |v_m^A - Rv_m|^2 \right) \approx R_A^B$$

where v_m^A is the inertial direction of the magnetic field in the locality where data is acquired. The weights λ_1 and λ_2 are chosen depending on the relative confidence in the sensor outputs. Due to the computational complexity of solving an optimization problem the reconstructed rotation is often obtained in a suboptimal manner where the constraints are applied in sequence; that is, two degrees of freedom in the rotation matrix are resolved by the acceleration readings and the final degree of freedom is resolved using the magnetometer. As a consequence, the error properties of the reconstructed attitude R_y can be difficult to characterize.

Moreover, if either magnetometer or accelerometer readings are unavailable (due to local magnetic disturbance or high acceleration maneuvers), then it is impossible to resolve the vectorial measurements into a unique instantaneous algebraic measurement of attitude.

11.3.3 Error Criteria for Estimation on $SO(3)$

Let $\hat{R}$ denote an estimate of the body-fixed rotation matrix $R = R_B^A$. The rotation $\hat{R}$ can be considered as coordinates for the estimator frame of reference $\{E\}$. It is also associated with the frame transformation:

$$\hat{R} = \hat{R}_E^A : \{E\} \rightarrow \{A\}.$$

The goal of attitude estimate is to drive $\hat{R} \rightarrow R$. The estimation error we propose to use is the relative rotation from body-fixed-frame $\{B\}$ to the estimator frame $\{E\}$:

$$\tilde{R} := \hat{R}^T R, \qquad \tilde{R} = \tilde{R}_B^E : \{B\} \rightarrow \{E\}. \tag{11.5}$$

The proposed observer design is based on Lyapunov stability analysis. The Lyapunov functions used are inspired by the cost function:

$$E_{tr} := \frac{1}{2} \| I_3 - \tilde{R} \|^2 = \frac{1}{2} \mathrm{tr}(I_3 - \tilde{R}) \tag{11.6}$$

One has that:

$$E_{tr} := \frac{1}{2} \mathrm{tr}(I - \tilde{R}) = (1 - \cos(\theta)) = 2 \sin(\theta/2)^2. \tag{11.7}$$

where θ is the angle associated with the rotation from $\{B\}$ to frame $\{E\}$. Thus, driving (11.6) to zero ensures that $\theta \rightarrow 0$.

11.4 Complementary Filters on $SO(3)$

A general framework for nonlinear complementary filtering on the special orthogonal group is introduced. The theory is first developed for the ideal case where $R(t)$ and $\Omega(t)$ are assumed to be known and used to drive the filter dynamics. Filter design for real world signals is considered in later Sections.

The goal of attitude estimation is to provide a set of dynamics for an estimate $\hat{R}(t) \in SO(3)$ to drive the error rotation (11.5) $\tilde{R}(t) \to I_3$. The kinematics of the true system is:

$$\dot{R} = R\Omega_\times = (R\Omega)_\times R \tag{11.8}$$

where $\Omega \in \{B\}$. The proposed observer equation is posed directly as a kinematic system for an attitude estimate $\hat{R}$ on $SO(3)$. The observer kinematics include a prediction term based on the Ω measurement and an innovation or correction term, $k_P\omega$, with $\omega := \omega(\tilde{R})$ derived from the error $\tilde{R}$ and $k_P > 0$ a positive gain. The general form proposed for the observer is:

$$\dot{\hat{R}} = (R\Omega + k_P\hat{R}\omega)_\times \hat{R}, \qquad \hat{R}(0) = \hat{R}_0. \tag{11.9}$$

The term $(R\Omega + k_p\hat{R}\omega) \in \{A\}$ is expressed in the inertial frame. The body-fixed-frame angular velocity is mapped back into the inertial frame $\Omega^A = R\Omega$. If no correction term is used ($k_P\omega \equiv 0$) then the error rotation $\tilde{R}$ is constant:

$$\dot{\tilde{R}} = \hat{R}^T(R\Omega)_\times^T R + \hat{R}^T(R\Omega)_\times R = \hat{R}^T\left(-(R\Omega)_\times + (R\Omega)_\times\right)R = 0. \tag{11.10}$$

The correction term $k_p\omega$ consists of a proportional gain $k_P > 0$ and the vector error term $\omega := \omega(\tilde{R}) \in \{E\}$, considered to be in the estimator frame of reference. The vector error term can be thought of as a nonlinear approximation of the error between R and $\hat{R}$. In practice, it will be implemented as an error between a measured estimate R_y of R and the estimate $\hat{R}$. The goal of the observer design is to find a simple expression for ω that leads to robust convergence of $R \to I$. Consider the choice:

$$\omega = \text{vex}(P_a(\tilde{R})). \tag{11.11}$$

Lemma 11.1 - Complementary filter: Consider the attitude kinematics (11.8) and assume that R and Ω are known. Choose ω according to

(11.11) and choose $k_P > 0$ a positive gain. Let $\hat{R}(t)$ denote the solution of (11.9) for initial condition $\hat{R}_0$. Then:

$$\dot{E}_{tr} = -2k_P \cos^2(\theta/2) E_{tr}$$

where E_{tr} is defined in (11.7). For any initial condition $\hat{R}_0$ such that $\mathrm{tr}(\tilde{R}_0) \neq -1$ then $\hat{R}(t) \to R(t)$ exponentially.

Proof: Deriving the Lyapunov function E_{tr} subject to dynamics (11.9) yields:

$$\begin{aligned}\dot{E}_{tr} &= -\frac{1}{2}\mathrm{tr}(\dot{\tilde{R}}) = -\frac{k_P}{2}\mathrm{tr}\left(\omega_\times^T \tilde{R}\right)\\ &= -\frac{k_P}{2}\mathrm{tr}\left[\omega_\times^T (P_s(\tilde{R}) + P_a(\tilde{R}))\right] = -\frac{k_P}{2}\mathrm{tr}\left[\omega_\times^T P_a(\tilde{R})\right]\\ &= -\frac{k_P}{2} << \omega_\times, P_a(\tilde{R}) >>\end{aligned}$$

Substituting for ω, (11.9) yields:

$$E_{tr} = -\frac{k_P}{2} \| P_a(\tilde{R}) \|^2 = -k_P |\omega|^2 .$$

Defining θ by the angle axis convention $\sin(\theta)a_\times = P_a(\tilde{R})$ for $|a| = 1$, one has $\| a_\times \| = 2$ and:

$$\begin{aligned}E_{tr} &= -k_P \sin^2(\theta) \| a_\times \|_\times^2 = -2k_P \sin^2(\theta)\\ &= -8k_P \sin^2(\theta/2)\cos^2(\theta/2) = -2k_P \cos^2(\theta/2) E_{tr}.\end{aligned}$$

The condition on the initial condition $\tilde{R}_0$ guarantees that $-\pi < \theta_0 < \pi$. The result follows from applying Lyapunov's direct method.

The filter (11.9) is called a *complementary filter on* $SO(3)$ since it recaptures the block diagram structure of a classical complementary filter.

In Figure 11.2, the '$\hat{R}^T$' operation is an inverse operation on *SO(3)* and it is equivalent to a '$-$' operation for a linear complementary filter. The '$\hat{R}^T R$' operation is equivalent to generating the error term '$y - \hat{x}$'. The two operations $P_a(\tilde{R})$ and $(R\Omega)_\times$ are maps from error space and velocity space into the tangent space of *SO(3)*; operations that are unnecessary on Euclidean space due to the identification $T_x R^n \equiv R^n$. The kinematic model is the Lie-group equivalent of a first order integrator.

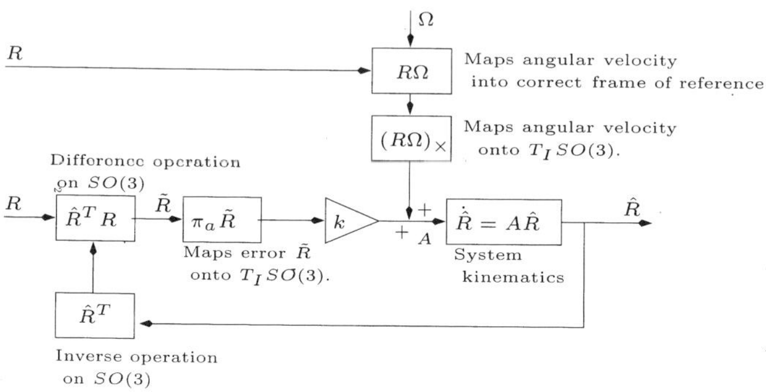

Fig. 11.2. Block diagram of the general form of a complementary filter on $SO(3)$.

To implement the complementary filter it is necessary to map the body-fixed-frame velocity Ω into the inertial frame. In practice, the 'true' rotation R is not available and an estimate of the rotation must be used. Two possibilities are considered:

Direct complementary filter: The constructed attitude R_y is used to map the velocity into the inertial frame $\Omega^A \approx R_y\Omega_y$. A block diagram of this filter design is shown in Figure 11.3. This approach can be linked to observers documented in earlier work [29] [32]. The approach has the advantage that it does not introduce an additional feedback loop in the filter dynamics, however, noise in the reconstructed attitude R_y will enter into the feed forward term of the filter.

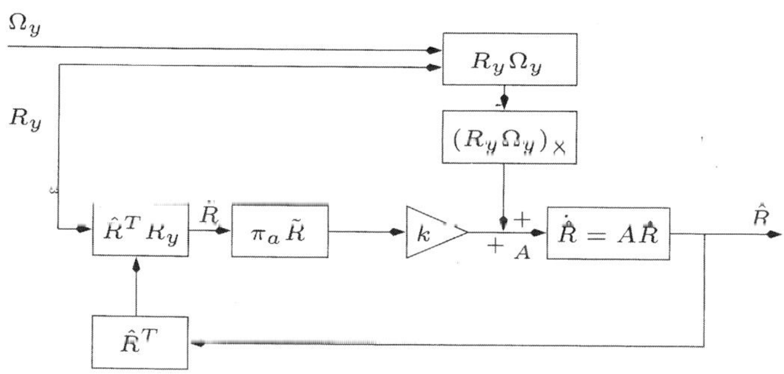

Fig. 11.3. Block diagram of the direct complementary filter on $SO(3)$.

Passive complementary filter: The filtered attitude $\hat{R}$ is used in the predictive velocity term $\Omega^A \approx \hat{R}\Omega_y$. A block diagram of this architecture is shown in Figure 11.4. The advantage lies in avoiding corrupting the predictive angular velocity term with the noise in the reconstructed pose. However, the approach introduces a secondary feedback loop in the filter and stability needs to be proved.

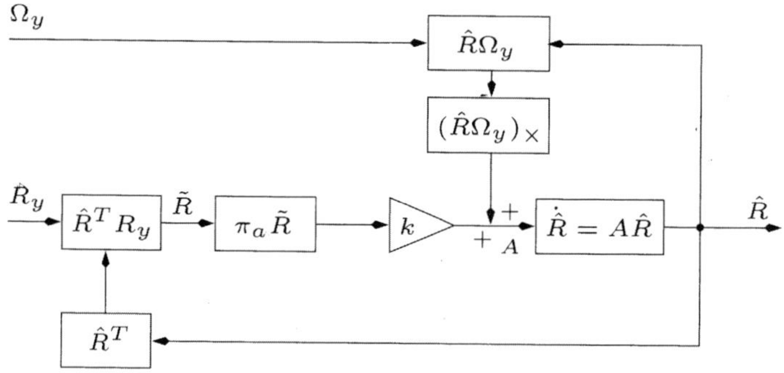

Fig. 11.4. Block diagram of the passive complementary filter on $SO(3)$.

Lemma 11.2. - Passive complementary filter: Consider the rotation kinematics (11.8) and assume that R and Ω are known. Let $k_P > 0$ and choose ω according to (11.11) Let the attitude estimate $\hat{R}(t)$ be given by the solution of:

$$\dot{\hat{R}} = \left(\hat{R}\Omega + k_P \hat{R}\omega\right)_\times \hat{R}, \qquad \hat{R}(0) = \hat{R}_0 \qquad (11.12)$$

Then:

$$\dot{E}_{tr} = -2k_P \cos^2(\theta/2) E_{tr}$$

where E_{tr} is defined in (11.7). For any initial condition $\hat{R}_0$ such that $\mathrm{tr}(\tilde{R}_0) \neq -1$, then $\hat{R}(t) \to R(t)$ exponentially.

Proof: Observe that:

$$\left(\hat{R}\Omega + k_P \hat{R}\omega\right)_\times \hat{R} = \hat{R}(\Omega + k_P\omega)_\times \hat{R}^T \hat{R} = \hat{R}(\Omega + k_P\omega)_\times$$

Differentiating E_{tr} subject to dynamics (11.12) yields:

$$E_{tr} = -\frac{1}{2}\operatorname{tr}(\dot{\tilde{R}}) = -\frac{1}{2}\operatorname{tr}(-(\Omega + k_P\omega)_\times \tilde{R} + \tilde{R}\Omega_\times)$$

$$= -\frac{1}{2}\operatorname{tr}([\tilde{R}, \Omega_\times]) - \frac{k_P}{2}\operatorname{tr}(\omega_\times^T \tilde{R})$$

$$= -\frac{k_P}{2} << \omega_\times, P_a(\tilde{R}) >>$$

since the trace of a commutator is zero, $\operatorname{tr}([\tilde{R}, \Omega_\times]) = 0$, a property of the passivity of the rotational kinematics on $SO(3)$. The remainder of the proof is identical to that of Lemma 11.1.

It is important to note that the direct and the passive complementary filters have different solutions even though the Lyapunov stability analysis appears identical. The different trajectories of $\hat{R}e_3$ are shown in Figure 11.5 for identical initial conditions and constant angular velocity Ω. The level sets of the Lyapunov function are the small circles of the hemisphere and the two trajectories always lie on the same small circle during the evolution of the filter.

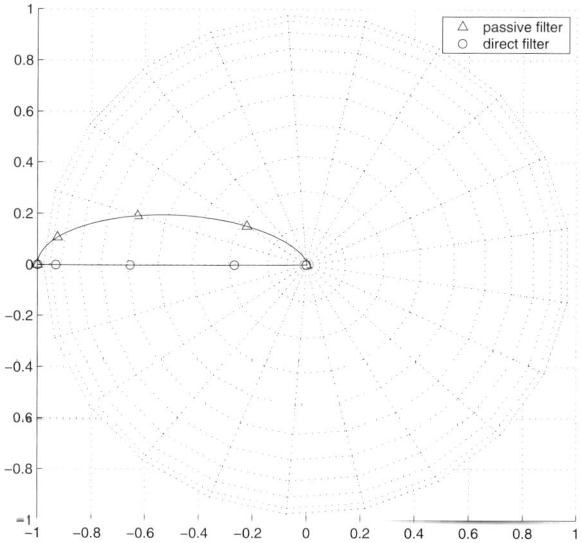

Fig. 11.5. Trajectories of $\hat{R}e_3$ for direct and passive complementary filters on the plan $\{e_1, e_2\}$ for initial deviation, $\tilde{R}_0$, corresponding to a rotation of $\frac{\pi}{2}$ rad around the e_2 axis and $\Omega = 0.3e_3$ rad/s.

If $\text{tr}(\tilde{R})(0) = -1$, then for both the direct and passive filter, it is easily verified that $\text{tr}(\dot{R}(t)) = 0$ for all $t \geq 0$. Hence, the set $U_0 = \{\tilde{R} \in SO(3) : \text{tr}(\tilde{R}) = -1\}$ is an invariant set of the error dynamics. This set corresponds to a maximum of the cost function E_{tr} and the descent condition in the Lyapunov arguments for both filters ensures that the invariant set is unstable.

There is no particular theoretical advantage to either the direct or the passive filter architecture in the case where exact measurements are assumed. However, it is straightforward to see that the passive filter (11.12) can be written as:

$$\dot{\hat{R}} = \hat{R}(\Omega_\times + k_P P_a(\tilde{R})). \tag{11.13}$$

This formulation suppresses entirely the requirement to represent Ω and $\omega = k_P P_a(\tilde{R})$ in the inertial frame and leads to the architecture shown in Figure 11.6. The passive complementary filter avoids coupling the reconstructed attitude noise into the predictive velocity term of the observer, has a strong Lyapunov stability analysis (Lemma 11.2), and provides a simple and elegant realization that will lead to the results in Section 11.6.

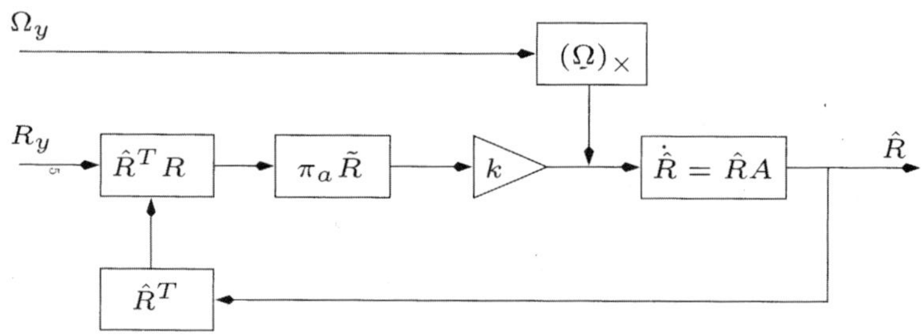

Fig. 11.6. Block diagram of the simplified form of the passive complementary filter.

11.5. Adaptive Gyro Bias Compensation

The direct and passive complementary filters on $SO(3)$ are extended to provide on-line estimation of time-varying bias terms in the gyroscope measurements.

For the following work it is assumed that a reconstructed rotation R_y and a biased measure of angular velocity Ω^y are available:

$$R_y \approx R, \qquad \text{valid for low frequencies,} \quad (11.14)$$

$$\Omega_y \approx \Omega + b, \qquad \text{for constant bias } b. \quad (11.15)$$

The approach taken is to add an integrator to the compensator term in the feedback equation of the complementary filter.

Let $k_P, k_I > 0$ be positive gains and define:

***Direct complementary filter with bias correction*:**

$$\dot{\hat{R}} = \left(R_y(\Omega^y - \hat{b}) + k_P \hat{R}\omega\right)_\times \hat{R}, \quad \hat{R}(0) = \hat{R}_0, \qquad (11.16)$$

$$\dot{\hat{b}} = -k_I \omega, \quad \hat{b}(0) = \hat{b}_0, \qquad (11.17)$$

$$\omega = \text{vex}(P_a(\tilde{R})), \quad \tilde{R} = \hat{R}^T R_y. \qquad (11.18)$$

***Passive complementary filter with bias correction*:**

$$\dot{\hat{R}} = \hat{R}\left(\Omega^y - \hat{b} + k_P \omega\right)_\times, \quad \hat{R}(0) = \hat{R}_0, \qquad (11.19)$$

$$\dot{\hat{b}} = -k_I \omega, \quad \hat{b}(0) = \hat{b}_0, \qquad (11.20)$$

$$\omega = \text{vex}(P_a(\tilde{R})), \quad \tilde{R} = \hat{R}^T R_y. \qquad (11.21)$$

The additional dynamics introduced for the $\hat{b}$ term can be thought of as an adaptive estimate of the gyro bias. The stability analysis is based on the same approach as the adaptive control Lyapunov function interpretation of the classical complementary filter presented in Section 11.2. The nonlinear state space $SO(3)$ introduces several complexities to the analysis, the most important following from the existence of local maxima of the Lyapunov function due to the topological structure of $SO(3)$. The proofs of the following two theorems are contained in the technical report [18]. Local stability results are given in [19].

***Theorem 11.1. - Direct complementary filter with bias correction*:** Consider the rotation kinematics (11.8) for a time-varying $R(t) \in SO(3)$ and with measurements given by (11.14), (11.15). Let $(\hat{R}(t), \hat{b}(t))$ denote the solution of (11.16) – (11.18). Define error variables $\tilde{R} = \hat{R}^T R$ and $\tilde{b} = b - \hat{b}$. Define $U \subseteq SO(3) \times R^3$ by:

$$U = \left\{ (\tilde{R}, \tilde{b}) \mid \mathrm{tr}(\tilde{R}) = -1, P_a(\tilde{b}_\times \tilde{R}) = 0 \right\}. \qquad (11.22)$$

Then:

- The set U is forward invariant and unstable with respect to the dynamic system (11.16) – (11.18);
- The error $(\tilde{R}(t), \tilde{b}(t))$ is locally exponentially stable to $(I, 0)$;
- For almost all initial conditions $(\tilde{R}_0, \tilde{b}_0) \notin U$ the trajectory $(\hat{R}(t), \hat{b}(t))$ converges to the trajectory $(R(t), b)$.

***Theorem 11.2. - Passive complementary filter with bias correction*:** Consider the rotation kinematics (11.8) for a time-varying $R(t) \in SO(3)$ and with measurements given by (11.14)-(11.15). Let $(\hat{R}(t), \hat{b}(t))$ denote the solution of (11.19)-(11.21). Define error variables $\tilde{R} = \hat{R}^T R$ and $\tilde{b} = b - \hat{b}$ and assume that $\Omega(t)$ is a bounded, absolutely continuous signal that is persistently exciting and uncorrelated to the error signal $\tilde{R} = \hat{R}^T R$. Define $U_0 \subseteq SO(3) \times R^3$ by:

$$U_0 = \left\{ (\tilde{R}, \tilde{b}) \mid \mathrm{tr}(\tilde{R}) = -1, \tilde{b} = 0 \right\}. \qquad (11.23)$$

Then:

- The set U_0 is forward invariant and unstable with respect to the dynamic system (11.19) – (11.21);
- The error $(\tilde{R}(t), \tilde{b}(t))$ is locally exponentially stable to $(I, 0)$;
- For almost all initial conditions $(\tilde{R}_0, \tilde{b}_0) \notin U_0$ the trajectory $(\hat{R}(t), \hat{b}(t))$ converges to the trajectory $(R(t), b)$.

Apart from the expected conditions inherited from Theorem 1 the key assumption in Theorem 11.2 is the persistence of excitation of $\Omega(t)$. The perturbation of the passive dynamics by the driving term Ω provides a disturbance that ensures that the adaptive bias estimate converges to the true gyroscopes' bias, a particularly useful property in practical applications.

11.5.1 Quaternion Versions of Direct and Passive Complementary Filters

Consider the rotation kinematics on $SO(3)$ (11.8). The associated quaternion kinematics are given by:

$$\dot{q} = \frac{1}{2} q \otimes \mathbf{p}(\Omega) \tag{11.24}$$

Let $q_y \approx q$ be a low frequency measure of q, and $\Omega_y \approx \Omega + b$ (for constant bias b) be the angular velocity measure. Let $\hat{q}$ denote the observer estimate and quaternion error $\tilde{q}$:

$$\tilde{q} = \hat{q}^{-1} \otimes q = \begin{bmatrix} \tilde{s} \\ \tilde{v} \end{bmatrix}$$

Note that:

$$2\tilde{s}\tilde{v} = 2\cos(\theta/2)\sin(\theta/2)a = \frac{1}{2}(\sin\theta)a = \mathrm{vex}(P_a(\tilde{R}))$$

where (θ, a) is the angle axis representation of $\tilde{R} = F(\tilde{q})$.

The direct complimentary filter is closely related to quaternion based attitude filters published over the last fifteen years [29] [34] [32]. The quaternion representation of the *direct complementary filter* (11.16)–(11.18) is:

$$\dot{\hat{q}} = \frac{1}{2} \hat{q} \otimes \mathbf{p}(\tilde{R}(\Omega_y - \hat{b}) + 2k_P \tilde{s}\tilde{v}) \tag{11.25}$$

$$\dot{\hat{b}} = -2k_I \tilde{s}\tilde{v} \tag{11.26}$$

There is a fifteen year history of using the quaternion representation and Lyapunov design methodology for filtering on $SO(3)$ (for example [34] [32] [29]). To the authors knowledge the Lyapunov analysis in all cases has been based around the cost function

$$\Phi(\tilde{q}) = (|\tilde{s}| - 1)^2 + |\tilde{v}|^2 .$$

Due to the unit norm condition it is straightforward to show that:

$$\Phi(\tilde{q}) = 2(1 - |\tilde{s}|) = 2\big(1 - |\cos(\theta/2)|\big)$$

The cost function proposed in this Chapter is $E_{tr} = (1 - \cos(\theta))$. It follows that the quadratic approximation of both cost functions around the point $\theta = 0$ is the quadratic $\theta^2/2$. The quaternion cost function Φ, however, is non-differentiable at the point $\theta = \pm\pi$ while the cost $\mathrm{tr}(I - \tilde{R})$ has a smooth local minimum at this point.

Almost all quaternion filters in the published literature have a similar flavor that dates back to the seminal work of Salcudean [29]. The closest published work to that undertaken in the present Chapter was published by

Thienel [33] [32]. The filter considered by Thienel et al. is given by:

$$\dot{\hat{q}} = \frac{1}{2}\hat{q} \otimes \mathbf{p}(\tilde{R}(\Omega_y - \hat{b} + k_P \mathrm{sgn}(\tilde{s})\tilde{v})) \qquad (11.27)$$

$$\dot{\hat{b}} = -k_I \mathrm{sgn}(\tilde{s})\tilde{v} \qquad (11.28)$$

The $\mathrm{sgn}(\tilde{s})$ term enters naturally in the filter design from the differential, $\frac{d}{dt}|\tilde{s}| = \mathrm{sgn}(\tilde{s})\frac{d}{dt}\tilde{s}$, of the absolute value term in the cost function Φ, during the Lyapunov design process.

Consider the observer obtained by replacing $\mathrm{sgn}(\tilde{s})$ in (11.27), (11.28) by $2\tilde{s}$. With this substitution, (11.28) is transformed into (11.26). To show that (11.27) transforms to (11.25) it is sufficient to show that $\tilde{R}\tilde{v} = \tilde{v}$. This is straightforward from:

$$2\tilde{s}\tilde{R}\tilde{v} = \tilde{R}(2\tilde{s}\tilde{v}) = \tilde{R}\mathrm{vex}(P_a(\tilde{R}))$$

$$= \mathrm{vex}(\tilde{R}P_a(\tilde{R})\tilde{R}^T) = \mathrm{vex}(P_a(\tilde{R})) = 2\tilde{s}\tilde{v}$$

It is shown that the quaternion filter (11.27), (11.28) is obtained from the standard form of the complimentary filter proposed in (11.16) – (11.18) with the innovation term (11.18) replaced by:

$$\omega_q = \mathrm{sgn}(\tilde{s})\tilde{v}, \quad \tilde{q} \in F^{-1}(\hat{R}^T R).$$

Note that the innovation defined in (11.18) can be written as $\omega = 2\tilde{s}\tilde{v}$. It follows that:

$$\omega_q = \frac{\mathrm{sgn}(\tilde{s})}{2\tilde{s}}\omega$$

The innovation term for the two filters varies only by the positive scaling factor $\mathrm{sgn}(\tilde{s})/(2\tilde{s})$. The quaternion innovation ω_q is not well defined for $\tilde{s} = 0$ (where $\theta = \pm\pi$) and these points are not well defined in the filter dynamics (11.27), (11.28). It should be noted, however, that $|\omega_q|$ is bounded at $\tilde{s} = 0$ and, apart from possible switching behavior, the filter can still be implemented on the remainder of $SO(3) \times R^3$. An argument for the use of the innovation ω_q is that the resulting error dynamics strongly force the estimate away from the unstable set U. An argument against its use is that, in practice, such situations will only occur due to extreme transients that would overwhelm the bounded innovation term ω_q in any case, and cause the numerical implementation of the filter to deal with a discontinuous argument. In practice, it is an issue of little significance since the filter will work sufficiently well to avoid any issues with the set

U for either choice of innovation. For $\tilde{s} \to 1$, corresponding to $\theta = 0$, the innovation ω_q scales to a factor of $1/2$ the innovation ω. A simple scaling factor like this is compensated for the in choice of filter gains k_P and k_I and makes no difference to the performance of the filter.

The quaternion representation of the *passive complementary filter* (11.19) – (11.21) is:

$$\dot{\hat{q}} = \frac{1}{2}\hat{q} \otimes \mathbf{p}(\Omega_y - \hat{b} + 2k_P\tilde{s}\tilde{v}) \tag{11.29}$$

$$\dot{\hat{b}} = -2k_I\tilde{s}\tilde{v} \tag{11.30}$$

Knowledge this version of the complementary filter on the quaternion group has not been considered in prior work. It is not surprising that the passive complementary filter has not been proposed by authors working purely in the quaternion representation since the passivity property is somewhat obscure in this representation.

11.6 Explicit Error Formulation of the Passive Complementary Filter

A weakness of the implementation of both the direct and passive complementary filters is the requirement for a reconstructed estimate of the attitude, R_y, to use as the driving term for the error dynamics. The reconstruction cannot be avoided in the direct filter implementation because the reconstructed attitude is also used to transform the measured angular velocity into the inertial frame. In this section, we show how the passive complementary filter may be reformulated in terms of direct measurements from the inertial unit.

Let $v_{0i} \in \{A\}$, $i = 1,\dots,n$, denote a set of n known inertial directions. The measurements considered are body-fixed-frame observations of the fixed inertial directions:

$$v_i = R^T v_{0i} + \mu_i, \quad v_i \in \{B\} \tag{11.31}$$

where μ_i is a noise process. Since only the direction of the measurement is relevant to the observer we assume that $|v_{0i}| = 1$ and normalize all measurements to ensure $|v_i| = 1$.

Let $\hat{R}$ be an estimate of R. Define:

$$\hat{v}_i = \hat{R}^T v_{0i}$$

to be the associated estimate of v_i. For a single direction v_i, the error considered is:

$$E_i = 1 - \cos(\angle v_i, \hat{v}_i) = 1 - < v_i, \hat{v}_i >$$

which yields:

$$E_i = 1 - \mathrm{tr}(\hat{R}^T v_{0i} v_{0i}^T R) = 1 - \mathrm{tr}(\tilde{R} R^T v_{0i} v_{0i}^T R)$$

For multiple measures v_i the following cost function is considered:

$$E_{mes} = \sum_{i=1}^{n} k_i E_i = \sum_{i=1}^{n} k_i - \mathrm{tr}(\tilde{R} M), \quad k_i > 0, \tag{11.32}$$

where:

$$M = R^T M_0 R \quad \text{with} \quad M_0 = \sum_{i=1}^{n} k_i v_{0i} v_{0i}^T \tag{11.33}$$

Assume linearly independent inertial direction $\{v_{0i}\}$ then the matrix M is positive definite $(M > 0)$ if $n \geq 3$. For $n \leq 2$ then M is positive semi-definite with one eigenvalue being zero. The weights $k_i > 0$ are chosen depending on the relative confidence in the measurements v_i. For technical reasons, assume additionally that the weights k_i are chosen such that M_0 has three distinct eigenvalues $\lambda_1 > \lambda_2 > \lambda_3$. A full proof of the following theorem may be found in [18], while a local stability proof may be found in [11].

Theorem 11.3. - Explicit complementary filter with bias correction: Consider the rotation kinematics (11.8) for a time-varying $R(t) \in SO(3)$ and with measurements given by (11.31) and (11.15). Assume that there are two or more, ($n \geq 2$) vectorial measurements v_i available. Choose $k_i > 0$ such that M_0 (defined by (11.33)) has three distinct eigenvalues. Consider the filter kinematics given by:

$$\dot{\hat{R}} = \hat{R}\left((\Omega^y - \hat{b})_\times + k_P(\omega_{mes})_\times\right), \quad \hat{R}(0) = \hat{R}_0 \tag{11.34}$$

$$\dot{\hat{b}} = -k_I \omega_{mes} \tag{11.35}$$

$$\omega_{mes} := \sum_{i=1}^{n} k_i v_i \times \hat{v}_i, \qquad k_i > 0. \tag{11.36}$$

and let $(\hat{R}(t),\hat{b}(t))$ denote the solution of (11.34) – (11.36). Assume that $\Omega(t)$ is a bounded, absolutely continuous signal that is persistently exciting and uncorrelated to the error signal $\tilde{R}=\hat{R}^T R$. Then:

- There are three unstable equilibriums of the filter characterized by $(\hat{R}_{*i},\hat{b}_{*i})=\left(U_0 D_i U_0^T R, b\right), i=1,2,3,$ where $D_1=\mathrm{diag}(1,-1,-1)$, $D_2=\mathrm{diag}(-1,1,-1)$ and $D_3=\mathrm{diag}(1,-1,-1)$ are diagonal matrices with entries as shown and $U_0\in SO(3)$ such that $M_0=U_0\Lambda U_0^T$ where $\Lambda=\mathrm{diag}(\lambda_1,\lambda_2,\lambda_3)$ is a diagonal matrix;
- The error $(\tilde{R}(t),\tilde{b}(t))$ is locally exponentially stable to $(I,0)$;
- For almost all initial conditions $(\tilde{R}_0,\tilde{b}_0)\neq(\hat{R}_{*i}^T R,b)$, $i=1,\ ...,3$, the trajectory $(\hat{R}(t),\hat{b}(t))$ converges to the trajectory $(R(t),b)$.

The quaternion representation of *explicit complementary filter* (11.34) – (11.36) is:

$$\omega_{mes}=-\mathrm{vex}\left(\sum_{i=1}^{n}\frac{k_i}{2}(v_i v_i^{\ T}-\hat{v}_i v_i^T)\right)... \tag{11.37}$$

$$\dot{\hat{q}}=\frac{1}{2}\hat{q}\otimes\mathbf{p}(\Omega_y-\hat{b}+k_P\omega_{mes}) \tag{11.38}$$

$$\dot{\hat{b}}=-k_I\omega_{mes}. \tag{11.39}$$

If $n=3$, the weights $k_i=1$, and the measured directions are orthogonal $(v_i^T v_j=0,\forall i\neq j)$ then $M=I_3$. The cost function E_{mes} becomes

$$E_{mes}=3-\mathrm{tr}(\tilde{R}M)=\mathrm{tr}(I_3-\tilde{R})=E_{tr}.$$

In this case, the explicit complementary filter (11.34) – (11.36) and the passive complementary filter (11.19) – (11.21) are identical.

If $n=2$, the two typical measurements obtained from an IMU unit are estimates of the gravitational, a, and magnetic, m, vector fields:

$$v_a=R^T\frac{a_0}{|a_0|},\qquad v_m=R^T\frac{m_0}{|m_0|}.$$

The cost function E_{mes} becomes:

$$E_{mes}=k_1(1-<\hat{v}_a,v_a>)+k_2(1-<\hat{v}_m,v_m>)$$

The weights k_1 and k_2 are introduced to weight the confidence in each measure. In situations where the IMU is subject to high magnitude accelerations (such as during takeoff or landing maneuvers) it may be wise to reduce the relative weighting of the accelerometer data ($k_1 << k_2$) compared to the magnetometer data. Conversely, in many applications the IMU is mounted in the proximity to powerful electric motors and their power supply busses leading to low confidence in the magnetometer readings (choose $k_1 >> k_2$). This is a very common situation in the case of mini aerial vehicles with electric motors. In extreme cases the magnetometer data is unusable and provides motivation for a filter based solely on accelerometer data.

11.6.1. Estimation from the Measurements of a Single Direction

Let v_a be a measured body fixed frame direction associated with a single inertial direction v_{0a}, $v_a = R^T v_{0a}$. Let $\hat{v}_a$ be an estimate $\hat{v}_a = \hat{R}^T v_{0a}$. The error considered is

$$E_{mes} = 1 - \mathrm{tr}(\tilde{R}M); \quad M = R^T v_{0a} v_{0a}^T R$$

A proof of the following corollary may be found in [18], while local stability is proven in [11].

Corollary 11.1: Consider the rotation kinematics (11.8) for a time-varying $R(t) \in SO(3)$ and with measurements given by (11.31) (for a single measurement $v_1 = v_a$) and (11.15). Let $(\hat{R}(t), \hat{b}(t))$ denote the solution of (11.34) – (11.36). Assume that $\Omega(t)$ is a bounded, absolutely continuous signal that is persistently exciting and uncorrelated to the error signal $\tilde{R} = \hat{R}^T R$. Define:

$$U_1 = \{(\tilde{R}, \tilde{b}) : v_{0a}^T \tilde{R} v_{0a} = -1, \tilde{b} = 0\}.$$

Then:

- The set U_1 is forward invariant and unstable under the closed-loop filter dynamics.
- The estimate $(\hat{v}_a, \hat{b})$ is locally exponentially stable to (v_a, b).
- For almost all initial conditions $(\tilde{R}_0, \tilde{b}_0) \notin U_1$ then $(\hat{v}_a, \hat{b})$ converges to the trajectory $(v_a(t), b)$.

An important aspect of the above Corollary is the convergence of the bias terms in all degrees of freedom. This ensures that (in the asymptotic limit) the drift in the attitude estimate around the unmeasured axis v_{0a} will be driven by a zero mean noise process rather than a constant bias term. In a practical setting, this makes the proposed filter a practical algorithm for most mAV applications.

11.7 Experimental Results

Experimental results are presented to demonstrate the performance of the proposed observers.

Experiments were undertaken on two real platforms to demonstrate the convergence of the attitude and gyro bias estimates. The first experiment was undertaken on a robotic manipulator with an IMU mounted on the end effector and supplied with synthetic estimates of the magnetic field measurement. The robotic manipulator was programmed to simulate the movement of a flying vehicle in hovering flight regime. The filter estimates are compared to orientation measurements computed from the forward kinematics of the manipulator. Only the passive and direct complimentary filters were run on this test bed. The second experiment was undertaken on the VTOL mAV HoverEye © developed by Bertin Technologies, see Figure 11.7. The VTOL belongs to the class of 'sit on tail' ducted fan VTOL mAV, like the iSTAR9 and Kestrel developed respectively by Allied Aerospace [15] and Honeywell [10]. It was equipped with a low cost IMU that consists of 3-axis accelerometers and 3-axis gyroscopes. Magnetometers were not integrated in the mAV due to perturbations caused by electrical motors. The explicit complementary filter was used in this experiment.

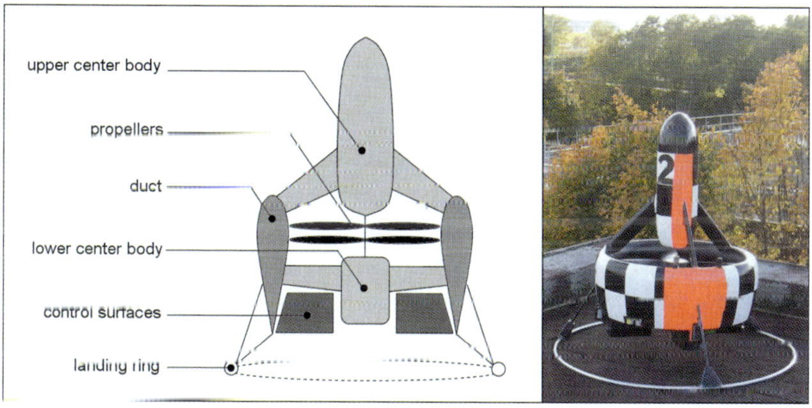

Fig. 11.7. The VTOL mAV HoverEye © of Bertin Technologies.

For both experiments the gains of the proposed filters were chosen to be: $k_P = 1\text{rd.s}^{-1}$ and $k_I = 0.3\text{rd.s}^{-1}$. The inertial data was acquired at rates of 25Hz for the first experiment and 50Hz for the second experiment. The quaternion version of the filters was implemented with first order Euler numerical integration followed by rescaling to preserve the unit norm condition.

Experimental results for the direct and passive versions of the filter are shown in Figures 11.8 and 11.9. In Figure 11.8 the only significant difference between the two responses lies in the initial transient responses. This is to be expected, since both filters will have the same theoretical asymptotic performance. In practice, however, the increased sensitivity of the direct filter to noise introduced in the computation of the measured rotation R_y is expected to contribute to slightly higher noise in this filter compared to the passive.

The response of the bias estimates is shown in Figure 11.9. Once again the asymptotic performance of the filters is similar after an initial transient. From this figure it is clear that the passive filter displays slightly less noise in the bias estimates than for the direct filter (note the different scales in the y-axis).

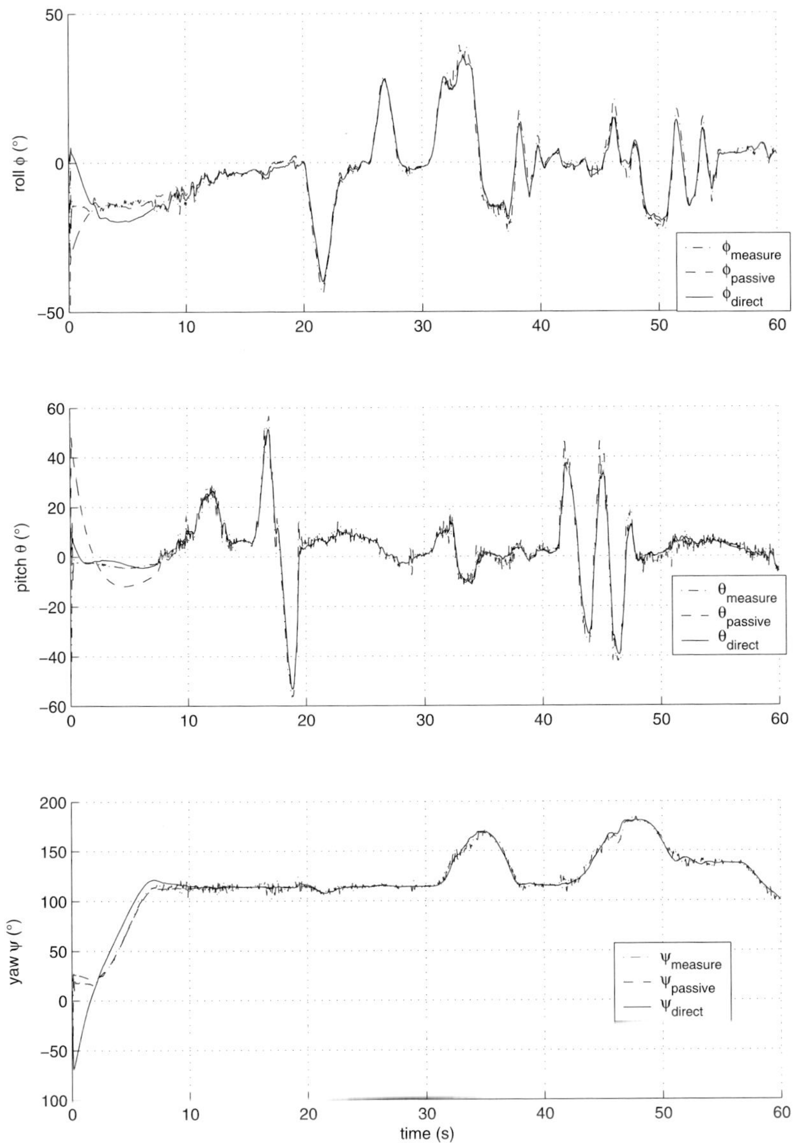

Fig. 11.8. Euler angles from direct and passive complementary filters.

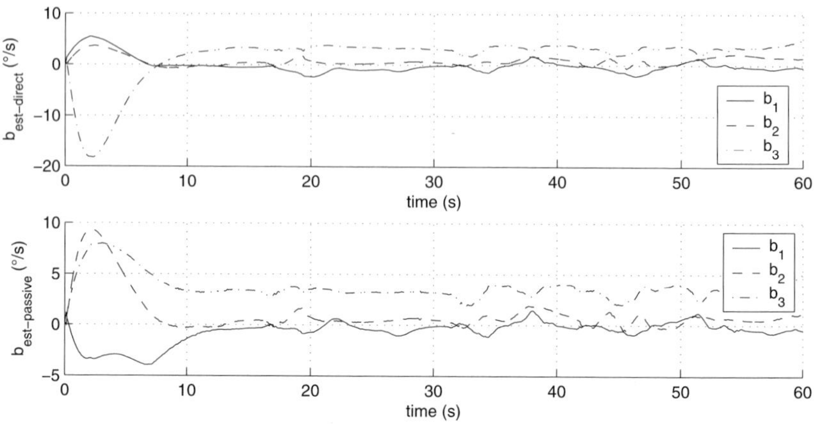

Fig. 11.9. Bias estimation from direct and passive complementary filters.

Figures 11.10 and 11.11 relate to the second experiment. The experimental flight of the mAV was undertaken under remote control by an operator. The experimental flight plan used was: First, the vehicle was located on the ground, initially headed towards $\psi(0)=0$. After take off, the vehicle was stabilized in hovering condition, around a fixed heading which remains close the initial heading of the vehicle on the ground. Then, the operator undertakes a 90° left turn maneuver, returns to the initial heading, and follows with a 90° right turn maneuver, before returning to the initial heading and landing the vehicle. After landing, the vehicle is placed by hand at its initial pose such that final and initial attitudes are the identical.

Figure 11.10 plots the pitch and roll angles (ϕ,θ) estimated directly from the accelerometer measurements against the estimated values from the explicit complementary filter. Note the large amounts of high frequency noise in the raw attitude estimates. The plots demonstrate that the filter is highly successful in reconstructing the pitch and roll estimates.

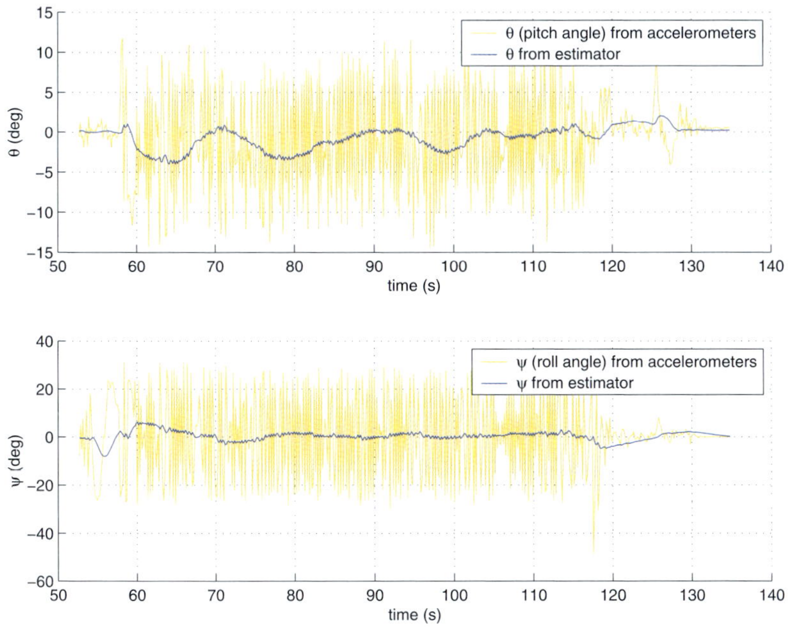

Fig. 11.10. Estimation results of the pitch and roll angles.

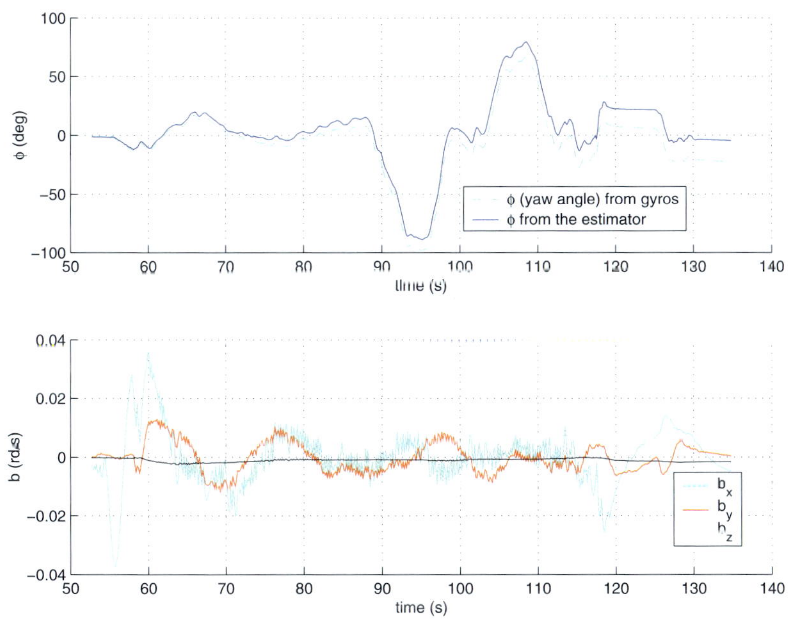

Fig. 11.11. Gyros bias estimation and influence of the observer on yaw angle.

Figure 11 presents the gyros bias estimation verses the predicted yaw angle (ϕ) based on open loop integration of the gyroscopes. Note that the explicit complementary filter here is based solely on estimation of the gravitational direction. Consequently, the yaw angle is the indeterminate angle that is not directly stabilized in Corollary 11.1. Figure 11.11 demonstrates that the proposed filter has successfully identified the bias of the yaw axis gyro. The final error in yaw orientation of the micro drone after landing is less than 5 degrees over a two minute flight. Much of this error would be due to the initial transient when the bias estimate was converging. Note that the second part of the figure indicates that the bias estimates are not constant. Although some of this effect may be numerical, it is also to be expected that the gyro bias on low cost IMU systems are highly susceptible to vibration effects and changes in temperature. Under flight conditions changing engine speeds and aerodynamic conditions can cause quite fast changes in gyro bias.

11.8 Conclusions

This Chapter has presented a general analysis of attitude observer design posed directly on the special orthogonal group. Three nonlinear observers have been proposed:

Direct complementary filter: A nonlinear observer posed on $SO(3)$ that is related to previously published nonlinear observers derived using the quaternion representation of $SO(3)$.

Passive complementary filter: A nonlinear filter equation that takes advantage of the symmetry of $SO(3)$ to avoid transformation of the predictive angular velocity term into the estimator frame of reference. The resulting observer kinematics is considerably simplified and avoid coupling of constructed attitude error into the predictive velocity update.

Explicit complementary filter: A reformulation of the passive complementary filter in terms of direct vectorial measurements, such as gravitational or magnetic field directions obtained for an IMU. This observer does not require on-line algebraic reconstruction of attitude and is ideally suited for implementation on embedded hardware platforms. Moreover, the filter remains well conditioned in the case where only a single vector direction is measured.

The explicit complementary filter was implemented as the primary attitude estimation system on several mAV vehicles world wide.

References

1. Bachmann E. R., Duman I. , Usta U. Y., McGhee R. B., Yun X. P., and Zyda M. J., "Orientation Tracking for Humans and Robots Using Inertial Sensors", *Proceedings, IEEE International Symposium Computational Intelligence in Robotics and Automation,* Monterey, CA, November 1999.
2. Baerveldt A-J., and Klang R., "A Low-Cost and Low-Weight Attitude Estimation System for an Autonomous Helicopter", *Intelligent Engineering Systems*, 1997.
3. Barshan B. and Durrant-Whyte H. F., "Inertial Navigation Systems for Mobile Robots", *IEEE Transactions on Robotics and Automation*, 44(4):751–760, 1995.
4. Bayard D. S., "Fast Observers for Spacecraft Pointing Control", *Proceedings, IEEE Conference on Decision and Control*, 4702–4707, Tampa, Florida, 1998.
5. Brown R. G., and Hwang P. Y. C., *Introduction to Random Signals and Applied Kalman Filtering*, John Wiley and Sons, 2nd Edition, 1992.
6. Corke P., Dias J. , Vincze M., and Lobo J. , "Integration of Vision and Inertial Sensors", *Proceedings, IEEE International Conference on Robotics and Automation*, W-M04, Barcelona, Spain, April 2005.
7. Corke P., "An Inertial and Visual Sensing System for a Small Autonomous Helicopter", *Journal of Robotic Systems*, 21(2):43–51, February 2004.
8. Creamer G., "Spacecraft Attitude Determination Using Gyros and Quaternion Measurements", *The Journal of Astronautical Sciences*, 44(3):357–371, July 1996.
9. Egland O., and Godhaven J. M., "Passivity-Based Adaptive Attitude Control of a Rigid Spacecraft", *IEEE Transactions on Automatic Control*, 39:842–846, April 1994.
10. Fleming J., Jones T., Gelhausen P., and Enns D., "Improving Control System Effectiveness for Ducted Fan VTOL UAVs Operating in Crosswinds", *Proceedings, 2nd Unmanned Unlimited System*, San Diego, CA, September 2003.
11. Hamel T., and Mahony R., "Attitude Estimation on $SO(3)$ Based on Direct Inertial Measurements", *Proceedings, International Conference on Robotics and Automation*, Orlando FL, April 2006.
12. Jun M., Roumeliotis S., and Sukhatme G., "State Estimation of an Autonomous Helicopter Using Kalman Filtering", *Proceedings, IEEE/RSJ International Conference on Intelligent Robots and Systems*, 1999.
13. Kim J.-H., and Sukkarieh S., "Airborne Simultaneous Localization and Map Building", *Proceedings, IEEE International Conference on Robotics and Automation*, 406–411, Taipei, Taiwan, September 2003.
14. Lefferts E. J., Markley F. L., and Shuster M. D., "Kalman Filtering for Spacecraft Attitude Estimation", *AIAA Journal of Guidance, Control and Navigation*, 5(5):417–429, September 1982.
15. Lipera L., Colbourne J. D., Tischler M. B., Hossein Mansur M., Rotkowitz M. C., and Patangui P., "The Micro Craft ISTAR Micro-Air Vehicle: Control

System Design and Testing", *Proceedings, 57th Annual Forum of the American Helicopter Society*, 1–11, Washington DC, May 2001.
16. Lobo J., and Dias J., "Vision and Inertial Sensor Cooperation Using Gravity as a Vertical Reference", *IEEE Transactions on Pattern Analysis and Machine Intelligence*, 25(12):1597–1608, December 2003.
17. Ma G.-F., and Jiang X.-Y., "Spacecraft Attitude Estimation from Vector Measurements Using Particle Filter", *Proceedings, 4th International Conference on Machine Learning and Cybernetics*, 682–687, Guangzhou, China, August 2005.
18. Mahony R., Hamel T., and Pflimlin J.-M., *Nonlinear Complementary Filters on the Special Orthogonal Group*, Technical Report, 2006.
19. Mahony R., Hamel T., and Pflimlin J.-M., "Complimentary Filter Design on the Special Orthogonal Group $SO(3)$", *Proceedings, IEEE Conference on Decision and Control*, Seville, Spain, December 2005.
20. Marins J. L., Yun X., Bachmann E. R., McGhee R. B., and Zyda M. J., "An Extended Kalman Filter for Quaternion-Based Orientation Estimation Using Marg Sensors", *Proceedings, IEEE/RSJ International Conference on Intelligent Robots and Systems*, 2003–2011, October / November 2001.
21. Metni N., Pflimlin J.-M., Hamel T., and Soueeres P., "Attitude and Gyro Bias Estimation for a Flying UAV", *Proceedings, IEEE/RSJ International Conference on Intelligent Robots and Systems*, 295–301, August 2005.
22. Metni N., Pflimlin J.-M., Hamel T., and Soueeres P., "Attitude and Gyro Bias Estimation for a VTOL UAV", *Control Engineering Practice*, 14(12):1511–1520, December 2006.
23. Pflimlin J-P., Hamel T., Souères P., and Metni N., "Nonlinear Attitude and Gyroscope's Bias Estimation for a VTOL UAV", *Proceedings, IFAC World Conference*, 2005.
24. Phillips R. E., and Schmidt G. T., *System Implications and Innovative Applications of Satellite Navigation*, Volume 207 of AGARD Lecture Series 207, Chapter GPS/INS Integration, 0.1–0.18, NASA Center for Aerospace Information, 1996.
25. Rehbinder H., and Ghosh B. K., "Pose Estimation Using Line-Based Dynamic Vision and Inertial Sensors", *IEEE Transactions on Automatic Control*, 48(2):186–199, February 2003.
26. Rehbinder H., and Hu X., "Nonlinear State Estimation for Rigid Body Motion with Low-Pass Sensors", *Systems and Control Letters*, 40(3):183–190, 2000.
27. Rehbinder H., and Hu X., "Drift-Free Attitude Estimation for Accelerated Rigid Bodies", *Automatica*, 4(4): 653–659, April 2004.
28. Roberts J., Corke P., and Buskey G., "Low-Cost Flight Control System for a Small Autonomous Helicopter", *Proceedings, Australasian Conference on Robotics and Automation*, Auckland, New-Zealand, 2002.
29. Salcudean S., "A Globally Convergent Angular Velocity Observer for Rigid Body Motion", *IEEE Transactions on Automatic Control*, 46, no 12:1493–1497, 1991.

30. Sukhatme G. S., Buskey G., Roberts J. M., Corke P. I., and Saripalli S., "A Tale of two Helicopters", *Proceedings, IEEE/RSJ International Conference on Intelligent Robots and Systems*, 805–810, Las Vegas, NV, October 2003.
31. Tayebi A., and McGilvray S., "Attitude Stabilization of a VTOL Quadrotor Aircraft", *IEEE Transactions on Control Systems Technology*, 14(3):562–571, May 2006.
32. Thienel J., and Sanner R. M., "A Coupled Nonlinear Spacecraft Attitude Controller and Observer with an Unknown Constant Gyro Bias and Gyro Noise", *IEEE Transactions on Automatic Control*, 48(11):2011 – 2015, November 2003.
33. Thienel J., *Nonlinear Observer/Controller Designs for Spacecraft Attitude Control Systems with Uncalibrated Gyros*, PhD Thesis, Faculty of the Graduate School of the University of Maryland, Department of Aerospace Engineering, 2004.
34. Vik B., and Fossen T., "A Nonlinear Observer for GPS and INS Integration", *Proceedings, 40th IEEE Conference on Decision and Control*, 2001.
35. Zimmerman M., and Sulzer W., "High Bandwidth Orientation Measurement and Control Based on Complementary Filtering", *Proceedings, Symposium on Robot Control*, Vienna, Austria 1991.

Chapter 12: Autonomous Solar UAV for Sustainable Flights[1]

This Chapter presents a historical review of solar aircrafts built and flown to date, followed by a design methodology for a solar unmanned aerial vehicle (UAV). The proposed methodology is applied to the prototype *Sky-Sailor* UAV. The UAV dynamic model, the navigation and control system are discussed along with potential applications.

12.1 Introduction

Development of a solar powered aircraft capable of continuous flight was still a dream a few years ago; this great challenge has become reality today. Significant progress has been made in the domains of flexible solar cells, high energy density batteries, miniaturized MEMS and CMOS sensors and powerful processors.

A solar powered UAV is equipped with solar cells covering its wing; it retrieves energy from the sun in order to supply power to its propulsion system and control electronics, and charges its battery with the energy surplus. During the night, the only available energy is from the battery, which discharges slowly until the next morning when a new cycle begins.

Major interdisciplinary efforts are needed to optimize and integrate concepts and technologies that will result in a fully functional system. A major issue is combination and sizing of different parts in order to optimize a certain criterion, for example endurance. Embedded payload is another parameter that needs be considered and optimized.

In 2004, the Autonomous Systems Lab at EPFL/ETHZ launched the *Sky-Sailor* project under a contract from the European Space Agency

---

[1] Written by A. North, R. Siegwart, W. Engel. This Chapter is based on Lecture Notes for the course *Aircraft and Spacecraft System Design* offered to students at ETH Zurich. A short version of the research appeared at the IEEE Robotics and Automation Magazine, Special Issue on Unmanned Aerial Vehicles, Vol. 13, No. 3, September 2006.

Kimon P. Valavanis (ed.), Advances in Unmanned Aerial Vehicles, 377–405.

(ESA). The objectives were the study and realization of a solar aircraft, fully autonomous in navigation and power generation, flying on Earth, thus evaluating indirectly the feasibility of a Mars dedicated version of such a UAV.

This Chapter presents mainly the methodology used for the global design of such a solar powered airplane capable of continuous flight on Earth. Even though the methodology has been applied to the first prototype of the *Sky-Sailor* UAV, it is rather general and it can be applied to small airplanes weighing less than a kilogram as well as to solar high altitude long endurance (HALE) platforms with a wingspan of tens of meters.

12.2 History of Solar Flight

12.2.1 Premises of Solar Aviation with Model Airplanes

The first flight of a solar powered aircraft took place on the 4th of November 1974, on the dry lake at Camp Irwin, California. *Sunrise I*, designed by R. J. Boucher from Astro Flight Inc. under a contract from ARPA, flew 20 minutes at an altitude of around 100 m during its inaugural flight. It had a wingspan of 9.76 m, weighed 12.25 kg and the power output of the 4096 solar cells was 450 W [2]. Flights that lasted three to four hours were conducted the winter, but *Sunrise I* was seriously damaged when flying in a sand storm. Thus, an improved lighter version with the same wingspan, *Sunrise II*, was built and tested on the 12th of September 1975. The new cells, with a higher efficiency of 14%, were capable of delivering 600 W. After many weeks of testing, this second version was also damaged due to a failure in the command and control system. Regardless, a historical event was accomplished, a solar powered flight was a reality and it was demonstration successfully.

On the other side of the Atlantic, Helmut Bruss was working in Germany on a solar model airplane in the summer of 1975 without knowing anything about Boucher's project. Unluckily, due to overheating of the solar cells level flight was never achieved, and it was his friend Fred Militky, who, one year later, completed the first flight with *Solaris*. Solaris completed three flights of 150 seconds reaching an altitude of 50 m on the 16th of August 1976 [3].

Since then, many different model airplane designers have tried to demonstrate flights using solar energy. Included in this list is Dave Beck with *Solar Solitude* (1996); Wolfgang Schaeper who set many records with *So-*

lar Excel in the 90's and Sieghard Dienlin with his tiny solar model *PicoSol* (1998). Obviously, endurance was improved from a few seconds to minutes, and then hours of flight [3].

12.2.2 The Dream of a Manned Solar Flight

After proving that it was feasible to fly solar powered model airplanes with sufficient illumination conditions, the new challenge that fascinated pioneers of the late 1970's was solar manned flights.

On the 19th of December 1978, Britons David Williams and Fred To launched *Solar One* on its maiden flight at Lasham Airfield [2] [21] and two years later, Larry Mauro flew for the first time his *Solar Riser* in California. For the two airplanes, the concept was to charge a battery on the ground using their solar panels and then achieve short duration flights.

The crucial stage of flying using as a single energy source the sun without any storage requirements was demonstrated by Dr. Paul B. MacCready and his company AeroVironment Inc. After the success of their human-powered airplanes from 1977 to 1979 when the *Gossamer Albatross* crossed the English Channel, the Dupont Company sponsored MacCready in an attempt to build a man-carrying solar plane. On the 18th of May 1980, the *Gossamer Penguin*, with MacCready's 13 year old son aboard, completed what can be considered as the world's first piloted, solar-powered flight.

However, the *Gossamer Penguin* was not safe for a pilot flying at more that a few feet. Hence, the *Solar Challenger* was built, a 14.2 m wingspan high-wing monoplane with 2500 W solar power. On July 7, 1981, it crossed the English Channel with solar energy as its sole power source having no additional energy storage system on-board.

The Challenger team was surprised to hear about a German competitor, Günter Rochelt, who was trying to achieve the same performance at the same time from another airport. His *Solair I*, a 16 m wingspan airplane built using Kevlar and covered by 2499 solar cells providing 1800 W [2] did not have enough energy to climb and, thus, incorporated a 22.7 kg battery. Rochelt did not cross the English Channel but on the 21st of August 1983 he flew *Solair I*, mostly on solar energy and also thermals, for 5 hours and 41 minutes.

Later on, in 1986, Eric Raymond started designing the *Sunseeker* in the United States; the success of the *Solar Riser* and the *Solar Challenger* two years later, as well as a meeting with Günter Rochelt convinced Raymond to build his own manned solar-powered aircraft. At the end of 1989, the

Sunseeker was test flown as a glider and during August of 1990, it crossed the United States in 21 solar-powered flights being 121 hours in the air.

In Germany, the town of Ulm organized regularly aeronautical competitions called Berblinger Contest. For the 1996 event, they offered attractive prizes to develop a real, practically usable solar aircraft that could stay in the air with at least half the solar energy a good summer day with clear sky could provide. This competition started activities around the world and more than 30 projects were announced; however, only one project was completed and it was ready for the final competition. On the 7th of July, the motor glider *Icaré 2* of Prof. Rudolf Voit-Nitschmann from Stuttgart University won the 100,000 DM price [24]. Two other interesting competitors were *O Sole Mio* from the Italian team of Dr. Antonio Bubbico and *Solair II* from the team of Prof. Günter Rochelt. Both projects were presented at an advanced stage of development, but they were not airworthy at the time of the competition. The first flight of *Solair II* took place two years later, in May 1998.

12.2.3 On the Way to HALE Platforms and Eternal Flight

After the success of the *Solar Challenger*, the United States government funded AeroVironment Inc. to study the feasibility of long duration, solar electric flight at high altitudes. In 1993, the *Pathfinder*, with its 30 m wingspan and 254 kg weigh was tested at low altitude; in 1994, it became part of NASA's Environmental Research Aircraft Sensor Technology (ERAST) program.

From 1994 to 2003, this program led to the construction of a series of three successive solar aircraft, *Pathfinder Plus*, *Centurion* and *Helios*. The latter was intended to be the ultimate "eternal airplane", incorporating energy storage for night-time flights. In 2001, *Helios* set an unofficial world-record altitude of 29,524 m (96,863 ft) but unfortunately, it never proved sustainable flight as it was destroyed when it fell into the Pacific Ocean on June 26, 2003 due to structural failures.

In Europe, many projects were also conduced based on HALE platforms. At the DLR Institute of Flight Systems, *Solitair* was developed as part of a study from 1994 to 1998 [23]. A 5.2 m wingspan proof-of-concept model was built and tested, while several other projects are still in progress as a continuation of the first scaled version [9]. The *Helinet* project, funded by a European Program, ran between January 2000 and March 2003 with the objective to study feasibility of a solar-powered HALE for broadband communications and Earth observation. The project involved ten European partners and led to the construction of a 24 m wingspan scale prototype. The Politecnico di Torino is still leading this research [17] [23].

QinetiQ, a British company, has been very active in the field of solar HALE platforms with *Zephyr*, an airplane which flew in July of 2006 for 18 hours, including 7 hours of flying in the dark. Weighing only 30 kg with 16 m wingspan, the aircraft used solar power for the ascent, reverting to battery power as dusk fell. *Zephyr* has recently been selected as the base platform for the Flemish HALE UAV remote sensing system *Mercator*, under the framework of the Pegasus project. The platform should fulfill missions like forest fire monitoring, urban mapping, coastal monitoring, etc.

The objective of *Helios* to prove feasibility of eternal flight for an unmanned airplane was accomplished on the 22nd of April 2005. Alan Cocconi, president and founder of AcPropulsion, flew his *Solong* during 24 hours and 11 minutes using only solar energy drawn from its solar panels and also thermals, currents of warm air rising from the desert floor. The 4.75 m wingspan and 11.5 kg airplane confirmed its capabilities two months later on the 3rd of June with a flight lasting 48 hours and 16 minutes.

The next dream to prove continuous flight with a pilot aboard will perhaps become a reality in the near future with *Solar-Impulse*, a project officially announced in Switzerland in 2003. A nucleus of twenty-five specialists, surrounded by some forty scientific advisors from various universities like EPFL, is working on the 80 m wingspan lightweight solar airplane. After manufacturing of a 60 m prototype in 2007-2008 and the final airplane in 2009-2010, an around-the-world flight is scheduled for May 2011 with a stopover in each continent.

12.3 Design Methodology for Level Flight

Whether used for surveillance at low altitudes or serve as a high altitude communication platform, a solar aircraft capable of continuous flight needs to fly at a constant altitude. For example, ground surveillance when a solar aircraft flies at high altitudes is not feasible, while for communication purposes flying at low altitude does not cover a sufficient area. Therefore, the focus is on level flight only.

In this case, energy and mass balances are the starting points of the design. In fact, the energy collected during the day by the solar panels has to be sufficient to power the motor, the on-board electronics and also charge the battery that provides enough power to fly from dusk to the next morning when a new cycle starts. Likewise, the lift force has to balance exactly the airplane weight so that altitude is maintained.

This leads finally to the hen and egg problem: the required power consumption dictates the choice and dimensions of various components like motor, solar cells, battery, but at the same time these parts determine the airplane gross weight used for calculating the required power. These relations are described next, while Figure 12.6 later on in this Chapter summarizes the overall design methodology.

12.3.1 Power Balance for Level Flight

The forces acting on the airplane during level flight are the lift L and the drag D defined as:

$$L = C_L \frac{\rho}{2} S V^2 \qquad D = C_D \frac{\rho}{2} S V^2 \tag{12.1}$$

where C_L and C_D are, respectively, the lift and drag coefficients, ρ is the air density, S the wing area and V the airplane relative speed, which is similar to the ground speed if one assumes no wind. C_L and C_D heavily depend on the airfoil, the angle of attack α, the *Re* number and *Mach* number. The drag coefficient is the sum of the airfoil drag C_{Da}, the parasitic drag of non-lifting parts that will be neglected here and the induced drag C_{Di} than can be estimated by:

$$C_{D_i} = \frac{C_L^{\,2}}{e\,\pi\,AR} \tag{12.2}$$

where e is Oswald's efficiency factor and AR the aspect ratio of the wing, the ratio between the wingspan and the chord. From (12.1) one can find the power for level flight:

$$P_{level} = \frac{C_D}{C_L^{3/2}} \sqrt{\frac{(m\,g)^3}{S}} \sqrt{\frac{2}{\rho}} \tag{12.3}$$

Using the relation between S, b and AR, one can rewrite:

$$P_{level} = \frac{C_D}{C_L^{3/2}} \sqrt{\frac{2\,AR\,g^3}{\rho}} \frac{m^{\frac{3}{2}}}{b} \tag{12.4}$$

Then, to obtain the total power consumption, efficiencies of the motor, its electronic controller, the gearbox and the propeller have to be taken into account, as well as the power consumption of the control and navigation system and the payload instruments.

12.3.2 Irradiance Model

A good model of irradiance depending on variables such as geographic position, time, solar panels orientation and albedo was developed based on [7]. This model was simplified for flat surfaces by a sinusoid as shown in Figure 12.1.

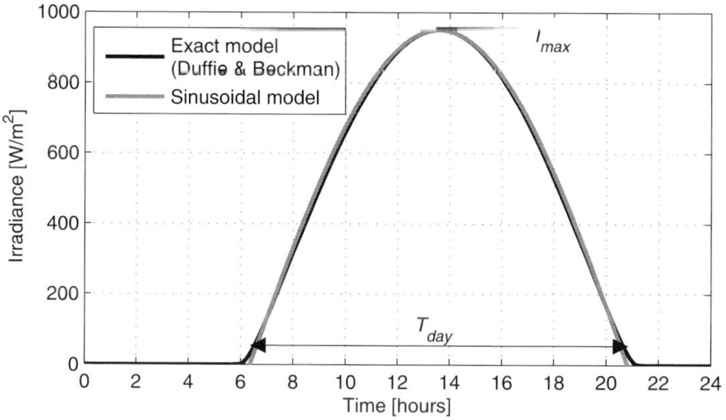

Fig. 12.1. Approximation of irradiance with a sinusoid (Lausanne, June 21).

The maximum irradiance I_{max} and the duration of the day T_{day}, which depend on the location and the date, allows one to compute the daily energy per square meter as depicted in (12.5). In order to take into account cloudy days, a constant is added with a value between 1 (clear sky) and 0 (dark). This constitutes a margin for the calculation as:

$$E_{day\ density} = \frac{I_{max}\, T_{day}}{\pi / 2} k_{solmargin} \tag{12.5}$$

12.3.3 Mass Estimation Models

For each part on the airplane, a good mass model is necessary in order to calculate the total mass m and use it in (12.4). Simple models will be briefly described, but not expressed in equation form.

The mass of the control and navigation system is considered fixed, just like the payload that is a requirement defined first. Concerning in the battery, its mass is directly proportional to the energy it needs to store, which is the product between power consumption and night duration, and inversely proportional to its energy density.

In the case of solar panels, one can find the area they cover by equating the total electric energy consumed each day with the total electric energy obtained from the sun:

$$P_{elec\ tot}\left(T_{day}+\frac{T_{night}}{\eta_{chrg}\,\eta_{dischrg}}\right)=\frac{I_{\max}\,T_{day}}{\pi/2}k_{solmargin}\,A_{solar}\,\eta_{cells}\,\eta_{mppt} \qquad (12.6)$$

The obtained area A_{solar} is then used to calculate the mass of the solar panels, taking into account the mass of the cells themselves and their encapsulation made of non-reflective sheets of polymer.

A special electronic device, called Maximum Power Point Tracker (MPPT) is required to regulate the voltage of the solar panels so that they provide the highest possible power. Its mass is proportional to the maximum power it has to convert, which may be calculated using the solar panel area previously calculated, based on the relation:

$$m_{mppt}=k_{mppt}\,P_{solmax}=k_{mppt}\,I_{\max}\,\eta_{cells}\,\eta_{mppt}\,A_{solar} \qquad (12.7)$$

The constant k_{mppt} is determined based on a study of existing high efficiency products as shown in Figure 12.2.

The mass of all electric cables, especially those connecting the solar panels to the MPPT, can be modeled according to the airplane wingspan and the electrical current. However, in order to avoid a too complex model, this mass is included in the on-board electronics.

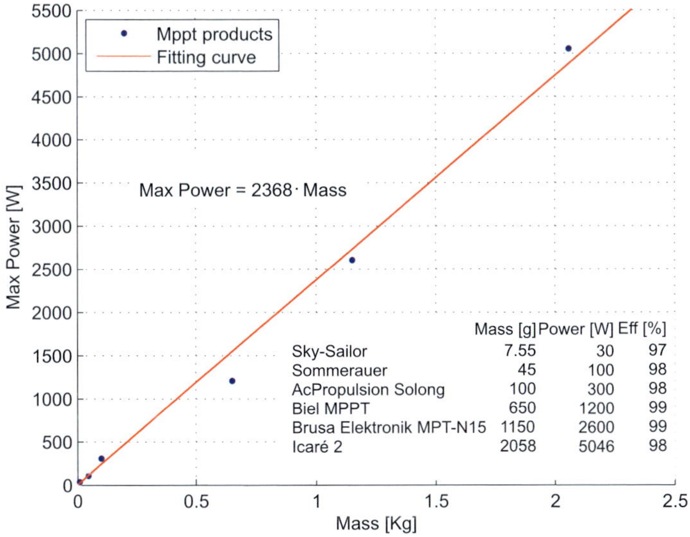

Fig. 12.2. Power density of high efficiency MPPTs.

Concerning the propulsion components, the motor, the gearbox and the propeller, a model has been proposed in [8] [9], adapted from civil aircraft to solar airplanes, which takes into account the number of blades, the propeller diameter and the power of the motor. Calculations show that estimation is far too optimistic for model aircraft. Further, in [18] and [25] very similar models have been proposed exclusively based on power, where the mass of the propulsion components is estimated as:

$$m_{prop} = 0.0045\, P_{prop} \qquad (12.8)$$

For real large scale solar airplanes like *Helios*, *Icaré 2* or *Solair II*, this factor is, respectively, 0.0033, 0.0012 and 0.0008 kg/W whereas the first experiments with *Sky-Sailor* showed a factor of around 0.010 kg/W. The reason is that for an airplane taking off from a runway, the difference between start power and mean power for level flight is small. On the contrary, in the case of a hand-launched model airplane that needs to increase its speed and gain altitude rapidly, the start power is far higher than the mean power required for level flight. Thus, the motor has to be oversized and its mass increases.

Finally, the mass of the airplane structure is the most difficult part to model and the two main approaches mainly used in literature for solar air planes appear inadequate. That is the reason this part is studied in detail and a new model is proposed.

The first approach by D.W. Hall [8] calculates separately the mass of all components of the airframe, i.e., spar, leading and trailing edge, covering, ribs, control surfaces, fuselage and tail, as functions of the total mass, aspect ratio and wing area. This approach was applied to an airplane with more than 60 m wingspan [6] but it is cannot be applied to model airplanes. The second approach proposed by W. Stender in 1969 [20], is based on statistical data for sailplanes with twin boom tails. The entire airframe weight is estimated parametrically as a function of aspect ratio, surface and number of boom tails:

$$W_{af} = 8.763\, n^{0.311}\, S^{0.778}\, AR^{0.467} \tag{12.9}$$

This simple model has been used to design solar airplanes [17] [18] [25]. In order to verify this model, a database containing wingspan, wing area, aspect ratio, structure weight and gross weight of 415 sailplanes of various dimensions was created. Sailplanes were divided into 92 radio-controlled unmanned models and 323 manned sailplanes. Figure 12.3 illustrates the weight of these samples as a function of the wing area and the aspect ratio. Equation (12.9) is obviously very optimistic for large scale sailplanes and too pessimistic for model airplane. Thus, using a least-square fitting method, a new equation may be derived based on the sailplane database described above:

$$W_{af} = 5.58\;\, S^{1.59}\, AR^{0.71} \tag{12.10}$$

Using the definition of aspect ratio, (12.10) may be also expressed as a function of wingspan:

$$W_{af} = 5.58 \cdot b^{3.18} \cdot AR^{-0.88} \tag{12.11}$$

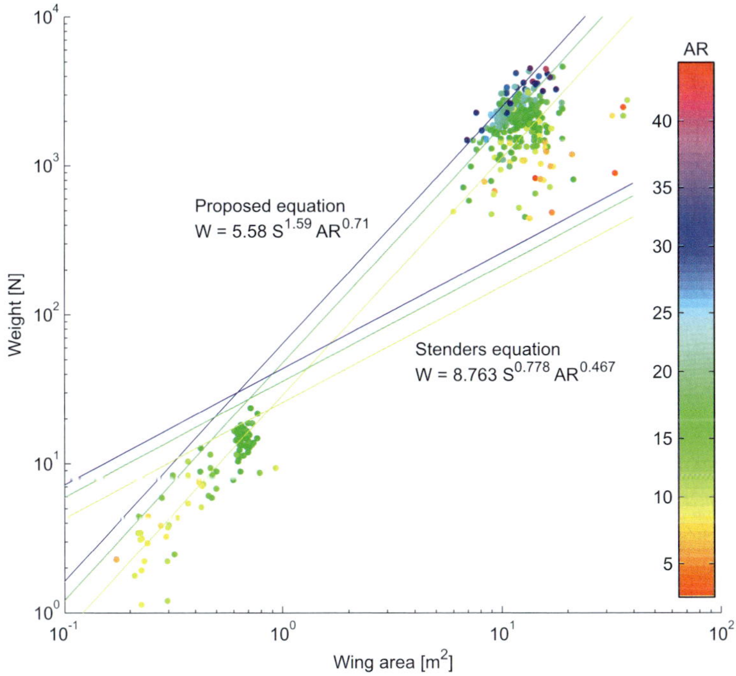

Fig. 12.3. Comparison of two airframe mass models with real data.

However, the above relations demonstrate on 'average' properties of the 415 sailplanes; but construction quality of each airplane varies. Being interested in a model of the highest quality sailplanes only, it is proposed that records be separated in two groups: one group includes sailplanes with lower actual weight compared to what would have been estimated using interpolation, while the second group includes the rest. Considering only the first group, and applying one more time the curve fitting process, after five iterations an equation is obtained that models the 5% best sailplanes:

$$W_{af} = 0.44 \ \ S^{1.55} \ \ AR^{1.3} \tag{12.12}$$

Equation (12.12) may be we-written using wingspan instead of surface as:

$$W_{af} = 0.44 \cdot b^{3.1} \cdot AR^{-0.25} \tag{12.13}$$

It is interesting to see the evolution of the constant and the two exponents during iterations when construction quality increases. The wing area is always related to the weight with a power exponent from 1.55 to 1.59;

this exponent does not change significantly. Influence of the aspect ratio increases with quality.

Several scientists have studied correlations between gross weight, wingspan, wing area and speed even from the hang glider to commercially flying airplanes, and also in the animal kingdom, from flies to albatross. A concise review of all these correlations may be found in [19]. A major contribution has been made by H. Tennekes in [22] in terms of correlations summarized in a log-log diagram named "The great flight diagram". The result is impressive: from the common fruit fly to the Boeing 747, all follow approximately a line corresponding to:

$$W / S = 47 \cdot W^{1/3} \tag{12.14}$$

The basis for (12.14) is the concept of geometric similarity. If one assumes geometric similarity among flying objects, the weight W is proportional to the cube of a characteristic length l, the surface S to the square and consequently the wing loading is linear with l and, thus, with $W^{1/3}$. It is interesting to observe that if the proposed model (12.12) is transformed to an equation similar to (12.14), the weight is linked to the wing loading with an exponent of 0.35, very close to the model proposed in [22]:

$$W_{af} / S = 0.59 \cdot W_{af}^{\ 0.35} \cdot AR^{0.84} \tag{12.15}$$

Figure 12.4 presents a combined diagram of "the great flight diagram", the 415 sailplanes in the database mentioned above, the mean and the top 5% of models, and a model by B.W. McCormick [12] for estimating the wing loading of manned airplanes based on square-cubing scaling laws. Observing Figure 12.4, a remarkable point is its asymptote at a weight of 1000 N that corresponds to the weight of a single human person in an incredibly lightweight airplane. The airplane 'closest' to this asymptote is the Gossamer Albatross, the human powered aircraft built by MacCready that crossed the English Channel in 1979.

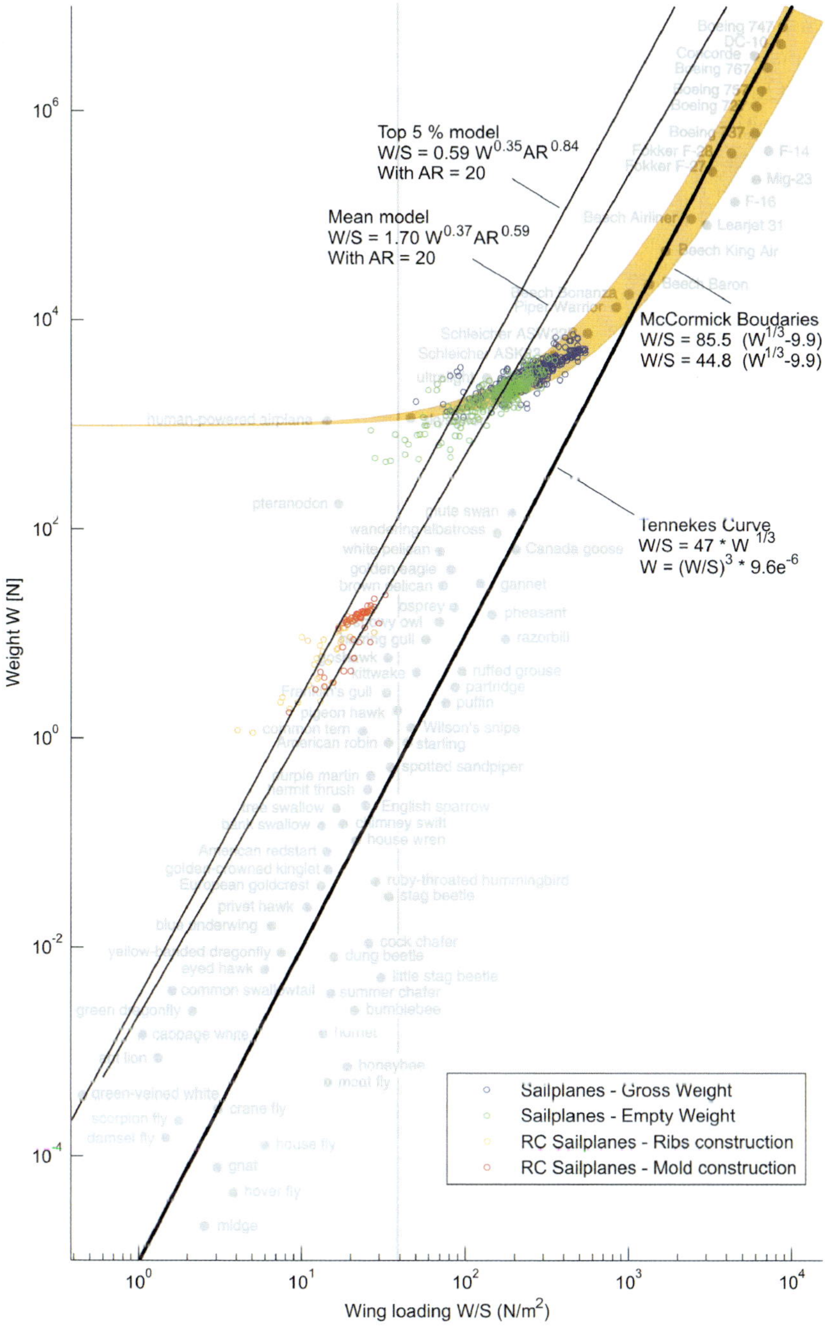

Fig. 12.4. The Great Flight Diagram [22] completed with statistics of the 415 sailplanes and McCormick's design.

Sixty-two of the most famous solar airplanes flown to date, from RC models to HALE platforms, are represented in the same way in Figure 12.5. One may observe that whereas small scale models are located in the same region with the non-solar ones, large scale solar airplanes are far away from the proposed and developed model.

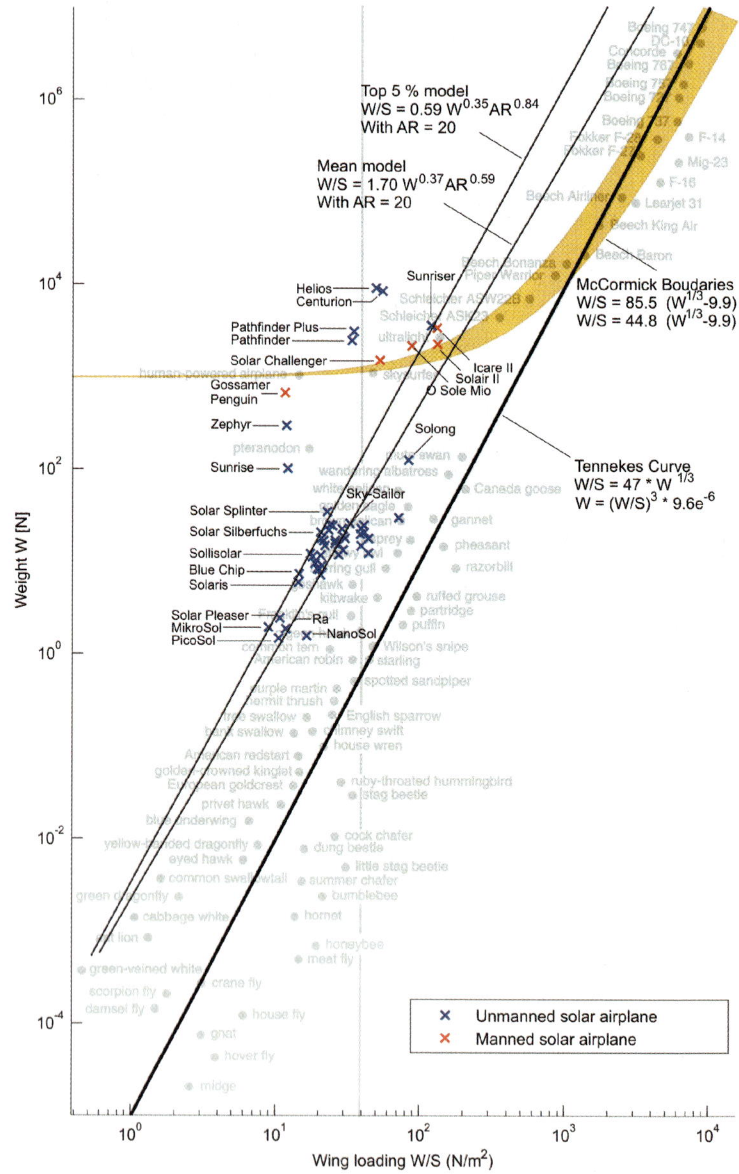

Fig. 12.5. The Great Flight Diagram [22] completed with 62 solar airplanes.

The reason is that *Helios*, *Centurion*, *Pathfinder*, *Pathfinder Plus* and *Zephyr* have a major difference in their configuration compared to all other models. In fact, their wing extremities are supported by several wheels, when not in flight. The flexion constraints on the wing are reduced, which allows for using a lighter construction method. These big models have impressive low weight but the direct consequence is their incredible fragility. That was the cause of the *Helios* crash in 2003. The five manned solar airplanes are slightly above McCormick upper boundary, except Gossamer Penguin, the solar version of human powered Gossamer Albatross.

12.3.4 Problem Summary

A schematic representation of the power balance and the mass balance is shown in Figure 12.6 where all relations previously derived are explicitly shown. In order simplify notation, the main expressions are replaced by variables a_0 to a_9.

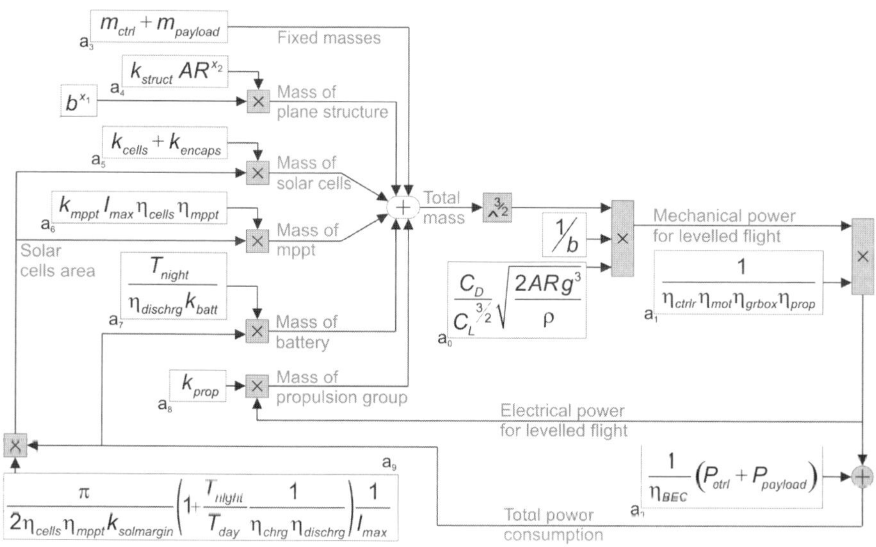

Fig. 12.6. Schematic representation of the design methodology.

This cyclic problem may be solved by using all these relations:

$$m = m_{ctrl} + m_{payload} + m_{struct} + m_{solar} + m_{batt} + m_{mppt} + m_{prop} \tag{12.16}$$

$$\underbrace{m - a_0 a_1 \left(a_7 + a_8 + a_9\left(a_5 + a_6\right)\right)}_{a_{10}} \frac{1}{b} m^{\frac{3}{2}} = \underbrace{a_2\left(a_7 + a_9\left(a_5 + a_6\right)\right) + a_3}_{a_{11}} + a_4\, b^{x_1} \qquad (12.17)$$

Reducing one more time equation complexity by using substitution variables, one obtains:

$$m - \underbrace{a_{10}\frac{1}{b}}_{a_{12}} m^{\frac{3}{2}} = \underbrace{a_{11} + a_4\, b^{x_1}}_{a_{13}} \qquad (12.18)$$

It can be shown that (12.18) has only a positive non-complex solution for m, which makes physically sense, if:

$$a_{12}^2\, a_{13} \le \frac{4}{27} \qquad (12.19)$$

For a given airplane configuration, feasibility is proved if this inequality is satisfied and at the same time if the surface of solar cells is smaller than the wing area.

In order to be able to extract meaningful information, it is necessary, among the thirty parameters that the proposed model contains, to distinguish between three different groups:

- The first group includes parameters linked to a technology and they are constant or they can be considered constant for a design. This may be the case of motor or propeller efficiencies that should be around 85 % when optimized for a special application.
- The second group of parameters is linked to the type of mission; they concern flight altitude, date and payload.
- The third group of parameters includes parameters that are varied during the optimization process in order to determine the airplane layout, for example the wingspan or the aspect ratio.

A complete list of these parameters is presented in Tables 12.1 to 12.3. The values that are mentioned have been used for the design of the *Sky-Sailor* first prototype.

Parameter	Value	Unit	Description
C_L	0.8	-	Airfoil lift coefficient
C_{Da}	0.013	-	Airfoil drag coefficient
E	0.9	-	Oswald's efficiency factor
I_{max}	950	[W/m^2]	Maximum irradiance
k_{batt}	190·3600	[J/kg]	Energy density of battery
k_{cells}	0.32	[kg/m^2]	Mass density of solar cells
k_{encaps}	0.22	[kg/m^2]	Mass density of encapsulation
k_{mppt}	0.00047	[kg/W]	Mass to power ratio of MPPT
k_{prop}	0.013	[kg/W]	Mass to power ratio of propulsion unit
k_{struct}	0.44/9.81	[kg/m^3]	Structural mass constant
m_{elec}	0.25	[kg]	Mass of navigation & control system
η_{bec}	0.7	-	Efficiency of step-down converter
η_{cells}	0.169	-	Efficiency of solar cells
η_{chrg}	0.98	-	Efficiency of battery charge
η_{ctrlr}	0.95	-	Efficiency of motor controller
$\eta_{dischrg}$	0.98	-	Efficiency of battery discharge
η_{grbox}	0.95	-	Efficiency of gearbox
η_{mot}	0.85	-	Efficiency of motor
η_{mppt}	0.97	-	Efficiency of MPPT
η_{prop}	0.85	-	Efficiency of propeller
P_{ctrl}	1	[W]	Power of navigation & control system
x_1	3.1	-	Structural mass area exponent
x_2	-0.25	-	Structural mass aspect ratio exponent

Table 12.1. Parameters that are constant or assumed constant.

Parameter	Value	Unit	Description
$k_{solmargin}$	0.7	-	Irradiance margin factor
$m_{payload}$	0.25	[kg]	Payload mass
$P_{payload}$	0.5	[W]	Payload power consumption
ρ	1.1655	[kg/m^3]	Air density (500 m)
T_{day}	14·3600	[s]	Day duration

Table 12.2. Parameters determined by mission.

Parameter	Value	Unit	Description
AR	12.9	-	Aspect ratio
B	3.2	[m]	Wingspan
M	2.6	[kg]	Total mass

Table 12.3. Variables linked to the airplane shape.

12.4 Application of the Proposed Methodology

Having defined mission requirements, the proposed method may be applied to evaluate possible shapes of a solar airplane. This will be illustrated with the example of the *Sky-Sailor* prototype. The objective is to embed a 250 g payload consuming 0.5 W and fly continuously at low altitude during two months in summer. These mission parameters allow for plotting the relations between the main variables, i.e., wingspan and aspect ratio, and the flight characteristics for all the possible configurations. This is shown in Figure 12.7 where one can first observe that the minimum airplane wingspan is around 2.5 m and that keeping the same construction quality the airframe will become too heavy beyond a certain point, thus, making continuous flight impossible. This means that for larger airplanes, the quality of the wing structure in terms of weight becomes more and more important. For a ten to twenty meters wingspan that corresponds to commercial sailplanes, construction methods should be improved considerably. It is also interesting to observe the mass distribution. Figure 12.8 shows that half of the weight is due to the battery. It is also shown that the structure itself is an important component whose weight increases with wingspan.

In this case, among the many possible configurations, the final choice has been based on considerations of the flight speed, size, and estimated production costs.

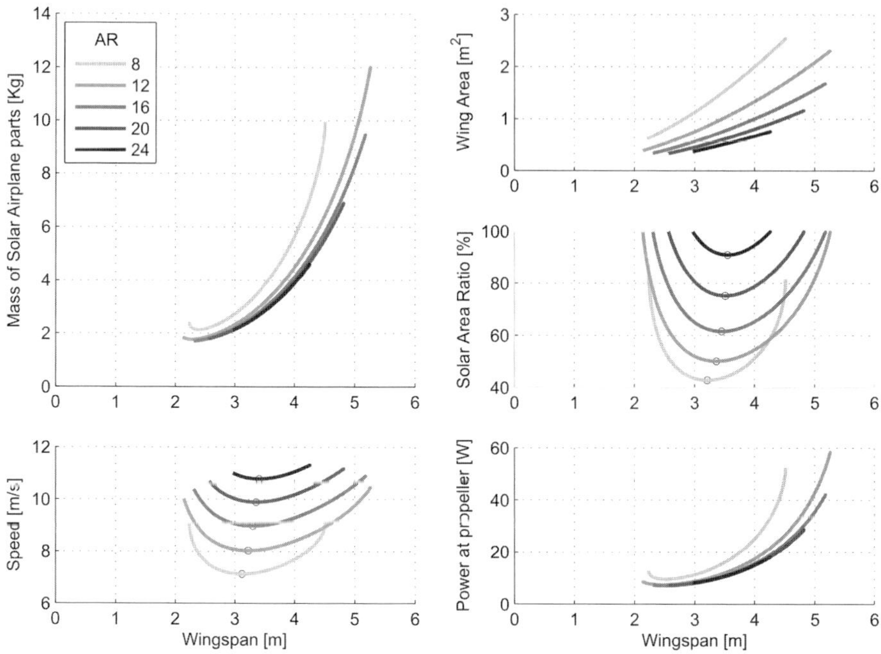

Fig. 12.7. Design plots of a solar airplane with a payload of 250 g.

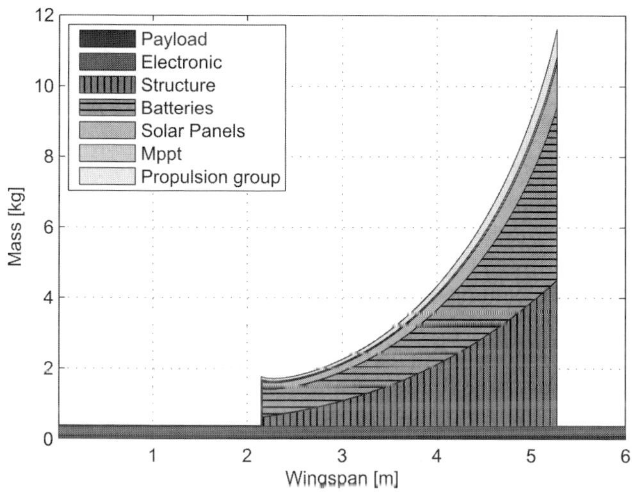

Fig. 12.8. Mass distribution with respect to wingspan assuming AR = 12.

By changing the value of parameters included in the first group, listed in Table 12.1, one may perform a cost-benefit analysis and evaluate trade offs technological changes may have on potential design improvements. For example, with regards to the choice of solar cells, one can compare the use of lightweight cells with low efficiency, compared to heavier ones that are twice more efficient. The impact of battery capacity on the feasibility of continuous flight at different altitudes can also be estimated. Figure 12.9 shows that in the case of the *Sky-Sailor*, doubling the capacity of the energy storage system results in flying at an altitude of 10,000 m.

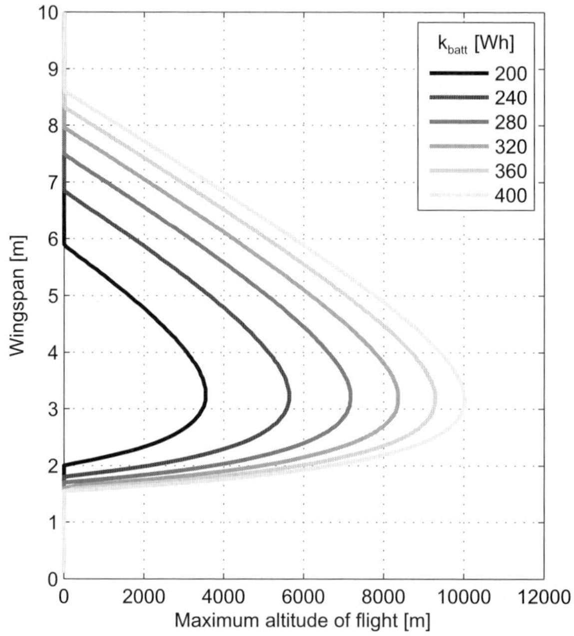

Fig. 12.9. Impact of the battery capacity on the flight altitude.

12.5 The Sky-Sailor Prototype

According to the results of the design study using the proposed methodology presented before, a fully functional prototype called *Sky-Sailor* was built. This is basically a model sailplane with an empty mass of 0.725 kg, a 3.2 m wingspan and a wing surface of 0.776 m^2, see Figure 12.10 [13].

Fig. 12.10. Sky-Sailor solar airplane.

The wing structure is essentially made of composite materials using a ribs construction method. The resulting total mass including motors, propeller, solar cells, batteries and controller is 2.6 kg. During nominal flight, the airplane speed is 8.2 m/s and the motor requires an electrical power of 16.25 W. Observing Figure 12.7 one can see that this fits quite well theoretical calculations.

12.5.1 Solar Generator System

The wing is covered by 216 silicon cells separated into three panels covering 0.51 m^2. These cells are preferable to gallium arsenide cells that have a higher efficiency but that are less flexible and heavier. They are encapsulated using a mechanically favorable symmetrical laminate combined with a fiber glass reinforced plastic coating what gives three non reflective highly flexible modules.

At maximum sun conditions, available power is 28 W per panel, which offers a total of 84 W. Three MPPTs were developed to ensure efficient use of the energy and a safe charge of the lithium polymer battery, which has a capacity of 230 Wh for a nominal voltage of 29.6 V.

12.5.2 Control and Navigation System

The *Sky-Sailor* flies autonomously using an on-board navigation and control system. High level commands are given from the ground control station (GCS). For this purpose, a lightweight and low-power consumption autopilot dedicated to this application was design and built. The system is mainly based on a DsPic interfaced with the different sensors and other electronic boards as shown in Figure 12.11.

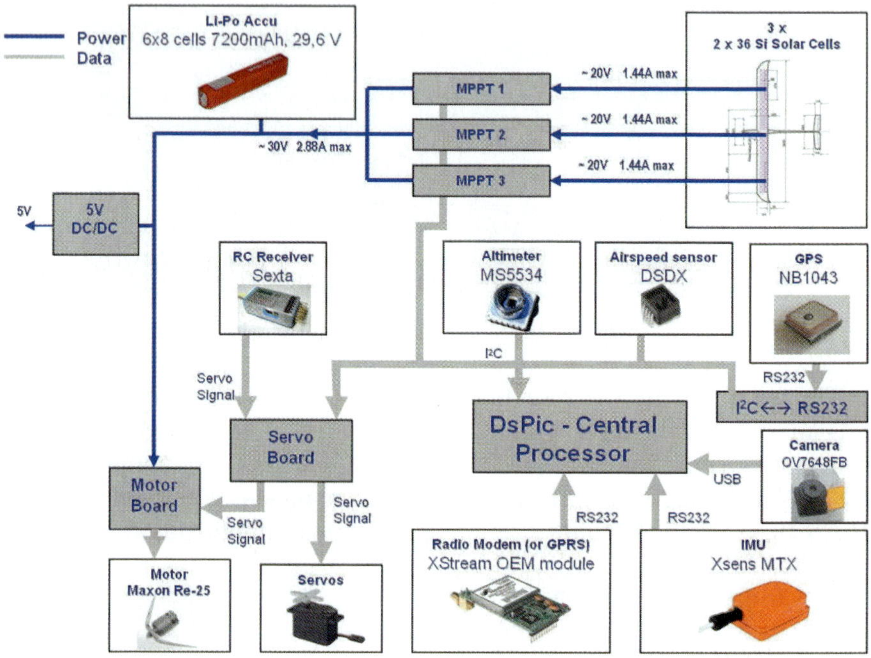

Fig. 12.11. Schematic view of the navigation and control system.

The sensor board contains an inertial measurement unit for measurement of the attitude, absolute and differential pressure sensors for altitude and air speed and a low power GPS. The DsPic issues commands to the actuators, i.e., the motor and the servo motors of the control surfaces. This is done through a servo board that also receives commands from the operator on the ground. This board is completely independent from the processor so that, in case of problem, the operator is able to take control of the airplane manually. Manual control is also used for take off, landing and during preliminary tests in order to tune the controller. The processor is also connected to the three MPPTs to obtain information about the power retrieved from the solar panels and the battery charge status. Finally, the autopilot board weighs 80 g and its power consumption is around 1.2 W.

The user can interact with the airplane using a GCS composed of a graphical user interface and a bidirectional radio modem for communication. The state of the airplane during flight is monitored and a visual feedback is given with virtual instruments and a 3-D map. Controller parameters may be tuned during flight experiments, while the operator may send high level commands or modify the trajectory.

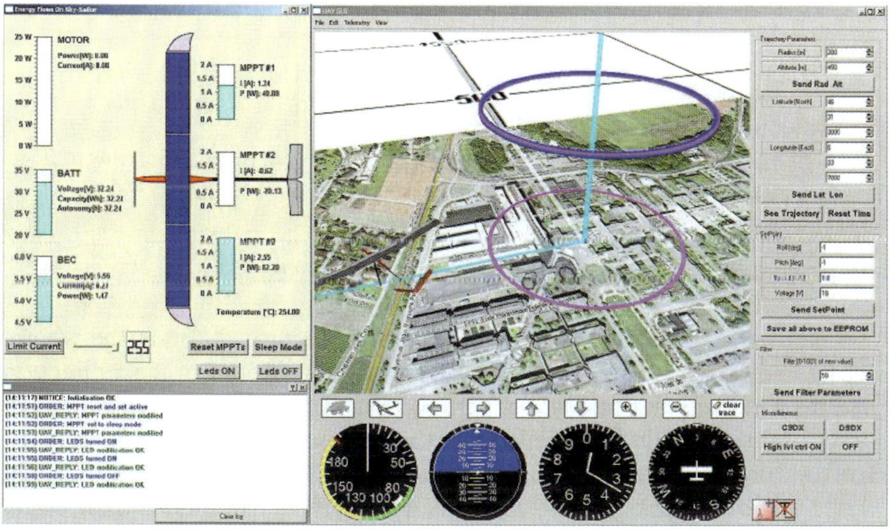

Fig. 12.12. *Sky-Sailor* graphical user interface.

For security purposes, all GCS commands are validated by the autopilot and retransmitted to the GCS again. Also, in case of communication loss, abnormal attitude or GPS problems, the interface warns the user so that fast action is taken according to the problem, reverting to remote control.

12.5.3 Modeling and Control

Dynamic modeling is an important step in developing and controlling a dynamic system. In fact, the model allows analyzing the system, determine functionality and behavior depending on different conditions. In the case of aerial robots where the risk of damage is very high, this is especially important in order to simulate and tune a controller before implementing it in real-time. Hence, a dynamic model of the *Sky-Sailor* was developed using Lagrange-Euler formalism as shown in Figure 12.13.

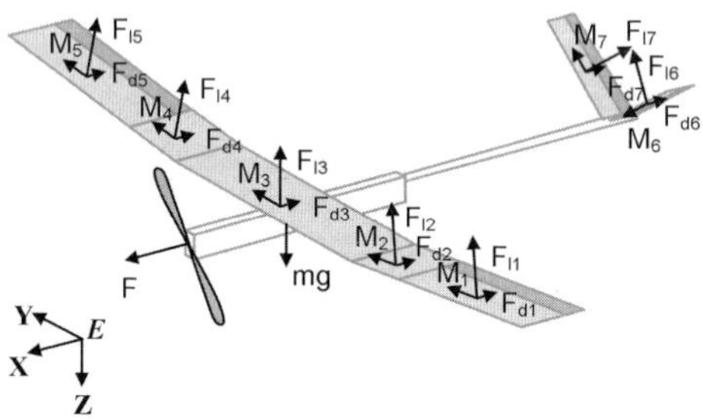

$$F_{tot} = F_{prop} + \sum_{i=1}^{7} F_{Li} + F_{di}$$

$$M_{tot} = \sum_{i=1}^{7} M_i + F_{Li} \times r_i + F_{di} \times r_i$$

$$F_{prop} = f(\dot{x}, U_1)$$

$$F_{li} = C_{li} \frac{\rho}{2} S_i v^2$$

$$F_{di} = C_{di} \frac{\rho}{2} S_i v^2$$

$$M_i = C_{mi} \frac{\rho}{2} S_i v^2 \cdot chord_i$$

$$\begin{aligned}
&[C_{l1}\ C_{d1}\ C_{m1}] = f(Aoa_i, U_2) \\
&[C_{li}\ C_{di}\ C_{mi}] = f(Aoa_i), \mathrm{i=2,3,4} \\
&[C_{l5}\ C_{d5}\ C_{m5}] = f(Aoa_i, U_3) \\
&[C_{l6}\ C_{d6}\ C_{m6}] = f(Aoa_i, U_4) \\
&[C_{l7}\ C_{d7}\ C_{m7}] = f(Aoa_i, U_5)
\end{aligned}$$

Fig. 12.13. Modeling of the *Sky-Sailor.*

The forces acting on the airplane shown in Figure 12.13 are the weight located at the center of gravity, the thrust of the propeller in the x direction and the aerodynamic forces, i.e., lift and drag of the seven subparts of the wing, and the *V*-tail. They depend on the relative airspeed, the angle of attack and the angle of the control surfaces, i.e., the ailerons, elevator and rudder. The moments are the aerodynamic moments and those induced by forces acting at a distance from the center of gravity. The final model was developed and implemented using *MATLAB* where it was validated with real experiments data.

The controller is composed of two different parts, an inner loop to maintain stability and an outer loop used to plan and follow a trajectory [11]. For low level control, an optimal linear state feedback control method following a Linear Quadratic Regulator has been implemented, using the model explained above. For high level, the choice was to adapt an algorithm proposed and tested for path tracking of a non-holonomic robot [4].

Simulations demonstrate good robustness under various conditions. Figure 12.14 shows a trajectory where the airplane flies with 8.2 m/s under wind turbulences of 2 m/s.

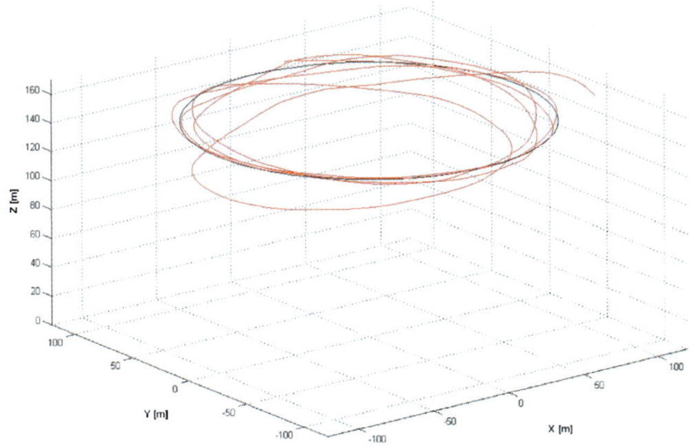

Fig. 12.14. Airplane trajectory with wind turbulence and farther initial position.

12.5.4 Current Project Status and Future Work

The first *Sky-Sailor* prototype was tested successfully and its flight characteristics were excellent in terms of stability and maneuverability. According to theoretical calculations during the design phase, the nominal speed and the power consumption were validated with experiments. The solar generator system composed of the solar panels and the MPPTs were successfully integrated and the charge of the battery using only solar power was proved to be efficient and safe, avoiding overcharge. Globally, the measurements showed that a 24 hours flight was feasible using the actual prototype.

Autonomous flights were also achieved using the proposed control and navigation system, the longest one being 5 hours. Further hardware improvements have been scheduled before attempting a long endurance flight. This milestone is scheduled for the summer of 2007. One further objective is to use the experience accumulated so far, in order to build a new prototype capable of larger payloads.

12.6 Potential Applications

The ability to fly autonomously for long hours even many days makes solar airplanes suitable for a wide range of civilian and military applications. Civil applications include coast or border surveillance, atmospheric and weather research and prediction, environmental, forestry, agricultural, and oceanic monitoring, imaging for the media and real-estate industries, etc.

Additional advantages of solar airplanes are their low cost and simplicity with which they can be launched. For example, in the case of potential forest fire risks during a warm and dry period, swarms of solar airplanes, easily launched by hand, could efficiently monitor a large surface, reporting rapidly any fire starts. This would allow for fast intervention and, thus, reduce the cost of such disaster in terms of human and material losses. At higher dimension, solar HALE platforms are expected to play a major role as communication relays and could replace advantageously satellites in a near future.

12.7 Conclusion

In this Chapter, a design methodology for solar airplanes aimed at flying continuously was presented. The methodology was based on the simple principles of mass and energy balances. The primary objective was not to limit applications to a certain dimension, but rather to be as generic as possible so that the methodology could be applied to different size solar planes.

Special attention was given to study how the mass of the various parts composing an airplane vary with scaling. These mass models constituted the basis of the discussed method and needed to be as precise as possible in order to produce correct results. The case of the airframe mass was investigated in detail and a new equation relating the aspect ratio and wingspan was proposed.

The *Sky-Sailor* solar airplane prototype was used to test the proposed method. The first experiments showed results that validated the efficiency of the proposed method. A dedicated navigation and control system was developed for autonomous flying that allowed the user on the ground to interact with the airplane using a graphical user interface.

Technology improvements in the fields of battery, solar cells, electronics and structure material will enhance the use of solar powered airplanes for diverse applications.

Nomenclature

Symbols	*Unit*	*Description*
A	[m/s]	speed of sound
A_{solar}	[m^2]	area of solar cells
AR	-	wing aspect ratio
b	[m]	wingspan
c	[m]	chord
CD	-	total drag coefficient
CD_a	-	airfoil drag
CD_i	-	induced drag coefficient
CL	-	lift coefficient
CM	-	moment coefficient
D	[N]	drag force
e	-	Oswald's efficiency factor
g	[m/s^2]	acceleration of Gravity
I_{max}	[W/m^2]	maximum irradiance
k_{batt}	[J/kg]	energy density of battery
k_{cells}	[kg/m^2]	mass density of solar cells
k_{encaps}	[kg/m^2]	mass density of encapsulation
k_{mppt}	[kg/W]	mass to power ratio of MPPT
k_{prop}	[kg/W]	mass to power ratio of propulsion unit
$k_{solmargin}$	-	irradiance margin factor
k_{struct}	[kg/m^3]	structural mass constant
L	[N]	lift force
m	[kg]	total mass
M	-	Mach number
M	[nm]	moment
P_{ctrl}	[W]	power of navigation and control system
$P_{payload}$	[W]	power consumption of payload
Re	-	Reynolds number
S	[m^2]	wing surface
T_{day}	[s]	duration of the day
T_{night}	[s]	duration of the night
V	[m/s]	speed
W	[N]	weight
W_{af}	[N]	weight of airframe
η	-	efficiency
μ	[kg/ms]	dynamic viscosity
ρ	[kg/m3]	air density

References

1. Berry P., "The Sunriser - A Design Study in Solar Powered Flight", *World Aviation Conference*, San Diego, CA, October 2000.
2. Boucher R. J., "History of Solar Flight", AIAA Paper 84-1429, 1984.
3. Bruss H., *Solar Modellflug Grundlagen, Enwicklung*, Praxis, Verlag für Technik und Handwerk, Baden-Baden, 1991.
4. Buccieri D., Mullhaupt P., Jiang Z., Bonvin D., "Velocity Scheduling Controller for a Nonholonomic Mobile Robot", *Proceedings, IEEE Chinese Control Conference*, 2006.
5. Colozza A. J., "Preliminary Design of a Long-Endurance Mars Aircraft", *Proceedings, 26th Joint Propulsion Conference*, AIAA 90-2000, Orlando, FL, July 1990.
6. Colozza A. J., *Effect of Power System Technology and Mission Requirements on High Altitude Long Endurance Aircraft*, NASA CR 194455, February 1994.
7. Duffie J. A., Beckman W. A., *Solar Engineering of Thermal Processes*, Second Edition, New York, Wiley-Interscience, 1991.
8. Hall D. W., Hall S. A., *Structural Sizing of a Solar Powered Aircraft, Lockheed Missiles and Space Company*, NASA Contractor Report 172313, 1984.
9. Keidel B., *Auslegung und Simulation von Hochfliegenden, Dauerhaft Stationierbaren Solardrohnen*, PhD Thesis, Lehrstuhl für Flugmechanik und Flugregelung, Technische Universität München, 2000.
10. MacCready P. B., Lissaman P. B. S., Morgan W. R., "Sun-Powered Aircraft Designs", *Journal of Aircraft*, Vol. 20, No. 6, June 1983.
11. Mattio A., *Modeling and Control of the UAV Sky-Sailor*, Master Project Report, Ecole Polytechnique Fédérale de Lausanne, Switzerland, 2006.
12. McCormick B. W., *Aerodynamics, Aeronautics and Flight Mechanics*, John Wiley & Sons, Inc. 1995.
13. Noth A., Engel W., Siegwart R., "Design of an Ultra-Lightweight Autonomous Solar Airplane for Continuous Flight", *Proceedings, Field and Service Robotics*, Port Douglas, Australia, 2005.
14. Noth A., Engel W., Siegwart R., "Flying Solo and Solar to Mars", *IEEE Robotics and Automation Magazine*, Special Issue on Unmanned Aerial Vehicles, Vol. 13, No. 3, September 2006.
15. Patel C., *The Design and Fabrication of a Solar Powered Model Aircraft*, B. Tech Thesis, Department of Aerospace Engineering, IIT Bombay, 2002.
16. Phillips W. H., "Some Design Considerations for Solar-Powered Aircraft", NASA Technical Paper 1675, 1980.
17. Romeo G., Frulla G., "HELIPLAT: High Altitude Very-Long Endurance Solar Powered UAV for Telecommunication and Earth Observation Applications", *The Aeronautical Journal*, 108, 277–293, 2004.
18. Rizzo E., Frediani A., "A Model for Solar Powered Aircraft Preliminary Design", *Proceedings, ICCES 04*, Vol. 1, 39-54, Madeira, Portugal, 2004.

19. Shyy W., Berg M., Ljungqvist D., "Flapping and Flexible Wings for Biological and Micro Air Vehicles", *Progress in Aerospace Science*, 35:455-506, 1999.
20. Stender W., *Sailplane Weight Estimation*, Organisation Scientifique et Technique Internationale du vol a Voile, 1969.
21. Stinton D., *The Design of the Aeroplane*, Second Edition, Blackwell Science, Oxford, UK, 2001.
22. Tennekes H., *The Simple Science of Flight, From Insects to Jumbo Jets*, MIT Press, 1996.
23. Tozer T. C., Grace D., Thompson J., and Baynham P., "UAVs and HAPs - Potential Convergence for Military Communications", *IEE Colloquium on "Military Satellite Communications"*, 6th June of 2000.
24. Voit-Nitschmann R., *Solar- und Elektroflugzeuge - Geschichte und Zukunft*, Jahrbuch aus Lehre und Forschung der Universität Stuttgart, Online Jahrbuch 2001.
25. Youngblood J. W., Talay T. A., Pegg R. J., "Design of Long-Endurance Unmanned Airplanes Incorporating Solar and Fuel Cell Propulsion", *Proceedings, AIAA / SAE / ASME 20th Joint Propulsion Conference*, 1984.

Chapter 13: The Integration of a Multimodal MAV and Biomimetic Sensing for Autonomous Flights in Near-Earth Environments[1]

This Chapter illustrates how integrating optic flow sensing for lateral collision avoidance with a novel miniature aerial vehicle (MAV) platform results in a vehicle that is well suited for flights in near-Earth environments. Section 13.2 discusses optic flow, reactive control experiments mimicking flying insects, and its limitations while Section 13.3 talks about the fixed-wing MAV with hovering capabilities. In Section 13.4, autonomous control of the aircraft's attitude during a hover is detailed. Section 13.5 concludes and outlines near future goals.

13.1 Introduction

Homeland security and disaster mitigation efforts are often taken place in unforeseen environments that include caves, tunnels, forests, cities, and even inside urban structures. Performing various tasks such as surveillance, reconnaissance, bomb damage assessment or search and rescue within an unfamiliar territory is not only dangerous but it also requires a large, diverse task force. Unmanned robotic vehicles could assist in such missions by providing situational awareness without risking the lives of soldiers, first responders, or other personnel. While ground-based robots have had many successes in search and rescue situations [6], they move slowly, have trouble traversing rugged terrain, and can still put the operator at risk. Alternatively, small unmanned aerial vehicles (UAVs) can provide soldiers and emergency response personnel with an "eye in the sky" perspective. On an even smaller scale, tiny bird-sized aircraft or micro air vehicles (MAVs) can be designed to fit in a backpack and can be rapidly deployed to provide surveillance and reconnaissance in and around buildings, caves, tunnels and other near-Earth environments. Navigating in these environments, however, remains a challenging problem for UAVs. In

[1] Written by W. Green, P. Y. Oh

Kimon P. Valavanis (ed.), Advances in Unmanned Aerial Vehicles, 407–430.

[7], promising results are shown for a rotorcraft equipped with a SICK laser scanner. However, because lift decreases with platform size, carrying this type of sensor on a MAV is not feasible.

For inspiration towards autonomous MAV navigation in near-Earth environments, the authors looked to flying insects. Flying insects, such as honeybees and fruit flies, use optic flow to navigate in complex and dynamic surroundings [2]. By mimicking insect behaviors as in [9], it was possible to demonstrate tasks such as collision avoidance and landing with a MAV inside an urban structure [4]. In these experiments, optic flow worked well for detecting obstacles on the sides of the MAV, but failed to avoid collisions when flying directly at low textured obstacles (e.g. walls). This is supported in [5] where in one experiment, optic flow sensors were used to calculate the distance to obstacles on the side of the MAV, but in the second experiment where the MAV started out by heading right towards a building, a laser ranger was used for obstacle detection. Furthermore, this experiment was performed outdoors with lots of flying space. This is not always the case when flying in near-Earth environments. To compensate, optic flow sensing was complemented with a secondary flight mode for a fixed-wing MAV. This secondary flight mode allows a fixed-wing MAV to hover and can be used to avoid imminent collisions and also to maneuver through tight spaces, see Figure 13.1.

Fig. 13.1. A fixed-wing MAV transitions to hovering mode to gently maneuver itself through a small opening of an urban structure. *Insert*: A surveillance image from the on-board wireless camera is also shown.

13.2 Optic Flow

Insects perform tasks such as collision avoidance and landing by perceiving the optic flow of their surroundings. Optic flow refers to the apparent motion of texture in the visual field relative to the insect's body. Through several experiments with honeybees [8] and fruit flies [10], it is suggested that flying insects avoid collisions by turning away from regions of high optic flow, see Figure 13.2. To mimic these navigation techniques, a 30 gram flying testbed was prototyped. Figure 13.3 shows this prototype which was designed to be small and fly at 2 m/s for extended reaction times in order to avoid detected obstacles.

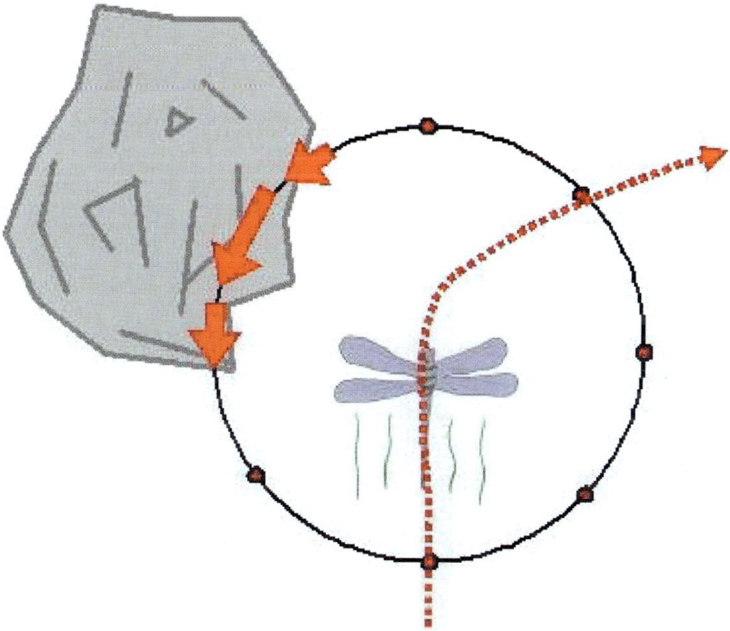

Fig. 13.2. Dragon fly moving away from regions of high optic flow in order to avoid a collision.

Fig. 13.3. The Drexel 30 gram prototype MAV that has a 60 cm wingspan. It flies at speeds of 2 m/s.

13.2.1 Collision Avoidance

Mimicking behaviors of flying insects required optic flow to be measured in front of the aircraft to detect oncoming collisions, see Figure 13.4. Figure 13.5 shows a one-dimensional optic flow sensor that was used in the experiments [1]. It is comprised of a mixed-mode vision chip which images the environment and performs low-level processing using analog VLSI circuitry. Then an off-the-shelf microcontroller performs mid and high-level processing using standard digital techniques. The resulting sensor, including optics, imaging, processing, and I/O weighs 4.8 grams. This sensor grabs frames at up to 1.4 kHz and measures optic flow up to 20 rad/s.

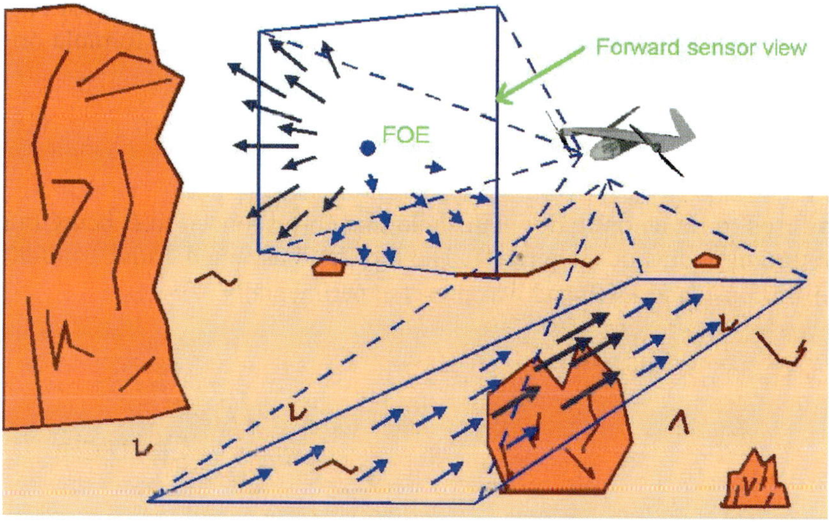

Fig. 13.4. Optic flow as seen by an aerial robot flying above the ground.

Fig. 13.5. Mixed-mode VLSI optic flow micro sensor. It is slightly bigger than a quarter.

Using two of these sensors angled at ±45° from the fuselage, optic flow fields were detected on each side of the aircraft. Optic flow is measured in rad/sec and it is a function of the MAV forward velocity, *V*, angular velocity, ω, distance D from an object, and the angle, α, between the direction of travel and the sensor's optical axis, see Figure 13.6. The formula originally derived in [13]:

$$OF = \frac{V}{D}\sin\alpha - \omega \tag{13.1}$$

was used to set an optic flow threshold that corresponded to D being twice the turning radius of the aircraft. The threshold assumed cruise conditions (i.e. V=const. and ω=0) and was preset experimentally.

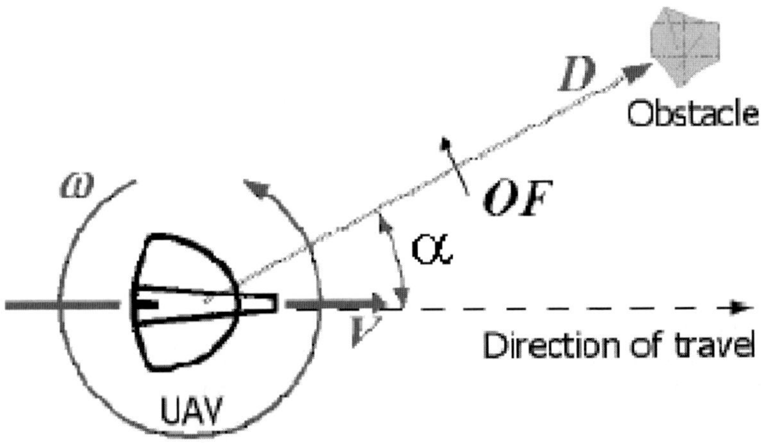

Fig. 13.6. One-dimensional optic flow during a MAV steady level flight.

The aircraft was then flown towards different obstacles and an approaching object on either side of the MAV would generate an increase in optic flow as seen in (13.1). The output of each of these sensors was fed into an on-board microcontroller. If the values from either of the sensors exceeded the threshold, the processor would apply full deflection to the rudder to avoid the collision. By implementing this reactive-based control scheme, autonomous collision avoidance was successfully demonstrated, see Figure 13.7.

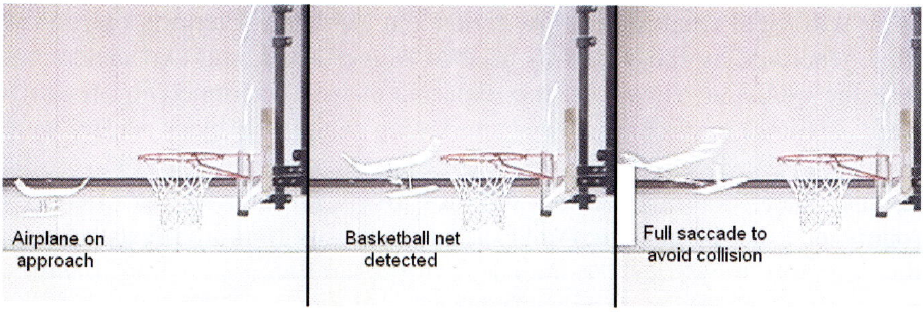

Fig. 7. Optic flow is used to sense when an obstacle is within two turning radii of the aircraft. The aircraft avoids the collision by fully deflecting the rudder.

13.2.2 Optic Flow Limitations

The proof of concept experiments showed promising results for using optic flow for lateral collision avoidance; however, there are some limitations when flying directly towards an object. For example, when two optic flow sensors are aligned at ±45° from the fuselage like in the experiments above, smaller objects such as poles could remain outside the sensor's field of view as shown in the left part of Figure 13.8. This is most likely why honeybees never fly a straight line towards a target, but rather make a slight zigzag pattern. This generates an artificial parallax that will yield optic flow values for smaller oncoming obstacles.

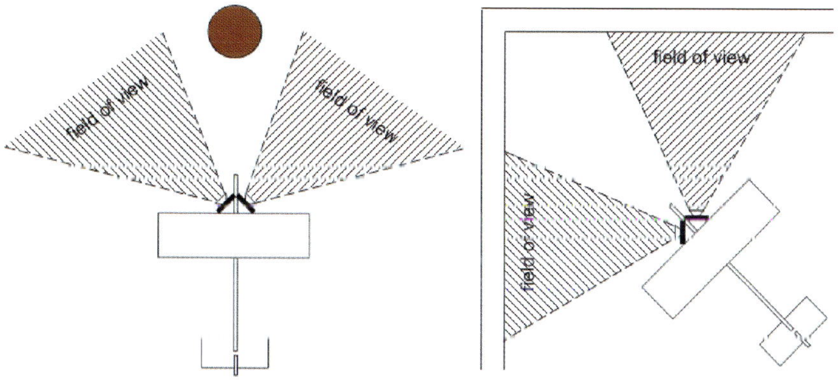

Fig. 13.8. Limitations of optic flow used for navigation.

Similarly, optic flow based collision avoidance is also insufficient when flying directly towards larger, homogenous obstacles such as walls (see the right part of Figure 13.8). Surfaces which are low in texture and uniform in color will yield small optic flow fields. In [14], the diverging optic flow field generated by the wall was used to trigger a warning two meters before the collision. However, the experiment was performed in an artificially textured environment (i.e. alternating white and black sheets were used as walls). Realistically, walls are often homogeneous and have little texture. Therefore, this method will most likely fail in more realistic environments. It has been shown in [11] that when fruit flies are presented with this scenario, they stick out their legs in preparation for landing. Landing on a wall is obviously not feasible for a MAV, however, a quick transition to a stationary attitude is. That is, a fixed-wing MAV can be designed to quickly transition to a hover in order to avoid collisions in these instances.

13.3 Fixed-Wing Hovering MAV

Integrating the endurance of fixed-wing aircraft with hovering capabilities of rotorcraft have recently been realized in the radio-controlled (RC) community through a maneuver known as *prop-hanging*. During a prop-hang, the longitudinal axis of the fuselage is completely vertical and the thrust from the motor balances the weight of the aircraft. Leveraging this maneuver, a prototype fixed-wing platform was developed with an additional flight mode for hovering [3]. Figure 13.9 shows the prototype in its hovering attitude. The prototype is constructed with a 3 mm depron foam core laminated with carbon fiber cloth. It has a 1 meter wingspan, weighs 600 grams, and can fly in cruise mode for 30 minutes on a 11.1 V, 1320 mAh lithium polymer battery. With a 6.6:1 gear ratio and a brushless motor which yielded 900 grams of thrust, the MAV has a thrust-to-weight (T/W) ratio of 1.5. This high T/W ratio was required in order to balance the weight of the aircraft and an extra 100 gram payload when in hover mode. In cruise flight (i.e. wings parallel to the ground), it has a speed range of 5-20 m/s.

Fig. 13.9. The Drexel hybrid prototype weighs 600 grams, has a 1 m wingspan, and has a speed range of 0 to 20 m/s. *Insert*: Wingtip motors are added to counter the rotation about the roll axis during a hover (i.e. torque roll).

13.3.1 Transition between Flight Modes

The most critical aspect of the hybrid design is the transition from cruise to hover flight, which will be used as a secondary collision avoidance maneuver, see Figure 13.10. During this phase, there exists an angle-of-attack, α, for which the wings are no longer a contributing factor to the lift component (i.e. stall). To achieve the transition, the aircraft has to leverage its momentum and essentially overpower its way through the stall regime. This requires a high T/W ratio so the momentum is not lost through the transition. Furthermore, as the aircraft is transitioning from cruise flight (minimum thrust) to the hovering flight mode, the throttle must be increased to balance the weight of the aircraft. The transition back to cruise mode is less complex. Vertical acceleration is first required to give the plane some momentum and then the elevator is deflected to pitch the aircraft forward into cruise mode.

Fig. 13.10. The MAV prototype with a 1 meter wingspan manually transitions from cruise flight (top left) through the stall regime (top middle) and into a hovering position (top right) to avoid a collision with a basketball net.

13.3.2 Hovering

After transitioning into the hovering mode, the attitude must be sustained by constantly adjusting four channels of a radio-controlled transmitter. The most critical task the expert human pilot has is to maintain the aircraft's vertical orientation by adjusting the rudder and elevator deflection angles. A secondary task involves modifying the throttle position to balance the weight of the aircraft. This is because once the stick position is found, it remains relatively constant as the aircraft is not gas powered and therefore maintains the same weight throughout the flight. Finally, the reaction of the MAV to the motor torque results in the plane rotating about the vertical axis when hovering. This is known as torque rolling and can sometimes be countered with aileron control. All of these efforts must be done simultaneously; this makes hovering a rather challenging task.

Rudder, elevator, and throttle control can be seen in more detail in Figure 13.11, which shows the forces acting on the MAV during a hover. The forces generated by rudder and elevator deflection regulate the aircraft's attitude while the thrust force balances the aircraft weight. Summing the forces in the vertical direction yields:

$$(T - D - F_E \sin\delta_E - F_R \sin\delta_R)\cos\psi\cos(\theta - 90) - W = ma_z \qquad (13.2)$$

where F_E and F_R are the elevator and rudder restoring forces, respectively, and are functions of the drag force, D, and control surface deflection angle, δ. When the aircraft is in a perfect hover (i.e. $\theta = 90$, $\psi = \delta_E = \delta_R = a_z = 0$) the thrust must equal both the weight and drag forces.

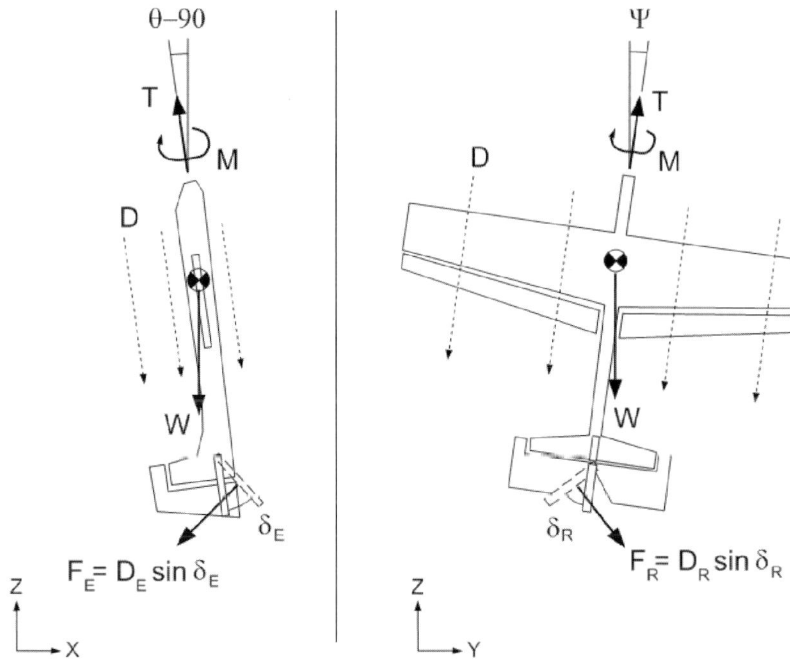

Fig. 13.11. When in a hovering attitude, the elevator and rudder control surfaces are used to regulate the pitch and yaw angles, respectively.

13.4 Autonomous Hovering

In order to autonomously avoid a collision by transitioning into hover mode, both the transition into hover and the hover itself must be automated. To regulate the attitude during a hover, data from a small and lightweight inertial measurement unit (IMU) is fed into an on-board control system. This data is captured during both manual and autonomous hovering and used to compare the controller performance to that of an expert human pilot. Furthermore, in order for the video from the MAV on-board camera to be useful in surveillance applications, the torque rolling during a hover must be eliminated. This is accomplished by mounting small motors with propellers on the tip of each aircraft wing.

13.4.1 Sensing and Control

Autonomous attitude control of this aircraft requires a sensor that can measure the vehicle's orientation when pitch angles approach and exceed ±90°. Figure 13.12 shows an IMU by Microstrain which outputs a gyroscopically stabilized four component quaternion describing the MAV orientation with respect to the fixed earth coordinate frame. It weighs just 30 grams (out of its protective casing) and is comprised of three triaxial accelerometers and angular rate gyros as well as three orthogonal magnetometers. The IMU, using RS232 communications, will transmit orientation data to the host computer at a clock cycle of around 10 ms. Therefore, embedding the sensor on the MAV platform will enable an on-board microcontroller to obtain the aircraft's orientation at a rate of 100 Hz.

Fig. 13.12. Microstrain's 30 gram IMU sensor was used to feedback attitude information to the on-board control system.

An on-board control system was designed using a PIC16F87 microcontroller and a RS232 converter chip to communicate serially with the IMU. The autonomous hovering algorithm begins by pinging the IMU for a measured quaternion, q_m, which corresponds to the MAV's attitude at that instant. The commanded quaternion, q_c, describing the desired attitude during hover mode is:

$$
\begin{aligned}
q_{c1} &= e_1 \sin(\theta/2) = 0.000i \\
q_{c2} &= e_2 \sin(\theta/2) = 0.707j \\
q_{c3} &= e_3 \sin(\theta/2) = 0.000k \\
q_{c4} &= \cos(\theta/2) = 0.707
\end{aligned}
\tag{13.3}
$$

Using the commanded and measured quaternions, the error quaternion, q_e, can be calculated using the following formula [12]:

$$q_e = q_c^* \otimes q_m \tag{13.4}$$

where q_c^* represents the conjugate of the commanded quaternion. The elevator and rudder control signals can be extracted from the error quaternion and proportional-derivative (PD) control is used to send pulse-width modulated (PWM) signals to the control surfaces. This, in turn, drives the aircraft orientation back to the hovering attitude. Figure 13.13 shows the control loop which repeats continuously and is synchronized with the IMU clock cycle (i.e. every 10 ms).

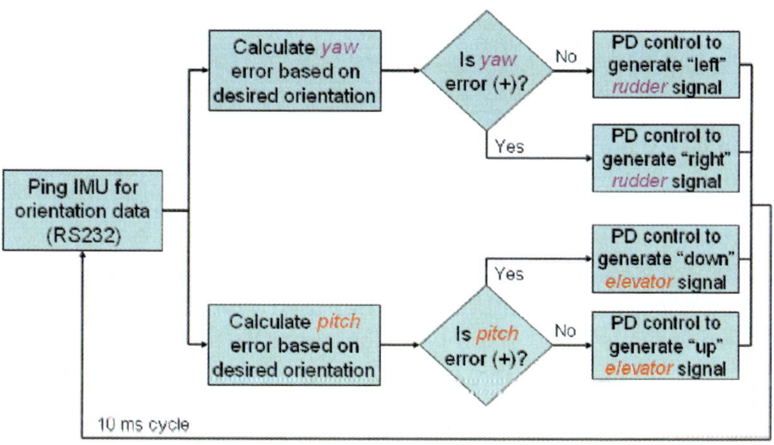

Fig. 13.13. Flow chart describing the proposed autonomous hovering algorithm.

13.4.2 Autonomous Hovering Experiments

The first autonomous hovering experiments were conducted inside an urban structure with limited flying space (i.e. 3m x 3m area) to demonstrate that hovering can be sustained within small areas. The MAV attitude is under full autonomous control through rudder and elevator inputs while the height is adjusted manually through throttle commands via the pilot until the aircraft's weight is balanced. Initial experiments demonstrated that the MAV was able to successfully hover in *hands-off* mode for several minutes before draining the battery, see Figure 13.14.

Fig. 13.14. MAV performing a *hands-off* autonomous hover in an urban structure. *Insert*: Also, a shot from the MAV camera is shown, which is slightly blurred because of the MAV torque roll.

Another experiment was performed to contrast hovering under both manual and autonomous control. The metrics used were duration of the hover before losing control and stability of the aircraft while in hovering mode. A skilled human pilot was initially given control of the aircraft and was instructed to fly around a gymnasium in cruise configuration, then transition from cruise to hover flight and attempt to hover the aircraft for as long as possible. The video stills[2] on the top of Figure 13.15 show the pilot struggling to keep the fuselage vertical, but is able to keep the aircraft positioned over a small area. Out of a few trials, the human pilot was able to sustain a hover for several minutes before draining the battery. However, the aircraft's pitch and yaw angles oscillated significantly as the pilot tried

[2] The video sequence shows three images extracted once a second for a period of three seconds. With the plane rotating at a rate of 0.25 revolutions per second, this is enough to show two quarter rotations.

to maintain the hover. This is supported with a portion of the captured flight data, labeled *human-controlled* shown in Figure 13.16. Next, the pilot was instructed to again fly in cruise configuration and manually transition from cruise to hover flight. However, instead of trying to hover the aircraft manually, the pilot flicked a switch on the transmitter which enabled the on-board control system. This time, the aircraft is fixed in a vertical position and is able to hover for more than 5 minutes before exhausting the battery (see bottom of Figure 13.15). Again, the flight data was captured and a fraction of it is shown in Figure 13.16.

Fig. 13.15. A skilled human pilot hovers a fixed-wing aircraft in a small gymnasium and struggles to maintain a vertical orientation (top). Under autonomous control, the same aircraft is able to sustain a hover while remaining fixed in the vertical position (bottom).

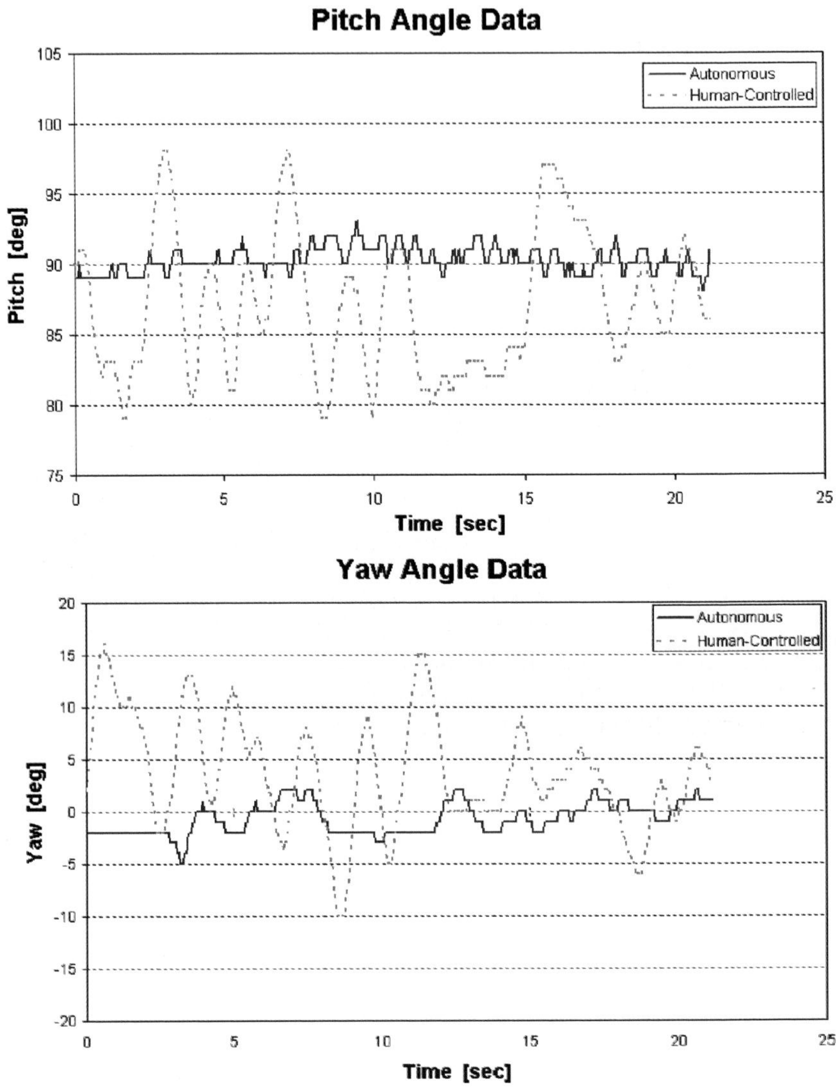

Fig. 13.16. Pitch and yaw angles captured during both human-controlled and autonomous hovering.

13.4.3 Torque Roll Control

As originally thought, the torque-roll did not affect the stability of the aircraft during a hover. That is, the MAV was still able to remain in the vertical position despite the rotations resulting from the motor torque. However, if this MAV was to be used in the field for surveillance and reconnaissance purposes, the view from the wireless camera on-board would have a dizzying effect as the plane was rotating at a rate of 20 rpm. Since the original aileron surface area did not create enough torque to counter the rotation when fully deflected, other alternatives had to be investigated. The first and most obvious was to increase the aileron surface area by lengthening them in the direction of the wing chord. However, this was not effective because the prop wash during a hover only flowed over approximately 40 percent of the ailerons. Furthermore, a longer aileron when fully extended caused some airflow to completely miss the tail. This significantly effected attitude regulation during a hover.

The second approach was to mount miniature DC motors and propellers on each wingtip. The motors and propellers were positioned to produce a thrust force in opposite directions which generated a rotational force countering the motor torque (see Figure 13.9). The wingtip motors are GWS EDP-20s which provide 23 grams of thrust with a 2510 direct drive propeller at 7.2 volts. With the same IMU used in the autonomous hovering experiments, the torque rolling rate (i.e. aircraft roll) can be fed back into the flight control system. Using this parameter, PD control was implemented on the error which determined the length of the PWM signal being output to the wingtip motors' brushed speed controller. A schematic of the setup for autonomous hovering with torque roll regulation is shown in Figure 13.17.

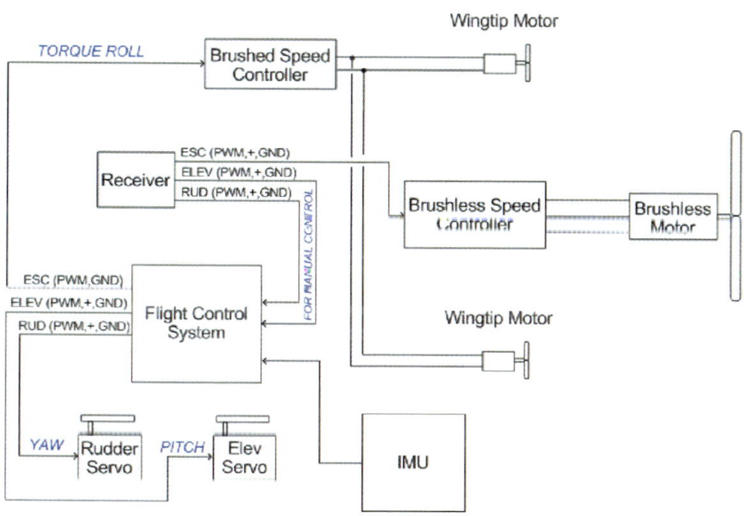

Fig. 13.17. Schematic showing the setup during an autonomous hover with torque roll control.

With the above setup, autonomous hovering experiments were conducted with and without torque roll control. The roll angle from the IMU was logged to the flight control system's external memory and it is shown in Figure 13.18. The top part of Figure 13.18 shows the case with no torque roll control. With the plane constantly rotating during a hover, a plot of the angular data would grow rapidly. To make each revolution more visible, the roll angle was bounded between -180° and 180°. Therefore, as the plane moves through one revolution the roll angle (in degrees) will go from 0 to 90, 90 to 180 (180 is equal to -180), -180 to -90, and -90 to 0. It can be seen on the top part of Figure 13.18 that the plane completes 7 full revolutions in 16 seconds, or about 26 rpm. The bottom part of Figure 13.18 shows the controlled condition and it can be seen that the plane remains at a relatively constant orientation.

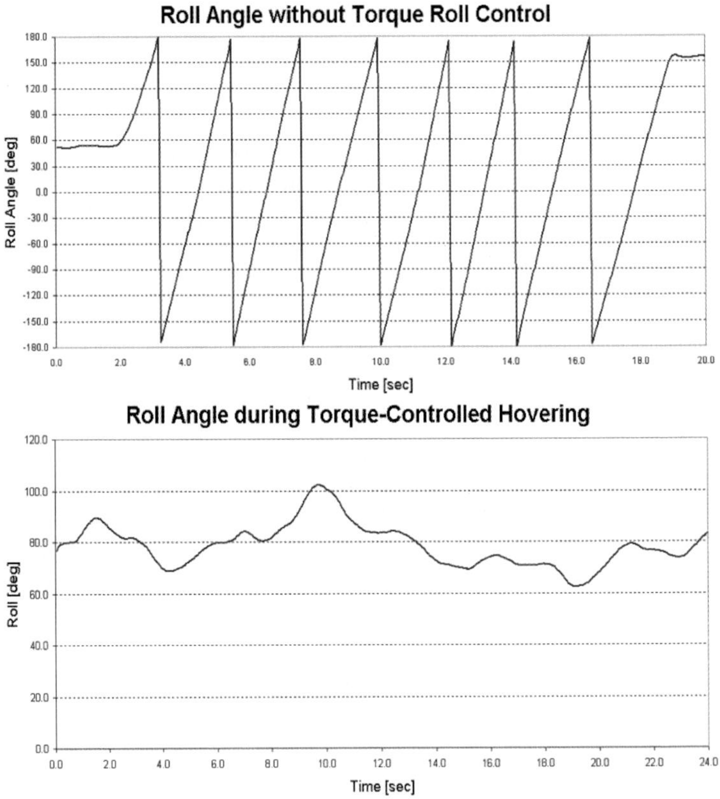

Fig. 13.18. Roll angle captured during autonomous hovering with and without torque roll control.

13.5 Autonomous Cruise-to-Hover Transition

During autonomous hovering, the commanded quaternion represents the aircraft in the vertical orientation with the belly facing north (using a north-east-down coordinate frame). When autonomous hovering is initiated, the aircraft is already close to the vertical orientation but may or may not be facing north. This will most likely result in small control effort from the elevator and rudder when compared to the control effort about the x-axis in the body frame. Essentially, the aircraft will get to the vertical orientation first and then roll until the belly faces north. As such, the commanded quaternion used for autonomous hovering, $q_c = 0.0i + 0.707j + 0.0k + 0.707$, is sufficient when releasing the aircraft in near-hovering orientation, but can be problematic for the transition from cruise to hover flight.

Imagine the MAV to be flying in cruise mode heading north (i.e. body and reference frames are aligned). If the commanded quaternion used for autonomous hovering was also used for the transition from cruise (heading north) to hover flight, the commanded, measured, and error quaternions would be:

$$\begin{aligned} q_c &= 0.0i + 0.707j + 0.0k + 0.707 \\ q_m &= 0.0i + 0.0j + 0.0k + 1.0 \\ q_e &= 0.0i + 0.707j + 0.0k + 0.707 \end{aligned} \tag{13.5}$$

Normalizing the error quaternion and finding the angular error about each axis results in a large pitch error and zero error about the roll and yaw axes. As desired, the controller would respond with large elevator deflection causing the aircraft to pitch up towards the vertical orientation. However, if the MAV was flying in cruise mode and heading south, the same commanded quaternion will not result in a pure pitch up maneuver. That is, if the commanded quaternion for hovering was used in this case, the commanded, measured and error quaternions would now be:

$$\begin{aligned} q_c &= 0.0i + 0.707j + 0.0k + 0.707 \\ q_m &= 0.0i + 0.0j + 1.0k + 0.0 \\ q_e &= 0.707i + 0.0j + 0.707k + 0.0 \end{aligned} \tag{13.6}$$

Normalizing q_e and finding the error about each axis results in large roll and yaw errors and zero pitch error. This is because a combination of roll and yaw will yield the minimum rotation to get the aircraft to the commanded quaternion (vertical with belly of aircraft facing north) rather than pitching up 90° and then rolling 180°. However, transitioning from cruise flight to hover mode when heading in any direction should only require a pitch up maneuver. That is, if the aircraft is in cruise mode heading south, the transition from cruise flight to hover mode should leave the aircraft in the vertical orientation with the belly facing south. To achieve this, a delta quaternion will be introduced which generates a commanded quaternion depending on the aircraft's orientation when the cruise-to-hover algorithm is initialized.

13.5.1 Cruise-to-Hover Algorithm

When the transition from cruise to hover is initialized, there is an angular error between the x axis of the aircraft and a vertical vector expressed in the reference frame. Assuming perfect cruise conditions (i.e. $\phi=0$, $\theta=0$, and $-180^o \leq \psi \leq 180^o$), this error is equal to ninety degrees about the pitch axis. As mentioned above, the conventional expression for the error quaternion cannot be used to solve for this angular error as it is dependent upon the aircraft heading. Instead, vector and quaternion mathematics will be used to generate a delta quaternion which represents a rotation from the initial aircraft attitude in quaternion form to the vertical orientation. Using the delta and measured quaternions, a commanded quaternion can then be calculated. Upon obtaining the commanded quaternion, the error quaternion can then be computed for each new measured quaternion (each control loop iteration), which will generate a pure pitching maneuver despite the initial aircraft heading.

Assuming the cruise to hover program has been initialized and the first measured quaternion has been acquired, the process to calculate the delta quaternion starts by defining the vertical vector in the reference (NED) frame, V_{vr}, and the aircraft's x axis in the body frame, V_b as shown in Figure 13.19:

$$\begin{aligned} V_{vr} &= 0.0i + 0.0j - 1.0k \\ V_b &= 1.0i + 0.0j + 0.0k \end{aligned} \tag{13.7}$$

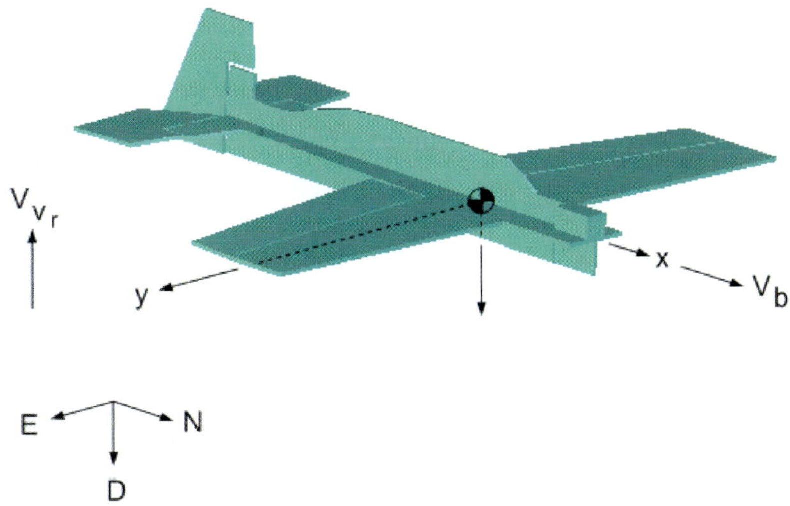

Fig. 13.19. The vertical vector in the reference frame is shown along with the x axis in the body frame.

It is desired to keep the error quaternion in the aircraft's body frame such that the angular errors can be used to proportionally control the aileron (or wingtip motor), elevator, and rudder surfaces. Therefore, all calculations will be performed in the body frame. As such, the first step is to transform the vertical vector from the reference frame (V_{vr}) to the body frame (V_{vb}) using the measured quaternion:

$$V_{vb} = q_m^* \; V_{vr} \; q_m \tag{13.8}$$

It should be noted that in the above equation, V_{vb} is first converted to a fourtuple by adding a zero to act as the scalar component. This is done to make it compatible with quaternion multiplication.

Now that the vertical vector and the aircraft's x-axis are both represented in the body frame, vector mathematics can be used to find a rotation axis and angle between the two vectors. The cross product is calculated to find the rotation axis, or the axis which is orthogonal to both vectors:

$$V_{rot} = V_b \times V_{vb} \tag{13.9}$$

Next the angle between the MAV's x-axis and vertical vector in the body frame can be found by:

$$\gamma = \cos^{-1}(V_b \cdot V_{vb}) \tag{13.10}$$

The rotation axis, V_{rot}, and angle, γ, that force the aircraft's x-axis to coincide with the vertical vector are now known and can be converted into a quaternion, which will be referred to as the delta quaternion, q_Δ:

$$
\begin{aligned}
q_{\Delta 1} &= V_{rot1} \sin(\gamma / 2) \\
q_{\Delta 2} &= V_{rot2} \sin(\gamma / 2) \\
q_{\Delta 3} &= V_{rot3} \sin(\gamma / 2) \\
q_{\Delta 4} &= \cos(\gamma / 2)
\end{aligned}
\tag{13.11}
$$

Now that the measured quaternion, q_m, and delta quaternion, q_Δ, are both known, the commanded quaternion can be calculated:

$$q_c = q_m \otimes q_\Delta \tag{13.12}$$

This entire process is performed once at the initialization of the cruise to hover maneuver. The resulting commanded quaternion remains constant and it is used along with a new measured quaternion at every iteration to compute the error quaternion. Finally, since the error quaternion is calculated in the body frame of the aircraft, the angular error about each axis can be used to control the aileron (or wingtip motor), elevator, and rudder surface deflection. Proportional-derivative (PD) control is performed on all control surfaces.

13.5.2 Autonomous Cruise-to-Hover Experiments

During the cruise-to-hover experiments, the human pilot loitered around a field in cruise mode and then flicked a switch on the transmitter to enable the on-board controller. This signaled the start of the autonomous transition and initially there was a large error between the desired and current attitude. This prompted full elevator deflection from the controller until the aircraft approached the hovering attitude. As the error decreased, so did the amount of elevator deflection. The transition takes about a second and it is shown in Fig. 13.20.

Fig. 13.20. The transition from cruise flight (left), through the stall regime (center), and into hover mode (right) is achieved autonomously.

13.6 Conclusions

Autonomously flying in caves, tunnels, and urban environments demands more than one sensing modality. This Chapter presented an optic flow based approach inspired by flying insects for avoiding collisions on the side of the MAV. However, there are a few realistic scenarios where optic flow will not be enough to detect oncoming collisions. This occurs when obstacles on approach are directly in front of the aircraft. Here, the diverging optic flow field will be minimal for low textured surfaces (i.e. homogenous walls). In this case, a MAV that can quickly transition into a hover in conjunction with a simple sonar or infrared sensor to trigger this maneuver can be used to avoid the otherwise fatal collision. Towards this, attitude feedback was used from an IMU mounted on-board the MAV to demonstrate autonomous hovering as well as an autonomous transition from cruise to hover flight modes. Furthermore, torque roll control was accomplished by mounting 5 gram brushed DC motors on each wingtip. The next step in this research is to mount a small proximity sensor on the front of the aircraft in order to trigger the autonomous transition from cruise to hover flight.

References

1. Barrows G., *Mixed-Mode VLSI Optic Flow Sensors for Micro Air Vehicles*, PhD Thesis, University of Maryland, 1999.
2. Gibson J. J., *The Ecological Approach to Visual Perception*, Houghton Mifflin, 1950.
3. Green W. E., Oh P. Y., "A MAV That Flies Like an Airplane and Hovers Like a Helicopter", *Proceedings, IEEE/RSJ International Conference on Advanced Intelligent Mechatronics*, Monterey, CA, 2005.

4. Green W. E., Oh P. Y., Barrows G., "Flying Insect Inspired Vision for Autonomous Aerial Robot Maneuvers in Near-Earth Environments", *Proceedings, IEEE International Conference of Robotics and Automation*, New Orleans, LA, 2004.
5. Griffiths S., Saunders J., Curtis A., Barber B., McLain T., Beard R., "Maximizing Miniature Aerial Vehicles", *IEEE Robotics and Automation Magazine*, Vol. 13, No, 3, 34-43, 2006.
6. Murphy R., Casper J., Hyams J., Micire M., Minten B., "Mobility and Sensing Demands in USAR", *Proceedings, IEEE Industrial Electronics Conference*, Vol. 1, 2000.
7. Shim D. H., Chung H., Sastry S. S., "Conflict-Free Navigation in Unknown Urban Environments", *IEEE Robotics and Automation Magazine*, Vol. 13, No. 3, 27-33, 2006.
8. Srinivasan M. V., Zhang S. W., Lehrer M., Collett T. S., "Honeybee Navigation En Route to the Goal: Visual Flight Control and Odometry", *Journal of Experimental Biology*, 237-243, 1996.
9. Srinivasan M. V., Chahl J. S., Weber K., Venkatesh S., Nagle M. G., Zhang S. W., "Robot Navigation Inspired By Principles of Insect Vision", *Robotics and Autonomous Systems*, Vol. 26, 203-216, 1999.
10. Tammero L. F., Dickinson M. H., "The Influence of Visual Landscape on the Free Flight Behavior of the Fruit Fly Drosophila Melanogaster", *Journal of Experimental Biology*, Vol. 205, 327-343, 2002.
11. Tammero L. F., Dickinson M. H., "Collision Avoidance and Landing Responses are Mediated by Separate Pathways in the Fruit Fly, Drosophila Melanogaster", *Journal of Experimental Biology*, Vol. 205, 2785-2798, 2002.
12. Wertz J. R., *Spacecraft Attitude Determination and Control*, Reidel Publishing Co, 1978.
13. Whiteside T. C., Samuel G. D., "Blur Zone", *Nature*, Vol. 225, 94-95, 1970.
14. Zufferey J. C., Floreano D., "Fly-Inspired Visual Steering of an Ultralight Indoor Aircraft", *IEEE Transactions on Robotics*, Vol. 22, No. 1, 137-146, 2006.

Chapter 14: Dynamic Localization of Air-Ground Wireless Sensor Networks[1]

This Chapter presents an air-ground localization approach in which ground nodes in collaboration with unmanned aerial vehicles (UAVs) generate optimal position estimates. The ground nodes support local sensing and communication, while unmanned aerial vehicles equipped with GPS may be used to localize the ground nodes network absolutely.

14.1 Introduction

The role of autonomous surveillance has proven to be important and applicable to a wide range of applications such as target location, map building, border security, pollution monitoring and control, and battle damage assessment. UAVs fit into the scenario of autonomous surveillance perfectly as they involve a low risk factor and facilitate technological advancements, making their use feasible in real world scenarios. UAVs are generally classified by their flight altitude, launch and recovery methods as detailed [23].

UAVs, together with randomly deployed stationary unattended ground sensors (UGS) can further enhance the performance of the autonomous surveillance tasks mentioned above. Since the information collected from UGS is of limited use if no information about the sensor position is available, the task of localizing the sensor nodes is of prime importance for sensor network applications. This Chapter discusses an air-ground localization scheme in which (UGS) nodes, with the aid of UAVs having on board GPS, generate their optimal position estimates.

Localization is classified in two categories: relative and absolute localization. In relative localization the sensor nodes are localized using the distances measured among the nodes with respect to an arbitrary internal co-

[1] Written by P. Dang, F. L. Lewis, D. O. Popa. This work was supported in part by the following grants: NSF Grant IIS-0326505, NSF Grant CNS-0421282, ARO Grant M-47928-CI-RIP-05075-1 and National Instrument Lead User Grant.

Kimon P. Valavanis (ed.), Advances in Unmanned Aerial Vehicles, 431–453.

ordinate system. Absolute localization, on the other hand localizes the network with respect to a known specified coordinate system.

An important development is an air-ground localization scheme, which performs relative and absolute localization of stationary UGS network with the aid of UAVs. The UGS nodes are simple and support local sensing, communication and computation. It becomes impractical to have GPS capability on UGS nodes due to energy and cost constraints. UAVs having GPS can be used to localize the UGS network absolutely.

The UGS node localization problem has been addressed by many researchers using various approaches. In [20] a distributed algorithm is proposed, where a fraction of nodes are already localized. Other schemes for localization such as SHARP, virtual force algorithm (VFA) and self-localization have been proposed in [1] [16] [24]. Other methods using fixed beacons with known positions are proposed in [24] [4] [19] [5]. Various approaches involving RSSI, TOA, AOA, signal pattern matching are explained in [6]. A detailed introduction to localization in sensor networks is given in [2]. Cooperative localization methods have been developed for relative localization in [17].

Air-ground localization schemes based on terrain aided navigation have been addressed in [10] [18] [12] [11]. These algorithms are known as Simultaneous Localization and Mapping (SLAM). Other airborne localization algorithms involving regular broadcast of UAV location have been proposed in [8]. These localization schemes depend on the path of UAV with respect to the deployed nodes and are restricted by power consumption and network congestion problems due to the regular broadcasts of UAV positions. In [15], problem of localizing vehicles where GPS signals are often unavailable is described.

With no a-priori terrain information available, the problem of localization is extremely challenging and various approaches based on an extended Kalman filter (EKF) have been proposed in [18] [12] [13]. The EKF is a recursive estimation technique based on first-order linearization of the nonlinear system, and can yield large estimation errors and even divergence of the filter. Thus, these techniques are sometime not very effective. For more on the divergence of the Kalman filter one may see [14].

This Chapter presents an alternative method of relative and absolute localization based on a potential field method [9]. Two algorithms are presented: relative localization algorithm and absolute localization algorithm. A dynamical model for each sensor node estimates the relative positions by employing a correction term based on a certain fictitious virtual force. In the relative localization algorithm the stationary UGS nodes are localized with respect to an internal coordinate frame. The relative localization algorithm proposed in the paper assumes that distance (i.e. range) measurements between sensor nodes are available. For absolute localization, it is assumed that some nodes have GPS absolute position information. Specifically, herein, the UGS nodes are localized with respect to a global frame provided using the absolute positions of several UAV with GPS as shown in Figure 14.1.

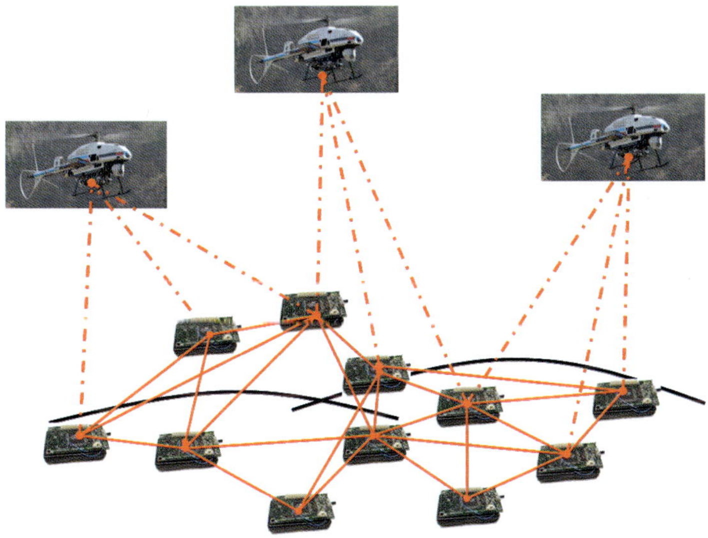

Fig. 14.1. Air-ground sensor network configuration.

The Chapter is organized as follows: Section 14.2 derives the position estimate dynamical model and a potential field function used for relative localization of a stationary UGS network. Section 14.3 discusses the algorithm used for relative localization and simulation results are presented. Section 14.4 extends the idea to absolute localization and presents the dynamical estimator model and a modified potential field function used for absolute localization. Section 14.5 discusses the algorithm used for absolute localization and simulation results are presented. Section 14.6 concludes the Chapter.

14.2 Virtual Node Dynamics for Relative Localization

A novel method is presented for relative localization of a network of stationary unattended ground sensors (UGS). It is assumed that distance (i.e. range) measurements between sensor nodes are available, specifically each sensor node measures the distance to at least three other nodes. The method uses a dynamical model for position estimates of each node that is driven by a fictitious virtual force based on range errors. These virtual dynamics have states which are the estimates of the relative positions, and reach a steady-state value that provides an optimal (in a sense to be made precise herein) estimate of the relative positions of all nodes in the network.

The UGS nodes do not physically move, but the virtual dynamics capture the available range information to dynamically compute the UGS relative position estimates. A certain potential field is introduced to generate optimal position locations in a least-squares sense. The potential field is used as a Lyapunov function and a Lyapunov proof shows how to generate appropriate virtual forces based on the gradient of the potential field.

14.2.1 System Description

The virtual dynamics used for generating position estimates of the stationary UGS nodes are described based on range information. The position estimate for the ith sensor node is given by:

$$X_i = [x_i \quad y_i]^T \tag{14.1}$$

where x_i and y_i are the x-y coordinates for the UGS node position estimate. The position estimation dynamics are given as:

$$\ddot{X}_i = \vec{f}_i \tag{14.2}$$

where $\vec{f}_i = [f_i^x \quad f_i^y]^T$ is the virtual force in the x and y directions to be specified. The state variable description form for the position estimate of the ith UGS node is by:

$$\begin{bmatrix} \dot{X}_i \\ \ddot{X}_i \end{bmatrix} = \begin{bmatrix} O_2 & I_2 \\ O_2 & O_2 \end{bmatrix} \begin{bmatrix} X_i \\ \dot{X}_i \end{bmatrix} + \begin{bmatrix} O_2 \\ I_2 \end{bmatrix} \begin{bmatrix} f_i^x \\ f_i^y \end{bmatrix} \tag{14.3}$$

where $O_2 = \begin{bmatrix} 0 & 0 \\ 0 & 0 \end{bmatrix}$ and $I_2 = \begin{bmatrix} 1 & 0 \\ 0 & 1 \end{bmatrix}$

14.2.2 Potential Field for Optimal Position Estimation

A potential field is now introduced to determine the virtual force $\vec{f}_i$ in (14.2) so that the position estimates reach a steady-state value that is an optimal estimate for the actual UGS node relative positions. Define a potential field as:

$$V_{ugs} = \sum_{i=1}^{N} \sum_{\substack{j=1 \\ i \neq j}}^{N} \frac{1}{2} K_{ij} (r_{ij} - \bar{r}_{ij})^2 \tag{14.4}$$

where $r_{ij} = [(x_i - x_j)^2 + (y_i - y_j)^2]^{1/2}$ is the estimated range and $\bar{r}_{ij}$ is the actual measured range between i^{th} and j^{th} UGS nodes.

Define the potential function for a single UGS node *i* by:

$$V_{i_{ugs}} = \sum_{\substack{j=1 \\ i \neq j}}^{N} \frac{1}{2} K_{ij} (r_{ij} - \bar{r}_{ij})^2 \tag{14.5}$$

The gradient of the potential with respect to the sensor node state is given by:

$$\frac{\partial V_{i_{ugs}}}{\partial X_i} = \vec{\nabla} V_{i_{ugs}} = \vec{\nabla} \sum_{\substack{j=1 \\ i \neq j}}^{N} \frac{1}{2} K_{ij} (r_{ij} - \bar{r}_{ij})^2 = \sum_{\substack{j=1 \\ i \neq j}}^{N} K_{ij} (r_{ij} - \bar{r}_{ij}) \vec{\nabla} (r_{ij} - \bar{r}_{ij}) \tag{14.6}$$

$$\frac{\partial V_{i_{ugs}}}{\partial X_i} = \sum_{\substack{j=1 \\ i \neq j}}^{N} K_{ij} (r_{ij} - \bar{r}_{ij}) \left[\frac{x_i - x_j}{\left\| X_i - X_j \right\|} \hat{x} \quad \frac{y_i - y_j}{\left\| X_i - X_j \right\|} \hat{y} \right] \tag{14.7}$$

where $\left\| X_i - X_j \right\| = r_{ij}$.

Theorem 14.1: Consider the position estimate dynamics (14.2) for each sensor node i in the network. Let the virtual force for i^{th} sensor node be given as:

$$\vec{f}_i = -\sum_{\substack{j=1 \\ i \neq j}}^{N} K_{ij} (r_{ij} - \bar{r}_{ij}) \frac{(X_i - X_j)}{\left\| X_i - X_j \right\|} - K_v X_i \tag{14.8}$$

Then the position estimates reach steady-state values that provide optimal estimates of the actual relative localization of the nodes in the sense that V_{ugs} is minimized.

Proof: Define the Lyapunov function:

$$L = V_{ugs} + \sum_{i=1}^{N} \frac{1}{2} \dot{X}_i^T \dot{X}_i \tag{14.9}$$

Differentiate to obtain:

$$\dot{L} = \sum_{i=1}^{N} \sum_{\substack{j=1 \\ i \neq j}}^{N} K_{ij} (r_{ij} - \bar{r}_{ij}) \dot{r}_{ij} + \sum_{i=1}^{N} \dot{X}_i^T \ddot{X}_i \tag{14.10}$$

One can compute:

$$\dot{r}_{ij} = \frac{\dot{X}_i^T (X_i - X_j)}{\left\| X_i - X_j \right\|} \tag{14.11}$$

and on substituting (14.2) and (14.11), obtain:

$$\dot{L} = \sum_{i=1}^{N} \sum_{\substack{j=1 \\ i \neq j}}^{N} K_{ij} (r_{ij} - \bar{r}_{ij}) \frac{\dot{X}_i^T (X_i - X_j)}{\left\| X_i - X_j \right\|} + \sum_{i=1}^{N} \dot{X}_i^T \vec{f}_i \tag{14.12}$$

Further, substituting the force (14.8) yields:

$$\dot{L} = -\sum_{i=1}^{N} \dot{X}_i^T K_v \dot{X}_i \tag{14.13}$$

Clearly, $\forall\ K_v > 0$, $\dot{L} \leq 0$ and the vector $\left[X_i \quad \dot{X}_i \right]^T$ is bounded which shows that the position estimate dynamics is SISL. Evaluating $\ddot{L}$ yields:

$$\ddot{L} = -2 \sum_{i=1}^{N} \dot{X}_i^T K_v \vec{f}_i \tag{14.14}$$

and on substituting (14.8) obtain:

$$\ddot{L} = -2 \sum_{i=1}^{N} \dot{X}_i^T K_v \left\{ -\sum_{j=1}^{N} K_{ij} (r_{ij} - \bar{r}_{ij}) \frac{(X_i - X_j)}{\left\| X_i - X_j \right\|} - K_v \dot{X}_i \right\} \tag{14.15}$$

Using the result obtained from Lyapunov analysis that the vector $\left[X_i \quad \dot{X}_i \right]^T$ is bounded yields that $\ddot{L}$ in (14.15) is also bounded. By Barbalat's Lemma [22] one deduces that $\dot{L} \to 0$ as $t \to \infty$, which yields $\dot{X}_i \to 0$

as $t \to \infty$. Therefore (14.2) shows that $\vec{f}_i$ goes to zero $\forall i$. Finally (14.7) and (14.8) show that $\frac{\partial V_{ugs}}{\partial X_i} \to 0, \forall i$, so V_{ugs} reaches a minimum.

14.3 Relative Localization Scheme for UGS Network

Using the system defined in (14.2) and the force control input defined as in (14.8) the wireless sensor network can be relatively localized. However, due to a nonlinear mapping between the range (r_{ij}) and the x-y coordinates, there are local minima in Lyapunov function in (14.9), which could lead to incorrect position estimates for the nodes. Though the Lyapunov term defined in (14.9) is positive definite in terms of range information, the Lyapunov derivative in (14.13) is independent of the range information term and is only dependent on the velocity of position estimates defined in (14.1). In this Section, the problem of local minima is studied and an algorithm is proposed to ensure that the network always converges to a unique solution.

14.3.1 Study of Local Minima

The next results show that a unique equilibrium point exists in a 3 sensor node configuration, but there exist local minima in a 4 sensor node configuration when the virtual force input is given by (14.8). Further in the Section, an algorithm is developed that always guarantees a unique equilibrium configuration for position estimate system in (14.2), for any number of UGS nodes.

Lemma 14.1: Let there be given 3 stationary UGS nodes (not in a straight line) and range measurements between all nodes. Then the position estimate dynamics in (14.2) with the virtual force in (14.8) converges to a unique steady-state value, which provides an optimal estimate for the relative position in terms of the minimum potential field in (14.4).

Proof: Assuming the first node to be at the origin and the second node to be along the *x-axis,* the coordinates of first and second nodes are given as *(0, 0)* and $(r_{12},0)$ respectively. Referring to Figure 14.2 the coordinates of the third node could be found, which are:

$$x_3 = r_{13}\cos(\theta); \quad y_3 = r_{13}\sin(\theta) \tag{14.16}$$

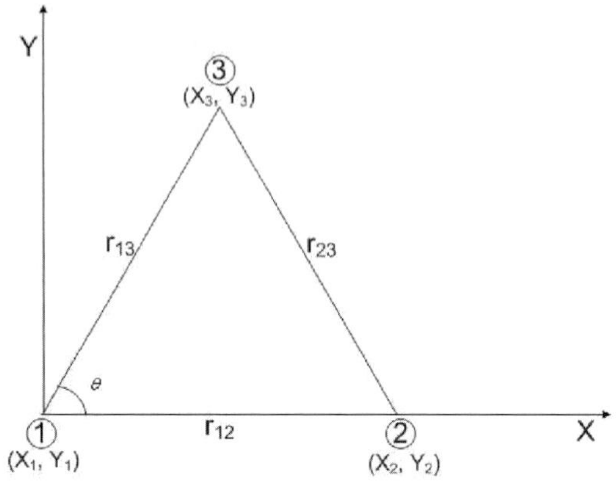

Fig. 14.2. Three UGS node case.

Using (14.8) to write down the force equation for node 1, one obtains:

$$f_1^x = -\frac{(r_{12} - \bar{r}_{12})}{r_{12}}(x_1 - x_2) - \frac{(r_{13} - \bar{r}_{13})}{r_{13}}(x_1 - x_3) \tag{14.17}$$
$$f_1^y = -\frac{(r_{12} - \bar{r}_{12})}{r_{12}}(y_1 - y_2) - \frac{(r_{13} - \bar{r}_{13})}{r_{13}}(y_1 - y_3)$$

and on substituting the coordinate values obtained in (14.16) yields:

$$f_1^x = -(r_{12} - \bar{r}_{12}) - (r_{13} - \bar{r}_{13})\cos(\theta) \tag{14.18}$$
$$f_1^y = -(r_{13} - \bar{r}_{13})\sin(\theta)$$

From the proof of Theorem 14.1, for the equilibrium point the forces $\vec{f}_1$ go to zero. Therefore, for $f_1^y = 0$, $r_{13} = \bar{r}_{13}$ is the only solution since $\sin(\theta) \neq 0$ under the constraint that the 3 nodes are not in a straight line. Looking back at (14.18) and in order to get $f_1^x = 0$, $r_{12} = \bar{r}_{12}$ is the only solution, which leads us to the unique solution of relative localization for 3 nodes.

Lemma 14.2: Let there be given 4 stationary UGS (no 3 of which are in a straight line) and range measurements between all nodes. Then the position estimate dynamics in (14.2) with the virtual force in (14.8) may not converge to a unique steady-state value.

Proof: Referring to Figure 14.3 and writing the coordinates for the 4 node case one obtains:

$$x_1 = 0, y_1 = 0;\ x_2 = r_{12}, y_2 = 0; \qquad (14.19)$$
$$x_3 = r_{13}\cos(\theta);\quad y_3 = r_{13}\sin(\theta)$$
$$x_4 = r_{14}\cos(\theta');\quad y_4 = r_{14}\sin(\theta')$$

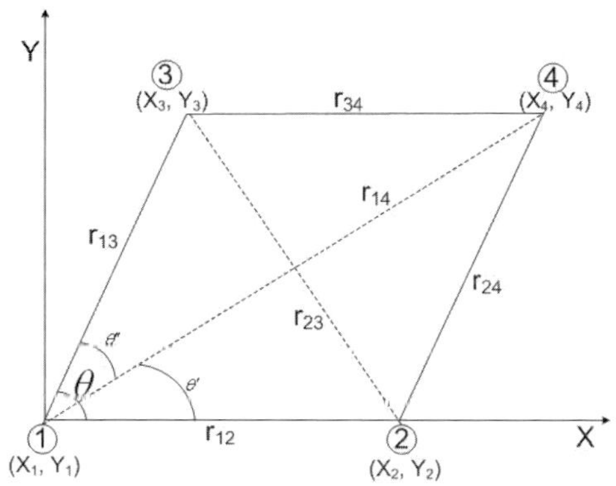

Fig. 14.3. Four UGS node case.

Using (14.8) to derive the force equation for node 1 results in:

$$f_1^x = -\frac{(r_{12}-\bar{r}_{12})}{r_{12}}(x_1 - x_2) - \frac{(r_{13}-\bar{r}_{13})}{r_{13}}(x_1 - x_3) - \frac{(r_{14}-\bar{r}_{14})}{r_{14}}(x_1 - x_4) \qquad (14.20)$$
$$f_1^y = -\frac{(r_{12}-\bar{r}_{12})}{r_{12}}(y_1 - y_2) - \frac{(r_{13}-\bar{r}_{13})}{r_{13}}(y_1 - y_3) - \frac{(r_{14}-\bar{r}_{14})}{r_{14}}(y_1 - y_4)$$

and on substituting the coordinate values from (14.19) yields:

$$f_1^x = -(r_{12}-\bar{r}_{12}) - (r_{13}-\bar{r}_{13})\cos(\theta) - (r_{14}-\bar{r}_{14})\cos(\theta') \qquad (14.21)$$
$$f_1^y = -(r_{13}-\bar{r}_{13})\sin(\theta) - (r_{14}-\bar{r}_{14})\sin(\theta')$$

The total force applied to node 1 is given as:

$$f_1^2 = (f_1^x)^2 + (f_1^y)^2 \qquad (14.22)$$

and by substituting (14.21) in (14.22) and simplifying the terms results in:

$$f_1^2 = a^2 + b^2 + c^2 + 2ab\cos(\theta) + 2ac\cos(\theta') + 2bc\cos(\theta'') \qquad (14.23)$$

where:

$$a = r_{12} - \bar{r}_{12};\quad b = r_{13} - \bar{r}_{13};\quad c = r_{14} - \bar{r}_{14} \qquad (14.24)$$

However, the force equation in (14.23) can also be obtained using the different configuration of nodes shown in Figure 14.4, in which the coordinates for the four nodes are given as:

$$x_1 = 0,\ y_1 = 0;\ x_2 = r_{12},\ y_2 = 0; \qquad (14.25)$$
$$x_3 = r_{13}\cos(\theta);\quad y_3 = r_{13}\sin(\theta)$$
$$x_4 = r_{14}\cos(\theta');\quad y_4 = -r_{14}\sin(\theta')$$

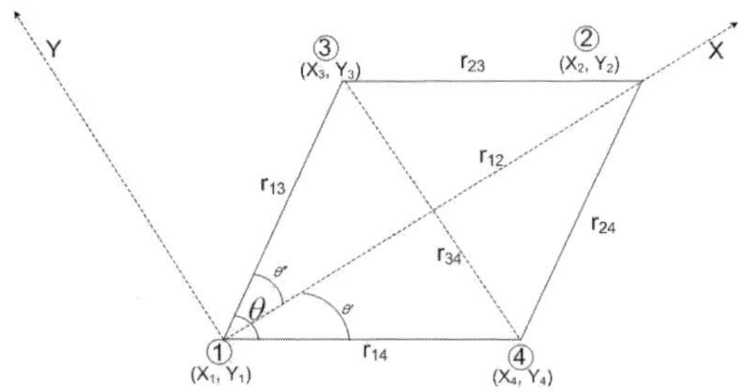

Fig. 14.4. Alternate configuration for four UGS nodes.

Thus, there are 2 minimum configurations for the sensor node to fall into local minima, one of which has $r_{ij} \neq \bar{r}_{ij}$.

14.3.2 Relative Localization Algorithm

Existence of local minima as mentioned in Lemma 14.2, with the potential field function defined in (14.4) can be resolved by the addition of only one node at a time to the sensor network when $N \geq 4$. According to Lemma 14.1, the first 3 nodes to arise in the UGS network attain a unique steady-state value for their position estimate. The final coordinates of these first 3 UGS nodes, already relatively localized, are used to calculate the initial starting point for the next UGS node to be added in the network.

The trilateration method [21] is used to obtain the initial starting point for the new UGS node. Use of only the trilateration process is inefficient for least squares localization of large wireless sensor nodes since, when a node is added, all other nodes may have to change their positions to reduce the potential occurred due to the measurement error. Here, trilateration is only used to get an initial position estimate for the new UGS node. Once the initial position estimate has been obtained, the algorithm is used to relatively localize the network with the control input in (14.8). This allows the relative position estimates of all the nodes in the network to be adjusted each time a node is added.

The starting point for UGS nodes when $N \geq 4$ is found using:

$$\begin{bmatrix} x_j \\ y_j \end{bmatrix} = \begin{bmatrix} 2(x_{j_1} - x_{j_3}) & 2(y_{j_1} - y_{j_3}) \\ 2(x_{j_2} - x_{j_3}) & 2(y_{j_2} - y_{j_3}) \end{bmatrix}^{-1} \begin{bmatrix} \bar{r}_{j_3 j_4}{}^2 - \bar{r}_{j_1 j_4}{}^2 + x_{j_1}{}^2 - x_{j_3}{}^2 + y_{j_1}{}^2 - y_{j_3}{}^2 \\ \bar{r}_{j_3 j_4}{}^2 - \bar{r}_{j_2 j_4}{}^2 + x_{j_2}{}^2 - x_{j_3}{}^2 + y_{j_2}{}^2 - y_{j_3}{}^2 \end{bmatrix} \tag{14.26}$$

where, (x_j, y_j) denotes the current estimate of the location of node j and $(x_{j_1}, y_{j_1}), (x_{j_2}, y_{j_2}), (x_{j_3}, y_{j_3})$ denotes the position estimates of the already localized nodes j_1, j_2, j_3 respectively.

14.3.3 Relative Localization Simulation Results

The relative localization algorithm mentioned above is simulated with seven UGS nodes in the sensor network, where each UGS node is assumed to have range measurement information with at least 3 other UGS nodes. The algorithm localizes the UGS nodes relative to each other and yields an optimal relative position estimates. The results are shown in Figure 14.5, where the estimates are plotted as functions of time given some incorrect initial conditions. One node is added at a time, and then all the estimates in the network are readjusted. The figure shows that the estimates converge to the true relative locations of the UGS.

Algorithm 1: Relative Localization Algorithm

1	Initialize $N = 3$ (number of UGS nodes in the network to start with the localization process);
2	Dynamically localize the network with $N = 3$ using (14.8) with range measurement information;
3	Increment N by 1 to keep a count of the number of sensor nodes;
4	Initialize the next sensor node position using (14.26);
5	Dynamically localize the network with the new sensor node using (14.8);
6	Repeat steps 3-5 until all the nodes have been localized.

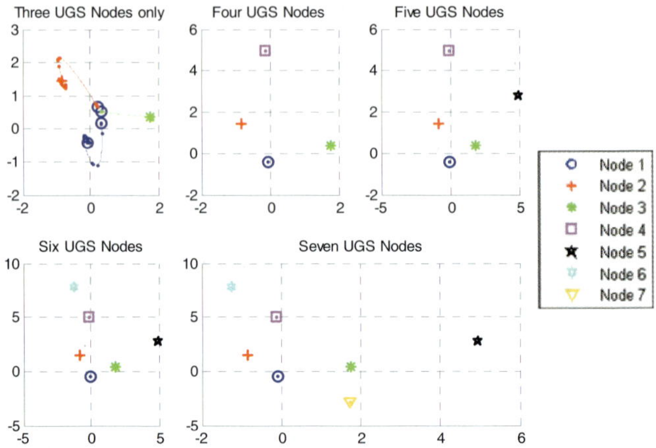

Fig. 14.5. Relative localization of 7 UGS nodes.

14.4 Absolute Position Localization for Air-Ground Wireless Sensor Networks

This Section discusses absolute localization of the UGS nodes with the help of UAVs. UAVs are assumed to have their absolute position information through on-board GPS or through contact with ground control station (GCS). Given the potential field defined in (14.4) for relative localization, a second potential field is added for absolute localization. The modified potential field is used as a Lyapunov function and a Lyapunov proof shows how to generate appropriate virtual forces based on the gradient of the modified potential field.

14.4.1 System Description

For the absolute localization algorithm the following assumptions are being made for UAVs, which are:

- UAVs have an altitude hold autopilot [7];
- UAVs operate in hover mode, so that they move over the stationary UGS network and assume a fixed position until the localization algorithm to be presented has been executed.

The virtual dynamics used for position estimates of the stationary UGS nodes are described based on range information and absolute position information of UAVs. Let the total number of UGS nodes and UAVs in the air-ground sensor network be N indexed by a set X_i; $i = \{1,2,3,.......N\}$. Let the number of UAVs with absolute position information be m, indexed by a set $X_{i_p}{}^a$; $\{p = 1,2,.....,m\}$ such that $X_{i_p}{}^a \subset X_i$ and the UGS nodes with no absolute position information be indexed by X_{i_p}; $\{p = m+1,.....,N\}$ such that $X_{i_p}{}^{ugs} \subset X_i$.

The position estimate for UGS nodes with no absolute position information is given as:

$$X_{i_p} = \begin{bmatrix} x_{i_p} & y_{i_p} \end{bmatrix}^T \tag{14.27}$$

where x_{i_p} and y_{i_p} are the x-y coordinates of the UGS node position estimates. The position estimation dynamics are given as:

$$\ddot{X}_{i_p} = \vec{f}_{i_p} \tag{14.28}$$

where, $\vec{f}_{i_p} = \begin{bmatrix} f_{i_p}{}^x & f_{i_p}{}^y \end{bmatrix}^T$ is a virtual force in the x and y directions to be specified.

The position estimate for a UAV with absolute position information is given as:

$$X_{i_p}{}^a = \begin{bmatrix} x_{i_p}{}^a & y_{i_p}{}^a \end{bmatrix}^T \tag{14.29}$$

where $x_{i_p}{}^a$ and $y_{i_p}{}^a$ are the x-y coordinates of the UAV position estimates.

The position estimation dynamics are given as:

$$\ddot{X}_{i_p}{}^a = \vec{f}_{i_p}{}^a \tag{14.30}$$

where:

$$\vec{f}_{i_p}{}^a = \begin{bmatrix} f_{i_p}{}^{ax} & f_{i_p}{}^{ay} \end{bmatrix}^T \tag{14.31}$$

is a virtual force in the x and y directions to be specified.

14.4.2 Potential Field for Optimal Position Estimation with Absolute Position Information

The potential field defined in (14.4) is now modified to incorporate the absolute position information available for UAVs. A new term is added to the already existing potential field for UGS nodes in (14.4) to obtain the potential field for UAVs, which is given as:

$$V_{uav} = \frac{1}{2}\sum_{p=1}^{m}\sum_{j=1}^{N} K_{i_p j}{}^{a}(r_{i_p j}{}^{a} - \bar{r}_{i_p j}{}^{a})^2 + \frac{1}{2}\sum_{p=1}^{m} K_{i_p}{}^{a}\left\|e_{i_p}{}^{a}\right\|^2 \tag{14.32}$$

where $e_{i_p}{}^{a} = \left[(x_{i_p}{}^{a} - \bar{x}_{i_p}{}^{a})^2 + (x_{i_p}{}^{a} - \bar{y}_{i_p}{}^{a})^2\right]^{1/2}$ and $\bar{X}_{i_p}{}^{a} = \left[\bar{x}_{i_p}{}^{a} \quad \bar{y}_{i_p}{}^{a}\right]$ is the known absolute position of UAV $i_p{}^{a}$. Therefore, the new potential field for the air-ground sensor network with UAVs and UGS nodes is now defined as:

$$V_p = V_{uav} + V_{ugs} \tag{14.33}$$

$$V_p = \frac{1}{2}\sum_{p=1}^{m} K_{i_p}{}^{a}\left\|e_{i_p}{}^{a}\right\|^2 + \frac{1}{2}\sum_{p=1}^{m}\sum_{j=1}^{N} K_{i_p j}{}^{a}(r_{i_p j}{}^{a} - \bar{r}_{i_p j}{}^{a})^2 + \frac{1}{2}\sum_{p=m+1}^{N}\sum_{j=1}^{N} K_{i_p j}(r_{i_p j} - \bar{r}_{i_p j})^2 \tag{14.34}$$

where $r_{i_p j}{}^{a} = [(x_{i_p}{}^{a} - x_j)^2 + (y_{i_p}{}^{a} - y_j)^2]^{1/2} = \left\|X_{i_p}{}^{a} - X_j\right\|$. Now, define the potential field for a single UAV $i_p{}^{a}$ with absolute position measurement by:

$$V_{i_p}{}^{a} = \frac{1}{2}K_{i_p}{}^{a}\left\|e_{i_p}{}^{a}\right\|^2 + \frac{1}{2}\sum_{\substack{j=1 \\ j \neq i_p}}^{N} K_{i_p j}{}^{a}(r_{i_p j}{}^{a} - \bar{r}_{i_p j}{}^{a})^2 \tag{14.35}$$

The potential V_{i_p} for a single UGS node i_p without absolute position measurement is the same as in (14.5). The gradient of the potential with respect to the UAV node states $i_p{}^{a}$ is given by:

$$\frac{\partial V_{i_p}{}^{a}}{\partial X_{i_p}{}^{a}} = K_{i_p}{}^{a}(X_{i_p}{}^{a} - \bar{X}_{i_p}{}^{a}) - \sum_{j=1}^{N} K_{i_p j}{}^{a}(r_{i_p j}{}^{a} - \bar{r}_{i_p j}{}^{a})\frac{(X_{i_p}{}^{a} - X_j)}{\left\|X_{i_p}{}^{a} - X_j\right\|} \tag{14.36}$$

whereas, the gradient of the potential $\left(\dfrac{\partial V_{i_p}}{\partial X_{i_p}}\right)$ with respect to the UGS node state i_p is same as in (14.7).

Based on the first assumption only x and y positions of the UAVs are considered, while the second assumption allows one to consider a UAV as a sensor node with absolute position information until the absolute localization algorithm has been executed.

Theorem 14.2: Consider the position estimate dynamics in (14.28) for each UGS node and (14.30) for each UAV with absolute position information. Let the virtual force for the UGS node and UAVs be given respectively as:

$$f_{i_p} = \sum_{j=1}^{N} K_{i_p j}(r_{i_p j} - \bar{r}_{i_p j})\frac{(X_{i_p} - X_j)}{\left\| X_{i_p} \quad X_j \right\|} - K_v \dot{X}_{i_p} \tag{14.37}$$

$$f_{i_p}{}^{a} = -K_{i_p}{}^{a}(X_{i_p}{}^{a} - \bar{X}_{i_p}{}^{a}) - \sum_{j=1}^{N} K_{i_p j}{}^{a}(r_{i_p j}{}^{a} - \bar{r}_{i_p j}{}^{a})\frac{(X_{i_p}{}^{a} - X_j)}{\left\| X_{i_p}{}^{a} - X_j \right\|} - K_v{}^{a}\dot{X}_{i_p}{}^{a} \tag{14.38}$$

Then the position estimates reach steady-state values that provide optimal estimates of the actual absolute localization of the nodes in the sense that V_p is minimized.

Proof: Define the Lyapunov function as:

$$L_p = V_p + \frac{1}{2}\sum_{p=1}^{m} \dot{X}_{i_p}{}^{a^T} \dot{X}_{i_p}{}^{a} + \frac{1}{2}\sum_{p=m+1}^{N} \dot{X}_{i_p}{}^{T} \dot{X}_{i_p} \tag{14.39}$$

Differentiate to obtain:

$$L_p = V_p + \sum_{p=1}^{m} \dot{X}_{i_p}{}^{a^T} \ddot{X}_{i_p}{}^{a} + \sum_{p=m+1}^{N} \dot{X}_{i_p}{}^{T} \ddot{X}_{i_p} \tag{14.40}$$

One can compute:

$$\dot{V}_p = \sum_{p=1}^{m} K_{i_p}{}^a \dot{X}_{i_p}{}^{a^T}(X_{i_p}{}^a - \bar{X}_{i_p}{}^a) + \sum_{p=1}^{m}\sum_{j=1}^{N} K_{i_pj}{}^a (r_{i_pj}{}^a - \bar{r}_{i_pj}{}^a)\dot{X}_{i_p}{}^{a^T}\frac{(X_{i_p}{}^a - X_j)}{\left\|X_{i_p}{}^a - X_j\right\|} \quad (14.41)$$
$$+ \sum_{p=m+1}^{N}\sum_{j=1}^{N} K_{i_pj}(r_{i_pj} - \bar{r}_{i_pj})\dot{X}_{i_p}{}^T \frac{(X_{i_p} - X_j)}{\left\|X_{i_p} - X_j\right\|}$$

and after further substitution one obtains:

$$\dot{L}_p = \sum_{p=1}^{m} K_{i_p}{}^a \dot{X}_{i_p}{}^{a^T}(X_{i_p}{}^a - \bar{X}_{i_p}{}^a) + \sum_{p=1}^{m}\sum_{j=1}^{N} K_{i_pj}{}^a (r_{i_pj}{}^a - \bar{r}_{i_pj}{}^a)\dot{X}_{i_p}{}^{a^T}\frac{(X_{i_p}{}^a - X_j)}{\left\|X_{i_p}{}^a - X_j\right\|} \quad (14.42)$$
$$+ \sum_{p=1}^{m} \dot{X}_{i_p}{}^{a^T} f_{i_p}{}^a + \sum_{p=m+1}^{N}\sum_{j=1}^{N} K_{i_pj}(r_{i_pj} - \bar{r}_{i_pj})\dot{X}_{i_p}{}^T \frac{(X_{i_p} - X_j)}{\left\|X_{i_p} - X_j\right\|} + \sum_{p=m+1}^{N} \dot{X}_{i_p}{}^T f_{i_p}$$

Moreover, looking at Theorem 14.1 and substituting the force inputs mentioned in (14.37) and (14.38) yields:

$$\dot{L}_p = -\sum_{p=1}^{m} \dot{X}_{i_p}{}^{a^T} K_v{}^a \dot{X}_{i_p}{}^a - \sum_{p=m+1}^{N} \dot{X}_{i_p}{}^T K_v \dot{X}_{i_p} \quad (14.43)$$

Clearly, $\forall\,(K_v, K_v{}^a) > 0$, $\dot{L}_p \le 0$ and the vector $\begin{bmatrix} X_{i_p} & \dot{X}_{i_p} \end{bmatrix}^T$ and $\begin{bmatrix} X_{i_p}{}^a & \dot{X}_{i_p}{}^a \end{bmatrix}^T$ is bounded which shows that the system is SISL. Evaluating $\ddot{L}_p$ results in:

$$\ddot{L}_p = -2\sum_{p=1}^{m} \dot{X}_{i_p}{}^{a^T} K_v{}^a \ddot{X}_{i_p}{}^a - 2\sum_{p=m+1}^{N} \dot{X}_{i_p}{}^T K_v \ddot{X}_{i_p} \quad (14.44)$$

which on substitution of (14.28) and (14.30) gives:

$$\ddot{L}_p = -2\sum_{p=1}^{m} \dot{X}_{i_p}{}^{a^T} K_v{}^a f_{i_p}{}^a - 2\sum_{p=m+1}^{N} \dot{X}_{i_p}{}^T K_v f_{i_p} \quad (14.45)$$

Using the result from the Lyapunov analysis that the vector $\begin{bmatrix} X_{i_p} & \dot{X}_{i_p} \end{bmatrix}^T$ and $\begin{bmatrix} X_{i_p}{}^a & \dot{X}_{i_p}{}^a \end{bmatrix}^T$ is bounded also yields that $\ddot{L}_p$ is also bounded. By Barbalat's Lemma [22] one deduces that $\dot{L}_p \to 0$ as $t \to \infty$, which yields $\dot{X}_{i_p}{}^a \to 0$ and $\dot{X}_{i_p} \to 0$ as $t \to \infty$. Therefore (14.28) shows that $\vec{f}_{i_p}$ goes to zero $\forall i_p$ with no absolute position information and (14.30) shows that $\vec{f}_{i_p}{}^{a}$ goes to zero $\forall i_p$ with absolute position information. Finally (14.37) and (14.38) shows that $\frac{\partial V_{i_p}}{\partial X_{i_p}} \to 0$ and $\frac{\partial V_{i_p}{}^{a}}{\partial X_{i_p}{}^{a}} \to 0$ respectively, so V_p reaches a minimum.

14.5 Absolute Position Algorithm

An algorithm is presented to dynamically localize the UGS network and uniquely determine the absolute position estimates of all the UGS nodes with the aid of at least three UAV with GPS. Before the algorithm is proposed, one should consider a potential problem of the UGS network being inverted (i.e. upside down) in its configuration.

In any network, if only one node has GPS information, the network can be uniquely localized with respect to an absolute coordinate frame modulo a rotation and an inversion (i.e. flip the net over). If two nodes have GPS, then the rotational uncertainty is removed, but the net can still be 'upside down'. Adding a third node with GPS removes this final uncertainty, and results in a correctly localized net in absolute coordinates.

To solve the 'upside down' problem for the sensor network, there must be at least 3 UAVs with absolute position information available. Unfortunately, when the third node with absolute position information is added to the wireless sensor network (WSN), the estimated positions in the WSN may already correspond to the inverted situation.

The upside down problem can be confronted as follows. According to the proposed algorithm, one UAV is added at a time to the air-ground sensor network. Let $i^{th}; j^{th}; k^{th}$ be the 3 UAVs to be added in that particular order to the network. Before the kth UAV is added, its initial position coordinates obtained from (14.26) using only range measurement is compared with its absolute position information and the error is defined as:

$$\varepsilon = \left\| X_{i_k}{}^{a} - \overline{X}_{i_k}{}^{a} \right\| \tag{14.46}$$

If $\varepsilon > \varepsilon_M$, (where ε_M is the known maximum range error depending on the types of system used), then the network is assumed to have improper position estimates due to the estimated network being inverted. To flip the estimated network positions upside down to the correct configuration, the orthogonal projection of all the UGS nodes already added to the network is taken on the line formed by the known positions of ith and j^{th} UAVs. The projection of the UGS nodes across that line [3] is given as:

$$\begin{bmatrix} x^o \\ y^o \end{bmatrix} = \frac{1}{a^2+b^2} \begin{bmatrix} b^2-a^2 & -2ab \\ -2ab & a^2-b^2 \end{bmatrix} \begin{bmatrix} x \\ y \end{bmatrix} - \begin{bmatrix} 2ac \\ 2ac \end{bmatrix} \tag{14.47}$$

where, (x, y) are the initial points and (x^o, y^o) are the final points after projection and:

$$a = y_j - y_i ; b = x_i - x_j ; c = (x_j - x_i) y_i - (y_j - y_i) x_i \tag{14.48}$$

Algorithm 2: Absolute Localization Algorithm

1 Relatively Localize the UGS nodes using Algorithm 1 for relative localization;
2 Initialize the UAV position using (14.26) from the range measurement information available;
  2.1 Increment i by 1 to keep a count of the number of UAVs with absolute position information;
    2.1.1 **if** $i = 3$
    2.1.2 use (46) to calculate ε
      2.1.2.1 **if** $\varepsilon > \varepsilon_M$
      2.1.2.2 Take orthogonal projection using (14.47) of UGS nodes on the line formed with first 2 UAVs having absolute position information;
      2.1.2.3 Dynamically localize the air-ground sensor network with (14.37) as the control input for UGS nodes and (14.38) as the control input for the third UAV;
      2.1.2.4 **end if**
    2.1.3 **else**
    2.1.4 Dynamically localize the air-ground sensor network with (14.37) as the control input for UGS nodes and (14.38) as the control input for UAV;
    2.1.5 **end if**
3 Repeat step 2 until all the 3 UAVs have been added.

14.5.1 Absolute Localization Simulation Results

Simulation results are presented for absolute localization. The air-ground sensor network model used for simulation consists of 7 UGS nodes and a single UAV with on-board GPS. Once the UGS nodes have been relatively localized using the algorithm mentioned in Section 14.3, the UAV hover over the terrain to absolutely localize the sensor network. The UAV stops at 3 different positions and at each position it measures the distance with at least 3 UGS nodes as shown in Figure 14.6.

Figure 14.7 shows the virtual movement of the UGS nodes when the UAV stops at 2 different positions. The UGS nodes move in a way to obtain its absolute coordinates. When the UAV moves to a third position, it estimates its position using (14.26) and then compares it with its position obtained from GPS. If the error obtained is more then ε_M then all the UGS nodes are reflected on the line formed by the points when UAV stopped at position 1 and 2 which is shown in Figure 14.8.

The reflection is done in order to solve the upside-down problem. Figure 14.9 shows the plot for the potential field function versus the number of iterations taken before the potential field converges to zero. The plot shows that the absolute localization algorithm estimates the UGS nodes position such that the desired distance is achieved.

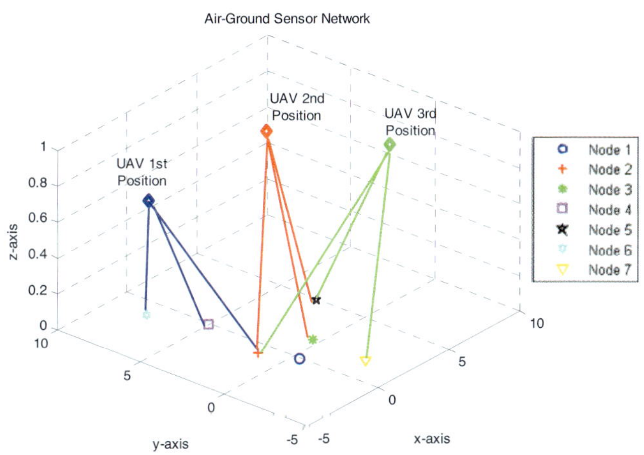

Fig. 14.6. Air-Ground sensor network.

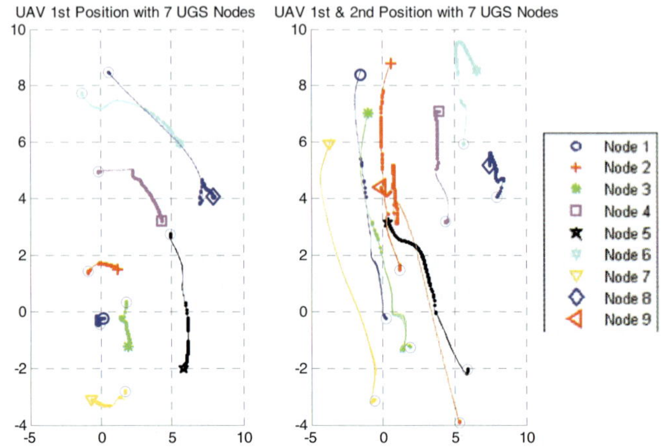

Fig. 14.7. UAV 1 and 2 with 7 UGS nodes.

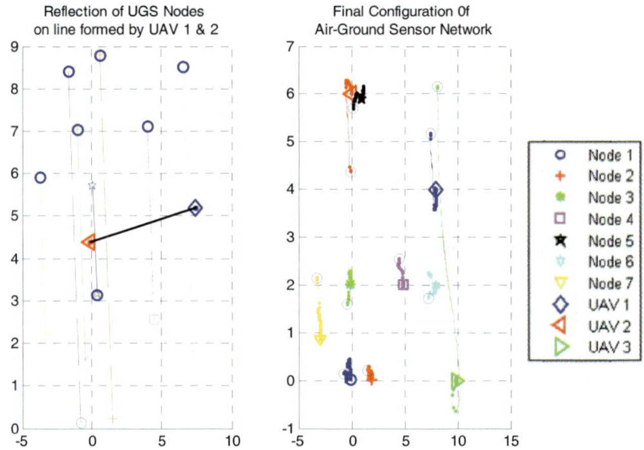

Fig. 14.8. Final configuration of UGS nodes.

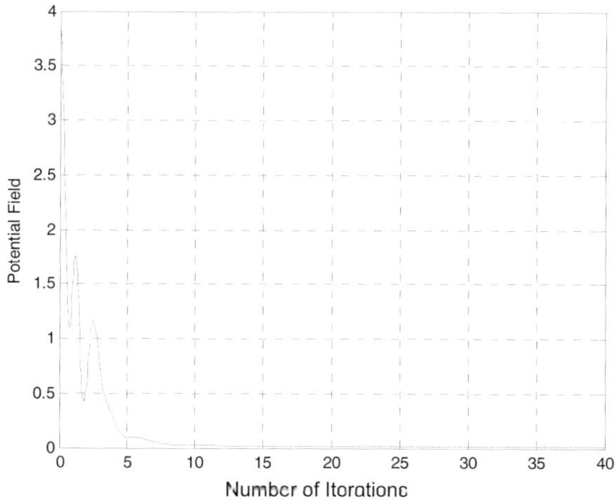

Fig. 14.9. Potential field function.

14.6 Conclusion

Efficient algorithms for relative and absolute localization are presented based on potential field methods, together with the mathematical analysis. The algorithms presented take care that the system does not fall into the local minima. The algorithm also takes into account the problem of the network being in an inverted configuration during absolute localization.

References

1. Ahmed A. A., Shi H., Shang Y., "SHARP. A New Approach to Relative Localization in Wireless Sensor Networks", *Proceedings, Distributed Computing Systems Workshops*, 892- 898, 2005.
2. Bachrach J., Taylor C., "Localization in Sensor Networks", *Handbook of Sensor Networks: Algorithms and Architectures*, Vol. 1, 2005.
3. Bronshtein I. N., Semendyayev K. A., Musiol G., Muehlig H., Mühlig H., *Handbook of Mathematics*, Springer; 4th Edition, 2003.
4. Bulusu N., Heidemann J., Estrin D., "GPS-less Low Cost Outdoor Localization for Very Small Devices", *Proceedings, 21st International Conference on Distributed Computing Systems*, Phoenix, AZ, 2001.

5. Bulusu N., Heidemann J., Estrin D., "Adaptive Beacon Placement", *Proceedings, 21st International Conference on Distributed Computing Systems*, Phoenix, AZ, 2001.
6. Bulusu N., Heidemann J., Estrin D., "GPS-less Low Cost Outdoor Localization for Very Small Devices", *IEEE Personal Communications Magazine*, Vol. 7, 28-34, 2005.
7. Castillo P., Lozano R, Dzul A., "Stabilization of a Mini Rotorcraft with Four Rotors", *IEEE Control System Magazine*, pp: 45-55, 2005.
8. Corke P., Peterson R., Rus D., "Coordinating Aerial Robots and Sensor Networks for Localization and Navigation", *Proceedings, 7th International Symposium Distributed Autonomous Robotic Systems*, Toulouse, France, 2004.
9. Khatib O., "Real-Time Obstacle Avoidance for Manipulators and Mobile Robots", *Proceedings, IEEE International Conference on Robotics and Automation*, Vol. 2, 500–505, 1985.
10. Kim J., Sukkarieh S., "Airborne Simultaneous Localization and Map Building", *Proceedings, Proceedings, IEEE International Conference on Robotics and Automation*, Vol. 1, 406–411, 2003.
11. Kim J., Sukkarieh S., "SLAM Aided GPS INS Navigation GPA Denied and Unknown Environments", *Proceedings, International Symposium on GNSS / GPS,* 2004.
12. Kim J., Ong L. L., Nettleton E., Sukkarieh S., "Decentralized Approach to Unmanned Aerial Vehicle Navigation: Without the Use of the Global Positioning System and Preloaded Maps", *Proceedings, I MECH E Part G Journal of Aerospace Engineering*, Vol. 218, Issue 6, 399-416 (18), 2004.
13. Langelaan J., Rock S., "Navigation of Small UAVs Operating in Forests", *Proceedings of the AIAA Guidance, Navigation and Control Conference*, 2004.
14. Lewis F. L., *Optimal Estimation*, John Wiley & Sons, 1986.
15. Luiz C., Grocholsky B., Keller J. F., Kumar V., Taylor C. J., "Experiments in Multirobot Air-Ground Coordination", *Proceedings, IEEE International Conference on Robotics and Automation*, New Orleans, LA, 2004.
16. Moses R. L., Krishnamurthy D., Patterson R. M., "A Self-Localization Method for Wireless Sensor Networks", *EURASIP Journal on Applied Signal Processing*, 2002.
17. Patwari N., Ash J. N., Kyperountas S., Hero A. O., Moses R. L., Correal N. S., "Locating the Nodes: Cooperative Localization in Wireless Sensor Networks", *IEEE Signal Processing Magazine*, Vol. 22, 54-69, 2005.
18. Paul A. S., Wan E. A., "Dual Kalman Filters for Autonomous Terrain Aided Navigation in Unknown Environments", *Proceedings, IEEE International Joint Conference on Neural Networks*, Vol. 5, 2784 -2789, 2005.
19. Savarese C., Rabaey J. M., Beutel J., "Locationing in Distributed Ad-hoc Wireless Sensor Networks", *Proceedings, IEEE International Conference on Acoustics, Speech, and Signal Processing*, 2001.
20. Semic S., Sastry S., *Distributed Localization in Wireless Sensor Networks*, Technical Report UCB/ERL M02/26, UC Berkeley, 2001.

21. Shi Q., Huo H., Fang T., Li D., "Using Space Distance Intersection for Node Location Computation in Wireless Sensor Networks", *Proceedings, International Workshop on Sensor Network and Applications*, 2005.
22. Slotine J., Li W., *Applied Nonlinear Control*, Prentice Hall.
23 Van Blyenburgh P., "UAVs: An Overview", *Air & Space Europe*, Vol. 1, Issue 5, 43-47 (5), Elsevier Science, 1999.
24 Zou Y., Chakrabarty K., "Sensor Deployment and Target Localization Based on Virtual Forces", *Proceedings, 22nd Annual Joint Conference of the IEEE Computer and Communications Societies*, Vol. 2, 1293 – 1303, 2003.

Chapter 15: Decentralized Formation Tracking of Multi-Vehicle Systems with Consensus-Based Controllers[1]

In the problem of formation tracking, multiple unmanned vehicles are required to follow spatial trajectories while keeping a desired inter-vehicle formation pattern in time. This Chapter considers vehicles with nonlinear dynamics that follow very general trajectories generated by some reference vehicles. Formations are specified using vectors of relative positions of neighboring vehicles and using consensus-based controllers in the context of decentralized formation tracking control. The key idea is to combine consensus-based controllers with the cascaded approach to tracking control, resulting in a group of linearly coupled dynamical systems. Two types of tracking controllers are proposed under different information flow topologies. Their stability properties are examined by using nonlinear synchronization theory. Simulation results are presented to illustrate the proposed method. The major advantage of the approach is that it is applicable to both unmanned ground vehicles, as well as aerial vehicles flying at a certain altitude. As such, the Chapter refers to 'unmanned mobile vehicles' in general.

15.1 Introduction

Control problems involving unmanned mobile vehicles have attracted considerable attention in the control community during the past decade. One of the basic motion tasks assigned to a mobile vehicle may be formulated as following a given trajectory [13] [25]. The trajectory tracking problem was globally solved in [20] by using a time-varying continuous feedback

---

[1] Written by L. Fang, P. J. Antsaklis

Kimon P. Valavanis (ed.), Advances in Unmanned Aerial Vehicles, 455–471.

law, and in [2] [12] [16] through the use of dynamic feedback linearization. The backstepping technique for trajectory tracking of nonholonomic systems in chained form was developed in [6] [10]. In the special case when the vehicle model has a *cascaded structure*, the higher dimensional problem can be decomposed into several lower dimensional problems that are easier to solve [17].

An extension to the traditional trajectory tracking problem is that of *coordinated tracking* or *formation tracking* as shown in Figure 15.1. The problem is often formulated as to find a coordinated control scheme for multiple unmanned vehicles that forces them to maintain some given, possibly time-varying, formation while executing a given task as a group. The possible tasks could range from exploration of unknown environments where an increase in numbers could potentially reduce the exploration time, navigation in hostile environments where multiple vehicles make the system redundant and thus robust, to coordinated path following. Detailed information may be found in recent survey papers [1] [21].

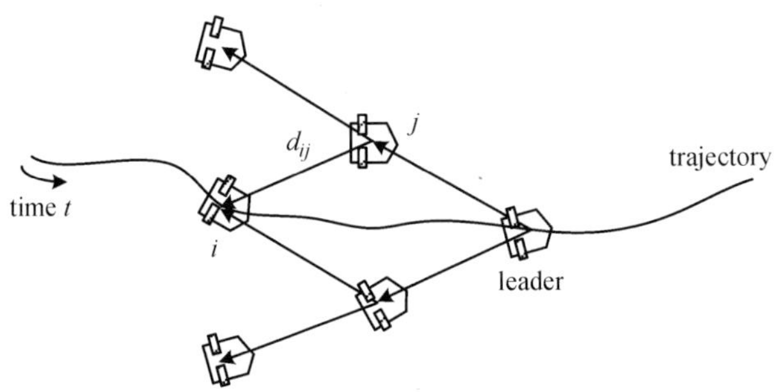

Fig. 15.1. Six unmanned vehicles perform a formation tracking task.

In formation control of multi-vehicle systems, different control topologies can be adopted depending on applications. There may be one or more leaders in the group, with other vehicles following one or more leaders in a specified way. In many scenarios, vehicles have limited communication ability. Since global information is often not available to each vehicle, distributed controllers using only local information are desirable. One approach to distributed formation control is to represent formations using the vectors of relative positions of neighboring vehicles and the use of consensus-based controllers with input bias [3] [11].

In this Chapter, the formation tracking problem for a group of vehicles is studied using the consensus-based controllers combined with the cascade approach [17]. The idea is to specify a reference path for a given, nonphysical point. Then a multiple vehicle formation, defined with respect to the real vehicles as well as to the nonphysical virtual leader, should be maintained at the same time as the virtual leader tracks its reference trajectory. The vehicles exchange information according to a communication digraph, *G*. Similar to the tracking controller in [17], the controller for each vehicle can be decomposed to two 'sub-controllers', one for positioning and one for orientation. Different from the traditional single vehicle tracking case, each vehicle uses information from its neighbors in the communication digraph to determine the reference velocities and stay at their designation in the formation. Based on nonlinear synchronization results [27], it is proven that consensus-based formation tracking can be achieved as long as the formation graph had a spanning tree and the controller parameters are large enough; they can be lower-bounded by a quantity determined by the formation graph.

Related work includes [4] [5] [9] [19] [22]. In [9], the vehicle dynamics were assumed to be linear and formation control design was based on algebraic graph theory. In [19], output feedback linearization control was combined with a second-order (linear) consensus controller to coordinate the movement of multiple mobile vehicles. The problem of vehicles moving in a formation along constant or periodic trajectories was formulated as a nonlinear output regulation (servomechanism) problem in [4]. The solutions adopted in [5] [22] for coordinated path following control of multiple marine vessels or wheeled vehicles built on Lyapunov techniques, where path following and inter-vehicle coordination were decoupled. Detailed information on consensus problems in networked systems may be found in [15] [18].

The proposed approach offers two key contributions: i) The consensus-based formation tracking controller for nonlinear vehicles is novel and its stability properties are examined using cascaded systems theory and nonlinear synchronization theory; ii) Global results allow one to consider a large class of trajectories with arbitrary (rigid) formation patterns and initial conditions.

Further, a novelty of this research that should not be overlooked is that the formation tracking in a 2-D setting studied in this Chapter includes hovercraft coordinating on a flat surface [7] or UAV flying at a constant altitude. Thus, the methodology proposed is easily extended and applied to UAV formation tracking in more general settings.

15.2 Preliminaries

15.2.1 Tracking Control of Unmanned Mobile Vehicles

A kinematics model of a hovercraft with two degrees of freedom is given by the following equations:

$$\dot{x} = v\cos\theta,\ \dot{y} = v\sin\theta,\ \dot{\theta} = \omega \tag{15.1}$$

where the forward velocity v and the angular velocity ω are considered as inputs, (x, y) is the center of the rear axis of the vehicle, and θ is the angle between heading direction and x-axis as shown in Figure 15.2.

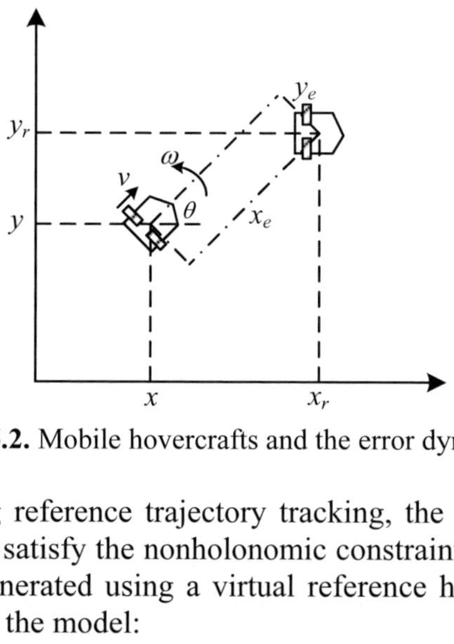

Fig. 15.2. Mobile hovercrafts and the error dynamics.

For time-varying reference trajectory tracking, the reference trajectory must be selected to satisfy the nonholonomic constraint. The reference trajectory is hence generated using a virtual reference hovercraft [8] which moves according to the model:

$$\dot{x}_r = v_r\cos\theta_r,\ \dot{y}_r = v_r\sin\theta_r,\ \dot{\theta}_r = \omega_r \tag{15.2}$$

where $[x_r\ y_r\ \theta_r]$ is the reference posture obtained from the virtual vehicle. Following [8] the error coordinates are defined as (Figure 15.2):

$$p_e = \begin{bmatrix} x_e \\ y_e \\ \theta_e \end{bmatrix} = \begin{bmatrix} \cos\theta & \sin\theta & 0 \\ -\sin\theta & \cos\theta & 0 \\ 0 & 0 & 1 \end{bmatrix} \begin{bmatrix} x_r - x \\ y_r - y \\ \theta_r - \theta \end{bmatrix} \tag{15.3}$$

It can be verified that in these coordinates the error dynamics become:

$$\dot{p}_e = \begin{bmatrix} \dot{x}_e \\ \dot{y}_e \\ \dot{\theta}_e \end{bmatrix} = \begin{bmatrix} \omega y_e - v + v_r \cos\theta_e \\ -\omega x_e + v_r \sin\theta_e \\ \omega_r - \omega \end{bmatrix}. \tag{15.4}$$

The aim of (single hovercraft) trajectory tracking is to find appropriate velocity control laws v and ω of the form:

$$\begin{aligned} v &= v(t, x_e, y_e, \theta_e) \\ \omega &= \omega(t, x_e, y_e, \theta_e) \end{aligned} \tag{15.5}$$

such that the closed-loop trajectories of (15.4) and (15.5) are stable in some sense (e.g., uniform globally asymptotically stable). As discussed in Section 15.1, there are numerous solutions to this problem in the continuous time domain. Here, the cascaded approach proposed in [17] is revisited. As a starting point, the notion of globally K-exponential stability is introduced.

Definition 15.1: A continuous function α: $[0, a) \to [0, \infty)$ is said to belong to class K if it is strictly increasing and $\alpha(0) = 0$.

Definition 15.2: A continuous function β: $[0, a)\times[0,\infty\ \infty) \to [0,\infty\ \infty)$ is said to belong to class KL if for each fixed s the mapping $\beta(r, s)$ belongs to class K with respect to r, and for each fixed r the mapping $\beta(r, s)$ is decreasing with respect to s and $\beta(r, s) \to 0$ as $s \to \infty$.

Definition 15.3: Consider the system:

$$\dot{x} = g(t, x),\ g(t, 0) = 0\ \forall t \geq 0 \tag{15.6}$$

where $g(t, x)$ is piecewise continuous in t and locally Lipschitz in x.

The system (15.6) is called *globally K-exponentially stable* if there exist $\xi > 0$ and a class K function $k(\cdot)$ such that:

$$\|x(t)\| \leq k(\|x(t)\|) e^{-\xi(t-t_0)}.$$

Theorem 15.1 ([17]): Consider the system (15.4) in closed-loop with the controller:

$$\begin{aligned} v &= v_r + c_2 x_e, \\ \omega &= \omega_r + c_1 \theta_e, \end{aligned} \tag{15.7}$$

where $c_1 > 0$ $c_2 > 0$. If $\omega_r(t)$, $\dot{\omega}_r(t)$, and $v_r(t)$ are bounded and there exist δ and k such that:

$$\int_t^{t+\delta} \omega_r(\tau)^2\, d\tau \geq k,\ \forall t \geq t_0 \tag{15.8}$$

then the closed-loop system (15.4) and (15.7), written compactly as:

$$\dot{p}_e = h(x_e, y_e, \theta_e)\big|_{v_r,\omega_r} = h(p_e)\big|_{v_r,\omega_r} \tag{15.9}$$

is globally K-exponentially stable.

In the above, the subscriptions for $h(\cdot)\big|_{v_r,\omega_r}$ mean that the error dynamics are defined relative to reference velocities v_r and ω_r. The tracking condition (15.8) implies that the reference trajectories should not converge to a point (or straight line).

This also relates to the well-known persistence-of-excitation condition in adaptive control theory. Note that control laws in (15.7) are linear with respect to x_e and θ_e. This is critical in designing consensus-based controller for multiple vehicle formation tracking as we shall see below.

15.2.2 Formation Graphs

Formations are considered that can be represented by acyclic directed graphs. In these graphs, the agents involved are identified by vertices and the leader-following relationships by (directed) edges. The orientation of each edge distinguishes the leader from the follower. Follower controllers implement static state feedback-control laws that depend on the state of the particular follower and the states of its leaders.

Definition 15.4 ([24]): A formation control graph $G = (V, E, D)$ is a directed acyclic graph consisting of the following.

- A finite set $V = \{v_1, \ldots, v_N\}$ of N vertices and a map assigning to each vertex a control system $\dot{x}_i = f_i(t, x_i, u_i)$ where $x_i \in \mathrm{R}^n$ and $u_i \in \mathrm{R}^m$.
- An edge set encoding leader-follower relationships between agents. The ordered pair $(v_i, v_j) \equiv e_{ij}$ belongs to E if u_j depends on the state of agent i, x_i.
- A collection $D = \{d_{ij}\}$ of edge specifications, defining control objectives (setpoints) for each j: $(v_i, v_j) \in E$ for some $v_i \in V$.

For agent j, the tails of all incoming edges to vertex represent leaders of j, and their set is denoted by $L_j \subset V$. Formation leaders (vertices of in-degree zero) regulate their behavior so that the formation may achieve some group objectives, such as navigation in obstacle environments or tracking reference paths.

Given a specification d_{kj} on edge $(v_k, v_j) \in E$, a set point for agent j can be expressed as $x_j^r = x_k - d_{kj}$. For agents with multiple leaders, the specification redundancy can be resolved by projecting the incoming edges specifications into orthogonal components:

$$x_j^r = \sum_{k \in L_j} S_{kj}\left(x_k - d_{kj}\right) \tag{15.10}$$

where S_{kj} are projection matrices with $\sum_k \text{rank}(S_{kj}) = n$. Then the error for the closed-loop system of vehicle j is defined to be the deviation from the prescribed set point $\tilde{x}_j = x_j^r - x_j$, and the formation error vector is constructed by stacking the errors of all followers:

$$\tilde{x} \equiv [\cdots \tilde{x} \cdots]^T,\ v_j \in V \setminus L_F .$$

15.2.3 Synchronization in Networks of Nonlinear Dynamical Systems

Definition 15.5: Given a matrix $V \in \mathrm{R}^{n \times n}$, a function $f(y, t) : \mathrm{R}^{n+1} \to \mathrm{R}^n$ is V-uniformly decreasing if $(y-z)^T V\left(f(y,t) - f(z,t)\right) \le -\mu \|y-z\|^2$ for some $\mu > 0$ and all $y, z \in \mathrm{R}^n$ and $t \in \mathrm{R}$.

Note that a differentiable function $f(y, t)$ is V-uniformly decreasing if and only if $V(\partial f(y)/\partial y) + \delta I$ for some $\delta > 0$ and all y, t. Consider the following synchronization result for the coupled network of identical dynamical systems with state equations:

$$\dot{x} = \left(f(x_1,t),\ldots,f(x_n,t)\right)^T + \left(C(t) \otimes D(t)\right)x + u(t), \tag{15.11}$$

where $x = (x_1, \ldots, x_N)^T$, $u = (u_1, \ldots, u_N)^T$ and $C(t)$ is a zero sums matrix for each t. $C \otimes D$ is the Kronecker product of matrices C and D.

Theorem 15.2 ([27]): Let $Y(t)$ be an n by n time-varying matrix and V be an n by n symmetric positive definite matrix such that $f(x, t) + Y(t)x$ is V-uniformly decreasing. Then the network of coupled dynamical systems in (11) synchronizes in the sense that $\|x_i - x_j\| \to 0$ as $t \to \infty$ for all i, j if the following two conditions are satisfied:

- $\lim_{t \to \infty} \|u_i - u_j\| = 0$ for all i, j;
- There exists an N by N symmetric irreducible zero row sums matrix U with nonpositive off-diagonal elements such that:

$$\left(U \otimes V\right)\left(C(t) \otimes D(t) - I \otimes Y(t)\right) \le 0 \tag{15.12}$$

for all t.

15.3 Basic Formation Tracking Controller

The control objective is to solve a formation tacking problem for N unmanned vehicles. This implies that each unmanned vehicle must converge to and stay at their designation in the formation while the formation as a whole follows a virtual vehicle. Equipped with the results presented in the previous Section, at first one should construct a basic formation tracking controller (FTC) from (15.7). Let $d_{ri} = [d_{x_{ri}} \;\; d_{y_{ri}}]^T$ denote the formation specification on edge (v_r, v_i). In virtue of linear structures of (15.7), the following basic FTC is proposed for vehicle i:

$$\begin{cases} v_i = v_r + c_2 x_{e_i} \\ \omega_i = \omega_r + c_1 \theta_{e_i} \end{cases} \tag{15.13}$$

where $c_1 > 0$, $c_2 > 0$ and:

$$p_{e_i} = [x_{e_i} \;\; y_{e_i} \;\; \theta_{e_i}]^T = \begin{bmatrix} \cos\theta_i & \sin\theta_i & 0 \\ -\sin\theta_i & \cos\theta_i & 0 \\ 0 & 0 & 1 \end{bmatrix} \begin{bmatrix} x_r - x_i - d_{x_{ri}} \\ y_r - y_i - d_{y_{ri}} \\ \theta_r - \theta_i \end{bmatrix} \tag{15.14}$$

Remark 15.1: It is not required to have constraints for every pair of vehicles. We need only a sufficient number of constraints which uniquely determine the formation.

Theorem 15 3: The basic FTC (15.13) and (15.14) solves the formation tracking problem.

Proof: By Theorem 15.1, every vehicle i follows the virtual (or leader) vehicle, thus the desired trajectory, with a formation constraint d_{ri} on edge (v_r, v_i). Therefore, all vehicles track the reference trajectory while staying in formation, which is specified by formation constraints d_{ri}'s as shown in Figure 15.3.

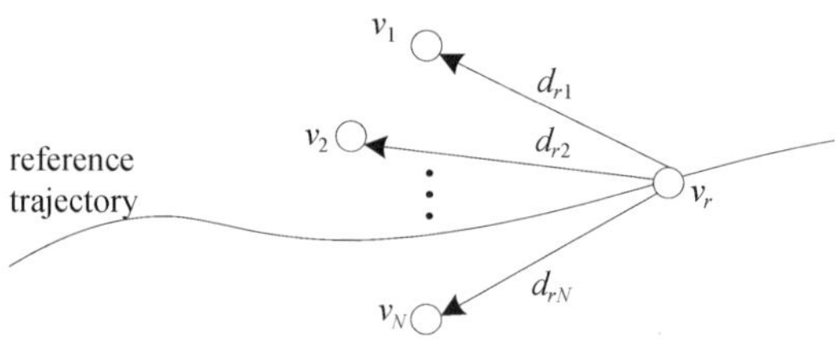

Fig. 15.3. Illustration of formation tracking using baseline FTC. The reference vehicle sends to vehicle i the formation specification d_{ri} as well as the reference velocities v_r and ω_r.

Corollary 15.1: Suppose only vehicle 1 follows the virtual vehicle. The composite system with inputs v_r and ω_r and states $\tilde{x}_1 = [x_{e_1}\ y_{e_1}\ \theta_{e_1}]$ is globally K-exponentially stable and therefore formation input-to-state stable (see Section 15.4).

Example 15.1-Basic FTC: Consider a system consisting of three vehicles, which are required to move in some predefined formation pattern. First, as in [4], consider the case of moving in a triangle formation along a circle. That is, the virtual (or reference) vehicle dynamics are given by:

$$x_r = v_r \cos(\omega r t) + x_{r0}, y_r = v_r \sin(\omega r t) + y_{r0}$$

where v_r is the reference forward velocity, ω_r the reference angular velocity, and $[x_{r0}\ y_{r0}]^T$ the initial offsets.

Assume that that parameters have the following values: v_r = 10, ω_r = 0.2, $[x_{r0}\ y_{r0}]^T = [-25\ 0]^T$. For simulation purposes, an isosceles right triangle was used with sides equal to $3\sqrt{2}$, $3\sqrt{2}$, and 6. Also fixed was the position of the virtual leader at the vertex with the right angle. Then, from the above constraints the required (fixed) formation specifications for the vehicles are given by $d_{r1} = [0\ \ 0]^T$, $d_{r2} = [3\ \ 3]^T$, $d_{r3} = [3\ \ -3]^T$.

For the basic FTC parameters were chosen as $c_1 = 0.3$ and $c_2 = 0.5$. Figure 15.4 shows the trajectories of the system for about 100 seconds. Initially the vehicles are not in the required formation; however, they form the formation quite fast (K-exponentially fast) while following the reference trajectory (solid line in the figure). Figure 15.5 shows the control signals v and ω for each vehicle.

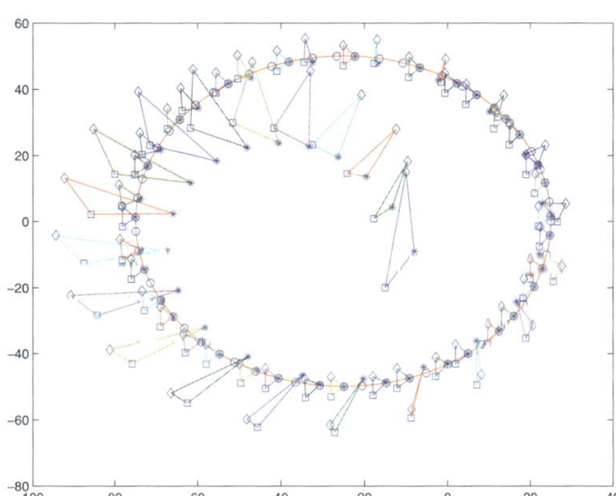

Fig. 15.4. Circular motion of three vehicles with a triangle formation. Initial vehicle postures are: $[-8\ -9\ 3\pi/5]^T$ for vehicle 1 (denoted as *); $[-15\ -20\ \pi/2]^T$ for vehicle 2 (□ - square); $[-10\ -15\ \pi/3]^T$ for vehicle 3 (◊ - diamond).

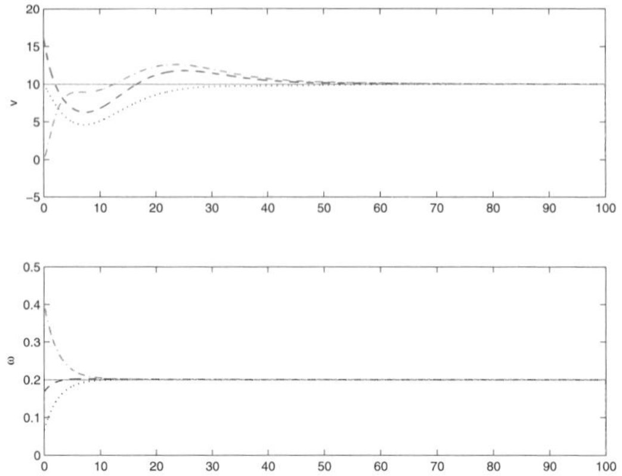

Fig. 15.5. Control signals v and ω for virtual vehicle: solid line; vehicle 1: dotted line; vehicle 2: dashed line; and vehicle 3: dot-dash line.

15.4 Consensus-Based Formation Tracking Controller

The basic FTC has the advantage that it is simple and leads to globally stabilizing controllers. A disadvantage, however, is that it requires every vehicle to get access to the reference velocities v_r and ω_r. This further implies that the reference vehicle needs to establish direct communication links with all other vehicles in the group, which may not be practical in some applications.

In a more general setting, one may assume that only a subset of vehicles (leaders) have direct access to the reference velocities. Other vehicles (followers) use their neighboring leaders' information to accomplish the formation tracking task. In this case, formation tracking controllers operate in a decentralized fashion since only neighboring leaders' information has been used.

Therefore, the consensus-based FTC for vehicle i is defined as follows:

$$\begin{cases} v_i = v_{r_i} + c_2 x_{e_i} + \sum_{j\in L_i} a_{ij}(x_{e_i} - x_{e_j}), \\ \omega_i = \omega_{r_i} + c_1 \theta_{e_i} + \sum_{j\in L_i} a_{ij}(\theta_{e_i} - \theta_{e_j}), \\ \dot{v}_{r_i} = \sum_{j\in L_i} a_{ij}(v_{r_j} - v_{r_i}), \\ \dot{\omega}_{r_i} = \sum_{j\in L_i} a_{ij}(\omega_{r_j} - \omega_{r_i}) \end{cases} \tag{15.15}$$

where:

$$p_{e_i} = \begin{bmatrix} x_{e_i} \\ y_{e_i} \\ \theta_{e_i} \end{bmatrix} = \begin{bmatrix} \cos\theta_i & \sin\theta_i & 0 \\ -\sin\theta_i & \cos\theta_i & 0 \\ 0 & 0 & 1 \end{bmatrix} \begin{bmatrix} x_i^r - x_i \\ y_i^r - y_i \\ \theta_i^r - \theta_i \end{bmatrix}.$$

and a_{ij} represents relative confidence of agent i in the information state of agent j.

Remark 15.2: As one can see from (15.15), the communication between vehicles is local and distributed, in the sense that each vehicle receives the posture and velocity information only from its neighboring leaders.

The following theorem is proven regarding the stability of the consensus-based FTC.

Theorem 15.4: The consensus-based FTC (15.15) solves the formation tracking problem if the formation graph G has a spanning tree and the controller parameters c_1, $c_2 > 0$ are large enough. Lower bounds for c_1 and c_2 are related to the Laplacian matrix for G.

Proof: Let L_G be the Laplacian matrix induced by the formation graph G and it is defined by:

$$(L_G)_{ij} = \begin{cases} \sum_{k=1,k\neq i}^{N} a_{ik}, & j = i \\ -a_{ij}, & j \neq i \end{cases}$$

with $P_e = [p_{e_1}, \ldots, p_{e_N}]^T \in \mathrm{R}^{3N}$, $[V_r \quad \Omega_r]^T = [v_{r_1}, \ldots, v_{r_N}, \omega_{r_1}, \ldots, \omega_{r_N}]^T \in \mathrm{R}^{2N}$. The closed loop system (15.15) - (15.4) for all vehicles can be expressed in a compact form as:

$$\dot{P}_e = \begin{bmatrix} h(p_{e_1})\big|v_{r_1}, \omega_{r_1} \\ \vdots \\ h(p_{e_N})\big|v_{r_N}, \omega_{r_N} \end{bmatrix} + (-L_G \otimes D)P_e, \tag{15.16}$$

$$\begin{bmatrix} \dot{V}_r \\ \dot{\Omega}_r \end{bmatrix} - (-L_G \otimes I_2)\begin{bmatrix} V_r \\ \Omega_r \end{bmatrix}, \tag{15.17}$$

where:

$$D = \begin{bmatrix} -1 & 0 & 0 \\ 0 & 0 & 0 \\ 0 & 0 & -1 \end{bmatrix} \tag{15.18}$$

describes the specific coupling between two vehicles.

It can be seen that (15.17) is in the form of linear consensus algorithms. Since the formation graph has a rooted spanning tree (with the root corresponding to the virtual vehicle), the reference velocities (coordination variables) $v_{r_i}(t)$ and $\omega_{r_i}(t)$ for any vehicle i in the group will approach $v_r(t)$ and $\omega_r(t)$, respectively, but with bounded tracking errors [14]. For an easy exposition, one may consider the tracking errors to be zero in this proof, and defer the discussion of its implication to the end of this Section.

Therefore, (15.16) may be re-written as:

$$\dot{P}_e = \begin{bmatrix} h(p_{e_1})|v_r,\omega_r \\ \vdots \\ h(p_{e_N})|v_r,\omega_r \end{bmatrix} + (-L_G \otimes D)P_e + \begin{bmatrix} \phi_1(t) \\ \vdots \\ \phi_N(t) \end{bmatrix} \tag{15.19}$$

and $\phi_i(t) \rightarrow 0$ as $t \rightarrow \infty$. The functions ϕ_i can be considered as residual errors that occurred when replacing v_{r_i} and ω_{r_i} in (15.16) with v_r and ω_r, respectively. Now (15.19) is in the same form with (15.12). Further, set $Y = \alpha D$ so that $h(p_e)+ \alpha Dp_e$ is V-uniformly decreasing (see Lemma 11 in [26]) provided that $c_1 - \alpha > 0$ and $c_2 - \alpha > 0$. Theorem 15.2 states that (15.19) synchronizes if there exists a symmetric zero row sums matrix U with non-positive off-diagonal elements such that $(U \otimes V)(-L_G \otimes D - I \otimes Y) \leq 0$. Since $VD \leq 0$ and $Y = \alpha D$, this is equivalent to:

$$U(-L_G - \alpha I) \geq 0\,. \tag{15.20}$$

Let $\mu(-L_G)$ be the supremum of all real numbers such that $U(-L_G - \alpha I) \geq 0$. It was shown in [28] that $\mu(-L_G)$ exists for constant row sum matrices and can be computed by a sequence of semi-definite programming problems. Choose c_1 and c_2 to be large enough such that:

$$\min\{c_1, c_2\} > -\mu(-L_G) \tag{15.21}$$

and the proof is complete.

In particular, an upper bound for $\mu(-L_G)$ is given by $\mu_2(-L_G) = \min Re(\lambda)$ where $Re(\lambda)$ is the real part of λ, the eigenvalues of $-L_G$ that do not correspond to the eigenvector e. It suffices to make $\min\{c_1, c_2\} > \mu_2(-L_G)$.

Example 15.2: In this example, virtual vehicle dynamics are of a sinusoidal form: $(x_r(t), y_r(t)) = (t, \sin(t))$. The acyclic formation graph with formation specifications is shown in Figure 15.6. The (un-weighted) Laplacian matrix that corresponds to Figure 15.6 is given by:

$$L_G = \begin{bmatrix} 1 & -1 & 0 & 0 \\ 0 & 2 & -1 & -1 \\ 0 & 0 & 1 & -1 \\ 0 & 0 & 0 & 0 \end{bmatrix}. \tag{15.22}$$

Since $\mu_2(-L_G) = -2$, consensus-based FTC (15.15) was used with positive c_1, c_2, say $c_1 = 0.3$ and $c_2 = 0.5$. As shown in Figure 15.7, successful formation tracking with a desired triangle formation is achieved. Vehicle control signals v_i's and ω_i's are shown in Figure 15.8.

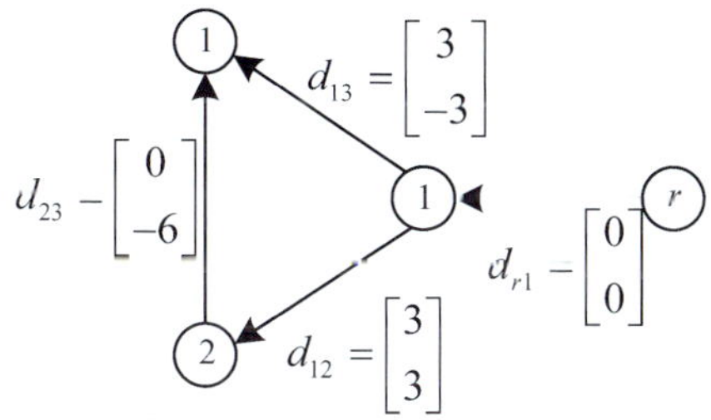

Fig. 15.6. A formation graph with formation specifications on edges: $d_{r1} = [0\ 0]^T$, $d_{12} = [3\ 3]^T$, $d_{13} = [3\ \text{-}3]^T$, $d_{23} = [0\ \text{-}6]^T$.

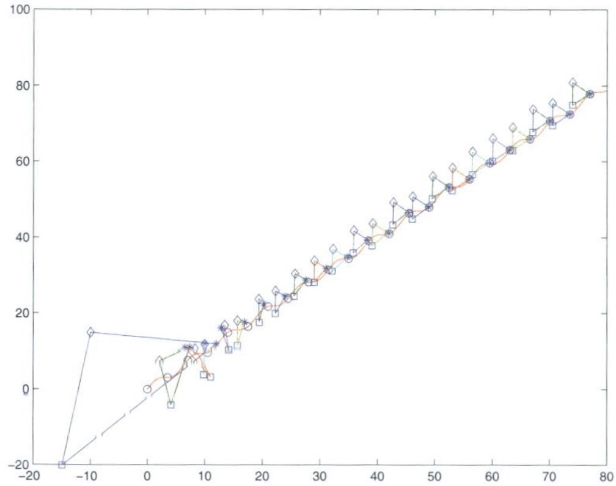

Fig. 15.7. Tracking a sinusoidal trajectory in a triangle formation. Initial vehicle postures are: $[12\ 12\ 0]^T$ for vehicle 1 (denoted as *); $[-15\ -20\ \pi/4]^T$ for vehicle 2 (□ - square); $[-10\ 15\ -\pi/4]^T$ for vehicle 3 (◊ - diamond).

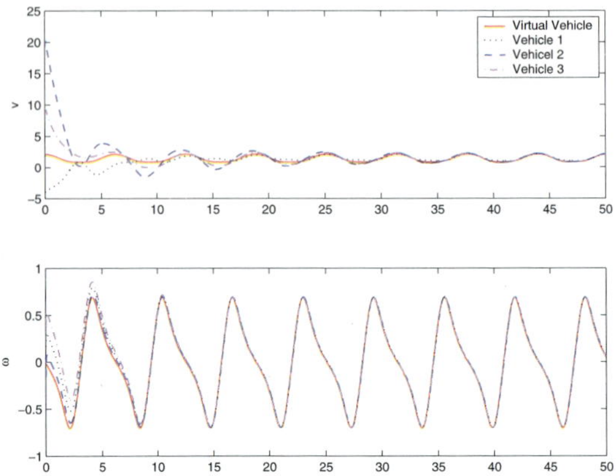

Fig. 15.8. Vehicle control signals v_i's and ω_i's.

15.4.1 Discussions on Formation ISS

In the proof of Theorem 15.4, it was assumed that the reference velocities $v_{r_i}(t)$ and $\omega_{r_i}(t)$ for any vehicle i in the group will eventually approach to $v_r(t)$ and $\omega_r(t)$. In fact, $v_{r_i}(t)$ and $\omega_{r_i}(t)$ cannot always follow time-varying $v_r(t)$ and $\omega_r(t)$ without errors, due to the low-pass nature of all consensus schemes. But the tracking errors between $v_r(t)$ and $v_{r_i}(t)$, $\omega_r(t)$ and $\omega_{r_i}(t)$ are known to be bounded, provided that:

- The formation graph has a spanning tree, and,
- $v_r(t)$ and $\omega_r(t)$ are uniformly bounded rate signals, i.e., $|\dot{v}_r(t)| \le m_1$ and $|\dot{\omega}_r(t)| \le m_2$ (see Proposition 2 in [14]).

A question that is raised naturally is the following: Does a variant of Theorem 15.4 hold with $|v_r - v_{r_i}| \le \varepsilon_1$, $|\omega_r - \omega_{r_i}| \le \varepsilon_2$, where ε_1 and ε_2 are reference velocities tracking errors? The answer is yes. To state this new result, one must introduce first the concept of leader-to-formation stability (LFS) [23].

Definition 15.6: A formation is called LFS if there exist a class *KL* function β and a class *K* function γ, such that for any initial formation error $\tilde{x}(0)$ and for any bounded inputs of the formation leaders, $\{w_l\}$ the formation error satisfies:

$$\|\tilde{x}(t)\| \le \beta(\|\tilde{x}(0)\|, t) + \sum_{l \in L_F} \gamma_l (\sup_{0 \le \tau \le t} \|w_l(\tau)\|) \tag{15.23}$$

As a variant of Theorem 15.4, the following theorem takes into account the effects of time-varying reference velocities on the formation stability.

Theorem 15.5: Consensus-based FTC (15) results in LFS if the formation graph has a spanning tree and the reference velocities are uniformly bounded rate signals.

Proof: The proof follows from Corollary 15.1 and the invariance property of LFS [23].

15.5 Conclusions and Future Work

This Chapter addressed the formation tracking problem for multiple mobile unmanned vehicles with nonholonomic constraints. A basic formation tracking controller (FTC) was developed as well as a consensus-based one using only neighboring leaders information. The stability properties of the multiple vehicle system in closed-loop with these FTCs were studied using cascaded systems theory and nonlinear synchronization theory. In particular, connections were established between stability of consensus-based FTC and Laplacian matrices for formation graphs. The simple formation tracking strategy holds great potential to be extended to the case of air and marine vehicles.

Collision avoidance and formation error propagation problems were not discussed. The proposed FTC does not guarantee avoidance of collisions and there is a need to consider them in future work. Theorem 15.5 showed that consensus-based FTC leads to LFS. The invariance properties of LFS under cascading could be explored to quantify the formation errors when individual vehicle's tracking errors are bounded. Formation tracking in a higher dimension is another interesting problem for future study.

References

1. Chen Y., and Wang Z., "Formation Control: A Review and a New Consideration", *Proceedings, IEEE/RSJ International conference on Intelligent Robots and Systems*, Canada, 2005.
2. d'Andre-Novel B., Bastin G., and Campion G., "Control of Nonholonomic Wheeled Mobile Vehicles by State Feedback Linearization", *International Journal of Robotic Research*, Vol. 14, No. 6, 543-559, 1995.
3. Fax J. A., and Murray R. M., "Information Flow and Cooperative Control of Vehicle Formations", *IEEE Transactions on Automatic Control*, Vol. 49, No. 9, 1465-1476, 2004.

4. Gazi V., "Formation Control of Mobile Vehicles Using Decentralized Nonlinear Servomechanism", *Proceedings, 12th Mediterranean Conference on Control and Automation*, Turkey, 2004.
5. Ghabcheloo R., Pascoal A., Silvestre C., and Kaminer I., "Coordinated Path Following Control of Multiple Wheeled Vehicles with Directed Communication Links", *Proceedings, 44th IEEE Conference on Decision and Control - ECC*, Spain, 2005.
6. Jiang Z.-P., and Nijmeijer H., "A Recursive Technique for Tracking Control of Nonholonomic Systems in Chained Form", *IEEE Transactions on Automatic Control*, Vol. 44, 265-279, 1999.
7. Jin Z., Waydo S., Wildanger E. B., etc., "MVWT-II: The Second Generation Caltech Multi-Vehicle Wireless Testbed", *Proceedings, American Control Conference*, Boston, 2004. .
8. Kanayama Y., Kimura Y., Miyazaki F., and Noguchi T., "A Stable Tracking Control Method for an Autonomous Mobile Robot", *Proceedings, IEEE International Conference on Robotics and Automation*, Cincinnati, OH, 1990.
9. Lafferriere G., Williams A., Caughman J., and Veerman J. P., "Decentralized Control of Vehicle Formations", *System and Control Letters*, Vol. 54, No. 9, 899-910, 2005.
10. Lee T.-C., Song K.-T., Lee C.-H., and Teng C.-C., "Tracking Control of Unicycle-Modeled Mobile Vehicles Using a Saturation Feedback Controller", *IEEE Transactions on Control Systems Technology*, Vol. 9, 305-318, 2001.
11. Lin Z., Francis B., and Maggiore M., "Necessary and Sufficient Graphical Conditions for Formation Control of Unicycles", *IEEE Transactions on Automatic Control*, Vol. 50, No. 1, 121-127, 2005.
12. Luca A. De., and Di Benedetto M. D., "Control of Nonholonomic Systems via Dynamic Compensation", *Kybernetica*, Vol. 29, No. 6, 593-608, 1993.
13. Luca A. De, Oriolo G., and Samson C., "Feedback Control of a Nonholonomic Car-like Robot", in *Robot Motion Planning and Control*, J.-P. Laumond, Editor, Springer, Lecture Notes in Computer and Information Sciences, Vol. 229, 171-253, 1998.
14. Olfati-Saber R., and Shamma J. S., "Consensus Filters for Sensor Networks and Distributed Sensor Fusion", *Proceedings, 44th IEEE CDC-ECC*, Seville, Spain, 2005.
15. Olfati-Saber R., Fax J. A., and Murray R. M., "Consensus and Cooperation in Multi-Agent Networked Systems", *Proceedings of the IEEE*, January 2007.
16. Oriolo G., Luca A. De, and Vendittelli M., "WMR Control via Dynamic Feedback Linearization: Design, Implementation, and Experimental Validation", *IEEE Transactions on Control Systems Technology*, Vol. 10, No. 6, 835-852, 2002.
17. Panteley E., Lefeber E., Loria A., and Nijmeijer H., "Exponential Tracking of a Mobile Car Using a Cascaded Approach", *Proceedings, IFAC Workshop on Motion Control*, Grenoble, France, 1998.
18. Ren W., Beard R. W., and Atkins E., "Information Consensus in Multi-Vehicle Cooperative Control: A Tutorial Survey of Theory and Applications", submitted to the *IEEE Control Systems Magazine*, 2005.

19. Ren W., and Atkins E., "Distributed Multi-Vehicle Coordinated Control via Local Information Exchange", submitted to the *International Journal on Robust and Nonlinear Control*, 2005.
20. Samson C., and Ait-Abderrahim K., "Feedback Control of a Nonholonomic Wheeled Cart in Cartesian Space", *Proceedings, IEEE International Conference on Robotics and Automation*, Sacramento, CA, 1991.
21. Scharf D. P., Hadaegh F. Y., Ploen S. R., "A Survey of Spacecraft Formation Flying Guidance and Control (Part II): Control", *Proceedings, American Control Conference*, Boston, MA, 2004.
22. Skjetne R., Ihle I.-A. F., and Fossen T. I., "Formation Control by Synchronizing Multiple Maneuvering Systems", *Proceedings, 6th IFAC Conference on Maneuvering and Control of Marine Craft*, Girona, Spain, 2003.
23. Tanner H. G., Pappas G. J., and Kumar V., "Input-to-State Stability on Formation Graphs", *Proceedings, 41st IEEE CDC*, Las Vegas, Nevada, 2002.
24. Tanner H. G., Pappas G. J., and Kumar V., "Leader-to-Formation Stability", *IEEE Transactions on Robotics and Automation*, Vol. 20, No. 3, 2004.
25. Canudas de Wit C., Khennouf H., Samson C., and Sordalen O. J., "Nonlinear Control Design for Mobile Vehicles", in *Recent Trends in Mobile Vehicles*, Vol. 11, Y. F. Zheng, Editor, World Scientific, 1993.
26. Wu C. W., and Chua L. O., "Synchronization in An array of Linearly Coupled Dynamical Systems", *IEEE Transactions on Circuits and Systems-I: Fundamental Theory and Applications*, Vol. 42, No. 8, 1995.
27. Wu C. W., "Synchronization in Networks of Nonlinear Dynamical Systems Coupled via a Directed Graph", *Nonlinearity*, Vol. 18, 2005.
28. Wu C. W., "On a Matrix Inequality and its Application to the Synchronization in Coupled Chaotic Systems", *Complex Computing-Networks: Brain-like and Wave-oriented Electrodynamic Algorithms*, Springer Proceedings in Physics, Vol. 104, pp. 279-288, 2006.

Chapter 16 "Hardware in the Loop" Tuning for a Volcanic Gas Sampling UAV[1]

In this Chapter, a newly designed partially autonomous unmanned aerial vehicle (UAV) is used for automatic sampling and analysis of gasses inside volcanic plumes. The UAV is equipped with a new autopilot and a gas sampling system. The 'hardware in the loop' methodology is used for tuning the autopilot without having to perform initial field trials. Even though simple PID controllers have been adopted in the actual autopilot, the system allows for implementation and tuning of more complex algorithms, if and when required.

16.1 Introduction

Significant advances have been made in recent years in volcanic eruption forecasting and in understanding the behaviour of volcanoes. A major requirement is improvement in the collection of field data using innovative methodologies and sensors. Collected data are typically used as input for computer simulations of volcanic activity, to improve forecasts for long-lived volcanic phenomena, such as lava flow eruptions and sand-rain.

Volcanology is of primary importance because of the huge impact that eruptions have in areas such as air-traffic safety, damage to buildings, climate change [1]. It has been estimated that 10% of the world population lives in areas directly threatened by volcanoes and only in the last 50 years several thousands of people have been killed by the effects of volcanic eruptions.

The DIEES at the University of Catania, Italy, has been involved in several research projects concerning volcanoes and related problems. Some

[1]Written by G. Astuti, D. Caltabiano, G. Giudice, D. Longo, D. Melita, G. Muscato, A. Orlando. This research was conducted in cooperation with OTe Systems Catania, Italy [16].

Kimon P. Valavanis (ed.), Advances in Unmanned Aerial Vehicles, 473–493.

activities relate to the *Robovolc* project, the aim of which is to develop robotic systems for exploration and analysis of volcanic phenomena [2]. As such, an unmanned ground vehicle (UGV) has been built and fully tested. An additional goal is the development of an autonomous underwater vehicle (AUV) for underwater volcano exploration, as well as the use of a UAV for volcanic gas sampling [3], as presented in this Chapter.

Analysis of the gas composition emitted by volcanoes is crucial for volcanologists to understand volcanic behaviour [4] [5]. Volcanic gas is quickly contaminated by the atmosphere and it is empirically worthless to collect it for analysis far from the eruptive vent. Indirect mechanisms to obtain required measures already exist; in fact, the chemical composition of the gas may be estimated using spectrographs, observing the plume from a safe place through a telescope [6]. Since such systems are not sufficiently reliable, direct measures of the most important variables of the plume (concentration of HCl, SO_2 and CO_2, pressure, temperature, wind velocity and direction) must be obtained.

This task can be achieved by using a UAV that carries a set of gas sampling sensors [7] [8] [9] [10]. There are two main barriers that do not allow a classic remote-controlled UAV to be used for such missions. The first is the required distance between a safe place and the volcanic plume (see Figure 16.1). Volcanic terrains are usually very rough in the proximity of the craters and it is not simple to carry the vehicle directly to the top of the volcano and find a suitable place for take-off and landing. For example, Figure 16.2 shows a typical rough terrain that can be found in the proximity of a crater.

Fig. 16.1. Plume on Mt. Etna.

Fig. 16.2. Typical terrain in the proximity of a crater.

Therefore, the distance required for teleoperations makes it very difficult to adopt a remote-controlled flying vehicle. The second obstacle relates to the fact that the gas within the plume is usually very dense; thus visual recognition of the vehicle is very difficult if not impossible. Consequently, the need to develop a UAV that could perform part of the mission autonomously is justified and it is of paramount importance.

The UAV mission should be to fly to the plume with an altitude drop of about 500 m, move to a predefined trajectory inside the plume and then come back to the base station. The total travel distance during the mission will be less than 6 km. Figure 16.3 depicts a typical UAV mission that is divided into five main phases:

a.) Take-off;
b.) Reaching the plume;
c.) Flying inside the plume (measurement phase);
d.) Approaching the landing area;
e.) Landing.

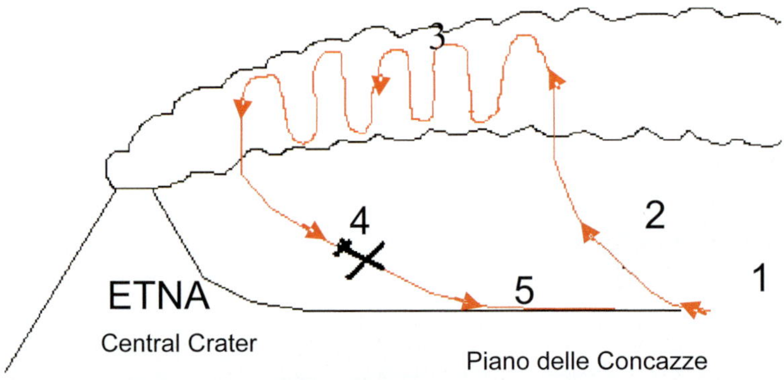

Fig. 16.3 A typical mission over Mt. Etna.

"Piano delle Concazze" is a plateau on Mt. Etna, about 2900 m above sea level, where it is possible for the aircraft to land and take-off. It must be observed that the main part of the mission is performed inside the plume, thus making it difficult for the operator to view the vehicle. Consequently, at least from phase b) to phase d) the mission must be performed autonomously by the UAV following a pre-planned trajectory.

An autopilot board with a suitable HMI (Human Machine Interface) has been developed. The HMI is useful for choosing the desired path, fixing waypoints on a map and performing other activities [3] while the control algorithm implemented on-board the autopilot guarantees the correctness of path tracking, heading, stability and other flight requirements.

Prerequisite to using the real UAV in field studies, and in order to avoid excessive field trials that could lead to dangerous consequences, a tuning procedure has been developed and implemented to test exhaustively the autopilot hardware in a virtual environment following the 'hardware in the loop' (HIL) technique [11] [12] [13] [14] [15].

The next Sections requirements for the designed UAV are briefly presented together with the system architecture and the HMI. Then the architecture of the proposed autopilot, its control algorithms and the HIL tuning methodology are described and results are presented.

16.2 The *VOLCAN* UAV

Mission requirements dictate UAV functionality. The objective is that of collecting and analyzing the composition of gas in volcanic plumes. Missions will be mainly performed over Mt. Etna (3300 m high); hence the UAV must be capable of flying up to 4000 m above sea level. At this altitude the wing lift is lower due to air rarefaction; the wing incidence must be emphasized and a more powerful engine is needed. Another important requirement is electric propulsion, essential to avoid contamination of the sampled gas by emissions of an internal combustion engine, which could happen due to the high sensitivity of the gas sensors adopted.

Gas analysis systems are often very slow, since they have to collect a sample of air and then execute a chemical analysis on it. The system adopted for this project has a sampling frequency of 1 Hz; for this reason, in order to collect several points while crossing through the plume, it is also very important to keep a low cruise speed, so a suitable airfoil is needed. A high payload is required in order to carry all analysis equipment. The main project specifications are:

- The adoption of an electric engine for propulsion;
- An autonomy of about 30 minutes;
- A payload of about 5 kg;
- A minimum cruise speed of about 40 km/h;
- A maximum cruise altitude of 4000 m;
- A working range of 3 km;
- Local and remote data log of obtained measures;
- An autonomous system (except for take-off and landing);
- Path planning through waypoints;
- Real-time visualization on a user-friendly GUI.

After several design trials and considerations the VOLCAN UAV designed is a V-Tail plane, with a configuration similar to the more famous Aerosonde [8] built using wood, carbon fiber and fibreglass. It has a wing span of 3 m, a total weight of 13kg, a 2000W brushless motor, Li-Po battery packs and a maximum cruise speed of 150km/h. It meets all specifications described above. Figure 16.4 is a snap shot of the UAV during preparation for take-off from a car on Mt. Etna. Take-off from a car may be needed in several sites where a suitable flat area for take-off is not available, also considering the small diameter of the aircraft wheels. For these trials the wheels are removed while a folding airscrew allows safe landing on the fuselage.

Fig. 16.4. The VOLCAN UAV being prepared for take-off from a car in a volcanic site (2900m).

Figure 16.5 shows a block diagram of the complete system. The autopilot is directly connected to the servos that actuate the mobile parts of the aircraft (aileron, flaps, V-tail rudder/elevator), and the motor power regulator (throttle), and receives as inputs data from a GPS and an AHRS (Attitude and Heading Reference System). The Base Station comprises a PC connected to the radio modem and an RC transmitter. From the Base Station it is always possible to switch from the Autopilot Mode to the PIC (Pilot in Command) remote-controlled mode.

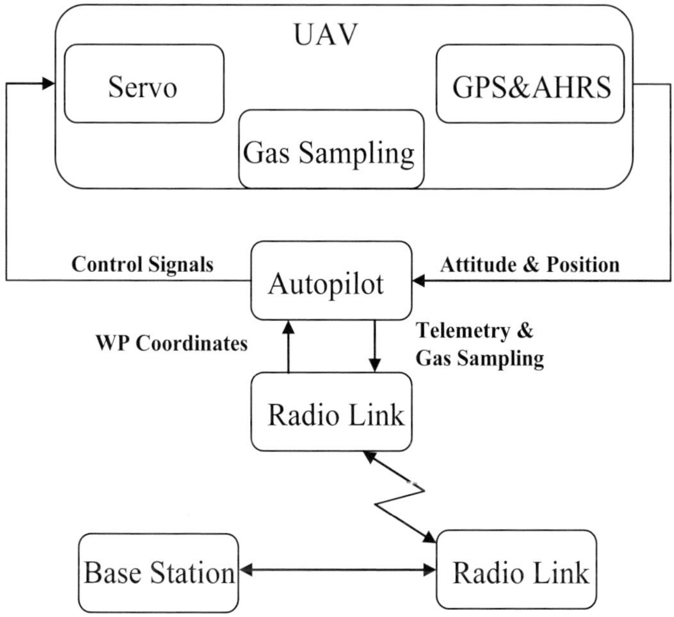

Fig. 16.5. Block diagram of the complete system.

16.3 Autopilot Architecture

The autopilot takes care of the aircraft stability and navigation; information received from the GPS and inertial systems allows the autopilot to reconstruct the attitude and the exact position of the UAV, so it can regulate the command signals to the aircraft servomotors.

It is necessary to choose from the Base Station (see Figure 16.5) the desired mission path via waypoints and to monitor the UAV parameters during its flight because of major changes in operating conditions during a typical mission that relate to changes in wind speed or direction, which modify the position of the plume.

The mission path is fixed through a sequence of waypoints; they are univocally determined assigning latitude, longitude and height above sea level; moreover, it is possible to set the flight velocity for each waypoint.

For the designed autopilot, a supervisor block acts as a trajectory generator and two different control algorithms are used to follow the mission course; the first is involved in minimization of the heading error; the second regulates the pitch angle to reach the desired altitude. Both use a cascade control scheme to set the position of the mobile parts because of the presence of subsystems with different dynamics. Figure 16.6 shows the architecture of the autopilot.

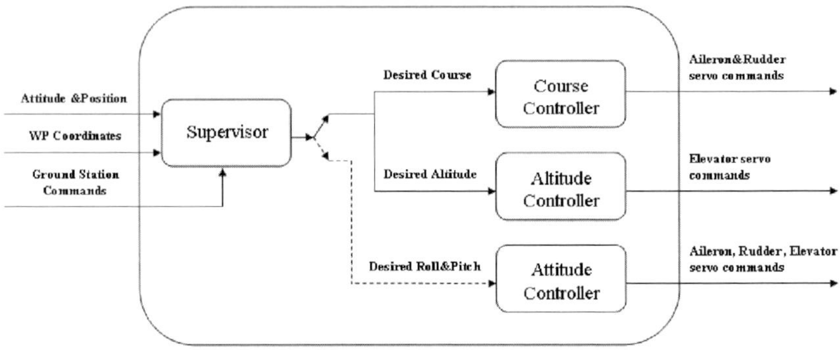

Fig. 16.6. Autopilot architecture; the two controllers involved in maintaining the course are shown. The dashed line illustrates the possibility to switch to the Attitude Controller mode.

The supervisor calculates set points for the controllers taking into account minimization of the *Cross Track Error*, that is, the perpendicular distance between the current position and the *Waypoints Line*, the imaginary line connecting two consecutive waypoints, as shown in Figure 16.7. Figure 16.8 shows the plane's real course obtained using *MATLAB*.

The autopilot internal loop control parameters can be finely tuned by activating the *Attitude Controller*; in this case the autopilot will generate servo commands to maintain the roll and pitch angles assigned by the base station while the Course and Altitude Controllers are turned off.

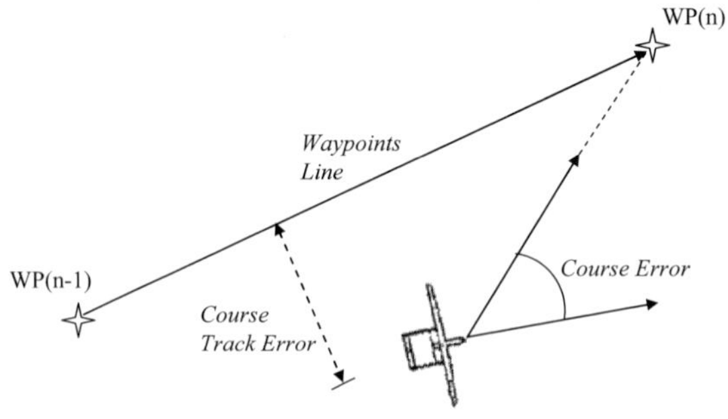

Fig. 16.7. Plane course error with respect to the next waypoint (wp).

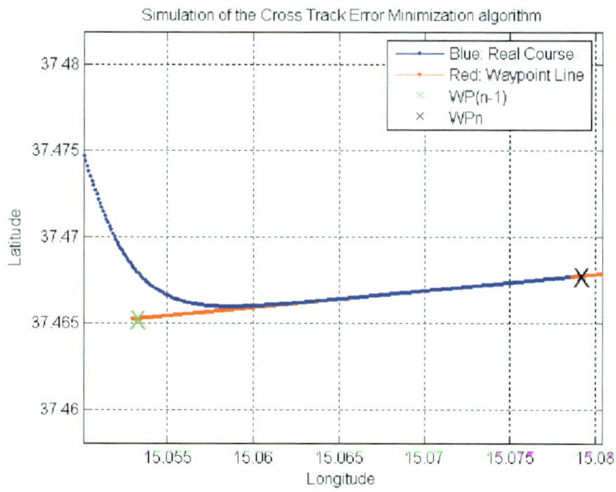

Fig. 16.8. Simulation of the cross track error minimization algorithm.

16.3.1 Heading Control Algorithm

The heading control algorithm computes the aileron and rudder positions from the course error (see Figure 16.7) that is, the angular difference between the actual trajectory, derived from two subsequent GPS measures, and the desired course, calculated taking into account GPS information and the latitude and longitude of the next waypoint.

The variables involved are:

- ΔLat_{uwp}, latitude difference between UAV and waypoint;
- ΔLon_{uwp}, longitude difference between UAV and waypoint;
- ΔLat_{u}, latitude difference between two consecutive UAV positions;
- ΔLon_{u}, longitude difference between two consecutive UAV positions;
- C_{des}, desired approaching course;
- C_{real}, actual route;
- C_E, difference between C_{des} and C_{real} (course error).

When GPS data, updated once per second, are received, the autopilot calculates first ΔLat_{uwp} and ΔLon_{uwp}, ΔLat_u and ΔLon_u, and then the desired course and the actual route $C_{des} = \text{atan2}(\Delta Lat_{uwp}, \Delta Lon_{uwp})$, $C_{real} = \text{atan2}(\Delta Lat_u, \Delta Lon_u)$, respectively.

Information about attitude, such as roll, pitch and yaw angles comes from the on-board Inertial Measurement Unit (IMU) and it is acquired every 12.5 ms.

During testing and tuning of the autopilot, simulation software was used to emulate the UAV flight, AHRS (Attitude and Heading Reference System) and GPS data, as described next.

Figure 16.9 shows the cascaded controller implemented for heading correction: the course error is used as a reference signal for a PID course regulator (PID_{Course}); the output of the latter is used as a reference signal for the roll angle PID regulation system (PID_{Roll}). The output of the PID_{Roll} is the set point for the servomotors that actuate the ailerons of the aircraft. These control signals are updated every 12.5 ms. A feedforward action, depending on the position of the ailerons, is used for the rudder servo motor. This strategy allows the yaw angle to be controlled taking into account the particular conformation of the tail of the VOLCAN UAV (V-Tail). The rudder deflection angle is a fraction of that of the ailerons, emphasizing the turning rate.

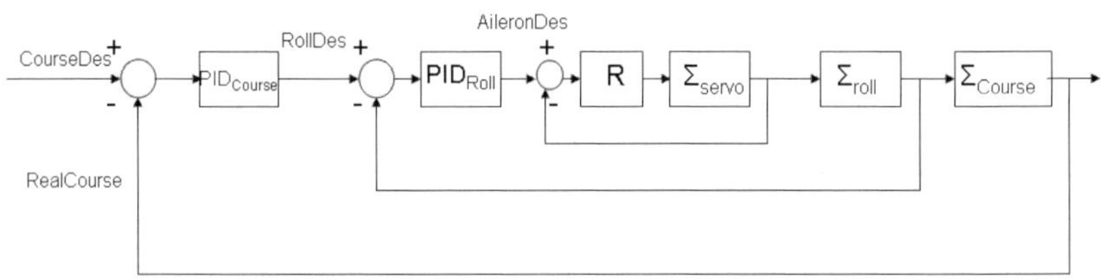

Fig. 16.9. Autopilot cascaded controller for heading regulation.

To avoid abrupt fluctuations in UAV attitude, variation of the control action of the PID_{Roll} is appropriately limited; moreover, the roll and pitch angles cannot exceed predefined security values for the VOLCAN aircraft.

The control strategy adopted guarantees both flight course holding and stability; in fact, the higher level regulator controls the approach route to the next waypoint; the lower level provides aircraft stability, regulating the flight parameters according to the reference signals generated by the higher-level control.

16.3.2 Altitude Control Algorithm

During flight, because of aerodynamic disturbances, a constant pitch angle does not lead to a constant altitude or climbing rate. Moreover, constant of elevator positions do not imply constant pitch angles. These considerations suggested the implementation of another cascaded controller for altitude regulation, shown in Figure 16.10. Altitude error is used as a reference signal for a PID altitude regulator (PID_{Alt}); the output of the latter is used as a reference signal for the pitch angle PID regulation system (PID_{Pitch}). The output of the PID_{Pitch} is the set point for the servomotors that actuate the elevators of the aircraft. These control signals are updated every 12.5 ms.

In this way, noise and disturbances in the faster dynamic system (elevator) do not affect regulation of the pitch angle, while disturbances in regulation of the pitch angle are not significant for the regulation of altitude.

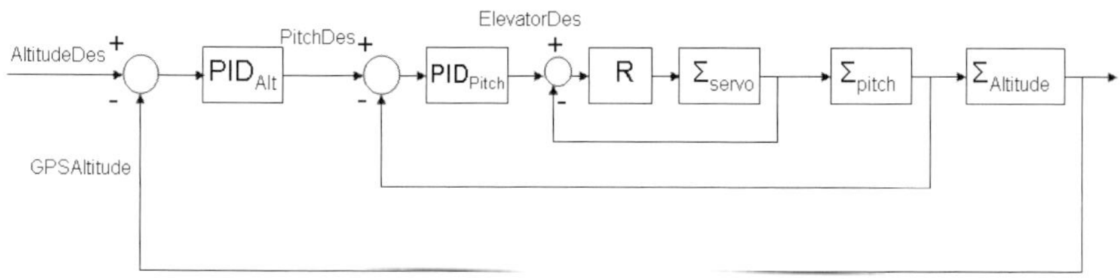

Fig. 16.10. Autopilot cascaded controller for altitude regulation.

16.3.3 Autopilot Hardware

The previously described architecture was implemented in hardware by using three modules connected together by means of the Can Bus, as is shown in Figure 16.11. One of these modules is the real autopilot (the smallest board), while the other two are the AHRS platform (with gyroscopes, accelerometers and pressure sensors) and the PWM generator to control the plane servomotors. The core of each module is based on a microcontroller. The autopilot electronics was designed and built by OTe Systems and further details can be found in [16].

Fig. 16.11. The autopilot system developed: from left to right, the real autopilot, the PWM generator module and the AHRS platform in comparison with a 2€ coin.

16.4 Autopilot Tuning: Hardware in the Loop Architecture

The HIL architecture represents a powerful and effective method to test and tune control systems [11] [12] [13] [14] [15]; the case of tuning an autopilot device is very critical, since experimental trials are performed with time consuming test flights and unsatisfactory results could lead to dangerous situations. A HIL simulator cannot fully replace field experiments, but it is very useful, especially in the preliminary phases, to discover and solve potential problems. Therefore, the major role of the HIL platform relates to improvement in development time, cost and risk reduction. Once performance characteristics and requirements of a given mission are satisfied, the same controller hardware can be directly connected to the real UAV.

Figure 16.12 shows the adopted architecture. The X-Plane flight simulator [13] by Laminar Research [17] has been used for all tests on the VOLCAN UAV. A model of the real aircraft was developed to determine the flight parameters; AHRS and GPS data are sent to the autopilot through a CAN Bus [18] using a dedicated plug-in and using the CANaerospace protocol [19], as described below. The block named "Autopilot" is the real electronic board while the HMI runs on a separate PC.

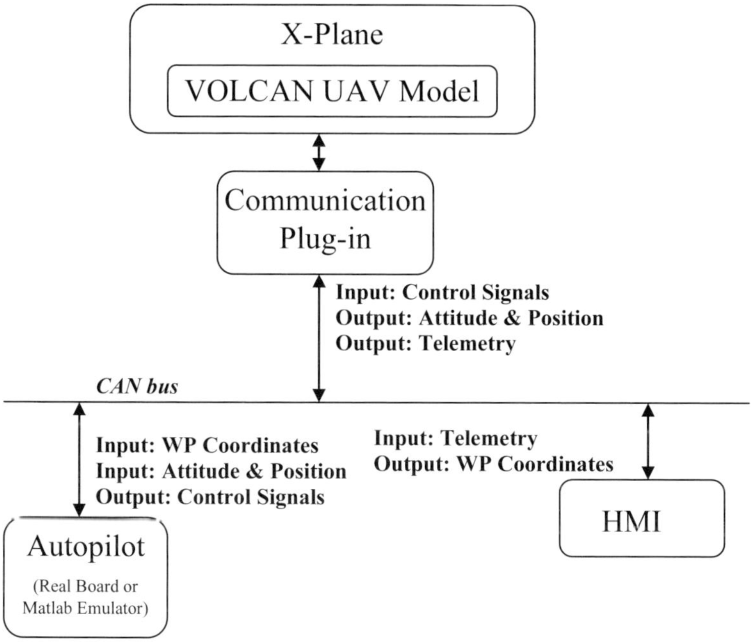

Fig. 16.12. The hardware in the loop architecture that is used during the autopilot tuning and testing.

This architecture is very flexible and its modularity allows better testing and tuning of devices involved; for example, the autopilot hardware could be initially replaced by a *MATLAB/Simulink* block to simulate navigation and stability algorithms. Moreover, the HIL allows the AHRS platform algorithms to be developed, tuned and tested. Raw noise-free IMU data (linear acceleration, angular velocity, air speed, barometric altitude, GPS) are obtained from the X-Plane simulator. Some noise, obtained from the sensor noise model used, is added to the simulator IMU data. By using these noisy data and extended Kalman filtering techniques, reliable AHRS data will again be available to the autopilot control system, see Figure 16.13. Once the filter is completely designed and tested, attitude and heading data coming from a real noisy AHRS module based on accelerometers and gyroscopes can be used. Once suitable performance for the autopilot and for the AHRS is obtained, it is possible to replace the X-Plane simulator with the real plane, performing field tests for fine parameter tuning.

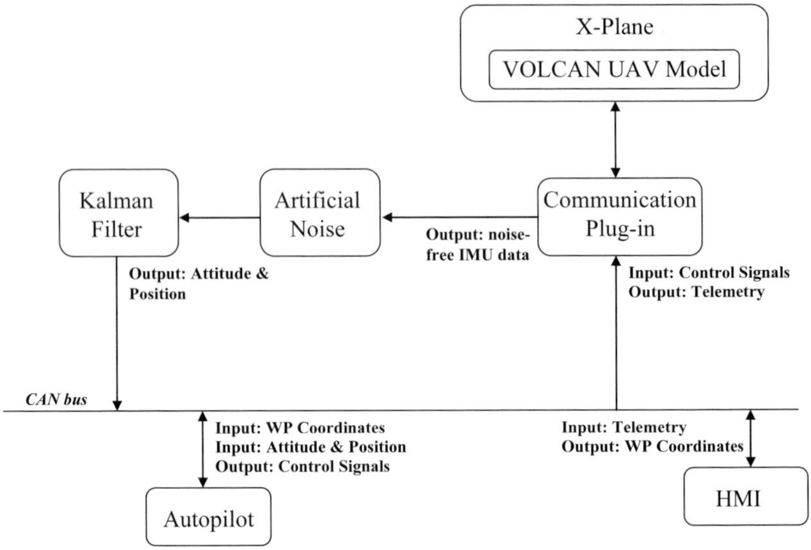

Fig. 16.13. The hardware in the loop architecture that is used during the AHRS platform tuning and testing.

16.4.1 X-Plane Simulator

X-Plane is approved by the Federal Aviation Administration (FAA) for training of pilots. X-Plane is a powerful simulator with characteristics [17]:

- High accuracy in flight model: X-Plane computes flight dynamics basing its algorithms on the geometry of the aircraft;
- Large database containing aircraft model and terrain coverage;
- Easily customized, allowing for easy creation of textures, sounds, and instrument panels;
- Complete navigation scenarios for the entire world;
- Powerful graphical engine for visualization.

Furthermore, X-Plane has two essential characteristics that are useful in realizing HIL architecture to test and tune an autopilot for aerial vehicles:

- Using special software (PlaneMaker and AirfoilMaker) it is possible to create customized aircrafts;
- X-Plane features can be extended by adding plug-in modules.

The last two properties have been used to realize a model of the Volcan UAV and to interact with the autopilot, allowing for data exchange and commands between the simulator and the electronic device developed.

16.4.2 The VOLCAN Model in X-Plane

To obtain optimal values for the autopilot parameters during the tuning phase, information received from the simulator must be the same to that provided by the real aircraft during a real flight. For this reason, a model of the VOLCAN UAV was created using the *Plane Maker* software; the plane was first drawn in *Rhinoceros*®, taking into account measures of the mobile and fixed parts of the aircraft. This is shown in Figure 16.14. This phase required particular attention and precision, since the graphical and geometrical accuracy affects flight dynamics.

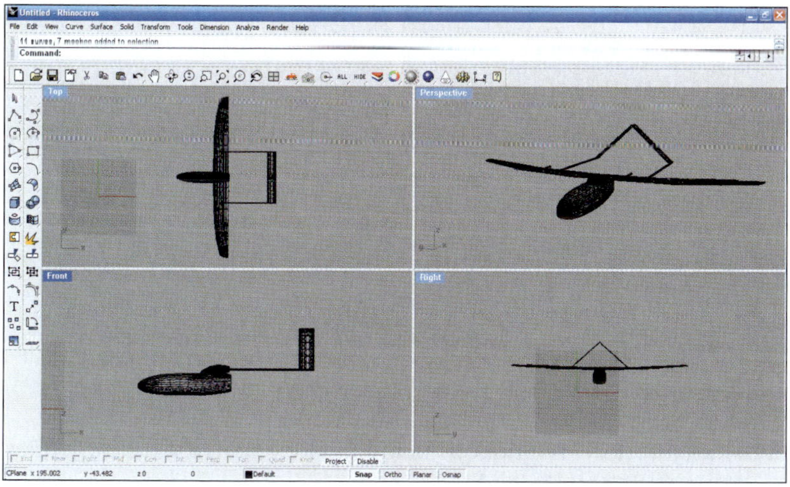

Fig. 16.14. 3-D drawing of the VOLCAN UAV realized by using Rhinoceros®.

The longitudinal sections of the fuselage can be imported in *Plane Maker*, specifying perimeter size through the coordinates of some boundary points, and the distance of each section from a fixed point; the software interpolates the surface, creating the model of the plane. For the mobile parts it is necessary to specify dimensions, deflection angles and the command signals associated with them.

The engine specifications section allows information about the propelling system to be provided: airscrew dimensions, kind of engine (in our case an electrical brushless DC motor was chosen), maximum motor rpm and maximum aircraft ground speed (20100 RPM and 150 km/h respectively). Moreover, plane weight and centre of gravity position have to be specified: VOLCAN UAV weighs 13 kg and the barycentre is near the motor, in the rear side of the aircraft. *AirfoilMaker* allows customized wings to be created, taking into account the characteristics of the real aircraft; the wing profile of VOLCAN UAV is classified as S3021. This airfoil is well suited for planes with a low cruise velocity. Figure 16.15 (a) shows the real aircraft while Figure 16.15 (b) shows the plane designed with *PlaneMaker*.

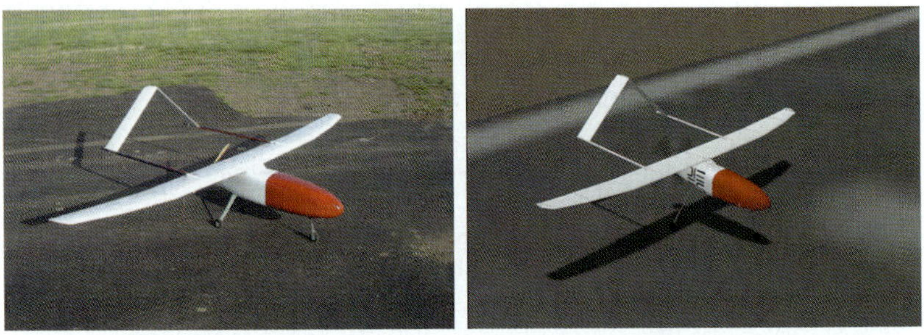

(a) (b)

Fig. 16.15. (a) The real UAV; (b) the one designed by using *PlaneMaker*.

16.4.3 X-Plane Plug-in

The HIL architecture requires a continuous exchange of information between the autopilot and the X-Plane: the simulator receives as input the control signals for the aircraft (coming from the real autopilot) and supplies the device with navigation data helpful for the control algorithms (see Figure 15.12).

Using a high-level language (Borland Delphi® in this specific case) a special plug-in was realized to expand X-Plane capabilities and gain access to the simulator data, sending them to the autopilot. Moreover, the software module developed reads control signals coming from the autopilot and writes data to the simulator, changing the positions of the mobile parts. It also makes a data log record for off-line analysis.

CAN Bus was adopted as the communication channel due to its real-time capabilities, high data rate (up to 1MBit/s), and extremely low probability of undetected data corruption, which makes it suitable for safety-critical applications. Moreover, the *CANAerospace* protocol was used since it is specifically designed for aerospace data exchange [19].

16.4.4 HMI

To follow the position of the plane position during a mission in real-time on a map, a user interface installed in the Base Station PC was realized, see figure 16.16. This GUI is typically used at the beginning of the mission to choose the desired path of the mission via waypoints, placing the position of each waypoint on a geo-referred map.

Fig. 16.16. The user interface: the map and the system instrumentation, the waypoints management area and the system parameter configuration area.

During the mission it is also possible to monitor UAV parameters such as battery levels, speed, position and orientation, the sensors measurements, the chemical composition of the air, the servo position, the current waypoint and so on. Moreover, HMI allows online setting of the navigation and stability algorithm parameters and data log recording.

16.5 Results

The HIL architecture proved to be a very useful system for autopilot tuning, with respect to finding optimal values for the control algorithm parameters to be found and aircraft stability and mission course holding to be reached.

To obtain the correct values for the PID parameters involved in the regulators a preliminary tuning was performed on the basis of a simplified dynamic model of the system and by *Simulink* simulations [3]. Then these parameters were tested on the HIL architecture; the flexibility to change parameters on-line through the user interface made this process faster and easier.

Figure 16.17 depicts the altitude response of the UAV after take-off; the first waypoint is 140 m above sea level. Figure 16.18 represents the roll trajectory for two references of +/-45° when only the attitude controller is active. Figure 16.19 shows an aerial view with results of an X-plane simulation; the HIL autopilot is activated and shown during a typical mission, when the same trajectory is followed twice. Figure 16.20 shows a graphical representation of the VOLCAN UAV during a flight.

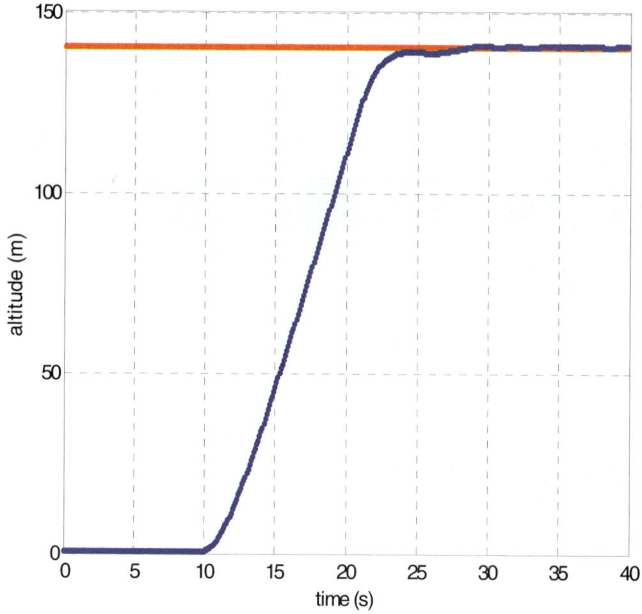

Fig. 16.17. Altitude trajectory (blue line) after take-off when the autopilot is engaged with an altitude reference signal of 140 meters (red line).

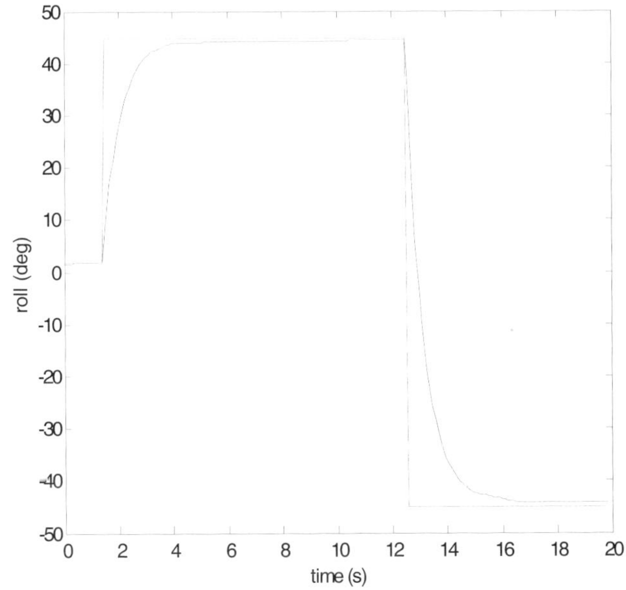

Fig. 16.18. Roll trajectory (blue line) for two references of +/-45° (red line), when only the attitude controller is active.

Fig. 16.19. An aerial view showing results of an X-Plane simulation with the HIL autopilot for two courses over a programmed path.

Fig. 16.20. The VOLCAN UAV during a simulated flight. Mt Etna is in the background.

16.6 Conclusions

This Chapter presented details of a UAV designed to carry real missions related to automatic sampling and analysis of gasses inside volcanic plumes. On-board the VOLCAN UAV is a new autopilot and gas sampling system. The HIL technique has been used to tune the autopilot avoiding the need to perform field trials. The modularity of this approach has allowed for tests to be performed on individual subsystems and extensive simulations have confirmed the suitability of the proposed methodology. Field measurement missions have been planned in order for fine-tuning of the controller and for volcanic gas analysis and sampling.

References

1. S. Alwyn, *La Catastrophe: The Eruption of Mount Pelee, the Worst Volcanic Disaster of the 20th Century*, Oxford University Press, 2002.
2. The ROBOVOLC project homepage: http://www.robovolc.dees.unict.it.

3. Caltabiano D., Muscato G., Orlando A., Federico C., Giudice G., Guerrieri S., "Architecture of a UAV for Volcanic Gas Sampling", *Proceedings, 10th IEEE International Conference on Emerging Technologies and Factory Automation*, Catania, Italy, September 2005
4. Aiuppa C., Federico A., Paonita G., Pecoraino M., Valenza S., "S, Cl and F Degassing as an Indicator of Volcanic Dynamics: the 2001 Eruption of Mount Etna", *Geophysics Research Letters*, 2002.
5. Symonds R., Rose W. I., Bluth G. J. S., Gerlach T. M., "Volcanic-gas Studies: Methods, Results and Applications", in: M. R. Carroll, J. R. Halloway (Eds.), *Volatiles in Magmas, Reviews in Mineralogy,* 30, 1-66, 1994.
6. Stix J., and Gaonac'h H., "Gas, Plume and Thermal Monitoring", in *Encyclopaedia of Volcanoes,* Edited by H. Sigurdsson, Academic Press, 1141-1164, 2000.
7. http://www.cas.usf.edu/~cconnor/vg@usf/uav.html.
8. http://www.aerosonde.com.
9. Kontitsis M. P., Tsourveloudis N. C., Valavanis K. P., "A UAV Based Automated Airborne Surveillance System", *Proceedings, 11th Mediterranean Conference on Control and Automation*, Rhodes, Greece, June 2003.
10. Spry S. C., Girard A. R., and Hedrick J. K., "Convoy Protection Using Multiple Unmanned Aerial Vehicles: Organization and Coordination", *Proceedings, American Control Conference*, Portland, OR, June 2005.
11. http://www.dspace.ltd.uk/ww/en/ltd/home/applicationfields/aerospace.cfm.
12. http://www.mit.edu/people/jhow/durip1.html.
13. McManus I., Greer D., Walker R., "UAV Avionics 'Hardware in the Loop' Simulator", *Proceedings, 10th Australian International Aerospace Congress*, Brisbane, Queensland, Australia, 2003.
14. Ashish G., Amitay I., Hemendra A., "Hardware-In-Loop Simulator for Mini Aerial Vehicle", *Proceedings, 6th Real-Time Linux Workshop*, Nanyang Technological University, Singapore, November 2004.
15. Johnson E. N., and Mishra S., "Flight Simulation for the Development of an Experimental UAV", *Proceedings, AIAA Modeling and Simulation Technologies Conference and Exhibit*, Monterey, California, August 2002.
16. http://www.otesystems.com.
17. http://www.X-Plane.com.
18. http://www.can.bosch.com.
19. http://www.canaerospace.com.

Chapter 17: A Modular On-board Processing System for Small Unmanned Vehicles[1]

This Chapter presents the overall concept for designing a modular on-board processing system for small unmanned vehicles (UVs). It is tested and implemented on a small unmanned helicopter as well as on a small unmanned ground vehicle. A detailed study describing key characteristics of on-board systems along with associated limitations is included. Implementations of two versions of a generic on-board system capable of vision processing and servo based control are presented along with detailed hardware specifications. Experimental data in a laboratory environment and outdoors validate design, functionality and key characteristics of the on-board processing systems, as well as their ability to serve multiple heterogeneous unmanned platforms.

17.1 Introduction

This Chapter has been motivated by the challenge to design and implement an on-board processing system that meets very limited payload capabilities of small unmanned aerial and ground vehicles while sacrificing minimal computational power and run time, still adhering to the low cost nature of commercial Radio Controlled (RC) equipment. Fundamental issues justifying the implementation of such an on-board system are increased autonomy, increased processing throughput, diversity in areas of application, and increased safety.

Small unmanned vehicles have shown great potential in research areas including, IED (Improvised Explosive Device) detection, search and res-

---

[1] Written by R. D. Garcia, K. P. Valavanis. This work has been supported partially by two Research Grants, ARO W911NF-06-1-0069 and SPAWAR N00039-06-C-0062.

Kimon P. Valavanis (ed.), Advances in Unmanned Aerial Vehicles, 495–529.

cue, reconnaissance, dynamic network repair, and traffic surveillance. The potential of unmanned vehicles is further increased due to their 'use-n-lose' ability that allows them to be used and never recovered. This ability is effective due to the low cost of each individual system.

Unmanned vehicles specifically designed for quick and easy deployment should be as efficient as possible. This includes efficiency of the platform itself, efficiency of all on-board components, and efficiency of the system as a whole. It may also include sensor and on-board processing system switching among vehicles. This allows for individual resources to share components resulting in stretching of overall resources without jeopardizing response time or effectiveness.

Payload is without question the most limiting factor in small UVs and it is the main boundary between full scale and small UVs. Full scale UVs that are proportional in size to their manned counterparts, are capable of carrying hundreds, and even thousands, of pounds of payload. This equates to vehicles that carry massive generators sacrificing horse power for electrical power, large sensor suites, multiple high power computers and multiple redundant components. Small UVs can sacrifice neither horsepower nor the payload loss required to generate electrical power or carry redundant components. Performed experiments have revealed realistic payload limitations around 8.5 pounds for a typical small UAV platform, creating major restrictions on devices placed on-board the platform. This is one of the main justifications for off-board processing, where UVs are typically equipped with only lightweight cameras and transmitters. Transmitted data are used by ground processing systems to perform all necessary computations. However, data transmission via wireless communication channels introduces both noise and data loss [1]. Transmitted video is commonly littered with static and color rearrangements as shown in Figure 17.1, while it is also typical to observe complete video dropout due to lost communication or bandwidth limitations. Wireless transmission also entails serious security issues with transmitted data being maliciously damaged or stolen. Software encryption only adds to computational demands and hardware encryption taxes the already limited payload of the platform. Hence, the preference to on-board processing.

On-board systems for UVs are a well researched area. However, there is still work to be done on small scale UVs [2] [3] [4] [15]. One example of a full scale on-board system is the one used by the Spartan Scout Unmanned Surface Vehicle, a rigid hull inflatable boat capable of semi-autonomous control, around 7 meters in length and capable of carrying payloads up to 3,200 lbs at ranges up to 14 miles for 3 hours [2] [5]. The Spartan Scout is controlled via a Graphical User Interface (GUI) from a nearby parent vessel.

Fig. 17.1. Examples of wireless video noise taken from a UAV.

The basic on-board equipment on the Spartan Scout is an electro-optical / infrared surveillance turret, surface radar, digital imagery transmission system, and an unmanned command and control system [6]. This equipment is standard for all Spartan vehicles and provides the system a concrete base of hardware to work with. This vessel is designed to integrate a multitude of modular pods allowing the platform to be quickly and easily customized for a specific task. The Spartan's modular pods include devices for Reconnaissance, Surveillance, and Target Acquisition (RSTA), Precision Strike (PS), Anti-Surface Warfare (ASuW), Force Protection (FP), and littoral Mine Warfare (MIW) [7]. However, the disadvantage is that approximately $30M USD has been spent for the development of four prototypes [2], a price tag that presents a very serious limitation in the non-military areas of research.

Smaller UAVs have also proven to be extremely effective platforms for on-board processing systems. One particularly notable platform is the Yamaha RMax consisting of a 2-stroke horizontally opposed 246cc engine mounted on a 3.63 m long frame [8]. It has a payload capacity of approximately 28 kg allowing it to accommodate very large on-board components containing multiple cameras, a radar altimeter, and complete desktop size computers [9].

Georgia Tech's Software Enabled Control (SEC) group has used the RMax platform (called the GTMax) along with a custom developed on-board system to assist in high performance autonomous control of the helicopter. The on-board system consists of a NovAtel RT-2 GPS receiver, sonar altimeter, HMR-2300 Magnetometer, ISIS-IMU, Radar Altimeter, two on-board computers, and an Aironet MC4800 wireless data unit [9]. The entire on-board system is powered by the RMax's on-board generator and control of the helicopter is handled by both Georgia Tech's on-board system and the Yamaha Attitude Control System (YACS) present on all standard RMax vehicles.

Small UVs are the most common and frequently used unmanned platforms today [14]. They allow for systems that are capable of long run times, reasonable payload capacities, and somewhat simplified storage and deployment. Above all, small platforms are popular due to their relatively inexpensive cost ranging from several thousand dollars to a few hundred thousand. This is mainly due to the systems ability to function using highly manufactured parts that do not require modification for size, weight and power consumption.

One notable small UAV platform, the USC AVATAR (Autonomous Vehicle Aerial Tracking and Reconnaissance), incorporates three firewire cameras, two IMUs, two PC-104 stacks (a stack of 5 and 6), two wireless transmitters, two solid state drives, and two power supplies [10]. This processing system is mounted on a Bergen Industrial Twin helicopter utilizing a 46cc twin cylinder two cycle engine with a 10 kg payload capacity.

The AVATAR system has been shown to be effective in both autonomous flight and visual identification of objects [11]. It has combined its visual recognition abilities with flight capabilities to perform vision assisted flights. This ability has been used to accomplish vision based autonomous landings and tracking of objects of interest. The platform has also been used in deployment of marsupial robots and the autonomous deployment and repair of sensor networks [12].

Prior research has shown the enormous benefits of on-board processing, but the migration of these systems to small highly agile platforms has yet to be fully explored. To effectively utilize the many benefits of miniature UVs, on-board processing comparable to those of larger platforms must be developed.

17.2 On-board System Development

Although the idea of an on-board processing system for small / miniature UVs appears to be a fairly straight forward design and implementation process, it does have several unforeseen pitfalls that need be explored before an effective and efficient design is accomplished. An adaptable on-board system must have generic abilities while considering platform limitations, hardware limitations, issues related to payload limitations, platform propulsion, operating environment, system power, and safety, all of which increase the complexity and difficulty of the on-board system design.

17.2.1 Generic Abilities

Adaptable systems must be generic enough to allow for functionality over a large domain but refrain from forcing the user into using hardware that may be either questionably useful or even hazardous to a task. To accomplish this, one must first identify and research the area of functionality of the adaptable system. In the case of UVs, familiarity with known platforms and types of tasks they are required to perform is needed along with identification of common aspects of these tasks.

Position Determination

Positional awareness is one of the most important aspects a UV must handle. Whether a UV is designed for indoor or outdoor environments it must have some idea of its position with respect to the environment. This may be accomplished with landmark based localization, dead reckoning, integration of velocity or acceleration, and GPS.

GPS is most widely utilized, particularly for outdoor position calculation. It computes UV positional data in three dimensions with reference to the Earth's coordinate system. This allows for robust and precise positional accuracy in most of the world. Although GPS is fairly robust it does have several issues: it cannot function indoors without using specially placed GPS repeaters and it must have a fairly unobstructed view of the sky for accurate position calculation. Positional data via GPS may even be corrupted by heavy cloud cover or frequency interference.

Robustness requires than an adaptable system be able to function in areas where GPS is not a realistic option: indoor environments and near buildings or large obstacles. This justifies the rational for vision based localization, dead reckoning, and integration of velocity or acceleration, which gives position with respect to the UV. Although these techniques are fairly inadequate by themselves, combinations of them have proven to be very effective [11].

To allow for positional accuracy both indoors and outdoors it is typical to see on-board systems equipped with multiple position sensors such as vision and GPS capabilities, or vision, GPS, and inertial sensors. This selection of sensors allows the system to function outdoors with positional accuracy provided by GPS, indoors with positional accuracy provided by both cameras and inertial sensors, and in transitions from both outdoor and indoor provided by all three sensors.

Orientation

Orientation plays a vital role in most UV designs. Position may provide partial information about the current state of the UV but it is insufficient when the vehicle attempts to transition to a new position or in some cases hold its current position. Typical platforms all require heading (yaw) to transition from a current position to a desired position. It is also imperative that UVs be able to accurately determine their roll and pitch position. This information is used to maneuver typical UAVs and is used for safety on most UGVs to prevent rollovers.

Orientation is sensed by electronic sensors or it is calculated based on passed information. Typically, calculation of orientation is limited to heading. This is usually accomplished by commanding a known movement, straight forward, and then using the previous and current position to calculate heading. This type of calculation can be very accurate if position information is very accurate or the calculation is performed over a large movement. Sensed orientation is typically accomplished via magnetometers which provide magnetic force readings on multiple axes. These readings use the magnetic field produced by the earth to determine roll, pitch, and yaw. Such readings are heavily influenced by magnetic fields produced by other objects including ferrous metals and electrical current. In dynamic systems these disturbances are typically filtered using gyroscopic readings on parallel axes.

Movement

Movement, although obvious, is crucial to any UV design. To be functional, a UV must have the ability to orient itself or a part of itself. This may be as simple as movement of a pan/tilt system or as complex as 3-D flight. Although there are extreme differences between the two previous examples they both have one fundamental similarity: they both control the position of an actuator or multiple actuators.

Examples of actuators include electric motors, thermal bimorphs, hydraulic pistons, relays, piezoelectric actuators, comb drives, and electroactive polymers. All of which transform some type of input signal into motion. In UV designs, this input signal is typically an electrical signal indicating the position and/or speed of the actuator.

The most common actuators utilized by small unmanned vehicles are servos. These actuators utilize Pulse Width Modulation (PWM) to determine the desired position of the motor. Typical RC servos use a range of 1000 to 2000 ms to describe every position within their range of motion and vary greatly in speed and torque.

Process Data

All of the above abilities are fairly useless without some level of data processing. Whether the processing is accomplished at a local ground station, on the UV, or by a human observer the data must be processed. This type of processing can be accomplished by small integrated hardware with minimal adaptabilities to massive multiprocessing machines. Processing systems range greatly in size, power consumption, heat dissipation, computational ability, and peripheral support. Examples of processing boards include basic stamp, PC-104, Advanced Technology Extended (ATX), ITX, and custom microprocessor designs. These boards allow for a multitude of input and outputs via various ports and support several variations of operating systems and peripheral devices.

When selecting a processing board one must first consider the location at which the processing system will be stored. Processing accomplished at a local ground station has the advantage of almost limitless computational and electrical power. Although this is very inviting the environment in which UVs typically operate (over long distances and typically not line of sight) and the medium by which they transfer data (802.11, serial modem, etc) is severely limiting.

17.2.2 Limitations

When designing an adaptable on-board processing system one must consider limitations related to payload, propulsion, operating environment, and electrical power, all of which add to the overall complexity of the on-board design.

Payload

Payload limitation is by far the most important limiting factor in small / miniature UVs. Such a limitation requires sacrifice of larger highly accurate sensors with smaller lighter less accurate sensors. It also limits the use of on-board equipment with high power consumption rates including high power processors, lasers range finders and radars. This is mainly due to the majority of platforms needing to carry all of the power required to operate the on-board system and platform. This creates an unforeseen payload decrease with every new piece of hardware. The designer is forced to consider both the actual weight of any hardware added to the on-board system and the weight of the extra power required to properly operate the hardware.

The limitation imposed on the size and weight of hardware added to any on-board system is always a trade off between the hardware's ability and the overall dimensions and weight. Any reduction in the ability of the hardware will most likely have to be overcome through software. However, extra strain placed on software may cause currently working software and hardware to fail.

Payload is also crucial when focusing on the dynamics and safety of a UV. Even payloads that fall under the maximum abilities of the vehicle may still create unforeseen complications. Any increase of the total weight of the vehicle affects overall dynamics, either positively or negatively depending on the hardware and platform. For example, a well placed weight on a UGV platform may lower the center of gravity decreasing the possibility of a roll over, or even decrease the overall vibration of the vehicle. It is also possible that this same weight could lower the ground clearance of the vehicle increasing the possibility of high centering. On the other hand, a seemingly small payload placed too far out on a fixed-wing aircraft could cause the wing to break under high wind or could cripple the ailerons in a side wind.

Propulsion

When designing an on-board system one must consider limitations imposed by the propulsion of the platforms to be utilized. In the area of miniature UV platforms, the types of propulsion are typically limited to jet, electric, methanol and gas. Discussion will focus on limitations that affect the design of the on-board processing system.

Electrical, methanol, gas and jet propulsion systems are very different; they each have some effect on any nearby or direct mounted object. For electrical propulsion this includes large magnetic fields created by the batteries and the electrical current used by the platform. This becomes apparent when one considers that many electric helicopters use nearly 40 Amps of current during flight. Current spikes may have adverse effects on unshielded wires and sensors (like electronic compasses). Methanol and gas propulsion systems typically expel a large amount of unburned oil and gas. This can be very hazardous to any electronics that are not environmentally protected. Jet propulsion exposes its surroundings to a great deal of heat and noise. This may cause damage to sensitive sensors or other hardware placed near the engine.

Platforms

All UV platforms have some type of limitation. Limitations may greatly cripple UV functionality and safety if they are not handled with care. Limiting factors include vibrations, freedom of movement, control difficulties, payload limitations, and safety.

Vibrations are of major concern mainly due to sensor noise caused by them. Many UV platforms rely on rate and acceleration measurements provided by gyroscopes and accelerometers for accurate vehicle functionality. One example of the severity of this noise is visible in Figure 17.2, where the level of noise from a static object is approximately 0.015 Gs compared to an object hard mounted on a small UAV in low idle, which is approximately 0.6 Gs. This is 40 times the amount of noise versus that in a static object. The level of severity is highly dependent on the mounting method, platform type, and propulsion type.

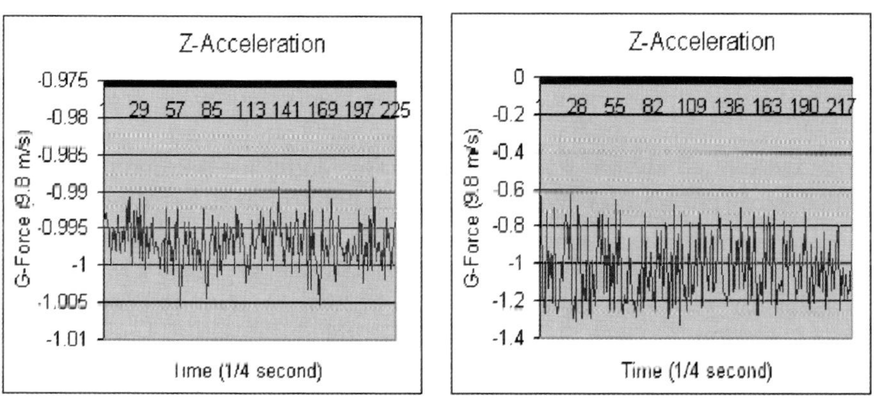

Fig. 17.2. Z Axis vibration from helicopter, engine off (left) and low idle (right).

Vibration is also an issue with the physical stability of the on-board system. Many electronic parts are built around the assumption that they will be used in a semi-static environment. However, when these electronic parts are exposed to high vibration/shock environments, their probability of failure increases. This may be overcome by using components specifically designed for high vibration environments or by reducing the amount of vibration felt by that component, for example using vibration reducing mounts in key areas. However, vibration reduction material should not cause amplification of vibration due to the frequency of the vibration.

Freedom of movement and control difficulties are also a concern when dealing with UV platforms. Major issues include the need to accurately and quickly determine position, orientation and rates in two / three dimensions. This may have a multitude of effects on the vehicle as clarified through an example. Consider what must be controlled when moving a UV forward: A UGV with Ackermann steering must assure that its turn angle is zero (steering control) and must have some forward rotation on the tires (acceleration control). A helicopter must assure that the vehicle does not loose altitude (collective control), that its main rotor turns (throttle control), that it does not roll to either side (aileron control), that it does not yaw left or right (heading control), and that it has some forward motion (pitch control).

Safety, although not entirely obvious, should be the most important of all concerns when dealing with any UV.

Environment

The environment in which a processing system functions has a great effect on the design of any such system. This effect is typically limited to the type of enclosure in which the on-board system is contained but can also reflect directly on the hardware itself. Specifically, hardware designed for a particular environment can alleviate constraints on the enclosure and improve the overall system's performance. This could include industrial designed hardware which is typically more tolerant of heat variations, moisture, and radiation.

Although some hardware may reduce constraints on the enclosure they are typically expensive and may go far beyond the requirements of the UV's operation. In these instances special attention should be taken to ensure that the enclosure can support all of the required operating environments. This includes environments that are exposed to chemicals, extreme heat and cold, radiation, moisture, pressure, interference etc. One must also assure that enclosure constraints do not directly conflict with on-board system's functionality. For example, an air tight enclosure will loose the ability to measure barometric pressure which is commonly used to measure altitude. One must even consider the type of material from which the enclosure is made. Materials that do not conduct heat will increase the overall temperature of the enclosed hardware, ferrous metals will have adverse effects on electric compasses, and some materials are too soft or rigid for a particular design.

Electrical Power

Power is a very limiting factor in any hardware design but especially limiting in miniature unmanned vehicles where payloads are highly limited. Most UVs require that all electrical power is carried on-board the platform. This requirement puts a great stain on the designer to assure that each piece of hardware is absolutely necessary and power efficient. It also forces the designer to consider power sources that have high power to weight ratios. Examples of such power sources would be lithium batteries (polymer and ion) and on-board generators.

Lithium polymer (LiPo) and ion batteries allow hardware to utilize power that is low in weight, high in power output, and rechargeable. Lithium batteries have a great advantage over Nickel Metal Hydride and Nickel Cadmium batteries due to their three and four times higher power to weight ratio respectively [17]. Although lithium batteries are very appealing to on-board system design, it does come at a price. Lithium batteries have very sensitive discharge and recharge ratios and are very sensitive to high impact shock. Incorrect care for these batteries can easily result in explosions and fire.

It is also appealing to allow a platform to supply its own electrical power via an on-board generator. Although this choice would seem optimal it does require several sacrifices. First, an on-board generator adds weight to the design pulling from an already taxed payload. Second, the power required to operate the generator is equal to or greater than power output by the generator. For example, a gasoline powered platform will use extra combustion to produce electrical power. This will increase the amount of fuel spent at any given time. Typically, an electrical generator will reduce to overall platform endurance.

17.3 Hardware and Operating System

This Section presents two prototypes of on-board processing systems developed for small UVs. These two systems have been chosen for their simplicity, use of Commercial off the Shelf Components (COTS), and adaptability. Both on-board systems have been developed at the USF Unmanned Systems laboratory.

17.3.1 On-board Processing System - Prototype Version 1 (V1)

The hardware components of the on-board system consist of:

- 1.2 GHz EPIA Embedded Processor
- Via motherboard
- Unibrain Firewire Camera
- Microstrain 3DM-G IMU
- 1 Gig 266 MHz RAM
- 1 Gig Compact Flash
- Compact Flash to IDE adapter
- Motorola M12+ GPS Receiver
- 8 Channel Servo Controller
- 200 W Power Supply
- 11.1 V LiPo Battery
- 802.11B Cardbus

This V1 configuration is chosen because of its high computational capabilities, various I/O ports, size, low heat emission, and cost. Figure 17.3 depicts the overall concept for the on-board processing system as well as connection descriptions. The on-board processing system is packaged into a 32x19x5 cm basswood box mounted on a lightweight aluminum sheet, see Figure 17.4. This sheet is mounted directly to the helicopter's skids via rubber insulated pipe clamps or to the UGV by rubber insulated aluminum sheets. The slim design of the enclosure allows for mounting of the hardware without modification to the standard carbon fiber skids of the helicopter and allows for a lower center of gravity on the UGV. The box is coated with a gas proof heat shrunk plastic typically used to coat model airplanes. Basswood is chosen for the enclosure due to its lightweight nature, lack of electrical conductance and ease of modification.

For the helicopter platform, the camera is shock mounted directly to a Lynxmotion pan/tilt unit, Figure 17.5. This unit is, in turn, hard mounted directly to the underside of the helicopter's servo tray. The pan/tilt system consists of two Futaba S3004 servos that are interconnected by 1/3cm laser cut Lexan. This setup allows the camera to pan and tilt up to 90°. Servo commands are issued by the eight channel servo control board located within the enclosure.

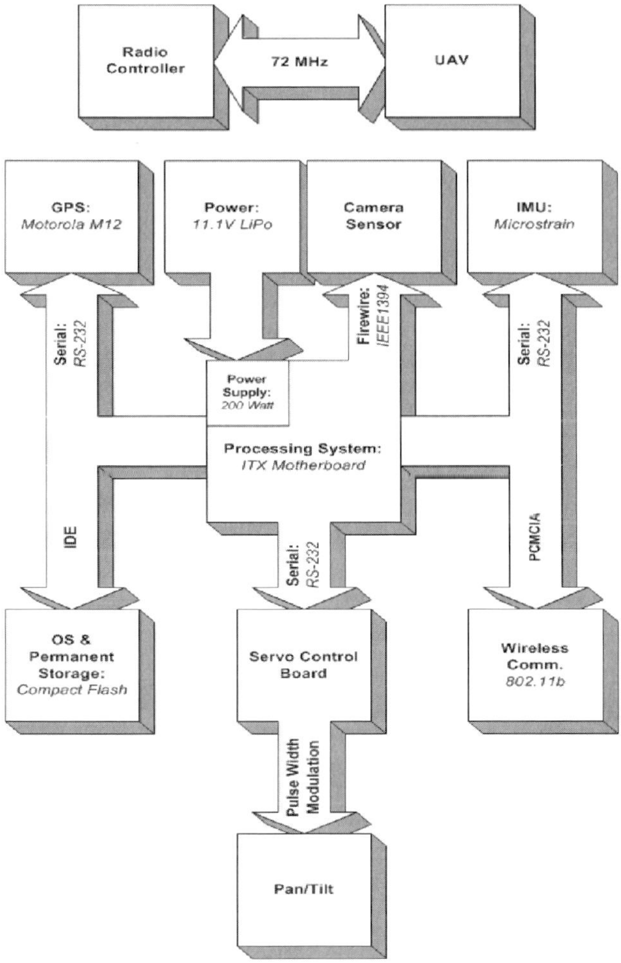

Fig. 17.3. Conceptual system diagram – V1.

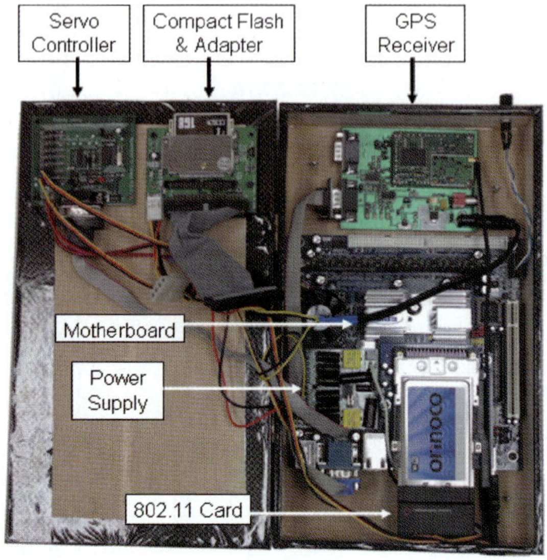

Fig. 17.4. On-board processing system in the enclosure – V1.

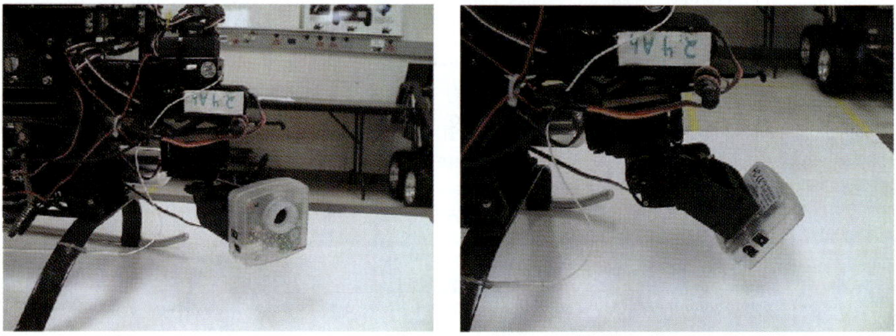

Fig. 17.5. Pan/Tilt camera mounted to the servo tray of a helicopter.

For the UGV platform, the camera is hard mounted to the front bumper of the vehicle and panning motions were assumed to be controlled by the direction of the vehicle.

To fully utilize the potential of the on-board system for the UGV the servo controller was directly connected to the speed controller and steering servo of the vehicle. This modification allows the entire movement of the platform to be controlled via the on-board processing system. This type of implementation was not considered an option on the helicopter platform due to safety concerns associated with the possibility of uncontrolled movements but was explored in the V2 prototype discussed later in this Section.

To satisfy the need for orientation data required by many software algorithms [13] a Microstrain 3-DMG was mounted to the UV. This device allows the on-board system access to the current orientation of the platform at up to 100Hz. The sensor is capable of sending both raw and gyro stabilized data and can supply the processing system with Euler angles, Quaternion vectors, roll rates, accelerations, and magnetic direction.

The on-board system is designed to receive GPS coordinates via the Motorola M12+ GPS receiver located within the enclosure and the active antenna mounted to either the horizontal fin of the VTOL or the top of the enclosure for the UGV. The horizontal fin is covered in an aluminum tape to assist in reception.

Power for the on-board system is supplied via the 11.1V 4Ah LiPo battery mounted on the lower front section of the boom for the VTOL and the undercarriage of the UGV. LiPo batteries were selected based on their high amperage, low weight, and small packaging. Power distribution to the hardware components is controlled by the 200 Watt ATX power supply. The power supply plugs directly into the motherboard allowing the unit to add nothing to the physical dimensions of the hardware.

The median for all peripherals of the on-board system is an EPIA VIA M2 motherboard. This 1.2GHz ITX motherboard provides multiple I/O interfaces, RAM and CPU on a single board. The most commonly used I/O interfaces along with the interface type and number available on the board are described in Table 17.1. The ITX motherboard also allows for a multitude of sensor suites and I/O devices to be added and removed from the on-board system with virtually no modification to the overall design due to low level integration of I/O ports. The ITX form motherboard also allows for an extremely thin design.

All communication with the on-board processing system is handled via 802.11 B. This is supported by an Orinoco Peripheral Component Micro-channel Interconnect Architecture (PCMCIA) card. This card interfaces directly with the motherboard via the supported PCMCIA slot. To support extended range this particular card is equipped with an external whip antenna. This antenna is mounted horizontally directly behind the enclosure for the helicopter and vertically at the front of the vehicle for the UGV.

The remaining hardware consists of a 1 Gig Compact Flash and IDE to Compact Flash (CF) adapter. The compact flash drive is responsible for the storage the operating system and hardware drivers. The CF adapter allows for a seamless interface between the software and the motherboard.

Port Type	# Available	Interface Type
IEEE1394	1	6 Pin Standard
USB	4	2x 5 Pin Standard, 2x Board Pinout
Serial	2	1x RS232, 1x Board Pinout
Card bus	1	Type I/II
Compact Flash	1	CF Slot
Ethernet	1	RJ45
S-Video Out	1	Standard 5 pin
Composite Video	1	RCA
LPT	1	Board Pinout
VGA	1	VGA
PS2	2	1x Keyboard, 1x Mouse
PCI	1	PCI Slot
IDE	2	40 Pin IDE
RAM	1	PC 233

Table 17.1. EPIA MII device support.

To select the operating system (OS) for the on-board system, several key requirements were identified like the IEEE 1394 and PCMCIA device support as well as installations that require less than 500 megabytes. These requirements were based on the need to support the Unibrain Firewire camera, Orinoco PCMCIA card, and the desire to have an installation that was less than half the size of the available RAM. Although the first two requirements are straight forward the third one does require further explanation.

CF cards are solid state storage that deteriorates with every write to the device. This becomes a considerable issue when one considers the number of writes made to permanent storage by the operating system. For this reason it was decided that the CF drive would only be used to load the operating system into memory. From that point all operations of the operating system would be performed in RAM. To allow the operating system to have a sufficient work area after being loaded into RAM the operating system had to be sufficiently smaller than the available RAM (1 Gig).

For the above reasons the Slackware 10.0 installation of Linux was chosen. This installation provides support for both PCMCIA and IEEE 1394 devices via its 2.4.26 kernel. The Slackware installation also provides support for low level customization during installation. Specifically, it provided the ability to remove all graphical content from the operating system allowing for a very small installation, less than 150 Megabytes compressed. Printer and sound drivers were also removed to bring the complete installation to approximately 92 Megabytes compressed.

17.3.2 On-board Processing System - Prototype Version 2 (V2)

The hardware components of the V2 on-board system consist of:

- 2.0 GHz Pentium M Processor
- Mini-ITX motherboard
- Sony Block Camera
- Microstrain 3DM-GX IMU
- 2 Gigs 333 MHz RAM
- 4 channel Frame Graber
- Superstar 2 GPS receiver
- Servo controller/safety switch
- 120 W Power Supply
- 11.1 V LiPo Battery
- 802.11B/G MiniPCI wireless card

This hardware configuration has been chosen because of its high computational capabilities, low power consumption, various I/O ports, size, low heat emission, and cost. Figure 17.6 depicts the overall concept for the on-board processing system as well as connection descriptions.

The on-board system is packaged into a 23x18x6 cm basswood box, as shown in Figure 17.7. This can be mounted in multiple configurations depending on the desired platform. The slim design of the enclosure allows for a mounting that minimally affects the center of gravity (CG) of the vehicle. The box is coated with a polyurethane coating to assist in repelling moisture and is lined with aluminum tape in an attempt to shield the GPS antenna from internal interference. Basswood has been chosen for the enclosure due to its lightweight nature, lack of electrical conductance, and ease of modification.

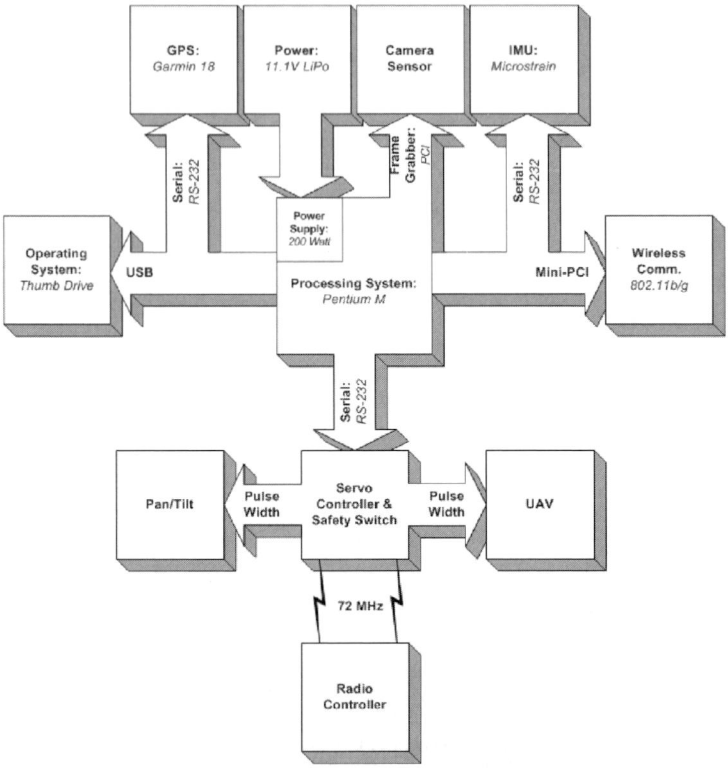

Fig. 17.6. Conceptual system diagram – V2.

Several variations of pan/tilt have been developed for this processing system all of which are controlled by two pulse width modulation (PWM) servos. This permits to control widely varying pan/tilts without hardware modification. Servo commands are issued by the Servo/Safety Switch located within the enclosure.

The system includes a custom designed breakout board to allow interfacing with the Servo/Safety switch. This was preferred over a breakout cable due to limiting of unused wires in the on-board system. The interface allows for nine PWM inputs, one of which must be the manual/computer designator switch, and nine PWM outputs from the servo/safety switch. The interface also allows the servo/safety switch to be powered by the mini-ITX power supply or a separate power source. This allows for some level of safety in the case that the mini-ITX power supply or power source fail during flight. Last the interface has both a regulated 5V and unregulated 12V output to allow powering of external sensors. The main purpose of the servo/safety switch is to allow manual takeover in the event that the vehicle becomes unstable under computer control.

To satisfy the need for orientation data required by many software algorithms [13] a Microstrain 3DM-GX is mounted on the platform. This device allows the on-board system access to the current orientation of the platform at up to 350Hz. The sensor is capable of sending both raw and gyro stabilized data and can return Euler angles, Quaternion vectors, roll rates, accelerations, and magnetic direction.

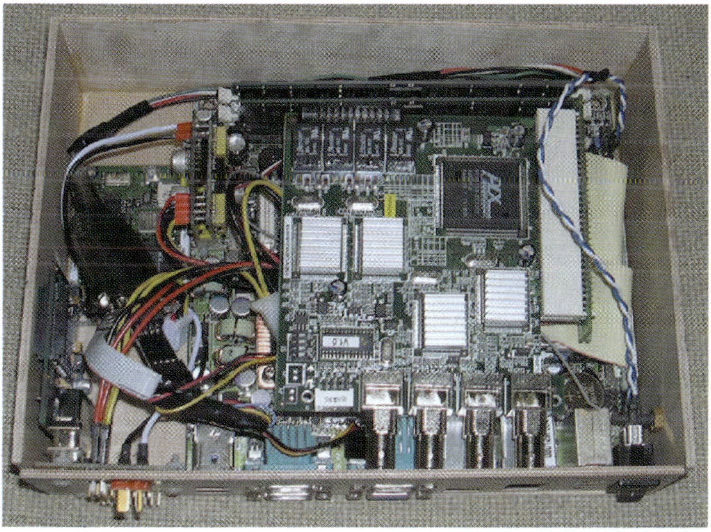

Fig. 17.7. On-board processing system in the enclosure – V2.

Power for the on-board system is supplied via the 11.1V 4.2Ah LiPo battery. LiPo batteries have been selected due to their high power to weight ratio, small packaging, and wide operating temperatures. Power distribution is controlled by the 120 Watt ATX power supply. The power supply plugs directly into the motherboard allowing the unit to add nothing to the physical dimensions of the on-board system.

The median for all peripherals of the on-board system is a G5M100-N motherboard. This Pentium M ITX motherboard provides multiple I/O interfaces, RAM, and CPU on a single board. The most commonly used I/O interfaces along with the interface type and number available on the board are described in Table 17.2.

Port Type	# Available	Interface Type
Ethernet	2	Rj45 Back Panel
USB	6	4x Back Panel, 2x Board Pinout
Serial	3	RS232 Back Panel
Mini PCI	1	MiniPCI slot
PCI	1	PCI Slot
IDE	2	1x 40 Pin IDE, 1x 44 Pin IDE
RAM	2	PC 2700
LCD Panel	1	LVDS
VGA	1	VGA
PS2	2	1x Keyboard, 1x Mouse

Table 17.2. G5M100-N device support.

The on-board system is designed to receive GPS coordinates via the Superstar 2 5Hz GPS receiver located within the enclosure and the externally mounted active antenna. Due to severe interference caused by the motherboard the GPS receiver is wrapped in a copper mesh which acts as a faraday cage shielding the sensor.

All communication with the on-board system is handled via the 802.ll B/G Pro 2200 Mini-PCI wireless card. This card interfaces directly with the motherboard via the supported Mini-PCI slot. To support extended range this card is wired to an external whip antenna. This antenna is mounted vertically to the front of the enclosure.

The remaining hardware consists of two serial to USB converters used to interface the GPS receiver and safety switch with the motherboard.

Several key requirements were identified before selecting the OS, including RS-232, USB, and Mini-PCI device support, as well as installations requiring less than 500 megabytes. These requirements were based on the need to support our desired hardware and any common hardware that may be utilized in the future.

One item neglected this far was that the on-board system does not include any permanent storage. All software is stored on a USB stick which is plugged in externally when the system is booted. During boot the entire OS is loaded into RAM where it can be accessed during the vehicles mission. After boot up is completed the USB stick is removed. Several issues supported the decision to use a USB stick for boot. First, the software had to be easily accessible and easily modifiable. A USB boot device allowed us to boot multiple variations of software including different operating systems and custom designed software for a particular platform by just using a different USB stick. Second, the on-board components needed to be minimal in both size and weight. Removing on-board storage allows the processing system to use fewer components and thus have less payload. Last, our previous statement that the system be "use-n-lose" requires that the system's software not be accessible to any one who might acquire the vehicle after it had been 'lost'. The vehicles lack of on-board storage, by definition, meets this requirement.

To adhere to our OS requirements the Slackware 10.2 installation of Linux has been chosen. This installation provides support for all of the currently required hardware via its 2.6.9 kernel, as well as support for low level customization during installation. Specifically, it provides the ability to remove all unnecessary content (graphics, printing, etc) from the OS allowing for a very small installation, less than 100 Megabytes compressed. Note that all the previously mentioned OS customizations were done during installation and added only about three minutes to the standard installation procedure.

17.4 Prototype V1/V2 Comparison

The most significant difference between the two prototype versions are the introduction of a safety switch and frame grabber, version V2, and removal of the CF and CF to IDE adapter, version V1. The safety switch is an obvious short coming of the first system. The lack of the safety switch limited the amount of control that could be accomplished with many platforms, specifically UAVs. The V1 design required that all control to the vehicle be routed through the on-board processing system. In the event of an on-board failure (software bug, hardware failure, battery failure, etc) the vehicle would be uncontrollable. This justifies the introduction of the safety switch into V2. The frame grabber is added to allow for greater flexibility of camera choice. IEEE1394 cameras were typically low in resolution, very high in cost, or very heavy. This presented several issues when operating in various environments and on various platforms. For these reason the frame grabber was incorporated in V2. Last, the CF and CF to IDE adapter were removed to take advantage of the BIOS support for USB 2.0 bootable devices. This allowed for the removal of components and thus weight while also removing two possible sources of failure. In V1, the CF also caused issues with updating the on-board software. Updates typically required removal of the CF which in turn required unmounting and disassembly of the enclosure. This is a very tedious and time consuming procedure and over time put a physical strain on the mounting hardware.

Less significant differences included the move from an EPIA embedded processor to a Pentium M processor, the increase from 1Gig of RAM to 2 Gigs, and the update to a 5Hz GPS. It is noteworthy to mention that the processor and motherboard upgrade allowed for a 25% decrease in amperage use when utilizing 100% of the processor, increased front side bus (FSB) speed, and increased processor frequency without increasing the size and weight of the motherboard.

Similarities between V1 and V2 include the use of a Mini-ITX motherboards, the utilization of a RAM drive OS, the use of an 11.1V LiPo battery, and the use of the Microstrain IMU. Each component was individually reevaluated to assure that it was the optimal choice given our requirements and that it was not prone to failure. Table 17.3 details a side by side comparison of the two prototype on-board processing systems.

Component	**Version 1 (V1)**	**Version 2 (V2)**
Motherboard	VIA MII	G5M100-N
Processor	EPIA 1.2 GHz	Pentium M 2.0 GHz*
Front Side Bus	266 MHz	400 MHz
L1 Cache	128 KB	64 KB
L2 Cache	64 KB	2 MB
RAM	1 Gigabyte	2 Gigabyte
Power Supply	200 Watt	120 Watt
RS-232	2	3
USB	3	6
PCI	1 (PCI)	1 (PCI),1(Mini-PCI)
Ethernet	1	2
IEEE 1394	1	0
BIOS USB Support	1.1 USB	1.1/2.0 USB
GPS	Motorola M12+ 1Hz	Superstar 2 5Hz
IMU	3DMG-X	3DMG-X
Enclosure Size	32x19x5 cm	23x18x6 cm
Weight**	3 lbs	2.5 lbs

* Maximum Processor Speed for the G5M100-N is 2.2 GHz

** Weight includes enclosure and all components within enclosure

Table 17.3. V1 & V2 comparison.

17.5 Experiments

Several experiments have been conducted to validate the on-board processing systems and quantify results. They included experiments for on-board system power consumption, ground versus on-board processing, vision tracking, platform payload limitations, overall system performance, teleoperation, waypoint navigation, and swarms.

17.5.1 Platforms

The first helicopter used for experiments is the Raptor 90 SE with the following characteristics:

Manufacturer:	Thunder Tiger
Rotor Diameter:	1580 mm (Symmetrical)
Dry Weight:	5.8 kg
Dimensions:	130x27x48cm (w/o Blades)
Payload Capacity:	4 kg
Endurance:	18 min
Battery:	4.8 V (2.6A) NiCad
Fuel:	30% Nitrous (Methanol)
Engine:	OS 0.91 C-Spec

This platform has been chosen due to its high power output and small size. The platform has been shown to have relatively low vibration and an ability to handle wind gust exceeding 15 mph.

The second helicopter used is the Maxi Joker 2 with the following characteristics:

Manufacturer:	Joker
Rotor Diameter:	1800 mm (Symmetrical)
Weight:	6.0 kg (with batteries)
Dimensions:	130x27x48cm (w/o Blades)
Payload Capacity:	3 kg
Endurance:	18 min
Battery:	4.8 V (2.6Ah) NiMh, 37 V (10Ah) LiPo
Engine:	Plettenberg HP370/40/A2

This platform has been chosen due to low vibrations and electric propulsion system.

To show wide applicability of the designed on-board processing system, additional experiments have been conducted using a UGV platform, an E-MAXX RC truck with the following characteristics:

Manufacturer:	Traxxas Corporation
Max Speed:	30 Mph
Drive system:	Shaft-drive 4WD
Dry Weight:	3.8 kg
Dimensions:	49x41x27cm
Endurance:	40 min
Battery:	Dual: 7.2V 8Ah LiPo
Motor:	Dual Titan™ 550
Speed Controller:	Novak HV-MAXX

This platform has been chosen due to its rugged nature, wide wheel base, adjustable suspension system, and low center of gravity.

Due to the sensitive dynamics of small unmanned helicopters, special attention has been taken to select and assemble all hardware. Helicopter roll and pitch movement is typically directed around the center of gravity (CG) [18]. This center of gravity is typically designed to reside on the main shaft of the platform approximately half way down the frame. This centrally located CG allows the helicopter to perform highly aggressive maneuvers in very confined areas.

To avoid obstruction of the helicopter's naturally aggressive abilities, extreme care was taken to select hardware that could be assembled and mounted in a manner that would minimally alter the CG. This involved a complete design that would weigh significantly less than the maximum payload of the platform, approximately 2.0 kg for both V1 and V2 (almost half the maximum payload). Minimal obstruction also included mounting the on-board system in a manner that would keep the CG centrally located, see Figure 17.8.

Although the dynamics of the UGV are not as sensitive as the helicopter's, special attention is given to assure that ground platforms are resilient to rollovers, high centering, and ground strikes, The UGVs are shown in Figure 17.9. To prevent rollovers, the on-board system is mounted as close to the platforms natural CG as physically possible and the stock shock mounts are moved away from the CG to increase the wheel base of the platform. To prevent high centering and ground strikes heavy duty springs were added to the suspension system. This forces the shocks to become stiff causing the suspension system to react more aggressively to vertical forces.

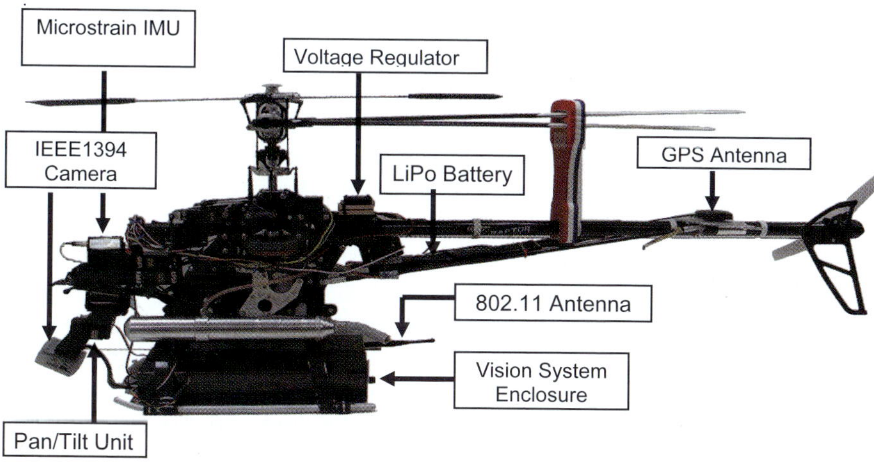

(a) Raptor 90 SE – V1

(b) Raptor 90 SE – V2

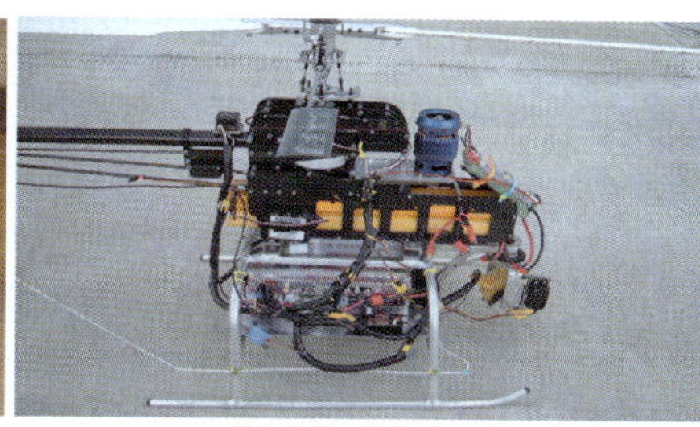

(c) Maxi Joker 2 – V2

Fig. 17.8. On-board systems mounted on the helicopters.

Fig. 17.9. E-MAXX with the on-board systems, V1 (left) and V2 (right).

17.5.2 Experiments with the V1 On-board System

The first experiments aim at verifying the endurance level of the on-board system with LiPo batteries. Due to the nature of LiPo cells, an 11.1V battery is considered completely spent when it reaches a voltage of 9V (this is 3V per LiPo cell). Lowering the voltage below 3V per cell will destroy the battery and may result in a fire [19].

To verify the run time of the on-board system, it was assembled in full and it was attached to a fully charged battery. The entire on-board system was then powered and was allowed to run in an idle state. Idle in this situation refers to the operation of system level processes only. This resulted in CPU utilization between 0% - 5%. During the experiment, GPS coordinates were transmitted by the receiver but ignored and the servos were command to a neutral state and held in position. The on-board system operated for approximately 2.0 hours before the battery voltage reached 9V.

Then, the on-board system was again attached to a fully charged battery and booted. The operating system immediately ran a user level process that grabbed and filtered images from the on-board camera. This process kept CPU utilization between 98% - 100%. The on-board system also served a wireless connection providing GPS coordinates to an external device. The on-board system operated continuously for 40 minuets before battery voltage was measured to be 9V.

The purpose of the second experiment was to quantify the processed frame rate that could be achieved and to compare this result to previous experiments using off-board processing [18]. The software utilized for processing the frames was tasked with identifying a simulated mine, black orb, in varying lighting and background. This was an exact copy of the software utilized in an off-board processing experiment.

Experiments showed frame rate acquisition and processing at a rate of 80 to 120 frames per second (fps) using image resolutions of *160x120* pixels. This exceeded camera limitations which could only grab frames at a rate of 30 fps. Experiments with an off-board processing system, utilizing a 900MHz video transmitter, showed a maximum realized frame rate of 15 fps using image resolutions of *320x240* pixels. This limitation was mainly due to the firewire driver for the video capture device which utilized DV format image, *720x480* pixels and color depth of 24 bit, at 30 fps which was down sampled to a usable lower resolution image [18]. It is also noteworthy to mention that ground processing resulted in a high number of false positive identifications caused by transmission noise and data loss. This type of false positive identification was removed with the use of the on-board system.

The on-board system was also programmed to track objects utilizing the helicopter pan/tilt system, see Figure 17.10. Specifically, software was written to identify objects within some threshold of a predetermined color and size [18]. Once an object was identified the center pixel of the object was approximated. Once the pixel was identified the code determined if the pixel was located within the center threshold of the image. The center threshold was determined to be ± 10 pixels. If the pixel was located within the center threshold both pan and tilt were held in place. If the pixel was not found to be within the center threshold it was determined if the pan, tilt, or both thresholds were broken and in which direction they were broken. This code was combined with servo controller code and used to move the pan/tilt one servo position per threshold violation.

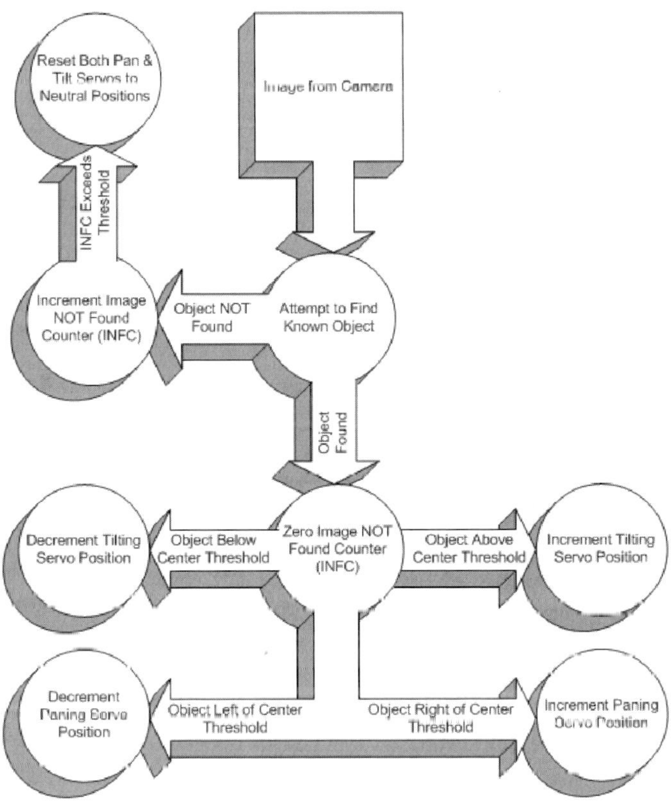

Fig. 17.10. Diagram of tracking software.

Once the tracking process is initiated it continues until the object no longer appears in frame or the hard limits of the pan/tilt are reached. If the object disappears from frame, as determined by the object recognition software, the pan/tilt holds position for up to 30 frames before returning to a neutral position. If the object reappears the tracking process continues. If the hard limits of the pan/tilt are reached the position is held until it disappears from frame or moves in a direction that does not violate the hard limits of the pan/tilt

The next experiment performed was to gain insight into the realistic payload capabilities of the helicopters. First, the helicopter was fitted with a small aluminum plate across the skids to which blocks of weighted aluminum would be added. The helicopter was then powered and flown at a starting payload of 2.5 lbs. Every consecutive flight increased the payload on the platform by 8 ounces. This continued until either the pilot deemed the vehicle unsafe to fly or the vehicle simply failed to lift the weight. At a payload of 10.5 lbs the helicopter was taken to a hover at approximately 10 ft where the vehicle was unable to sustain altitude for longer than 2 min. To ensure personal safety and longevity of the equipment the maximum payload set for this vehicle was 8.5 lbs. This was deemed the optimal payload by the pilot based on vehicle responsiveness.

Next, experiments were performed at the Naval Surface Warfare Center in Panama City. The helicopter was tasked with identifying a target object (black orb) and presenting an estimated GPS coordinate for that object to a UGV in the area, Figure 17.11.

Fig. 17.11. Helicopter and UGVs searching for a simulated mine.

The helicopter was first teleoperated through a series of six GPS coordinates at an altitude of approximately five meters. This altitude was chosen based on the field of view of the camera and to prevent false positive identifications experienced at lower altitudes from grass color and shadows. Each GPS coordinate was approximately fifteen meters from the previous

GPS coordinate and arranged in a raster scan configuration. This resulted in a search area of approximately 450 square meters. The desired object was then randomly placed within the search area. Upon visual detection of the designated object the helicopter was teleoperated to a hover and remained in position until a ground robot arrived. The hovering position of the helicopter was utilized as the GPS estimation of the object. This was deemed a valid estimation due to the almost vertical positioning of the on-board camera.

Identification of the object was handled by on-board vision algorithms utilizing the color and size of the object [18]. Upon identification of the object an internal flag was set. This flag was passed to the ground station upon status request, typically once per second. After receipt of the flag the ground station tasked a local ground robot to the estimated position. Figure 17.12 shows a screenshot of the helicopter requesting help from a UGV after visual detection of a possible 'mine'. Upon arrival at the estimated GPS coordinate, the ground robot began a spiral search for the desired object and the helicopter was released for further searching.

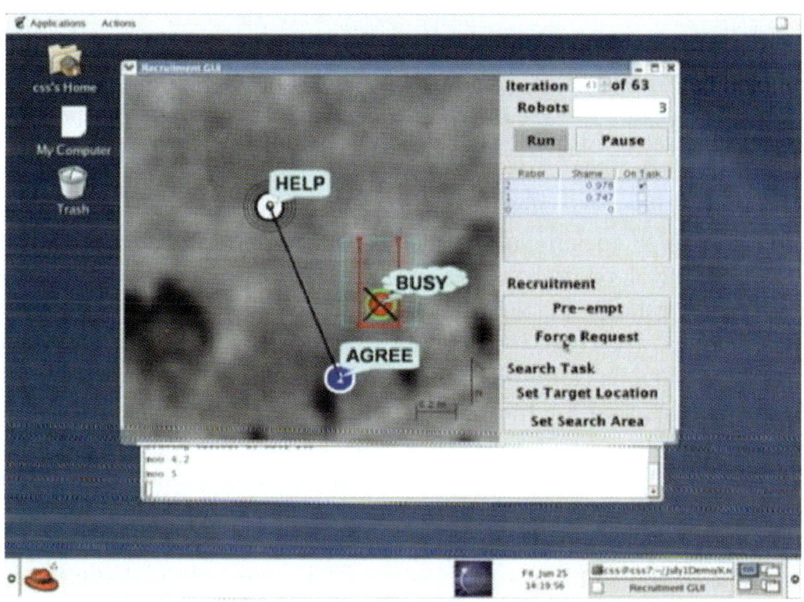

Fig. 17.12. Ground station GUI visualizing helicopter's request for a UGV.

The last UAV experiment was geared towards understanding the potential and problems with traffic surveillance. This was accomplished by utilizing the on-board processing system and the helicopter under radio controlled mode, to retrieve aerial video of traffic.

Video produced by the processing system revealed several issues. Distortion in the lens created a 'rounded' effect on images, see Figure 17.13. Roadway that was undoubtedly flat appeared curved in the image. This also caused distortion to the vehicles traveling on the roadway and made automated vehicle identification somewhat difficult. Second, the video was very out of focus. Although it seems that a minor adjustment could fix it, it is almost impossible to know the altitude and angle at which the helicopter will reside while monitoring traffic. Hence, it is very difficult to focus the lens before flight, suggesting that an auto focus lens or on-board controlled focus will prove useful.

Fig. 17.13. Camera distortion and poor iris control (left); poor focus (right).

Captured images revealed issues based around iris control. The Fire-I camera attempted to simulate iris control through software; however, it bases this control on initial measurements or when light entry exceeds some large threshold. Since the camera is typically only inches from the ground when powered on, it is heavily shadowed by itself and the helicopter. As the helicopter gains altitude, more light enters into the iris but typically does not exceed the preset threshold. This results in images that lose distinction in both color and clarity (Figure 17.13).

Other issues noted during these experiments were the amount of aerial obstacles present around roadways, including power lines, tree lines, light post and signs along with the difficulty involved in finding emergency landings in areas.

The first experiment performed on the UGV platform was teleoperated control. This was done to validate the claim that the on-board system was both generic and highly adaptable.

The on-board system was first mounted to the UGV with one minor modification: all platform servos (speed, gear selection, and steering) were connected directly to the servo controller. This removed the control from the standard radio controller and gave it to the on-board processing system. Code was then implemented that gave command of the vehicle to any machine with login permissions. The user was then able to drive the vehicle, via the keypad, using a remote machine. The user was also able to utilize the same software that was tested and implemented for the VTOL including video and status passing as well as GPS and IMU data.

It is noteworthy to mention that the time required removing the on-board system from the VTOL, mount it to the UGV, and have the on-board system fully operational is about 15 minutes.

The last experiment performed was waypoint navigation of the UGV. This was performed to validate the claim that the on-board system possesses the ability to effectively control a miniature vehicle.

The on-board system was first given a list of desired GPS waypoints. The on-board system was then command to move the platform to these waypoints stopping at the last one. This was accomplished by comparing the current GPS coordinate of the UV to the next waypoint. These two positions were then used to calculate the easterly and northerly error. These two errors were used to calculate the angle from north from the UV to the waypoint. The heading of the UV was then requested from the IMU and subtracted from the error angle. This angle was used as the steering angle of the UV's front wheels.

Make note that due to the limitations of Ackermann steering and the design of the E-MAXX the vehicles turning angle was limited to 45°. Any calculated angle above 45° or below -45° was adjusted to this maximum in that direction.

The speed of the UV was controlled by both the distance from the waypoint and the turning angle of the vehicle. The larger the distance of the UGV from the waypoint the faster the UGV was command to go. This was limited by a maximum speed of approximately 10 Mph. This speed was further reduced based on the turning angle of the front tires. This was to avoid roll over of the vehicle caused by high speed turns. The UGV was also lower limited in speed to assure that the vehicle did not stop in the event that uneven terrain was reached. The UGV navigated successfully through several waypoint patterns on uneven terrain through heavy grass; two illustrations are depicted in Figure 17.14. Autonomous navigation videos can be viewed at www.csee.usf.edu/~rdgarcia/Videos/EMAXX/.

Fig. 17.14. E-MAXX autonomously navigating waypoints.

17.5.3 Experiments with the V2 On-board System

To assure that the V2 system could also sustain endurances that allow for sufficient usability on a UV, several experiments were also performed.

First the V2 on-board system was attached to a fully charged battery, powered, and allowed to run in an idle state. During this experiment all servos were held at neutral positions, the camera, video transmitter, GPS, and IMU were all powered and operational. The on-board system operated for approximately 1.4 hours before the 11.1V 4.2 Ah battery voltage reached 9V.

Next, the on-board system was again attached to a fully charged battery and it was booted. The operating system immediately ran a user level process that collected and transferred data from both the IMU and GPS to a remote shell. This allowed the system to utilize between 98% - 100% of the CPU as well as the wireless device which transferred approximately 500 bytes of data per second. The on-board system operated continuously for 1.15 hours before battery voltage was measured to be 9V.

For simplicity and safety, experiments utilizing the V2 on-board processing system were first performed on UGVs. Waypoint code from V1 was extracted and modified to work on V2. After validation, the safety switch was tested; the system was sent through a series of waypoints and control was randomly removed from the on-board system. The vehicle was then human-controlled to a new position where control was relinquished to the UGV. The UGV successfully navigated through all waypoints with no visible side effects caused by the human's intervention.

After validation of the safety switch ability to allow a human to takeover control of the vehicle, the on-board system was incorporated into a swarm of four UGV's. The first swarm experiment performed was 'follow the leader'. Three of the four vehicles were assigned a leader to follow. Once a leader began to move its GPS coordinates were transferred to any robot requesting its position. These points were stored as waypoints on the following vehicle and used as the path. Once the leader had moved a safe distance away the following vehicle would begin to move. The speed of the following vehicle was determined based on its path distance to its leader.

The experiment was performed by allowing a human to drive the lead car. Once the lead vehicle had moved approximately 15 feet its follower began to move. This continued until all vehicles were moving.

It is noteworthy to mention two obvious problems with this technique. First, GPS is typically inaccurate at slow speeds. Thus, any error in GPS is transmitted to all subsequent robots. Second, any error or failure seen on a vehicle is inevitably visible in all following vehicles. Thus if robot #2 missed a waypoint and circled around, each vehicle behind robot #2 would also circle around even if they did not miss a waypoint.

Next, experiments were performed to create automated vehicle formations throughout waypoint navigation. The vehicles were given a set of waypoints to maneuver through and each vehicle used it's position, the positions of the other vehicles, known/sensed obstacles, and the formation desired and calculated direction & velocities for the vehicle. Utilizing this method the vehicles were able to form multiple configurations including ellipse, square, circle, and line.

Although the full capabilities of the system have not been utilized on the helicopter, there have been several initial experiments with it. The first experiment performed was to verify that the system could coordinate with the ground vehicles. This was accomplished by incorporating the helicopter in the 'follow the leader' scenario. The helicopter was used as the lead vehicle in the swarm and teleoperated by a human pilot. This experiment verified that communication and coordination between the ground and air vehicles was implemented into the swarm, as shown in Figure 7.15.

Fig. 17.15. "Follow the leader" swarm following a helicopter with V2.

17.6 Summary

It is clear that developing a robust on-board processing system for small unmanned vehicles is not trivial. Much attention must be paid to assure that the on-board system integrates all the necessary components without inhibiting the vehicle's ability to utilize new hardware as it is developed. There must also be a thorough understanding of the vehicles that will utilize the system and the overall objectives the vehicles will be expected to perform.

The presented on-board processing systems have shown promise as candidate systems to achieve autonomous navigation. They are modular, flexible, easily integrated with several platforms. Most important of all, they follow the plug-in plug-out concept; they are very light weighted and much cost effective.

References

1. Aida H., Tamura Y., Tobe Y., Tokuda H., "Wireless Packet Scheduling with Signal-to-Noise Ratio Monitoring", *Proceedings, 25th Annual IEEE Conference on Local Computer Networks*, November 2000.
2. Vergun D., "Spartan Unmanned Surface Vehicle Envisioned for Array of High-Risk Missions", *Sea Power*, May 2003.
3. "*New Generation Autonomous Helicopter to Create New Era of Human Safety*", CSIRO, Australia. Available on line at: http://www.csiro.au/index.asp?type=mediaRelease&id=Prhelicopter.

4. Amidi O., *An Autonomous Vision-Guided Helicopter*, Master's Thesis, Carnegie Mellon University, 1996.
5. http://www.defenseworld.net/World-Military-News.asp/var/6979-DefenseAerospacePressnews-1, "Republic of Singapore Navy Hosts IMDEX Asia 05 Ship Display", Defense World, 2004.
6. "Robotic Naval Ships", News about Naval Forces, 23 December 2003.
7. http://www.naval-technology.com/projects/littoral/, Littoral Combat Ship Core Capabilities, 2005.
8. http://www.yamaha-motor.co.jp/global/business/sky/lineup/rmax/index.html.
9. Johnson E., and Mishra S., "Flight Simulation for the Development of an Experimental UAV," *Proceedings, AIAA Modeling and Simulation Technologies Conference and Exhibit*, August 2002.
10. http://www-robotics.usc.edu/~avatar/vision_hw.htm, *Vision Hardware*, University of Southern California, 2004.
11. Mejias L., Saripalli S., Sukhatme G., Cervera P., "Detection and Tracking of External Features in an Urban Environment Using an Autonomous Helicopter," *Proceedings, IEEE International Conference on Robotics and Automation*, April 2005.
12. http://www-robotics.usc.edu/~avatar/history.htm, History, *USC Autonomous Flying Vehicle Project*, 2004.
13. Ruffier F., and Franceschini N., "Visually Guided Micro-Aerial Vehicle: Automatic Take off, Terrain Following, Landing and Wind Reaction", *Proceedings, IEEE International Conference on Robotics and Automation*, April – May 2004.
14. Brock K., *Development of an Autonomous Robotic Aerial Vehicle*, The University of Arizona, 2002.
15. Storvik M., *Guidance System for Automatic Approach to a Ship*, Master's Thesis, Norwegian University of Science and Technology, 2003.
16. Garcia R., Valavanis K., Kontitsis M., "A Multiplatform On-board Processing System for Miniature Unmanned Vehicles", *Proceedings, IEEE International Conference on Robotics and Automation*, May 2006.
17. http://www.ezonemag.com/pages/faq/a300.shtml.
18. Gavrilets V., Mettler B., Feron E., "Nonlinear Model for Small-Size Acrobatic Helicopter," *Proceedings, AIAA Guidance, Navigation, and Control Conference and Exhibit*, August 2001.
19. http://www.thunderpower-batteries.com/images/THPSafetyWarnings.pdf, *Important Safety Instructions and Warnings*, Thunder Power Batteries, 2004.

PART V

Epilogue

Chapter 18: Conclusions and the Road Ahead[1]

This Chapter is the epilogue that brings closure to the book. It summarizes UAV abilities, functionality and usefulness; it tabulates current UAV limitations and bottlenecks, paves the way for future applications and recommends what needs be done in the future that will make UAVs the unique and integral part of military and civilian applications. The contents of this Chapter may be considered as 'heavily biased' because they follow directives, roadmaps and publications from the US DOD and related study groups. However, since the vast majority of UAV applications, missions and tasks still refer to the military domain, observations and suggestions are sound.

18.1 Why UAVs?

The answer to the simple question "*Why UAVs*?" is more than obvious when one focuses on military applications. The focus on and the perspective from the military point of view (overlooking civilian applications) is justified by the fact that even today UAVs are basically and mostly used in the battlefield. Therefore, the answer to the posed question is given in terms of the UAV definition and types of military missions they are suitable for.

According to Joint Publication 1-02, DOD Dictionary [3], the definition of a UAV or unmanned aircraft (UA), reads: "*A powered aerial vehicle that does not carry a human operator, uses aerodynamic forces to provide vehicle lift, can fly autonomously or be piloted remotely, can be expendable or recoverable, and can carry a lethal or non-lethal payload. Ballistic or semi ballistic vehicles, cruise missiles, and artillery projectiles are not considered unmanned aerial vehicles.*"

The statement that UAVs are best suited for '*dull, dirty and dangerous*' missions has merit and it is supported because:

[1] Written by K. P. Valavanis, G. J. Vachtsevanos, P. J. Antsaklis

Kimon P. Valavanis (ed.), Advances in Unmanned Aerial Vehicles, 533–543.

- Dull operations that require more than 30- or 40- hour missions are best carried out using UAVs, since crew members in manned aircraft are used to much shorter duty cycles. Before the 1990's crews were used to four-hour sorties and missions. Fatigue and mission duration compromise proficiency and functionality of crew members, thus, the UAV alternative prevails.
- Dirty operations may require that UAVs fly into nuclear clouds (as happened in 1946–1948) immediately after bomb detonation, a mission that is clearly dangerous to human crews and threatens human lives.
- Dangerous operations like reconnaissance over enemy territory may result in loss of human lives, thus UAVs are preferred.

As stated in [2] and [3], UAVs are preferred over manned aircraft not only because of downsizing risk and increasing confidence in mission success avoiding at the same time the human cost of losing lives if the mission is unsuccessful, but also because unmanned vehicles have better and sustained alertness over humans during dull operations.

Even when considering UAVs for civilian applications, for example traffic monitoring, fire detection and border patrol, advantages are also obvious justifying their preference over their manned counterparts.

18.2 Discussion and Observations

It has been very clear that unmanned fixed- and rotary- wing aircraft have played a major role to the success of military operations (surveillance, reconnaissance and rescue).

It is also true that the Joint Strike Fighter (JSF) is the last manned aircraft to be produced by the Armed Forces. As such, future combat operations will rely basically on unmanned aircraft operating under very challenging and unusual conditions, for example, in urban warfare environments where rules of engagement are very unconventional compared to the traditional battlefield.

However, based on the current state-of-the-art, still limited reliability, reduced autonomy and sensor technology limitations, cost-benefit analysis particularly for civilian applications, platform miniaturization, payload limitations and operator workload requirements of current unmanned vehicles present major roadblocks to their success and full utilization.

To be more specific, future surveillance and reconnaissance missions and tasks will need advanced sophisticated modeling, planning and control technologies, as well as full vehicle autonomy. In short, looking ahead, metrics for UAV success will be very different and much more demanding compared to today's standards.

Future operations, military and civilian, will also shift from the single to multiple UAV concepts. Operations will require multiple UAVs functioning in a cooperative mode, sharing resources and complementing other air, ground, sea-surface and underwater assets.

Thus, it is essential to abstract from current implemented approaches and considerations, and view an ensemble of multiple and heterogeneous unmanned vehicles as a 'system of systems', where a single UAV is functioning as a 'sensor' or as an 'agent' or as a 'node'. That is, multiple UAVs engaged in an assigned mission will function as a swarm, constituting a network of distributed (in the spatial-temporal sense) sensors that must be coordinated to complete a complex mission [4]. Achieving the ultimate goal of full autonomy for a swarm of vehicles executing a complex surveillance and reconnaissance mission still remains a major challenge.

This generalized role of a UAV as part of a complex system dictates that not only networking but also computing and communication issues, fault tolerance, coupling, collision avoidance, real-time reconfiguration and replanning must be considered as UAVs are tasked to perform diverse missions (like surveillance, reconnaissance, forest fire detection, rescue operations, pipeline monitoring).

In more detail, the role UAVs will play in the future may be summarized in terms of the following (not exclusive) observations:

- Future urban warfare, search and rescue, border patrol, homeland security forest fire detection, traffic monitoring, and other applications will utilize an unprecedented level of automation in which human-operated, autonomous, semi-autonomous air and ground platforms will be linked through a coordinated control system to perform complex missions.
- The main challenge in future operations will relate to networked UAVs with adaptable operational procedures, planning and asset de-confliction, increased situation awareness, coupled with cutting edge technologies to realize autonomous collaborative operations.
- Technical challenges will stem from real-time sensing, computing and communication requirements, environmental and operational uncertainty, hostile threats and the emerging need for improved UAV and UAV team autonomy and reliability.
- Significant challenges will also relate to inter-UAV communications, links to command and control, contingency management, etc.

Central to the above observations is the common requirement and emerging need for improved UAV autonomy. This is also a major requirement stated in the DOD UAV roadmap [2] [3] as depicted in Figure

18.1 that shows the time table for the autonomous control level trend. The same roadmap details the need for new technologies that will address single vehicle and multi-vehicle autonomy issues. In essence, Figure 18.1 tabulates unmanned aircraft sophistication levels from the DOD perspective, which cover the whole spectrum from the teleoperated and preprogrammed flight of a single aircraft to self-actuating and fully autonomous group vehicle flights.

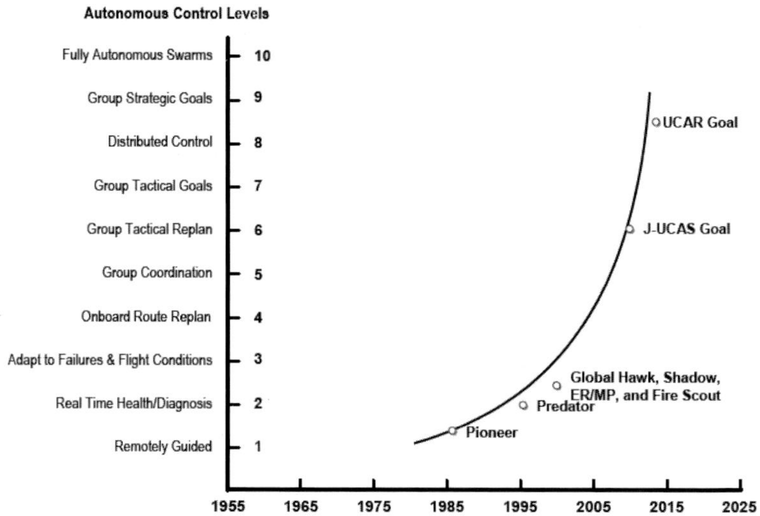

Fig. 18.1. Autonomous control level trend (taken from [2] [3]).

Coupled with the autonomous control level trend is the autonomous control level chart shown in Figures 18.2 (a) and (b). Challenges increase significantly as one move up the hierarchy of the chart from single to multi-vehicle coordinated control. Only moderate success has been currently reported in meeting the lower echelon challenges, leaving open the whole field for subsequent developments.

Technically, to meet stated challenges, innovative coordinated planning and control technologies such as distributed artificial intelligence (DAI), multi agent System (MAS) theory, computational intelligence and soft computing, generalized system theory, as well as game theory and dynamic optimization, coupled with sophisticated hardware and software architectures will be needed. Even though related approaches and methodologies have been investigated intensively in recent years, most work has been focused on solving particular problems, such as formation control and autonomous search, while less attention has been paid to the overall 'system architecture' concept, especially from an implementation and integration point of view.

Technical challenges and related enabling technologies are presented next.

Autonomous Control Level (ACL) Chart

Note: As ACL increases, capability includes, or replaces, items from lower levels

Level	Level Descriptor	Observe Perception / Situational Awareness	Orient Analysis / Coordination	Decide Decision Making	Act Capability
10	Fully Autonomous	Cognizant of all within Battlespace	Coordinates as necessary	Capable of total independence	Requires little guidance to do job
9	Battlespace Swarm Cognizance	Battlespace inference – Intent of self and others (allies and foes). Complex Intense environment – on-board tracking	Strategic group goals assigned. Enemy Strategy inferred.	Distributed tactical group planning. Individual determination of tactical goal. Individual task planning/execution. Choose tactical targets	Group accomplishment of strategic goal with no supervisory assistance
8	Battlespace Cognizance	Proximity inference – Intent of self and others (allies and foes). Reduced dependence upon off-board data.	Strategic group goals assigned Enemy tactics inferred ATR	Coordinated tactical group planning Individual task planning / execution Choose targets of opportunity	Group accomplishment of strategic goal with minimal supervisory assistance
7	Battlespace Knowledge	Short track awareness – History and predictive battlespace data in limited range, timeframe and numbers.	Tactical group goals assigned. Enemy Trajectory estimated.	Individual task planning / execution to meet goals	Group accomplishment of tactical goal with minimal supervisory assistance
6	Real-Time Multi-Vehicle Cooperation	Ranged awareness – Onboard sensing for long rang, supplemented by off-board data	Tactical group goals assigned Enemy location sensed / estimated	Coordinated trajectory planning and execution to meet goals – Group optimization	Group accomplishment of tactical goal with minimal supervisory assistance
5	Real-Time Multi-Vehicle Coordination	Sensed awareness – Local sensor to detect others, fused with off-board data	Tactical group plan assigned RT Health Diagnosis; Ability to compensate for most failures and flight conditions; Ability to predict onset of failures (e.g. Prognostic Health Mgmt) Group diagnosis and resource management	On-board trajectory replanning – Optimize for current and predictive conditions Collision avoidance	Group accomplishment of tactical plan as externally assigned Air collision avoidance Possible close air space separation for AAR, formation in non-thread conditions

(a)

Autonomous Control Level (ACL) Chart

Note: As ACL increases, capability includes, or replaces, items from lower levels

Level	Level Descriptor	Observe Perception / Situational Awareness	Orient Analysis / Coordination	Decide Decision Making	Act Capability
4	Fault / Event Adaptive Vehicle	Deliberate Awareness – Allies communicate data	Tactical plan assigned Assigned Rules of Engagement RT Health Diagnosis. - Ability to compensate for most failures and flight conditions – inner loop changes reflected in outer loop performance	On-board trajectory replanning – Event driven, Self resource management, Deconfliction	Self accomplishment of tactical plan as externally assigned Medium vehicle airspace separation
3	Robust Response to Real-Time Faults / Events	Health / Status history and models	Tactical plan assigned RT Health Diagnostics Ability to compensate for most control failures and flight conditions	Evaluate status vs. required mission capabilities Abort / RTB if insufficient	Self accomplishment of tactical plan as externally assigned
2	Changeable Mission	Health / Status sensors	RT Health diagnosis Off-board replan	Execute preprogrammed or uploaded plans in response to mission and health conditions	Self accomplishment of tactical plan as externally assigned
1	Execute Preplanned Mission	Preloaded mission data Flight Control and Navigation Sensing	Pre / Post Flight BIT Report status	Preprogrammed mission and abort plans	Wide airspace separation requirements
0	Remotely Piloted Vehicle	Flight Control sensing Nose camera	Telemetered data Remote pilot commands	N/A	Control by remote pilot

(b)

Fig. 18.2. The autonomous control level chart.

18.3 Technical Challenges and Enabling Technologies

Judging from the current state-of-the-art in UAV technology, evaluating their utilization in military and civilian domains, and considering that the future trend moves from the 'single system' to the 'system of systems' paradigm, a potential list of technical challenges that need be overcome includes (but it is not limited to) the following:

- Modeling: Spatial-temporal modeling paradigms will be essential to improve real-time planning and control of heterogeneous networked systems of UAVs.

- Control: Hierarchical/Intelligent control methodologies suitable for multiple networked systems will be required. New reasoning paradigms for tracking, pursuit-evasion, surveillance/reconnaissance, coordinated control, planning and scheduling, obstacle avoidance, and deconfliction need be derived and implemented.
- Networking and communications: Inter- and intra- system coupling needs be considered between assets, robustness and reliability should be investigated along with secure communication protocols. The need for joint command, control and communication capabilities will be essential. A well defined set of criteria that improves Quality of Service requirements, mostly beyond the transport level, will be essential to evaluate systems of UAVs.
- Computing: Increased on-board processing and on-board computational requirements will enhance unmanned aircraft autonomicity. Enhanced on-board capabilities should be coupled with developing new hardware and software architectures, eventually leading to open system architecture designs.
- Sensors and sensing strategies: Computationally efficient, reliable and accurate sensor fusion algorithms will be needed, coupled with relevant hardware and software requirements. Effectiveness metrics for networked sensors should be defined that fuse and process data in real-time leading to decision making.
- Performance metrics, verification and validation: Formal methods for verification and validation purposes (on top of hardware-in-the-loop and software-in-the-loop verification) will be needed, along with well defined metrics to measure success.

Those challenges are coupled with relevant enabling technologies that need be enhanced and/or developed in parallel to arrive at desirable levels of functionality and operability. Enabling technologies that correspond to the above technical challenges are, respectively:

- New modeling techniques will be required to capture coupling between individual system/sensor dynamics, communications, etc., with 'system of systems' behaviors. Hybrid system approaches will play a key role in capturing complex behaviors and defining the means to represent and manage uncertainty, including spatial-temporal models of distributed sensors to integrate system and motion dependencies, contingency planning and situation awareness.
- Intelligent and hierarchical/distributed control concepts must be developed and expanded to address 'system of systems' configurations. Game theoretic approaches and optimization algorithms running in

real-time to assist in cooperative control and adversarial reasoning will be needed. Comprehensive approaches to control of networks of dynamic agents will be essential to tackle coordination and cooperation issues.

- Networking and communications will deviate from traditional communication protocols and standards, requiring novel solutions to overcome jamming, consider UAVs as decoys, as communication relays and as hubs, when necessary. Security metrics will need to be redefined.
- Computing technologies will be based on embedded processing requirements, leading to new, reliable, and fault-tolerant computing platforms. Software reliability issues should be revisited and redefined.
- Sensors and sensing strategies will need innovative technologies and solutions in wireless communications. This will be coupled with building improved, reliable and cost-effective sensor suites, as well as 'smart' sensors, leading to better sensing strategies for massive data processing, data mining and sensor fusion.
- The need for defining new 'system of systems' performance and effectiveness metrics for verification, validation and assessment of networked systems is more than obvious. Multiple UAV systems are much more demanding than single UAV ones.

18.4 The Road Ahead

During the last fifteen years, we have witnessed significant advances in the unmanned aircraft and unmanned systems state-of-the-art. Several advances have been presented in this book.

Major research and development projects, sponsored primarily by the military sector, have contributed towards development and deployment of aerial, undersea and ground unmanned systems with new designs surfacing at an increasing rate. Common objectives and goals of any such designs are improved autonomy through innovative sensing and control strategies, enhanced reliability and endurance through advanced materials and power plants.

Yet, new and novel concepts and technologies are required for a more widespread use of such critical assets, not only for military but also for commercial and other applications such as homeland security, rescue operations, forest fire detection, delivery of goods, to name just a few applications.

References [2] and [3] published by the US DOD present significant challenges to the designer, the technology developer, the OEM and the end user that must be addressed for the user community to gain sufficient confidence in UAV utility.

Perhaps the most important challenge on the road ahead is associated with what needs be accomplished at the unmanned vehicle level. The following recommendations are justifiable and very valid:

- UAVs must become 'smaller and smarter'. Micro- and organic- vehicles must be developed that are capable of carrying sufficient payloads of sensing, computing and communication capabilities to insure reliable and robust intelligence, surveillance, reconnaissance capabilities, communicate useful information to command and control stations, and cooperate with other unmanned assets.
- UAVs must operate with enhanced mobility and improved autonomy. They must posses maneuvering capabilities required to execute difficult missions in cluttered urban environments. Operator load must be reduced through increased autonomy.
- UAVs must be able to function as fixed-wing and rotary-wing unmanned aircraft as described in Chapter 13, depending on mission requirements.
- UAVs must become more fault-tolerant; that is, they must be designed to monitor their own performance and sustain an acceptable level of performance in the presence of non-catastrophic failure modes or large disturbances. Fault tolerant control technologies are currently being investigated and some initial successful implementations have been reported. The absence of an on-board human pilot requires new and innovative technologies for designing and deploying such high confidence systems.
- UAVs may be biologically inspired, that is, possess design and behavioral attributes that imitate as strongly as possible those of biological species. Larger class vehicles should carry significant payloads and must be design to operate at high altitude with long endurance through new and perhaps 'exotic' power plants. Biologically inspired UAVs must possess attributes of stealthiness, while noise levels associated with such vehicles must be reduced within acceptable limits.
- Teams or groups of heterogeneous unmanned systems must exhibit attributes of autonomy so that complex mission scenarios can be achieved without requiring an extensive operator workload. Aspects of multi-vehicle cooperative technologies are currently being researched and, to a limited extent, they have been demonstrated.

➢ Software platforms for single and multiple vehicle formations must be further developed and improved to accommodate open system architectures, standards and protocols that will facilitate the expedient, effective and reliable introduction of unmanned assets to the military and civilian sectors.

18.5 Epilogue

Federated systems consisting of multiple unmanned aerial vehicles performing complex missions present new challenges to the control community. UAVs must possess attributes of autonomy in order to function effectively in a 'system of systems' configuration.

Coordinated and collaborative control of UAV swarms demands new and novel technologies that integrate modeling, control, communications and computing concerns into a single architecture. Typical application domains include reconnaissance and surveillance missions in an urban environment, target tracking and evasive maneuvers, search and rescue operations, homeland security, etc.

Major technological challenges remain to be addressed for such UAV swarms, or similar federated system of systems configurations to perform efficiently and reliably. Excessive operator load, autonomy issues and reliability concerns have limited thus far their widespread utility. The systems and controls community is called upon to play a major role in the introduction of breakthrough technologies in this exciting area.

References

1. *Unmanned Vehicles Handbook 2002*, The Shepard Press, 2002.
2. *Unmanned Aircraft Systems Roadmap 2005-2030*, Office of the Secretary of Defense, August 2005.
3. *OSD UAV Roadmap 2002-2027*, Office of the Secretary of Defense Acquisition, Technology, & Logistics, Air Warfare, December 2002.
4. Vachtsevanos G., Tang L., and Reimann J. "An Intelligent Approach to Coordinated Control of Multiple Unmanned Aerial Vehicles," *Proceedings of the American Helicopter Society 60th Annual Forum*, Baltimore, MD, June 2004.
5. Vachtsevanos G., Kim W., Al-Hasan S., Rufus F., Simon M., Schrage D., and Prasad J. V. R., "Mission Planning and Flight Control: Meeting the Challenge with Intelligent Techniques", *Journal of Advanced Computational Intelligence*, Vol. 1, (1), pp. 62-70, October 1997.

6. Vachtsevanos G., Ludington B., Reimann J., Antsaklis P., Valavanis K., "Modeling and Control of Unmanned Aerial Vehicles – Current Status and Future Directions", CD-ROM Proceedings, *13th Mediterranean Conference on Control and Automation*, Cyprus, June 2005.

International Series on
INTELLIGENT SYSTEMS, CONTROL, AND AUTOMATION: SCIENCE AND ENGINEERING

Editor: Professor S. G. Tzafestas, *National Technical University, Athens, Greece*

---

1. S.G. Tzafestas (ed.): *Microprocessors in Signal Processing, Measurement and Control*. 1983 ISBN 90-277-1497-5
2. G. Conte and D. Del Corso (eds.): *Multi-Microprocessor Systems for Real-Time Applications*. 1985 ISBN 90-277-2054-1
3. C.J. Georgopoulos: *Interface Fundamentals in Microprocessor-Controlled Systems*. 1985 ISBN 90-277-2127-0
4. N.K. Sinha (ed.): *Microprocessor-Based Control Systems*. 1986 ISBN 90-277-2287-0
5. S.G. Tzafestas and J.K. Pal (eds.): *Real Time Microcomputer Control of Industrial Processes*. 1990 ISBN 0-7923-0779-8
6. S.G. Tzafestas (ed.): *Microprocessors in Robotic and Manufacturing Systems*. 1991 ISBN 0-7923-0780-1
7. N.K. Sinha and G.P. Rao (eds.): *Identification of Continuous-Time Systems*. Methodology and Computer Implementation. 1991 ISBN 0-7923-1336-4
8. G.A. Perdikaris: *Computer Controlled Systems*. Theory and Applications. 1991 ISBN 0-7923-1422-0
9. S.G. Tzafestas (ed.): *Engineering Systems with Intelligence*. Concepts, Tools and Applications. 1991 ISBN 0-7923-1500-6
10. S.G. Tzafestas (ed.): *Robotic Systems*. Advanced Techniques and Applications. 1992 ISBN 0-7923-1749-1
11. S.G. Tzafestas and A.N. Venetsanopoulos (eds.): *Fuzzy Reasoning in Information, Decision and Control Systems*. 1994 ISBN 0-7923-2643-1
12. A.D. Pouliezos and G.S. Stavrakakis: *Real Time Fault Monitoring of Industrial Processes*. 1994 ISBN 0-7923-2737-3
13. S.H. Kim: *Learning and Coordination*. Enhancing Agent Performance through Distributed Decision Making. 1994 ISBN 0-7923-3046-3
14. S.G. Tzafestas and H.B. Verbruggen (eds.): *Artificial Intelligence in Industrial Decision Making, Control and Automation*. 1995 ISBN 0-7923-3320-9
15. Y.-H. Song, A. Johns and R. Aggarwal: *Computational Intelligence Applications to Power Systems*. 1996 ISBN 0-7923-4075-2
16. S.G. Tzafestas (ed.): *Methods and Applications of Intelligent Control*. 1997 ISBN 0-7923-4624-6
17. L.I. Slutski: *Remote Manipulation Systems*. Quality Evaluation and Improvement. 1998 ISBN 0 7932 4822 2
18. S.G. Tzafestas (ed.): *Advances in Intelligent Autonomous Systems*. 1999 ISBN 0-7932-5580-6
19. M. Teshnehlab and K. Watanabe: *Intelligent Control Based on Flexible Neural Networks*. 1999 ISBN 0-7923-5683-7
20. Y.-H. Song (ed.): *Modern Optimisation Techniques in Power Systems*. 1999 ISBN 0-7923-5697-7
21. S.G. Tzafestas (ed.): *Advances in Intelligent Systems*. Concepts, Tools and Applications. 2000 ISBN 0-7923-5966-6
22. S.G. Tzafestas (ed.): *Computational Intelligence in Systems and Control Design and Applications*. 2000 ISBN 0-7923-5993-3
23. J. Harris: *An Introduction to Fuzzy Logic Applications*. 2000 ISBN 0-7923-6325-6

International Series on
INTELLIGENT SYSTEMS, CONTROL, AND AUTOMATION: SCIENCE AND ENGINEERING

24. J.A. Fernández and J. González: *Multi-Hierarchical Representation of Large-Scale Space.* 2001 ISBN 1-4020-0105-3
25. D. Katic and M. Vukobratovic: *Intelligent Control of Robotic Systems.* 2003 ISBN 1-4020-1630-1
26. M. Vukobratovic, V. Potkonjak and V. Matijevic: *Dynamics of Robots with Contact Tasks.* 2003 ISBN 1-4020-1809-6
27. M. Ceccarelli: *Fundamentals of Mechanics of Robotic Manipulation.* 2004 ISBN 1-4020-1810-X
28. V.G. Ivancevic and T.T. Ivancevic: *Human-Like Biomechanics.* A Unified Mathematical Approach to Human Biomechanics and Humanoid Robotics. 2005 ISBN 1-4020-4116-0
29. J. Harris: *Fuzzy Logic Applications in Engineering Science.* 2005 ISBN 1-4020-4077-6
30. M.D. Zivanovic and M. Vukobratovic: *Multi-Arm Cooperating Robots.* Dynamics and Control. 2006 ISBN 1-4020-4268-X
31. V.G. Ivancevic and T. Ivancevic: *Geometrical Dynamics of Complex Systems.* A Unified Modelling Approach to Physics, Control, Biomechanics, Neurodynamics and Psycho-Socio-Economical Dynamics. 2006 ISBN 1-4020-4544-1
32. V.G. Ivancevic and T.T. Ivancevic: *High-Dimensional Chaotic and Attractor Systems.* A Comprehensive Introduction. 2007 ISBN 1-4020-5455-6
33. K.P. Valavanis (ed.): *Advances in Unmanned Aerial Vehicles.* State of the Art and the Road to Autonomy. 2007 ISBN-978-1-4020-6113-4